VOLUME EIGHTY EIGHT

ADVANCES IN
BOTANICAL RESEARCH

Plant Epigenetics Coming of Age for Breeding Applications

ADVANCES IN BOTANICAL RESEARCH

Series Editor

Jean-Pierre Jacquot
Professor, Membre de L'Institut Universitaire de France, Unité Mixte de Recherche INRA, UHP 1136 "Interaction Arbres Microorganismes", Université de Lorraine, Faculté des Sciences, Vandoeuvre, France

VOLUME EIGHTY EIGHT

ADVANCES IN BOTANICAL RESEARCH

Plant Epigenetics Coming of Age for Breeding Applications

Edited by

MARIE MIROUZE
Institut de Recherche pour le Développement (IRD), UMR232, Laboratoire Genome et Developpment des Plantes, Perpignan, France

ETIENNE BUCHER
IRHS, Université d'Angers, INRA, AGROCAMPUS-Ouest, Université Bretagne Loire, Angers, France

PHILIPPE GALLUSCI
UMR EGFV, Bordeaux Sciences Agro, INRA, Université de Bordeaux, Villenave d'Ornon, France

ELSEVIER

ACADEMIC PRESS
An imprint of Elsevier

Academic Press is an imprint of Elsevier
125 London Wall, London EC2Y 5AS, United Kingdom
525 B Street, Suite 1650, San Diego, CA 92101, United States
50 Hampshire Street, 5th Floor, Cambridge, MA 02139, United States
The Boulevard, Langford Lane, Kidlington, Oxford OX5 1GB, United Kingdom

First edition 2018

ISBN: 978-0-12-815403-8
ISSN: 0065-2296

For information on all Academic Press publications visit our website at https://www.elsevier.com/books-and-journals

Publisher: Zoe Kruze
Acquisition Editor: Sam Mahfoudh
Editorial Project Manager: Teresa Pons-Ferrer
Production Project Manager: James Selvam
Cover Designer: Victoria Pearson

Typeset by TNQ Technologies

CONTENTS

CONTRIBUTORS

Isabel Allona
Centro de Biotecnología y Genómica de Plantas, Universidad Politécnica de Madrid (UPM), Instituto Nacional de Investigación y Tecnología Agraria y Alimentaria (INIA), Madrid, Spain

Christophe Ambroise
Laboratoire de Mathématiques et Modélisation d'Évry, LaMME, UMR 8071, CNRS-Université d'Evry Val d'Essonne, Evry, France

Mario A. Arteaga-Vazquez
Laboratory of Epigenetics and Developmental Biology, Instituto de Biotecnología y Ecología Aplicada, Universidad Veracruzana, Xalapa, Mexico

Margot M.J. Berger
UMR EGFV, Bordeaux Sciences Agro, INRA, Université de Bordeaux, Villenave d'Ornon, France

Etienne Bucher
IRHS, Université d'Angers, INRA, AGROCAMPUS-Ouest, Université Bretagne Loire, Angers, France

Marco Catoni
The Sainsbury Laboratory, University of Cambridge, Cambridge, United Kingdom

Amanda Malvessi Cattani
Graduate Program in Cell and Molecular Biology, Centro de Biotecnologia, Universidade Federal do Rio Grande do Sul, Porto Alegre, Brazil

Daniel Conde
Centro de Biotecnología y Genómica de Plantas, Universidad Politécnica de Madrid (UPM), Instituto Nacional de Investigación y Tecnología Agraria y Alimentaria (INIA), Madrid, Spain

Sandra Cortijo
The Sainsbury Laboratory, University of Cambridge, Cambridge, United Kingdom

Vítor da Silveira Falavigna
AGAP, Univ. Montpellier, CIRAD, INRA, Montpellier SupAgro, Montpellier, France

Paulo Ricardo Dias de Oliveira
Centro Nacional de Pesquisa de Uva e Vinho, Empresa Brasileira de Pesquisa Agropecuária, Bento Gonçalves, Brazil

Ana E. Dorantes-Acosta
Laboratory of Epigenetics and Developmental Biology, Instituto de Biotecnología y Ecología Aplicada, Universidad Veracruzana, Xalapa, Mexico

Régis Fichot
Laboratoire de Biologie des Ligneux et des Grandes Cultures, LBLGC, EA 1207 USC 1328, INRA-Université Orléans, Orléans, France

Ryo Fujimoto
Graduate School of Agricultural Science, Kobe University, Rokkodai, Nada-ku, Kobe, Japan

Philippe Gallusci
UMR EGFV, Bordeaux Sciences Agro, INRA, Université de Bordeaux, Villenave d'Ornon, France

Svetlana Gribkova
Laboratoire de Probabilités et Modèles Aléatoires, LPMA, UMR 8001, CNRS-Université Paris Diderot, Paris, France

Daniel Grimanelli
Institut de Recherche pour le Développement (IRD), UMR232, Université de Montpellier, Montpellier, France

Tzung-Fu Hsieh
Department of Plant and Microbial Biology, North Carolina State University, Raleigh, NC, United States; Plants for Human Health Institute, North Carolina State University, North Carolina Research Campus, Kannapolis, NC, United States

Yu-Hung Hung
Department of Plant and Microbial Biology, North Carolina State University, Raleigh, NC, United States; Plants for Human Health Institute, North Carolina State University, North Carolina Research Campus, Kannapolis, NC, United States

Mathieu Ingouff
Institut de Recherche pour le Développement (IRD), UMR232, Université de Montpellier, Montpellier, France

Véronique Jorge
Biologie intégrée pour la valorisation de la diversité des arbres et de la forêt, BioForA, UMR 0588, INRA-ONF, Orléans, France

Sunil Kumar Kenchanmane Raju
Department of Plant Biology, Michigan State University, East Lansing, MI, United States

Junhua Kong
UMR EGFV, Bordeaux Sciences Agro, INRA, Université de Bordeaux, Villenave d'Ornon, France

Sophie Lanciano
Institut de Recherche pour le Développement (IRD), UMR232, Laboratoire Génome et Développement des Plantes, Perpignan, France

Olivier Leblanc
Institut de Recherche pour le Développement (IRD), UMR232, Université de Montpellier, Montpellier, France

Grégoire Le-Provost
Biodiversité, Gènes et Communautés, BIOGECO, INRA, Univ. Bordeaux, Cestas, France

Fang Liu
Plants for Human Health Institute, North Carolina State University, North Carolina Research Campus, Kannapolis, NC, United States; State Key Laboratory for Conservation and Utilization of Subtropical Agro-bioresources and College of Agriculture, Guangxi University, Nanning, China

Hiroki Maeji
Department of Bioengineering, Nagaoka University of Technology, Nagaoka, Niigata, Japan

Stéphane Maury
Laboratoire de Biologie des Ligneux et des Grandes Cultures, LBLGC, EA 1207 USC 1328, INRA-Université Orléans, Orléans, France

Peter Meyer
The University of Leeds, Centre for Plant Sciences, Leeds, United Kingdom

Marie Mirouze
Institut de Recherche pour le Développement (IRD), UMR232, Laboratoire Génome et Développement des Plantes, Perpignan, France

Naomi Miyaji
Graduate School of Agricultural Science, Kobe University, Rokkodai, Nada-ku, Kobe, Japan

Chad E. Niederhuth
Department of Plant Biology, Michigan State University, East Lansing, MI, United States

Taisuke Nishimura
Department of Bioengineering, Nagaoka University of Technology, Nagaoka, Niigata, Japan

Omar Oltehua-Lopez
Laboratory of Epigenetics and Developmental Biology, Instituto de Biotecnología y Ecología Aplicada, Universidad Veracruzana, Xalapa, Mexico

Luc Pâques
Biologie intégrée pour la valorisation de la diversité des arbres et de la forêt, BioForA, UMR 0588, INRA-ONF, Orléans, France

Christophe Plomion
Biodiversité, Gènes et Communautés, BIOGECO, INRA, Univ. Bordeaux, Cestas, France

Diogo Denardi Porto
Centro de Pesquisa Agropecuária do Trópico Semiárido, Empresa Brasileira de Pesquisa Agropecuária, Petrolina, Brazil

Luís Fernando Revers
Graduate Program in Cell and Molecular Biology, Centro de Biotecnologia, Universidade Federal do Rio Grande do Sul, Porto Alegre, Brazil; Centro Nacional de Pesquisa de Uva e Vinho, Empresa Brasileira de Pesquisa Agropecuária, Bento Gonçalves, Brazil

Jérôme Salse
Génétique, Diversité et Écophysiologie des Céréales, GDEC, UMR 1095, INRA-Université Clermont Auvergne, Clermont-Ferrand, France

Léopoldo Sanchez-Rodriguez
Biologie intégrée pour la valorisation de la diversité des arbres et de la forêt, BioForA, UMR 0588, INRA-ONF, Orléans, France

Tiago Sartor
Graduate Program in Cell and Molecular Biology, Centro de Biotecnologia, Universidade Federal do Rio Grande do Sul, Porto Alegre, Brazil

Vincent Segura
Biologie intégrée pour la valorisation de la diversité des arbres et de la forêt, BioForA, UMR 0588, INRA-ONF, Orléans, France

Carolina Pereira Silveira
Centro Nacional de Pesquisa de Uva e Vinho, Empresa Brasileira de Pesquisa Agropecuária, Bento Gonçalves, Brazil

Mamadou Dia Sow
Laboratoire de Biologie des Ligneux et des Grandes Cultures, LBLGC, EA 1207 USC 1328, INRA-Université Orléans, Orléans, France

Emeline Teyssier
UMR EGFV, Bordeaux Sciences Agro, INRA, Université de Bordeaux, Villenave d'Ornon, France

Michael Thieme
Institute of Plant and Microbial Biology, University of Zürich, Zollikerstrasse, Zürich, Switzerland

Jörg Tost
Laboratoire Épigénétique et Environnement, LEE, CNRGH CEA, Evry, France

Wenyan Xiao
Department of Biology, St. Louis University, St. Louis, MO, United States

Xiang-Qian Zhang
Plants for Human Health Institute, North Carolina State University, North Carolina Research Campus, Kannapolis, NC, United States; Guangdong Engineering Research Center of Grassland Science, College of Forestry and Landscape Architecture, South China Agricultural University, Guangzhou, China

PREFACE

Over the last century, a remarkable yield increase has been achieved for many crop plants. Multiple factors have contributed to this achievement, but undoubtedly breeding has been a major driving force that has resulted in the improvement of traits of agronomical importance including growth and vigor, increased yields and better tolerance to environmental constraints and to agronomical practices. So far, crop improvement has relied on the genetic variability found within species or in wild relatives and on the introgression of traits of agronomical importance into elite crops. The recent completion of many major crop genomes has greatly facilitated precision breeding by revealing gene variants associated with specific traits. These approaches that include genome-wide association studies (GWAS), genomic selection and other genetic strategies have provided new ways to capture a substantial proportion of the sequence-based heritable variation underlying phenotypic traits (Morrell, Buckler, & Ross-Ibarra, 2012). There is still, however, a significant amount of heritability that cannot be explained by genetic methods, the "missing heritability" (Maher, 2008), that in humans can represent a substantial part of the heritability as recently shown in the case of human height (Bourrat, Lu, & Jablonka, 2017; Manolio et al., 2009; Visscher, 2008). Various mechanisms that could account for the "missing heritability" have been suggested. Possible explanations include the combination of rare alleles, variations in gene copy number or the contribution of additional gene variants that remain to be discovered and occur at very low frequency in populations (Bourrat et al., 2017; Bandyopadhyay, Chanda, & Wang, 2017).

As far as plants are concerned, GWAS appears to be a more efficient approach as illustrated by the recent demonstration that a significant proportion of flowering time variation can simply be explained by sequences diversity, however, still leaving 45% of the phenotypic diversity of this trait unexplained (Brachi, Morris, Borevitz, & Benjamin Brachi, 2011; Li, Huang, Bergelson, Nordborg, & Borevitz, 2010). This suggests that variation in gene sequences is unlikely to explain, in plants also, the complete range of the observed phenotypic diversity. This is even more true when the genetic diversity of crops is limited, for example in tomato (Lin et al., 2014), or when clonal propagation is an important method for plant multiplication (Verhoeven & Preite, 2014).

Epigenetics has been defined as the study of heritable variations of phenotypes that do not involve alterations in the genetic information (Berger et al, 2009; Dupont et al., 2009) and consequently could also be involved in determining phenotyping diversity independently of DNA sequence variations. This has been illustrated in several studies showing that epigenetic variations can lead to stable modifications of plant traits of agronomical importance, for example fruit ripening (Manning et al., 2006), plant height (Miura et al., 2009), or fruit nutrient content (Quadrana et al., 2014). In addition, an exciting aspect of epigenetic mechanisms is their potential contribution to the so-called "soft inheritance" (Richards, 2006), a term proposed by Mayr's to explain that "*the genetic basis of characters could be modified either by direct induction by the environment, or by use and disuse, or by an intrinsic failure of constancy, and that this modified genotype was then transmitted to the next generation*" (Mayr et al., 1980). Epigenetic processes could therefore contribute to the heritable phenotypic diversity through mechanisms other than DNA sequence variation (Bourrat et al., 2017). It is the focus of the present book to discuss the extent to which epigenetic diversity could contribute to stable phenotypic diversity in plants and therefore becomes part of new breeding strategies aiming at facing the numerous challenges raised by climate change and thus affecting food security in the upcoming century.

The first chapter provides an historical perspective of epigenetics that has been intensively studied over the last 20 years in plants and since 1950 in animals leading to the demonstration that they play essential functions in controlling many aspects of cell-specific gene expression programs and cell differentiation processes. In the second chapter, molecular mechanisms underlying the epigenetic processes are described highlighting the fundamental role of DNA methylation, histone posttranslational modifications (PTMs) and small noncoding RNAs in determining the conformational state of the chromatin and thereby its transcriptional state. Indeed, the heritability, stability and diversity of epigenetic marks are an essential aspect to consider for potential breeding applications. Of particular interest is the question of the relevance of locus-specific or genome-wide epigenetic diversity in determining phenotypic variations as discussed in Chapter 3. In addition, the demonstration that epigenetic recombinant-inbred *A. thaliana* lines (epiRILs) generated from isogenic parents with divergent DNA methylome segregate tens of stable parental differentially methylated regions that are stably inherited and act as epigenetic quantitative traits, highlights the relevance of DNA methylation as a stable epigenetic mark, as discussed in Chapter 4. This chapter shows that variation in DNA methylation impacts on plant phenotypes over generations and provides a new paradigm in which DNA methylation could shape phenotypic variability, contributing to the so-called "missing heritability" and could therefore be used in breeding strategies. Indeed, a precise analysis of how epigenetic marks, including DNA methylation, histone marks and small RNAs can be transmitted through sexual and asexual plant reproduction is a central issue when considering the potential role of epigenetic heritability in breeding strategies. Chapter 5 presents the current knowledge of epigenetic marks, their stability and transmission both in sexual and asexual reproduction, and thereby provides an insight into the potentialities and limits of the use of epigenetic variability in plant improvement.

An important function of epigenetic regulation is its contribution to the stability of genomes, with a special role in the control of transposable element (TE) mobility. Indeed TEs have long been considered a major risk for genome integrity as they can generate deleterious mutations, but recent work highlights their fundamental role in plant stress responses and in crop domestication. Therefore, TEs could be important factors to consider in breeding strategies (Chapter 6). Over the last century, both grafting and heterosis have contributed to the significant increase in crop yield by improving plant vigor, yield or resistance to pathogens. Although the mechanisms underlying these processes are still far from being understood, recent work indicates that epigenetic mechanisms play an important function in these processes (Chapters 7 and 8). Chapters 9 and 10 focus on the essential roles of epigenetic mechanisms in controlling traits of agronomical importance as flowering time and bud break in perennial plants and the regulation of fleshy fruit development and quality, respectively. Finally the last chapters discuss the importance of epigenetic mechanisms in cereal yield and quality (Chapter 11) and forest tree management (Chapter 12).

Philippe Gallusci
Etienne Bucher
Marie Mirouze

ACKNOWLEDGMENTS

We would like to thank Pr Graham Seymour for helpful comments and manuscript editing.

REFERENCES

Bandyopadhyay, B., Chanda, V., & Wang, Y. (2017). Finding the sources of missing heritability within rare variants through simulation. *Bioinformatics and Biology Insights, 11*. http://dx.doi.org/10.1177/1177932217735096.

Berger, S. L., Kouzarides, T., Shiekhattar, R., & Shilatifard, A. (2009 Apr 1). An operational definition of epigenetics. *Genes Dev, 23*(7), 781–783. http://dx.doi.org/10.1101/gad.1787609.

Bourrat, P., Lu, Q., & Jablonka, E. (2017). Why the missing heritability might not be in the DNA. *Bioessays: News and Reviews in Molecular, Cellular and Developmental Biology, 39*(7), 1–2. http://dx.doi.org/10.1002/bies.201700067.

Brachi, B., Morris, G. P., Borevitz, J. O., Benjamin Brachi, G. P. M., & Borevitz, J. O. (2011). Genome-wide association studies in plants: the missing heritability is in the field. *Genome Biology, 12*(232), 232. http://dx.doi.org/10.1186/gb-2011-12-10-232.

Dupont, C., Armant, D. R., & Brenner, C. A. (2009). Epigenetics: definition, mechanisms and clinical perspective. *Seminars in reproductive medicine, 27*(5), 351–357.

Li, Y., Huang, Y., Bergelson, J., Nordborg, M., & Borevitz, J. O. (2010). Association mapping of local climate-sensitive quantitative trait loci in *Arabidopsis thaliana*. *Proceedings of the National Academy of Sciences, 107*(49), 21199–21204. http://dx.doi.org/10.1073/pnas.1007431107.

Lin, T., Zhu, G., Zhang, J., Xu, X., Yu, Q., Zheng, Z., et al. (2014). Genomic analyses provide insights into the history of tomato breeding. *Nature Genetics, 46*(11), 1220–1226. http://dx.doi.org/10.1038/ng.3117.

Maher, B. (2008). Personal genomes: The case of the missing heritability. *Nature, 456*, 18–21. http://dx.doi.org/10.1038/456018a.

Manning, K., Tor, M., Poole, M., Hong, Y., Thompson, A. J., King, G. J., et al. (2006). A naturally occurring epigenetic mutation in a gene encoding an SBP-box transcription factor inhibits tomato fruit ripening. *Nature Genetics, 38*(8), 948–952. http://dx.doi.org/10.1038/ng1841.

Manolio, T. A., Collins, F. S., Cox, N. J., Goldstein, D. B., Hindorff, L. A., Hunter, D. J., et al. (2009). Finding the missing heritability of complex diseases. *Nature, 461*(7265), 747–753. http://dx.doi.org/10.1038/nature08494.

Mayr, E. (1980). In E. Mayr, & W. B. Provine (Eds.), *The Evolutionary Synthesis* (1–48). Cambridge, Massachusetts; London, England: Harvard Univ. Press.

Miura, K., Agetsuma, M., Kitano, H., Yoshimura, A., Matsuoka, M., Jacobsen, S. E., et al. (2009). A metastable DWARF1 epigenetic mutant affecting plant stature in rice. *Proceedings of the National Academy of Sciences of the United States of America, 106*(27), 11218–11223. http://dx.doi.org/10.1073/pnas.0901942106.

Morrell, P. L., Buckler, E. S., & Ross-Ibarra, J. (2012). Crop genomics: Advances and applications. *Nature Reviews Genetics, 13*(2), 85–96. http://dx.doi.org/10.1038/nrg3097.

Quadrana, L., Almeida, J., Asís, R., Duffy, T., Dominguez, P. G., Bermúdez, L., et al. (2014). Natural occurring epialleles determine vitamin E accumulation in tomato fruits. *Nature Communications, 5*(May), 3027. http://dx.doi.org/10.1038/ncomms5027.

Richards, E. J. (2006). Inherited epigenetic variation, revisiting soft inheritance. *Nature Reviews Genetics*, 7, 395–401. http://dx.doi.org/10.1038/nrg1834 10.1038/nrg1834.

Verhoeven, K. J. F., & Preite, V. (2014). Epigenetic variation in asexually reproducing organisms. *Evolution; International Journal of Organic Evolution, 68*(3), 644–655. http://dx.doi.org/10.1111/evo.12320.

Visscher, P. M. (2008). Sizing up human height variation. *Nature Genetics, 40*, 489–490. http://dx.doi.org/10.1038/ng0508-489/10.1038.

CHAPTER ONE

Epigenetics — A Historical Perspective

Peter Meyer
The University of Leeds, Centre for Plant Sciences, Leeds, United Kingdom
E-mail: p.meyer@leeds.ac.uk

Contents

Abstract

The term 'epigenetics' refers to heritable and reversible changes to a gene that do not alter its DNA sequence but that affect its competence or efficiency of expression. The three contexts in which epigenetic changes are predominantly studied today are their contribution to the developmental program of an organism, their part in the response of an organism to changing environmental conditions and their contribution to evolutionary flux via the production of new epigenetic variants.

Jean-Baptiste Lamarck's theory about the inheritance of acquired characteristics initiated the discussion about the role of heritable epigenetic changes in evolution, and Conrad Waddington's epigenetic landscape model provided a new definition of epigenesis, the development of the phenotype, as the combination of genetic expression and tissue interaction. Plant research has played a crucial role in the identification of epigenetic concepts and mechanisms, especially via the analysis of plant transposons and transgenes that helped to discover the importance of chromatin states and small RNAs.

Advances in Botanical Research, Volume 88
ISSN 0065-2296
https://doi.org/10.1016/bs.abr.2018.08.003

1. EPIGENETIC CONCEPTS IN EVOLUTION THEORIES

From the 3rd century before Christ until the early 19th century, species were considered the product of a divine act of creation and therefore distinct and unchangeable. This theory was in accordance with Christian dogmata but its foundation lies in Aristotle's metaphysical theory (Tredennick, 1933) that had replaced Heraklit's earlier doctrine of 'flux', which had argued in favour of continuous change and development of objects, probably representing the earliest model for an evolutionary theory. In contrast, Plato's favourite pupil Aristotle argued that an eternal and immutable, immaterial Mind or God pursues the work of creation. The predominance of this view is exemplified by Carl von Linné's statement in 1735 'there are as many species as there have been created by the eternal being' or by Louis Agassiz' comment in 1859 that 'all plant and animal species represent god's embodied doctrine of creation' (Haeckel, 1909). A paramount principle in Aristotle's view on creation is that it results in perfection, an aspect that was eagerly defended by clerics against the alternative Pythagorean view that perfection could only be achieved at the end of a developmental process (Rev. M'Mahon, 1896).

The first to deviate from the ancient doctrine of creation, was Jean-Baptiste Lamarck with his definition of evolution as a natural order of species constituting a series, which traversed from the most imperfect species to the most perfect ones. Nature did not bring them all into existence at the same moment but had to form them in turn, beginning with the simplest and ending with the most complex organisation. Lamarck also introduced a new concept of adaptive evolution in response to an organism's activity and environment when he proposed that (i) in animals the extended use or lack of use of specific organs will cause these to become enlarged or diminished over time, and that (ii) such changes, if experienced by both sexes, are passed on to and are preserved in individuals of successive generations (Lamarck, 1809).

In contrast to Darwin, Lamarck did not have access to the empirical tools and theories that only emerged during the first half of the 19th century, exemplified in particular by Carl Ernst von Baer's *Developmental history of animals* in 1828, which established the foundation of comparative embryology, Johannes Müller's textbook of physiology in 1833, which introduced new experimental tools, especially microscopic, chemical and physiological analysis, and Schleiden and Schwann's cell theory in 1838 (Haeckel, 1909).

Interestingly, 40 years before Schleiden and Schwan had published their cell theory, Lamarck had already proposed that all plants and animals are essentially composed of cellular tissue (Lamarck, 1809). Unfortunately, even after this statement was rediscovered in 1922 (Gerould, 1922), Lamarck was never properly credited for it.

In his work on *The variation of animals and plants under domestication* (Darwin, 1868), Darwin appeared to support Lamarck's concept of acquired characteristics when he acknowledged the role of changing environmental conditions in the evolution of heritable traits. His description of early hatching of the Aylesbury ducks could be considered an example of transgenerational inheritance of a genetic trait that had been altered by changes in environmental conditions. Aylesbury ducks were kept in houses and were reared as early as possible for the London market. When ducks that had hatched from eggs of Aylesbury ducks, were kept together with other ducks under identical conditions, they hatched their first brood in January, about two months earlier than the other ducks. In grandchildren of the Aylesbury ducks, hatching was restored again to March, similar to all other ducks. This showed that the early period of hatching had been inherited for one generation before being reversed again in the next generation. As another example of epigenetic inheritance, Darwin refers to Monnier's experiments with winter- and summer-varieties of wheat. Monnier had sown winter-wheat in spring. Only 4% of the plants produced ripe seeds, but when these were sown and re-sown in spring for three years, all seeds ripened. A converse experiment growing summer-wheat in autumn, produced similar results, demonstrating that winter-varieties can be converted into summer-varieties and *vice versa*.

Despite this recognition of the influence of environmental conditions in generating (epi)genetic variation, Darwin was unjustly critical of Lamarck's theory stating that one half of Lamarck's arguments was obsolete in view of the scientific progress made since 1809, while the other half was erroneous or defective. Darwin especially considered Lamarck's suggestion about the cause of the gradual modification of species as 'inapplicable to the whole vegetable world' (Darwin, 1887). Contrary to the criticism that Lamarck received from Darwin and some of Darwin's supporters, Lamarck did belief in a tendency towards perfection. He did acknowledge the existence of a natural order in the successive development of life but he proposed that external conditions gave it infinite variety by directing an infinite number of evolutionary branches (Gardiner, 1897).

Lamarck's theories had been vividly attacked by many of his contemporaries, in particular by George Cuvier, the influential proponent of catastrophism, who defended the concept of the preservation of species. When Lamarck died in 1829, blind and in poverty, he had been rated a discredited author and most of his extensive writings had been destroyed. It took another 80 years before Lamarck's significant contribution as a pioneer of evolution was recognised by the erection of a monument in the Jardin des Plantes in Paris (A MONUMENT TO LAMARCK, 1907). Darwin's dismissal of Lamarck's work may have contributed to a rather unnecessary dogmatic conflict between the followers of Lamarck's and Darwin's evolutionary theories. While one group allocated a role to the environment as driver for genetic variation, the other group focuses on its role in selecting the most favourable individuals among a pool of random mutants. August Weismann's theory of the continuity of the germplasm (Weismann, 1892) cemented this division as it was interpreted as proof for an exclusion of environmental conditions in altering genetic information that could be passed on to the next generation.

2. PAUL KAMMERER – CHARLATAN OR GRANDFATHER OF EPIGENETICS?

The 20th century saw a number of attempts to resurrect the concept of the selection of heritable characteristics. The first and one of the most controversial advocate was Paul Kammerer who investigated the adaptation of salamanders, toads, lizards and sea quirts to changing environmental conditions (Kammerer, 1909). While Kammerer's work is praised by some as a milestone in epigenetic research representing the first example of heritable germline modification in direct response to environmental change (Vargas, Krabichler, & Guerrero-Bosagna, 2017), it is also tainted by a long dispute about the accuracy and reproducibility of his data. This is illustrated by debates about his experiments on the midwife toad *Alytes obstetricans* (Kammerer, 1906) and the yellow sea squirt *Ciona intestinalis* (Kammerer, 1923a). While the midwife toad experiments dominate the debate about his work until today, Kammerer favoured his sea squirt experiments, which he referred to as his *experimentum crucis*, over his *Alytes* experiments as proof for the inheritance of acquired characteristics (Kammerer, 1923a).

Sea squirts have two tube-like oral siphons protruding from their heads. When he cut off the siphone, Kammerer observed that they grew back to be longer than in their original form. More importantly, the offspring inherited longer than normal siphons, which Kammerer interpreted as an example of acquired characteristics. This work was challenged by colleagues who could not repeat his experiments, and who claimed that it was well-known that abnormally long siphons could be induced by keeping the animal in suspension of abundant food, which should therefore have been taken into account by proper controls (Fox, 1923). In response to this criticism, Kammerer emphasised that the effect of 'super-regeneration' was reproducible as it had already been observed by others before his experiments but that the effect may only occur in some *Ciona* populations (Kammerer, 1923b). While this unresolved dispute may be explained by the sensitivity of epigenetic experiments to experimental conditions and by a variation in responsiveness to environmental conditions among species or populations, Kammerer's midwife toad data were criticized by some in the context of negligent or even intentional misrepresentation.

Male toads that breed in water, develop black nuptial pads on their palms and fingers to clasp the female during the mating season. Midwife toads mate on land and males therefore do not need to develop nuptial pads. By increasing the temperature, Kammerer forced land-based midwife toads to breed in water, where a small number of eggs survived. The offspring of this experiment bred in water, irrespective of the temperature. Moreover, from the third generation onwards, offspring that were bred in a moist environment, developed nuptial pads as an indicator for an inherited characteristic. The nuptial pads first only appeared on the innermost finger, later spreading to other regions with a high degree of unpredictable variability (Kammerer, 1923a). An interesting aspect of many of the phenotypes Kammerer reported, is their erratic and stochastic appearance, a hallmark of many epigenetic phenotypes that may have contributed to the irritation among Kammerer's peers who were predominantly used to phenotypes based on monogenetic traits controlled by wildtype and mutants alleles that represented functional or non-functional states, respectively. As the terrestrial midwife toad evolved from an aquatic form that originally had pads, the emergence of nuptial pads may have been induced by the random reestablishment of an ancestral trait, for example by reactivation of an existing but silenced gene.

One of the fiercest critics of Kammerer's work was William Bateson who complained about the unusually poor quality of the pictures in Kammerer's publication, which made it impossible to verify the conclusions, and who expressed concerns that Kammerer had not been able to provide specimens or photomicrographs to the community (Bateson, 1919). This criticism triggered an exchange of accusations between Kammerer and his critics, which ended with the discovery that Kammerer's specimen had been manipulated by the injection of ink (Noble, 1926), followed by Kammerer's suicide six weeks later. In a letter to the Moscow Academy of Science, Kammerer acknowledged that he had found several of his samples at the Vienna institute to have been manipulated but that he had played no part in this falsification, which had placed his whole life's work into doubt (Kammerer, 1926). Kammerer's suicide took the attention away from his complicated experimental studies towards a political debate that painted Kammerer either as an attention-seeking fraudster (Weissmann, 2010) or as a heroic victim of the scientific establishment that had deliberately sabotaged his work (Koestler, 1972).

3. POLITICAL DOGMATA

It certainly did not help that Kammerer and some of his critics had found themselves on different sides of a war, and that the subject was further dragged into the political arena when Kammerer established an affiliation with the Communist regime in Russia, which looked very favourably upon the inheritance of acquired characteristics. In most Western countries, Kammerer's case had discredited the concept of Lamarckian inheritance, which made it difficult for contemporary scientists to propose any models that linked heritable changes in phenotype to environmental conditions. In contrast, in the Soviet Union, such claims were received with open arms. The horticulturist Ivan Vladimirovich Mitchurin argued that a coddled 'bourgeois' treatment of plants prevented them of their capacity for 'self-activity'. He therefore grew fruit trees under 'Spartan conditions' at high density and without fertilisers on poor soil until fruit set, to force the development of novel hybrid varieties that were well adapted to their new environment. He proposed that new characteristics acquired in this way, could be transmitted to future generations (Michurin, 1952).

The most prominent advocate for acquired characteristics, Trofim Denisovich Lysenko, was also the most infamous character who successfully used his political influence to silence his critics. He rejected Mendelian genetics as incompatible with Marxist dialectics, contrasting it with his theory of development by stages. He argued that successive stages of development can be affected by changing environmental conditions, which are transmitted to future generations, thus creating species that are better adapted to these changing environmental conditions. He also propagated the idea that by grafting one plant onto another, genetic traits could be transmitted by transfusion. In 1940, Lysenko became Director of the Institute of Genetics of the Institute of Science of the U.S.S.R. and President of the Lenin All-Union Academy of Agricultural Sciences. In 1948, Lysenko's theory was fully endorsed at a meeting of the Lenin Agricultural Academy as the only correct theory, and classical genetics was denounced as contrary to Darwinism. Lysenko used his political influence to remove colleagues from office who had supported Mendelian genetics, some of which died in labour camps (Caspari & Marshak, 1965).

Lysenko's downfall in 1965 was rightly hailed as a liberation of science from political dogma, but it was also falsely interpreted as proof that all theories that argued for an environmental influence on heritable genetic changes were nonsensical and worthless. The Lamarckian theory was declared obsolete as any evidence in its support had turned out to be 'either invalid or fake' (Caspari & Marshak, 1965). Very few scientists at the time were prepared to consider that the environmental influence was more complex than geneticists were willing to admit, and that, in this context, Lysenko was worth listening to (Teich, 2007).

One lesson the scientific community should have learned from the various religious and political aspects that crept into the evaluation of evolutionary concepts, is to become more sensitive to interest groups occupying scientific theories, which would certainly also improve the quality of our current discussions about climate change or about the various applications of transgene technologies. The fact that Stalin supported Lysenko's theories does not discredit experiments that demonstrate heritable environmental influences on gene function. Equally, the possibility that Kammerer's samples could have been manipulated by his enemies does not automatically verify his experiments. It seems short-sighted to define epigenetics in contrast to Mendelian genetics as it rather represents an extended view on how genetic information can be used. Experimental data should certainly not be influenced by what scientists or their funders believe or expect. With regard to their reproducibility, we have to consider that epigenetic systems are susceptible to environmental conditions that are often not well defined and therefore subject to a high degree of variation. This can make them more complicated to study but it is certainly worth the effort.

4. CONRAD WADDINGTON

Conrad Waddington coined the term 'epigenetics' to illustrate the role of gene activity in epigenesis, i.e. in the development of the embryo with its specific cell types from an undifferentiated structure. He defined the 'epigenotype' as a complex of developmental processes that connect the genotype of the fertilised egg with the phenotype of the organism as the result of the use of the genotype. Any disturbance of this process at an early stage will cause significant changes in many different tissue types (Waddington, 1942b).

In contrast to the common contemporary perception that the concept of acquired characteristics was invalid, Conrad Waddington's work on genetically assimilated traits demonstrated that evolution does not exclusively depend on the selection of chance events but that genotypes can develop in response to changing environmental conditions. Waddington placed his theory between the selectionist theory and the Lamarckian concept of acquired characteristics, by linking environmental influences to genetic factors, which we may today refer to as epigenetic states. He proposed that in the early members of an evolutionary chain, phenotypic changes are induced by environmental conditions but that in later stages of evolution, these changes can be maintained by internal genetic factors, independent of the environmental stimulus (Waddington, 1942a). The genetic assimilation model does no longer depend on random mutations as the only pool for selection, but it also does not follow Lamarck's concept as it does not allow environmental conditions to induce any kind of genetic change. Instead it acknowledges that environmental conditions can induce heritable adaptive changes but it limits these effects to a group of susceptible genes that canalise the development of adaptive characters.

Waddington exemplified this canalisation in his epigenetic landscape concept (Waddington, 1940), which depicts a ball rolling down a valley as an illustration of a cell's development, which becomes more and more committed towards its ultimate fate, just like that ball commits to enter a path that more and more restricts its mobility (Fig. 1). At the beginning of its journey the ball represents a pluripotent state that can be channelled into different paths down the hill. The genetic assimilation model assumes that the route the ball takes is not random but can be influenced by environmental conditions and that its passage can induce heritable changes that influence which passage the ball takes in subsequent generations. Waddington

Figure 1 An illustration of Conrad Waddington's epigenetic landscape model. A ball rolling down a landscape of ridges and valleys will commit to a specific pathway, illustrating the journey of an undifferentiated cell that develops into a specific cell type.

illustrated this concept by applying a heat shock treatment to *Drosophila* pupae, which induced a crossveinless wing phenotype in some adults, which was not observed in the control group. When flies with the induced crossveinless phenotype were bred over several generations, the phenotype did appear even without the heat treatment. It was unlikely that the heat treatment had simply induced mutations in genes responsible for the polygenic phenotype. On the other hand, there was no indication that the crossveinless wing phenotype provided any selective advantage to cope with increased temperature. Waddington therefore suggested that the increased temperature treatment, which was most efficient 21 to 23 h after puparium formation, caused a significant level of variation in different phenotypes, including the crossveinless wing phenotype (Waddington, 1953).

Waddington's concept of canalisation identified a build-in flexibility of the genotype that allows it to adapt to environmental changes, highlighting the interaction between environmental conditions and the genome in establishing cell linages with defined epigenetic states that are faithfully maintained during mitosis. While the genome still determined the phenotype, the epigenome controlled how the available genetic information contributed to the phenotype. Within an organism, this results in the establishment of different epigenomes, which share the same genome but differ in how the genetic information is used. With regard to the heritable changes observed, Waddington suggested that the variation in phenotypes that could be induced in response to environmental stress, reflected the presence of a pool of already existing alleles that were combined in the assimilated populations (Waddington, 1956).

5. BARBARA MCCLINTOCK

In 1951, Barbara McClintock published her studies of newly arisen mutable loci in maize (McClintock, 1951), which illustrated the activity of transposable elements that could switch between transcriptionally active and silent states. Earlier work on position-effect phenomena in *Drosophila* (Muller, 1930) and *Oenothera* (Catcheside, 1939) had highlighted that a gene's translocation to a new chromosomal position could alter its expression. As this effect was heritable but also reversible, it represented an epigenetic influence of the chromosomal environment on the gene's state of expression. McClintock argued that gene activity is a reflection of the structural organisation of the genome. Her interpretation that 'nuclear components are formed into organized functional units' and that 'changes in pattern-controlling mechanisms would serve as a primary source of potential variability of genic expression without changes in the genes themselves' (McClintock, 1951) turned out to be a rather accurate model for chromatin-based epigenetic control mechanisms described many decades later (Lorch, Maier-Davis, & Kornberg, 2010). Her proposition that inactive or suppressed genes are covered by 'nongenic chromatin materials' and that genes are activated when they become exposed due to a physical change in the covering material (McClintock, 1951) was later confirmed by numerous studies that demonstrated a link between gene function and accessibility (Weintraub & Groudine, 1976) (Van Blokland, Ten Lohuis, & Meyer, 1997).

McClintock had coined the term "controlling elements" for transposable elements to highlight their ability to influence the expression of nearby genes (McClintock, 1953). Transposable elements are activated in response to biotic (Mottinger, Dellaporta, & Keller, 1984) or abiotic stress (Walbot, 1999), which let McClintock to propose that this was part of an adaptation process of the genome to environmental stress. If in response to stress, transposable elements could move into or near a new gene locus, this could modify their expression. McClintock pointed out that some of the changes induced by activated elements were only detected in later generations

(McClintock, 1984), which suggested that these changes were heritable, in accordance with Lamarck's hypothesis for the inheritance of acquired characteristics.

McClintock's work had highlighted the presence of a powerful new driver for genetic variation, which was not mediated by stable genetic mutation but based on reversible and heritable changes in gene expression. A few of her colleagues embraced this concept at the time (Stadler, 1954), while others struggled to accept that mobile elements could have a controlling effect on genes. One criticism she received was that she was confusing cause and effect, and that development controlled transposition (Comfort, 2002). The scientific community quickly acknowledged the concept of transposition as a driver for genetic mutation but it took considerably longer before the significance of transposable elements was acknowledged as a source of allelic regulatory variation in gene response to stress (Makarevitch et al., 2015) (Forestan et al., 2016).

6. THE ROLE OF TRANSGENIC PLANTS IN IDENTIFYING EPIGENETIC PATHWAYS

Following up on McClintock's work on transposable elements in maize, the plant research community continued to make essential contributions to the elucidation of epigenetic phenomena and mechanism. The arrival of plant transformation technologies in the early 1980s provided transgenes associated with distinct phenotypes that could be used to visualize epigenetic effects, in particular the inactivation of transgenes. In contrast to endogenous traits, the representation of many transgenes in subsequent generations did not strictly follow Mendelian rules. One reason for the under-representation of introduced genes appeared to be their inactivation via DNA methylation, which had been observed in crown gall tumours that displayed heritable silencing of all or some hormone-producing genes. These genes were activated again after application of the demethylation agent 5-azacytidine (Hepburn, Clarke, Pearson, & White, 1983). Early studies used methylation-sensitive restrictions enzymes to compare active and silenced states in crown gall tumour lines and in transgenic plants, demonstrating a link between gene silencing and methylation of cytosines (Amasino, Powell, & Gordon, 1984) in CG and CNG sequence contexts (Peerbolte, Leenhouts, Hooykaas-van Slogteren, Wullems, & Schilperoort, 1986). Bisulfite sequencing revealed that, in contrast to mammals where methylated cytosines were predominantly found in CG contexts, plants contained methylated cytosines in all sequence contexts (Meyer, Niedenhof, & ten Lohuis, 1994).

The correlation between DNA methylation and structural changes was confirmed as methylated and silenced transgenes showed a reduced sensitivity to endonucleases reflecting a higher level of chromatin condensation (Van Blokland et al., 1997). This suggested that transcriptional silencing involved structural changes similar to heterochromatinisation and that DNA methylation changes were either the cause or consequence of structural changes. In accordance with Mc Clintock's reports on the influence of the chromosomal environment on the activity of transposons, a similar effect could be observed for transgenes, which were prone to become methylated and transcriptionally silenced when they integrated into a densely methylated genomic region (Prols & Meyer, 1992).

7. HOMOLOGY-DEPENDENT GENE SILENCING

Especially in transgenic lines with multiple transgene copies, fewer than expected progeny plants contained active transgenes (Deroles & Gardner, 1988), suggesting that the presence of multiple homologous transgenes triggered transgene silencing. This 'homology-dependent gene silencing' effect could be activated by crossing two lines that contained homologous transgenes. Two resistance marker constructs that shared a common promoter were silenced when they were combined in the same plant. Suppression correlated with methylation within a common promoter region that both genes shared. Both the inactivation and promoter methylation of the two genes were reversed when the two homologous transgenes were segregated in backcrossed progeny (Matzke, Primig, Trnovsky, & Matzke, 1989). This demonstrated that DNA methylation was dependent on the presence of a second homologous copy, which explained the predisposition of transformants with multiple transgene copies to become inactive.

Another homology-dependent silencing effect, which was, however, not based on transcriptional gene silencing (TGS) via DNA methylation, was reported in two independent studies that had expressed the floral pigmentation pathway *genes dihydroflavonol-4-reductase (DFR)* and *chalcone synthase A (chsA)*, respectively, in petunia (Napoli, Lemieux, & Jorgensen, 1990) (Krol, Leon, Beld, Mol, & Stuitje, 1990). In a large number of transformants, both the transgene and its endogenous homologous version had become inactive, resulting in a loss-of-function phenotype with a drastic decrease in mRNA levels, both for the transgene and the homologous endogene. This effect, which was coined 'co-suppression', was controlled by the transgene as the endogene became re-activated when it was segregated from the transgene (Napoli et al., 1990). Nuclear run-on studies showed that, unlike homology-based silencing effects that involved hypermethylation of promoters, co-suppression was a post-transcriptional silencing (PTGS) event that involved the joint degradation of endogenous and transgenic *chsA* transcripts. Initially, it was unclear if transgene expression was a requirement for co-suppression as even promoter-less *chsA* transgene constructs could induce co-suppression (Van Blokland, Van der Geest, Mol, & Kooter, 1994). Co-suppression studies of nitrate reductase in tobacco, did finally demonstrate the requirement of transgene expression for PTGS, as a transcriptionally silenced nitrate reductase transgene lost its ability to induce co-suppression (Vaucheret, Nussaume, Palauqui, Quilléré, & Elmayan, 1997). Co-suppression was not restricted to transgenic plants as it resembled effects in flowers with pigmented sectors observed in some natural petunia varieties. These phenotypes were later identified as the consequence of tissue-specific silencing effects mediated by dominant *chsA* copies (Della Vedova et al., 2005; Koseki, Goto, Masuta, & Kanazawa, 2005).

8. RNA AS A MEDIATOR OF SILENCING

Early mechanistic models for homology-based gene silencing phenomena had focussed on a transfer of repressive DNA methylation marks following the association of methylated and unmethylated homologous DNA regions. These DNA association models, were, however, not suitable to explain co-suppression effects that did not involve changes in DNA methylation but were mediated by post-transcriptional degradation. Alternative models therefore focused on a role of RNA molecules as signal molecules that mediate gene silencing of homologous DNA and RNA regions.

Evidence for a role of RNA molecules in DNA methylation and TGS came from the analysis of a transgenic tobacco line that contained a transgene cDNA copy of the Potato spindle tuber viroid (PSTVd). Infection of this line with PSTVd, which resulted in the accumulation of viroid RNA in the nucleus, also triggered DNA methylation of the homologous PSTVd sequences of the T-DNA, suggesting that viroid RNA was causal to the methylation of homologous DNA sequences (Wassenegger, Heimes, Riedel, & Sanger, 1994). The authors favoured an RNA/DNA hybrid model to explain the targeting of DNA methylation to DNA regions with homology to the PSTVd RNA. This model was later corrected following the discovery of small RNAs but it was confirmed two decades later in mammals that RNA/DNA hybrids could actually trigger DNA methylation at rDNA loci (Schmitz, Mayer, Postepska, & Grummt, 2010).

Viral research also prepared the ground to identify the significance of RNA for PTGS. Early strategies to design virus-resistant plants involved the expression of viral coat protein genes with the expectation that the abundant presence of its coat protein would prepare a plant against the invasion of viral RNA. This concept actually appeared to work as transgenic coat-protein lines became resistant to RNA virus infection. Surprisingly, however, this resistance could even be observed in transgenic lines that expressed non-coding or truncated sense transcripts of viral coat proteins (Lindbo & Dougherty, 1992). While these transgenes were transcribed at high levels, the plants showed very low steady-state levels of transgenic RNA. This suggested that virus resistance was mediated by a PTGS mechanism that was triggered by the presence of homologous RNA copies that could form a dsRNA with the viral RNA. The authors proposed that PTGS was triggered by the production of small mRNA segment from the transgene DNA, which hybridised to homologous regions of the viral RNA, providing targets for amplification by RNA-dependent RNA polymerases and for degradation by dsRNA-specific RNAses (Lindbo, Silva-Rosales, Proebsting, & Dougherty, 1993). Within a few years, this model was confirmed as a correct prediction of the mechanisms involved in PTGS.

In petunia lines with *chsA* co-suppression and in white flower sectors of wildtype lines, aberrant *chsA* fragments were detected, including polyA-free 3′ RNA fragments, which suggested that *chsA* endogene and transgene transcripts were degraded by endonucleolytic cleavage at different positions. This supported a concept of transcript cleavage after pairing with an aberrant homologous transcript by a dsRNA-specific RNase (Metzlaff, Odell, Cluster, & Flavell, 1997). Direct evidence that transcript degradation was mediated by dsRNA formation came from the introduction of transgenes that produced dsRNAs and successfully generated resistance against homologous viral RNA (Waterhouse, Graham, & Wang, 1998). Mutant selection against PTGS in *Neurospora crassa* identified the first example of a RNA-dependent RNA polymerase (RdRP) as a component in the PTGS pathway (Cogoni & Macino, 1999).

A breakthrough for our understanding of the mechanistic role of double-stranded and aberrant RNAs in epigenetic gene regulation, came from the discovery of a group of small RNA molecules in plants with post-transcriptionally silencing transgenes (Hamilton & Baulcombe, 1999). These lines contained ~25nt RNA fragments in antisense orientation to the introduced transgenes, which derived from reverse transcription of a sense RNA template by an RdRP. The association of silencing with a small and potentially mobile RNA offered a plausible explanation for previous observations of a systemic gene silencing signal that moved from a local infiltration area into adjacent cells (Voinnet & Baulcombe, 1997), or from a silenced stock to a non-silenced scion (Palauqui, Elmayan, Pollien, & Vaucheret, 1997). Following the identification of genes encoding functions required for the production of small RNAs, the analysis of mutants of these genes confirmed the long-range mobility of different classes of small RNAs (Molnar et al., 2010).

9. OUTLOOK

The observations that environmental conditions can cause changes in DNA methylation and gene expression (Ten Lohuis, Galliano, Heidmann, & Meyer, 1995) (Kovarik, Koukalová, Bezdek, & Opatrn, 1997) and that DNA methylation changes can be propagated over several generations (Cubas, Vincent, & Coen, 1999) (Silveira et al., 2013) have revived Lamarck's concept of adaptive evolution in response to environmental conditions. The mobility of small RNAs that control DNA methylation or post-transcriptional silencing, offers an attractive mechanistic model for the transmission of epigenetic states within a plant but even to subsequent generations if small RNAs could move from somatic tissue into gametophytes. Heritable methylation states that are not controlled by small RNAs (Singh, Zubko, & Meyer, 2008) (Watson, Hawkes, & Meyer, 2014), may be propagated by recruiting DNA methyltransferase to heritable chromatin states or structural features.

As attractive as epigenetic models have recently become, especially in the context of the climate change debate, it will be important to avoid general claims about the inheritance of environmentally induced traits. Each individual case for epigenetic inheritance will require proof of a causal link between a gene and a specific phenotype, and evidence for the heritable transmission of epi-alleles of this gene with defined epigenetic, phenotypic and expression states.

REFERENCES

Amasino, R. M., Powell, A. L. T., & Gordon, M. P. (1984). Changes in T-DNA methylation and expression are associated with phenotypic variation and plant regeneration in a crown gall tumor line. *Molecular and General Genetics, 197*(3), 437–446.

Bateson, W. (1919). Dr. Kammerer's testimony to the inheritance of acquired characters. *Nature, 103*(2592), 344.

Caspari, E. W., & Marshak, R. E. (1965). The rise and fall of Lysenko. *Science, 149*(3681), 275–278.

Catcheside, D. G. (1939). A position effect in Oenothera. *Journal of Genetics, 38*(1), 345–352.

Cogoni, C., & Macino, G. (1999). Gene silencing in Neurospora crassa requires a protein homologous to RNA-dependent RNA polymerase. *Nature, 399*(6732), 166–169.

Comfort, N. C. (2002). Barbara McClintock's controlling elements: Premature discovery or stillborn theory? In E. B. Hook (Ed.), *Prematurity in scientific discovery* (pp. 175–199). Berkeley: University of California Press.

Cubas, P., Vincent, C., & Coen, E. (1999). An epigenetic mutation responsible for natural variation in floral symmetry. *Nature, 401*(6749), 157–161.

Darwin, C. R. (1868). *The variation of animals and plants under domestication.* London: John Murray.

Darwin, F. (1887). *The life and letters of Charles Darwin* (Vol. I). D.Appleton and Company.

Della Vedova, C. B., Lorbiecke, R., Kirsch, H., Schulte, M. B., Scheets, K., Borchert, L. M., et al. (2005). The dominant inhibitory chalcone synthase allele C2-Idf (Inhibitor diffuse) from Zea mays (L.) acts via an endogenous RNA silencing mechanism. *Genetics, 170*(4), 1989–2002.

Deroles, S. C., & Gardner, R. C. (1988). Analysis of the T-DNA structure in a large number of transgenic petunias generated by agrobacterium-mediated transformation. *Plant Molecular Biology, 11*(3), 365–377.

Forestan, C., Aiese Cigliano, R., Farinati, S., Lunardon, A., Sanseverino, W., & Varotto, S. (2016). Stress-induced and epigenetic-mediated maize transcriptome regulation study by means of transcriptome reannotation and differential expression analysis. *Scientific Reports, 6*, 30446.

Fox, H. M. (1923). Dr. Kammerer's Ciona experiments. *Nature, 112*, 653.

Gardiner, J. (1897). Lamarck and the 'perfecting tendency'. *Science, 6*(157), 995–997.

Gerould, J. H. (1922). Lamarck, mirbel and the cell theory. *Science, 55*(1425), 421–422.

Haeckel, E. (1909). *Das Weltbild von Darwin und Lamarck.* Leipzig: Alfred Kröner Verlag.

Hamilton, A. J., & Baulcombe, D. C. (1999). A species of small antisense RNA in posttranscriptional gene silencing in plants. *Science, 286*(5441), 950–952.

Hepburn, A. G., Clarke, L. E., Pearson, L., & White, J. (1983). The role of cytosine methylation in the control of nopaline synthase gene expression in a plant tumor. *Journal of Molecular and Applied Genetics, 2*(3), 315–329.

Kammerer, P. (1906). Experimentelle Veränderung der Fortpflanzungstätigkeit bei Geburtshelferkröte (*Alytes obstetricans*) und Laubfrosch (*Hyla arborea*). *Archiv für Entwicklungsmechanik der Organismen, 22*, 48–140.

Kammerer, P. (1909). Vererbung erzwungener fortpflanzungsanpassungen. *Archiv für Entwicklungsmechanik der Organismen, 28*(4), 447–545.

Kammerer, P. (1923a). Breeding experiments on the inheritance of acqired characters. *Nature, 111*, 637–640.

Kammerer, P. (1923b). Experiments on Ciona and Alytes. *Nature, 112*, 826–827.

Kammerer, P. (1926). Paul Kammerer's letter to the Moscow Academy. *Science, 64*(1664), 493–494.

Koestler, A. (1972). *The case of the midwife toad.* New York: Random House Inc.

Koseki, M., Goto, K., Masuta, C., & Kanazawa, A. (2005). The star-type color pattern in petunia hybrida 'red star' flowers is induced by sequence-specific degradation of chalcone synthase RNA. *Plant & Cell Physiology, 46*(11), 1879–1883.

Kovarik, A., Koukalová, B., Bezdek, M., & Opatrn, Z. (1997). Hypermethylation of tobacco heterochromatic loci in response to osmotic stress. *Theoretical and Applied Genetics, 95*(1–2), 301–306.

Krol, A. R. v. d., Leon, A. M., Beld, M., Mol, J. N. M., & Stuitje, A. R. (1990). Flavonoid genes in petunia: Addition of a limited number of gene copies may lead to a suppression of gene expression. *The Plant Cell, 2*(4), 291–299.

Lamarck, J. B. P. A. (1809). *Philosophie Zoologique ou exposition des considerations relatives a l'histoire naturelle des animaux*. Paris: J.B. Bailliere.

Lindbo, J. A., & Dougherty, W. G. (1992). Untranslatable transcripts of the tobacco etch virus coat protein gene sequence can interfere with tobacco etch virus replication in transgenic plants and protoplasts. *Virology, 189*(2), 725–733.

Lindbo, J. A., Silva-Rosales, L., Proebsting, W. M., & Dougherty, W. G. (1993). Induction of a highly specific antiviral state in transgenic plants: Implications for regulation of gene expression and virus resistance. *The Plant Cell Online, 5*(12), 1749–1759.

Lorch, Y., Maier-Davis, B., & Kornberg, R. D. (2010). Mechanism of chromatin remodeling. *Proceedings of the National Academy of Sciences, 107*(8), 3458–3462.

Makarevitch, I., Waters, A. J., West, P. T., Stitzer, M., Hirsch, C. N., Ross-Ibarra, J., et al. (2015). Transposable elements contribute to activation of maize genes in response to abiotic stress. *PLoS Genetics, 11*(1). e1004915.

Matzke, M., Primig, M., Trnovsky, J., & Matzke, A. J. M. (1989). Reversible methylation and inactivation of marker genes in sequentially transformed tobacco plants. *EMBO Journal, 8*, 643–649.

McClintock, B. (1951). Chromosome organization and genic expression. *Cold Spring Harbor Symposia on Quantitative Biology, 16*, 13–47.

McClintock, B. (1953). Induction of instability at selected loci in maize. *Genetics, 38*(6), 579–599.

McClintock, B. (1984). The significance of responses of the genome to challenge. *Science, 226*(4676), 792–801.

Metzlaff, M., Odell, M., Cluster, P. D., & Flavell, R. B. (1997). RNA-mediated RNA degradation and chalcone synthase A silencing in Petunia. *Cell, 88*(6), 845–854.

Meyer, P., Niedenhof, I., & ten Lohuis, M. (1994). Evidence for cytosine methylation of non-symmetrical sequences in transgenic Petunia hybrida. *The EMBO Journal, 13*, 2084–2088.

Michurin, I. V. (1952). *Some problems of method*. Moscow: Foreign Languages Publishing House.

Molnar, A., Melnyk, C. W., Bassett, A., Hardcastle, T. J., Dunn, R., & Baulcombe, D. C. (2010). Small silencing RNAs in plants are mobile and direct epigenetic modification in recipient cells. *Science, 328*(5980), 872–875.

A MONUMENT TO LAMARCK. (1907). *Science, 25*(646), 795.

Mottinger, J. P., Dellaporta, S. L., & Keller, P. B. (1984). Stable and unstable mutations in aberrant ratio stocks of maize. *Genetics, 106*(4), 751–767.

Muller, H. J. (1930). Types of visible variations induced by X-rays inDrosophila. *Journal of Genetics, 22*(3), 299–334.

Napoli, C., Lemieux, C., & Jorgensen, R. (1990). Introduction of a chimeric chalcone synthase gene into petunia results in reversible Co-Suppression of homologous genes in Trans. *The Plant Cell Online, 2*(4), 279–289.

Noble, G. K. (1926). Kammerer's Alytes. *Nature, 118*(2962), 209–211.

Palauqui, J.-C., Elmayan, T., Pollien, J.-M., & Vaucheret, H. (1997). Systemic acquired silencing: Transgene-specific post-transcriptional silencing is transmitted by grafting from silenced stocks to non-silenced scions. *The EMBO Journal, 16*, 4738–4745.

Peerbolte, R., Leenhouts, K., Hooykaas-van Slogteren, G. M. S., Wullems, G. J., & Schilperoort, R. A. (1986). Clones from a shooty tobacco crown gall tumor II: Irregular T-DNA structures and organization, T-DNA methylation and conditional expression of opine genes. *Plant Molecular Biology, 7*(4), 285–299.
Prols, F., & Meyer, P. (1992). The methylation patterns of chromosomal integration regions influence gene activity of transferred DNA in Petunia hybrida. *The Plant Journal, 2*(4), 465–475.
Rev. M'Mahon, J. H. (1896). *The metaphysics of Aristotle*. London: George Bell & Sons.
Schmitz, K.-M., Mayer, C., Postepska, A., & Grummt, I. (2010). Interaction of noncoding RNA with the rDNA promoter mediates recruitment of DNMT3b and silencing of rRNA genes. *Genes & Development, 24*(20), 2264–2269.
Silveira, A. B., Trontin, C., Cortijo, S., Barau, J., Del Bem, L. E. V., Loudet, O., et al. (2013). Extensive natural epigenetic variation at a *de novo* originated gene. *Plos Genetics, 9*(4). e1003437.
Singh, A., Zubko, E., & Meyer, P. (2008). Co-operative activity of DNA methyltransferases for maintenance of symmetrical and non-symmetrical cytosine methylation in *Arabidopsis thaliana*. *The Plant Journal, 56*, 814–823.
Stadler, L. J. (1954). The gene. *Science, 120*(3125), 811–819.
Teich, M. (2007). Haldane and Lysenko revisited. *Journal of the History of Biology, 40*(3), 557–563.
Ten Lohuis, M., Galliano, H., Heidmann, I., & Meyer, P. (1995). Treatment with propionic and butyric acid enhances expression variegation and promoter methylation in plant transgenes. *Biological Chemistry Hoppe-seyler, 376*(5), 311–320.
Tredennick, H. (1933). *Aristotle - the metaphysics books*. London: Willian Heinemann.
Van Blokland, R., Ten Lohuis, M., & Meyer, P. (1997). Condensation of chromatin in transcriptional regions of an inactivated transgene: An indication for an active role of transcription in gene silencing. *Molecular and General Genetics, 257*, 1–13.
Van Blokland, R., Van der Geest, N., Mol, J. N. M., & Kooter, J. M. (1994). Transgene mediated suppression of chalcone synthase expression in Petunia hybrida results from an increase in RNA turnover. *The Plant Journal, 6*, 861–877.
Vargas, A. O., Krabichler, Q., & Guerrero-Bosagna, C. (2017). An epigenetic perspective on the midwife toad experiments of Paul Kammerer (1880–1926). *Journal of Experimental Zoology Part B: Molecular and Developmental Evolution, 328*(1–2), 179–192.
Vaucheret, H., Nussaume, L., Palauqui, J. C., Quilléré, I., & Elmayan, T. (1997). A transcriptionally active state is required for post-transcriptional silencing (Cosuppression) of nitrate reductase host genes and transgenes. *The Plant Cell Online, 9*(8), 1495–1504.
Voinnet, O., & Baulcombe, D. C. (1997). Systemic signalling in gene silencing. *Nature, 389*, 553.
Waddington, C. H. (1940). *Organisers and genes*. Cambridge: Cambridge University Press.
Waddington, C. H. (1942a). Canalization of development and the inheritance of acquired characters. *Nature, 150*, 563.
Waddington, C. H. (1942b). The epigenotype. *Endeavour, 1*, 18–20.
Waddington, C. H. (1953). Genetic assimilation of an acquired character. *Evolution, 7*(2), 118–126.
Waddington, C. H. (1956). Genetic assimilation of the bithorax phenotype. *Evolution, 10*(1), 1–13.
Walbot, V. (1999). UV-B damage amplified by transposons in maize. *Nature, 397*(6718), 398–399.
Wassenegger, M., Heimes, S., Riedel, L., & Sanger, H. L. (1994). RNA-directed de novo methylation of genomic sequences in plants. *Cell, 76*(3), 567–576.
Waterhouse, P. M., Graham, M. W., & Wang, M.-B. (1998). Virus resistance and gene silencing in plants can be induced by simultaneous expression of sense and antisense RNA. *Proceedings of the National Academy of Sciences, 95*(23), 13959–13964.

Watson, M., Hawkes, E., & Meyer, P. (2014). Transmission of epi-alleles with MET1-dependent dense methylation in *Arabidopsis thaliana*. *PLoS One, 9*(8). e105338.
Weintraub, H., & Groudine, M. (1976). Chromosomal subunits in active genes have an altered conformation. *Science, 193*(4256), 848–856.
Weismann, A. (1892). *Das Keimplasma: eine Theorie der Vererbung*. Jena: Gustav Fischer.
Weissmann, G. (2010). The midwife toad and alma mahler: Epigenetics or a matter of deception? *The FASEB Journal, 24*(8), 2591–2595.

CHAPTER TWO

Epigenetic Mechanisms in Plants

Hiroki Maeji and Taisuke Nishimura[1]
Department of Bioengineering, Nagaoka University of Technology, Nagaoka, Niigata, Japan
[1]Corresponding author: E-mail: tnishi@vos.nagaokaut.ac.jp

Contents

Abstract

Gene expression is associated with chromatin status, which is often determined by epigenetic marks such as DNA methylation, histone modification, and histone variants. Genetic and genomic studies mainly in *Arabidopsis thaliana* have revealed the functions of these epigenetic marks and identified many of the factors involved in the regulation of epigenetic marks. Based on the findings obtained from these studies, new molecular breeding techniques will undoubtedly be developed and applied to commercially important plant species. In this chapter, the mechanisms and functions of DNA methylation, histone modification, and of histone variants are outlined.

1. INTRODUCTION

In eukaryotes, gene regulation is closely associated with chromatin status. In most cases, chromatin status is independent of nucleotide sequence and is maintained during cell division. Therefore, chromatin dynamics are considered to be under epigenetic regulation. Chromatin is formed of

Advances in Botanical Research, Volume 88
ISSN 0065-2296
https://doi.org/10.1016/bs.abr.2018.09.014

nucleosomes which are composed of dimers of the histone core proteins, H2A, H2B, H3 and H4, surrounded by DNA; linker DNAs bound to H1 protein are inserted between the nucleosomes. Chromatin is generally classified as euchromatin or heterochromatin. Euchromatin is a lightly packed chromatin and genes within these regions are transcriptionally active. Heterochromatin is a more tightly packed chromatin that is transcriptionally inactive because transcriptional activators cannot access target genes.

Chemical modifications of DNA and histone proteins influence chromatin features by affecting interactions between DNA and histone proteins, and by altering recruitment of protein factors to chromatin. In mammals and plants, methylation at the fifth position of cytosine often affects gene expression; by contrast, gene regulation by cytosine methylation is less important in model organisms such as yeasts, nematodes, and fruit flies (Goll & Bestor, 2005). Histone proteins, especially the so-called histone tails located at the N-terminal residues, can undergo a range of post-transcriptional modifications including acetylation, methylation, ubiquitination, and phosphorylation; these modifications determine the features of the chromatin and affect gene transcription (Kouzarides, 2007). Another factor that influences chromatin dynamics is histone variants. There are several variants of each histone core protein, except for H4 protein. Some of these variants differ in only a few amino acids; however, these differences alter the characteristics of the protein. Nucleosomes can vary depending on the composition of histone variants (Talbert & Henikoff, 2017).

Molecular genetic analyses of *Arabidopsis thaliana* (hereafter Arabidopsis) mutants identified mechanisms that regulate DNA methylation, histone modification, and histone variants, and identified their physiological functions in plants. Some features of these mechanisms are common to those of mammals, although others are unique to plants. In this chapter, we focus on the epigenetic regulation mechanisms characterized in these studies.

2. DNA METHYLATION

2.1 Functional Characteristics of DNA Methylation in Plants

In mammals, cytosine methylation plays an important role in gene expression during development and environmental responses; it is also important to silencing of Transposable Elements (TEs), genomic imprinting, and X chromosome inactivation (Bird, 2002). In contrast, cytosine

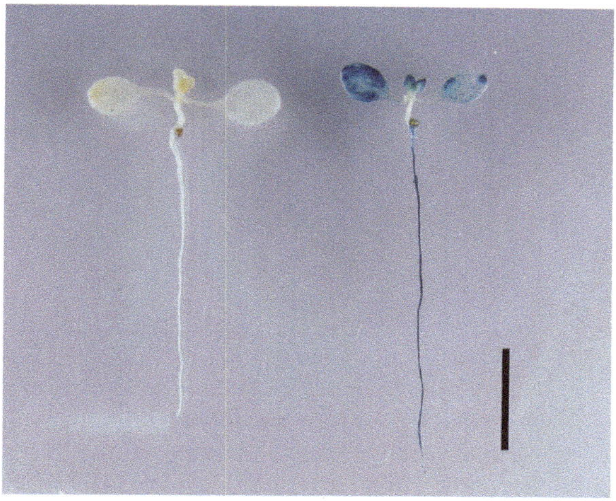

Figure 1 ***Loss of DNA methylation releases transgene silencing.*** Left: Arabidopsis L5 line into which a GUS reporter gene has been transformed (Morel et al., 2000). The transgene is methylated and expression is repressed. Right: Arabidopsis L5 line with a *met1* background in which DNA methylation is severely compromised. Repression of transgene expression is released. Bar = 5 mm.

methylation in plants mainly functions in genome stability such as formation of heterochromatin for TE silencing, repeat sequences, and transgenes (Chan, Henderson, & Jacobsen, 2005) (Fig. 1). However, cytosine methylation can also function in other ways in plants. For example, the Arabidopsis DNA methyltransferase mutant, *dna methyltransferase1* (*met1*), in which cytosine methylation throughout the genome is largely lost, shows developmental abnormalities (Finnegan, Peacock, & Dennis, 1996; Reinders et al., 2009; Ronemus, Galbiati, Ticknor, Chen, & Dellaporta, 1996) and defects in environmental responses (Dowen et al., 2012; Reinders et al., 2009) (Fig. 2). This suggests that cytosine methylation may have other roles in the control of gene expression during plant development and for environmental responses. How cytosine methylation acts in these pathways is still largely unknown.

Cytosine methylation has been observed in heterochromatic chromosomal regions in mammals and plants, which is consistent with its role in heterochromatin formation. In the past decade, genome-wide methylation analyses have shown that cytosine methylation also frequently occurs in coding regions of active genes (Bewick & Schmitz, 2017). This type of methylation is termed gene body methylation and has been observed in both plant and mammalian genomes, although its function is still unknown.

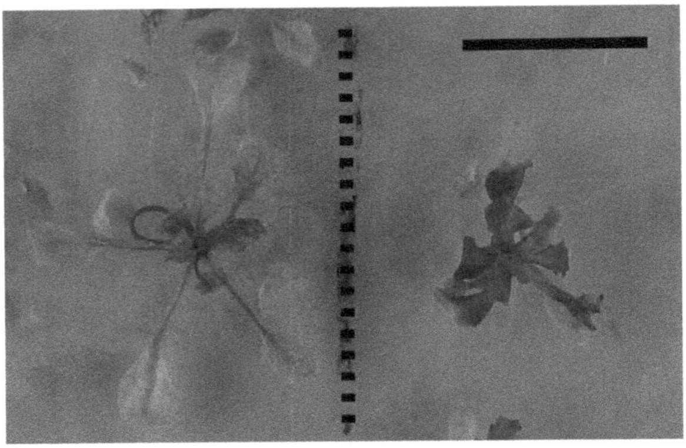

Figure 2 ***Loss of DNA methylation leads to resistance against pathogenic bacteria.*** Left: Wild-type plant infected with *Pseudomonas syringae* pv. *tomato* strain DC3000 by Flood-inoculation Method (Ishiga, Ishiga, Ichinose, & Mysore, 2017). Right: *met1* mutant infected with the same bacterium. A resistance phenotype is observed in the *met1* mutant. Photograph was taken by Ami Hoshino. Bar = 1 cm.

Interestingly, some plant species can completely lose gene body methylation in their genomes without any visible defects (Bewick et al., 2016). This indicates that gene body methylation may not have conserved roles in plants.

In the human genome, most cytosine methylation occurs in a symmetric CpG context (Lister et al., 2009). However, in the Arabidopsis genome, non-CpG contexts, such as CpHpG and CpHpH (H = A, T or C), are often methylated in addition to CpG methylation (Cokus et al., 2008; Lister et al., 2008). Non-CpG methylation is also found in mammalian embryonic stem cells in which 25% of methylcytosine is found in non-CpG contexts (Lister et al., 2009).

The pattern of DNA methylation can be maintained stably into daughter cells during cell division. In plants, some cytosine methylation, especially CpG methylation, is maintained during meiosis in contrast to mammals in which global cytosine methylation is erased at early embryogenesis and reprogrammed during later development (Feng, Jacobsen, & Reik, 2010; Paszkowski & Grossniklaus, 2011). This means that a change to the DNA methylation pattern, called epi-allele, in plants can behave like a mutation of nucleotide sequences and cause phenotypic changes (Cubas, Vincent, & Coen, 1999; Jacobsen & Meyerowitz, 1997; Manning et al., 2006; Soppe et al., 2000). CpHpH methylation is erased at early stages of plant development (Calarco et al., 2012; Kawakatsu, Nery, Castanon, & Ecker, 2017; Slotkin et al., 2009); reprogramming of CpHpH methylation occurs mainly in TEs, and the significance of this for plant development is not known although suggested as a mechanism of TE surveillance.

2.2 Maintenance of DNA Methylation

During DNA replication, the newly synthesized DNA strand does not possess a methyl group at cytosines and the double stranded DNA is hemimethylated. In mammals, a maintenance DNA methyltransferase, called Dnmt1, can restore the hemimethylated state of CpG site to a fully methylated one immediately after the DNA synthesis by DNA polymerase (Goll & Bestor, 2005). In contrast to mammals, independent mechanisms requiring three different classes of DNA methyltransferases are known to act on maintenance of DNA methylation in plants (Law & Jacobsen, 2010) (Fig. 3). The domain structures of each class of the DNA methyltransferases are highly conserved among plant species (Pavlopoulou & Kossida, 2007).

An Arabidopsis Dnmt1 orthologue, MET1, maintains CpG methylation in a manner similar to mammals (Law & Jacobsen, 2010; Pavlopoulou & Kossida, 2007) (Fig. 4). The MET1 protein has a catalytic domain for DNA methyltransferase activity in its C-terminus. In the N-terminus region,

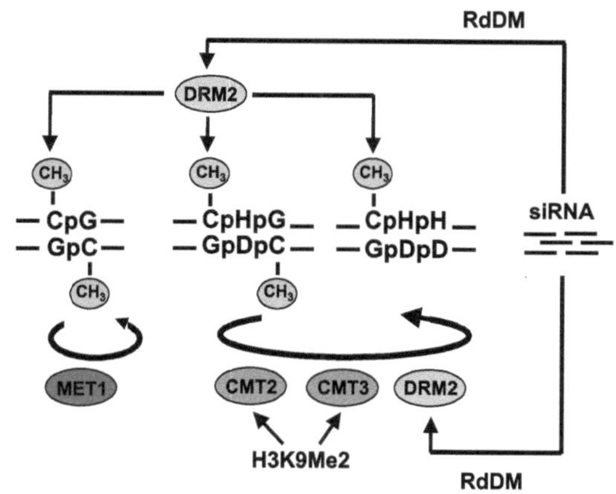

Figure 3 ***Enzymes for DNA methylation and a working model in Arabidopsis.*** In plants, DNA methylation occurs in three sequence contexts, CpG, CpHpG and CpHpH. DNA methylation in Arabidopsis is catalyzed by three classes of DNA methyltransferase. Class one: MET1 maintains CpG methylation. Class two: CMT2 and CMT3 mainly maintain CpHpG and CpHpH methylation, respectively, and methylation by CMT2 and CMT3 is facilitated by H3K9Me2. Class three: DRM2 establishes DNA methylation in all sequence contexts through the RdDM pathway. In addition, DRM2 maintains CpHpG and CpHpH methylation in a similar manner.

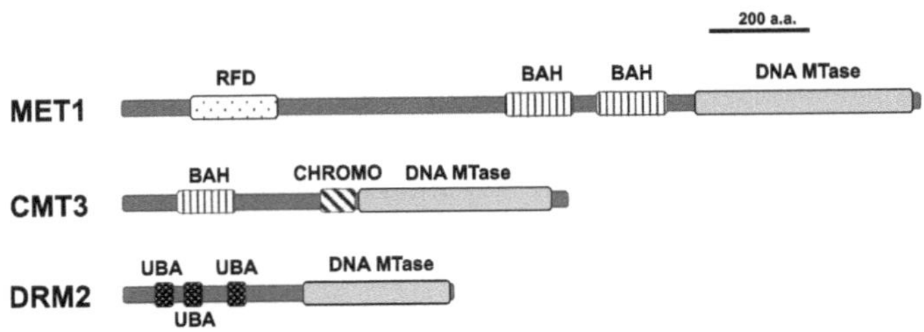

Figure 4 ***Structures of three classes of DNA methyltransferases in Arabidopsis.*** See text for the details.

MET1 protein has two replication foci domains (RFDs) for targeting of the protein to the replication foci. Two bromo-adjacent homology (BAH) domains, which are often found among the proteins involved in chromatin regulation and predicted to act on protein-protein interaction, are located between the RFD and the DNA methyltransferase domains. In Arabidopsis, VIM1, an orthologue of mammal Uhrf1, can bind to hemi-methylated CpG via the SRA domain, which is required for maintenance of DNA methylation by MET1 (Kim, Kim, Richards, Chung, & Woo, 2014; Yao, Song, He, Kumaran, & Dunn, 2012).

The remaining two classes of DNA methyltransferases maintain non-CpG methylation. The second class of DNA methyltransferase is CHROMOMETHYLASE (CMT) which is found only in the plant kingdom (Pavlopoulou & Kossida, 2007). In addition to the DNA methyltransferase domain and BAH domain, CMTs possess a chromatin organization modifier (CHROMO) domain which recognizes di-methylation of lysine at the ninth position of histone H3 (H3K9Me2) with BAH domain (Du et al., 2012) (Fig. 4). In Arabidopsis, the maintenance of DNA methylation in CpHpG context is catalyzed by CMT3, one of CMT proteins. CMT3 can bind methylated histone H3 tails via recognition of H3K9Me2 by the CHROMO and BAH domains, and maintains CpHpG methylation in the targeted regions (Du et al., 2012; Lindroth et al., 2004). Loss of H3K9Me2 in the mutant of *KRYPTONITE (KYP)/SUPRESSOR OF VARIEGATION 3-9 HOMOLOG 4 (SUVH4)*, encoding the histone methyltransferase largely responsible for H3K9Me2, results in a drastic decrease of CpHpG methylation (Jackson, Lindroth, Cao, & Jacobsen, 2002). In addition to the histone methyltransferase domain, KYP/SUVH4 has an SRA domain which binds methylated DNA (Du et al., 2014; Johnson et al., 2007). These studies suggest that CMT3 and KYP/SUVH4 form a self-reinforcing loop between DNA methylation and histone methylation.

INCREASE IN BONSAI METHYLATION1 (IBM1) has a catalytic activity of demethylation of H3K9 (Saze, Shiraishi, Miura, & Kakutani, 2008). In *ibm1* mutants, ectopic H3K9 methylation and CMT3-dependent CpHpG hypermethylation occurs at protein coding genes, suggesting that IBM1 prevents the repression of these protein coding genes via H3K9 demethylation and non-CG DNA demethylation (Inagaki et al., 2010; Miura et al., 2009). CMT2, another CMT protein in Arabidopsis, mainly maintains CpHpH methylation of TEs in histone H1-containing heterochromatin (Zemach et al., 2013). An *in vitro* study revealed that CMT2 can bind to H3K9Me2 and methylate both CpHpG and CpHpH sites (Stroud et al., 2014).

DOMAIN REARRANGED METHYLTRANSFERASE (DRM), an orthologue of human Dnmt3 *de novo* DNA methyltransferase, belongs to the third class of DNA transferases (Pavlopoulou & Kossida, 2007). The DRM2 protein, a DRM methyltransferase acting mainly in Arabidopsis, is responsible for the maintenance of non-CpG methylation (Cao et al., 2003). Although DRM2 can maintain methylation of all sequence context, the Ubiquitin Associated (UBA) domain of DRM2 protein is required for non-CpG methylation, especially CpHpH methylation in addition to the methyltransferase domain (Henderson et al., 2010; Zhong et al., 2014) (Fig. 4). DRM2 can be recruited to the target regions in a small RNA-dependent manner which is called the RNA-dependent DNA methylation (RdDM) pathway (Herr, Jensen, Dalmay, & Baulcombe, 2005; Kanno et al., 2005; Onodera et al., 2005; Pontier et al., 2005) (Fig. 5). In the first step of the RdDM pathway, a plant specific RNA polymerase, Pol IV, transcribes RNAs from the target gene loci. Pol IV targeting requires its binding protein, SHH1, which can bind to histone H3 tail via the chromatin-binding SAWADEE domain recognizing both unmethylated K4 and methylated K9 (Law et al., 2013). From the Pol IV-transcribed RNAs, double stranded RNAs (dsRNAs) are synthesized by RNA-DEPENDENT RNA POLYMERASE 2 (RDR2) and digested into 24 nt small interference RNAs (siRNAs) by DICER LIKE3 (DCL3) (Xie et al., 2004). After the 3′ end methylation by HUA ENHANCER 1 (HEN1), 24 nt siRNAs are loaded into ARGONAUTE (AGO) proteins, primarily AGO4 or 6 (Li, Yang, Yu, Liu, & Chen, 2005; Zheng, Zhu, Kapoor, & Zhu, 2007; Zilberman, Cao, & Jacobsen, 2003). AGO-siRNA complexes interact with non-coding RNAs, which are transcribed from the target loci by another plant specific RNA polymerase, Pol V, via base-paring between siRNA and non-coding RNA (Wierzbicki, Haag, & Pikaard, 2008; Wierzbicki, Ream, Haag, &

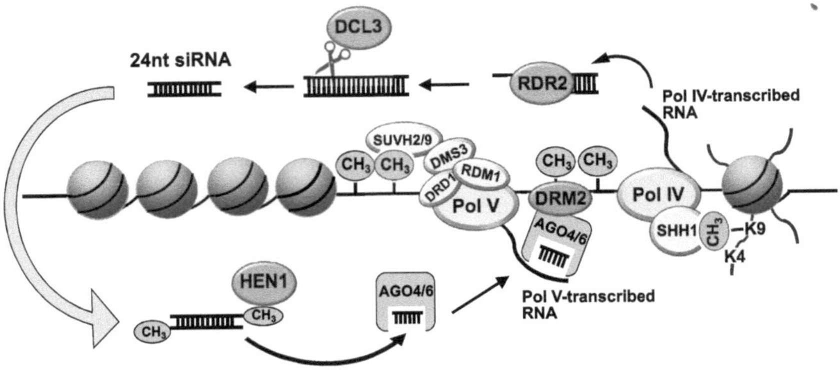

Figure 5 ***Hypothetical model of the RdDM (RNA-directed DNA methylation) pathway.*** Pol IV transcribes RNA from the target region via the binding of the Pol IV-interacting protein, SHH1, to unmethylated K4 and methylated K9 of histone H3. RDR2 synthesizes dsRNA from the Pol IV-transcribed RNA and DCL3 cleaves dsRNA to 24 nt siRNAs; the 3′ ends of these siRNAs are methylated by HEN1. The methylated 24 nt siRNAs are loaded onto AGO4 or AGO6 to form a complex. AGO4/6-siRNA complexes interact with Pol V-transcribed non-coding RNAs, and the complexes recruit DRM2 to the target region. SUVH2 and SUVH9 recognize methylated DNA to recruit Pol V to the target region via the DDR complex (DRD1, RDM1 and DMS3).

Pikaard, 2009). Since DRM2 binds to AGO4 *in vivo*, it is proposed that the interaction of AGO-siRNA complexes with Pol V-transcribed non-coding RNAs recruits DRM2 methyltransferase to target loci (Zhong et al., 2014). Recently it was reported that AGO4 can interact with DNA directly at target loci, although how the AGO4-DNA interaction acts on the recruitment of DRM2 methyltransferase is still unknown (Lahmy et al., 2016). Association of Pol V with target regions involves SUVH2 and SUVH9 which bind to methylated DNA via their SRA domain and then interact with the DDR complexes; DEFECTIVE IN RNA-DIRECTED DNA METHYLATION1 (DRD1), DEFECTIVE IN MERISTEM SILENCING3 (DMS3) and RNA-DIRECTED DNA METHYLATION1 (RDM1); the DDR complexes bind directly to Pol V which then associates with its target regions (Johnson et al., 2014; Liu et al., 2014). The recruitment of Pol V to methylated regions by SUVH2 and SUVH9 is proposed to form a self-reinforcing loop.

2.3 *De Novo* DNA Methylation

In Arabidopsis, *de novo* DNA methylation of previously unmethylated DNA sequences can be induced by DRM2, an orthologue of mammal *de novo*

DNA methyltransferase, in all three sequence contexts: CpG, CpHpG and CpHpH (Cao & Jacobsen, 2002) (Fig. 3). Consistent with the requirements of the RdDM pathway for DRM2 activity described in the previous section, mutations affecting the RdDM pathway cause defects in *de novo* methylation of transgenes (Kanno et al., 2005). In mutants defective in DNA methylation, TEs in which DNA methylation is lost can be re-methylated by the restoration of DNA methylation activity. This DNA re-methylation is also mediated through the RdDM pathway (Teixeira et al., 2009). This type of *de novo* DNA methylation partly involves components for post-transcriptional gene silencing (PTGS), a mechanism that functions in antiviral and antibacterial defenses (Cuerda-Gil & Slotkin, 2016). In PTGS, the enzymes DCL2 or DCL4 process dsRNA synthesized by RNA pol II and RDR6 to produce 21–22 nt siRNAs that then associate with AGO1 or AGO2 to degrade target RNAs (Pumplin & Voinnet, 2013). Re-methylation of some TEs requires 21–22 nt or 24 nt siRNAs produced by the PTGS components, Pol II and RDR6 (Mari-Ordonez et al., 2013; Nuthikattu et al., 2013). Alternative RdDM pathways, so-called non-canonical RdDMs, have been reported to be involved in the initiation of gene silencing, but the detailed mechanisms of these processes are still largely unknown (Cuerda-Gil & Slotkin, 2016). During egg cell development, reprogramming of CpHpH methylation is controlled by non-canonical RdDM which is independent of Pol IV activity (Ingouff et al., 2017).

2.4 DNA Demethylation

Although the pattern of genomic DNA methylation is stably maintained throughout plant growth in Arabidopsis, it can be reset and reprogrammed at specific developmental stages, such as pollen and seed formation (Calarco et al., 2012; Ingouff et al., 2017; Kawakatsu et al., 2017; Slotkin et al., 2009). Genome-wide DNA demethylation occurs in plant germ cells that are not of embryonic origin, specifically, the vegetative nucleus of pollen and the central cell of the female gametophyte (Calarco et al., 2012; Ibarra et al., 2012; Schoft et al., 2011; Slotkin et al., 2009). Similar to mammals, genetic imprinting is established by DNA methylation, and the pattern of expression of imprinted genes is determined by DNA demethylation (Ikeda & Nishimura, 2015).

DNA demethylation can be achieved by passive and active mechanisms in plants and mammals. Passive mechanisms involve repression of DNA methylation activities and cause genome-wide reductions in DNA methylation levels during DNA replication. During seed germination in

Arabidopsis, CpHpG and CpHpH methylation levels at RdDM- and CMT2-targeted TEs decrease without the involvement of known DNA demethylases (Kawakatsu et al., 2017). It has been reported that transcriptional repression of Arabidopsis MET1 is associated with genome-wide DNA demethylation in the central cells (Jullien et al., 2008). However, a recent study showed that MET1 repression does not involve global reduction of CpG methylation in the central cells (Park et al., 2016). The role of MET1 repression in passive DNA demethylation during female gametogenesis is still controversial.

Expression of imprinted genes involves active DNA demethylation in Arabidopsis (Ikeda & Nishimura, 2015). In mammals, active DNA methylation is mediated by the TET oxidase which converts 5-methylcytosine to 5-hydroxymethylcytosine (He et al., 2011; Ito et al., 2011). Hydroxymethylcytosine is then removed by the base excision repair (BER) pathway and/or by a passive mechanism as Dnmt1 does not recognize hydroxymethylcytosine. As the Arabidopsis genome does not include a gene encoding TET oxidase, other enzymes, such as DEMETER (DME), REPRESSOR OF SILENCING 1 (ROS1), DEMETER-LIKE 2 (DML2) and DML3, are involved in BER and have a role in DNA demethylation (Choi et al., 2002; Gong et al., 2002; Penterman et al., 2007). The latter enzymes have bifunctional activities, namely, as DNA glycosylases and apurinic/apyrimidinic (AP) lyases, and remove methylated cytosines from the DNA backbone to generate a single-nucleotide gap (Agius, Kapoor, & Zhu, 2006; Gehring et al., 2006; Morales-Ruiz et al., 2006). Eventually, the single-nucleotide gap is refilled with an unmethylated cytosine by an unidentified DNA polymerase and the DNA ligase AtLIG1 in the BER pathway (Andreuzza et al., 2010; Li, Duan, Zhu, Qian, & Zhu, 2015). The *DME* gene is expressed in central cells in which repression of imprinted genes is released; as expected, *dme* mutants show defects in the regulation of imprinted genes (Choi et al., 2002).

3. HISTONE MODIFICATIONS

In eukaryotes, the amino acid residues of the N-terminal tails of the histone core proteins H2A, H2B, H3 and H4 can show several types of modification, for example methylation, acetylation, phosphorylation, and ubiquitination. The pattern of modification on the histone tail, the so-called histone code, is closely associated with chromatin status and gene expression.

Genome-wide studies using chromatin immunoprecipitation (ChIP)-seq showed that the effects of histone modifications on gene expression are often conserved between animals and plants (Roudier et al., 2011; Sequeira-Mendes et al., 2014). Some histone modifications are involved in the memory of developmental and environmental cues in plants, indicating the need for mechanisms to maintain information on histone modification patterns during DNA replication as well as for DNA methylation (Avramova, 2015; Berry & Dean, 2015; Lämke & Bäurle, 2017). In plants, enzymes that regulate histone modifications have been identified through the analyses of Arabidopsis mutants (Table 1). In the remainder of this section, we describe the regulatory mechanisms for each type of histone modification, with the exception of phosphorylation which acts on cell division rather than gene expression (Rossetto, Avvakumov, & Cote, 2012).

3.1 Histone Methylation

In eukaryotes, the mechanism of histone methylation has been intensely studied. Histone methylation involves the modification of the alkaline amino acids, lysine and arginine. Methylation at lysine residues has two opposite effects on gene expression depending on the position of the modified residues. Methylation of lysine at the fourth (H3K4) and the 36th (H3K36) positions is associated with transcriptional activation, while methylation at the ninth (H3K9) and 27th (H3K27) positions is associated with transcriptional repression (Roudier et al., 2011; Sequeira-Mendes et al., 2014). Each residue can be methylated to different extents, namely, mono-, di-, and tri-methylated, by different mechanisms. Histone lysine methylation is recognized by specific domains of "reader" proteins that regulate chromatin status and gene expression. The identified domains include chromo, tudor, malignant brain tumor (MBT), PWWP, plant homeo domain (PHD), and WD40 (Xiao, Lee, & Wagner, 2016; Yun, Wu, Workman, & Li, 2011).

Several histone lysine methyltransferases and demethylases are classified as "writers" and "erasers", respectively, with respect to lysine methylation at different positions (Liu, Lu, Cui, & Cao, 2010; Xiao et al. 2016) (Table 1). Histone lysine methyltransferases have a SET catalytic domain that is conserved among eukaryotes. A number of genes encoding SET domain-containing proteins, called SET domain-containing groups (SDGs), have been identified in the Arabidopsis genome (Baumbusch et al., 2001; Ng et al., 2007). These SDGs are classified into four subgroups based on their

Table 1 Writers and Erasers of Histone Modifications in Arabidopsis

Modifications	Writers	Erasers	References
Histone Methylation			
H3K4Me1	SDG25/ATXR7	FLD, LDL1, LDL2, JMJ14, JMJ15	(1), (2), (3), (4), (5), (6)
H3K4Me2	SDG2/ATXR3, SDG4/ASHR3, SDG25/ATXR7, SDG30/ATX2	FLD, LDL1, LDL2, JMJ14, JMJ15, JMJ18	(1), (2), (3), (4), (5), (6), (7), (8), (9), (10)
H3K4Me3	SDG2/ATXR3, SDG4/ASHR3, SDG25/ATXR7, SDG26/ASHH1, SDG27/ATX1	JMJ14, JMJ15, JMJ18	(1), (5), (6), (7), (8), (9), (10), (11),
H3K9Me1	SDG9/SUVH5, SDG23/SUVH6,	IBM1	(12), (13), (14), (15)
H3K9Me2	SDG31/SUVR4,	IBM1	(12), (13), (14), (15)
H3K9Me3	SDG33/KYP/SUVH	ND	(12), (13), (14)
H3K27Me1	SDG15/ATXR5, SDG34/ATXR6	ND	(16)
H3K27Me2	SDG1/CLF, SDG5/MEA, SDG10/SWN	JMJ11/ELF6, JMJ12/REF6, JMJ30, JMJ32	(17), (18), (19), (20), (21), (22)
H3K27Me3	SDG1/CLF, SDG5/MEA, SDG10/SWN	JMJ11/ELF6, JMJ12/REF6, JMJ30, JMJ32	(17), (18), (19), (20), (21), (22)
H3K36Me1	SDG4/ASHR3	ND	(8)
H3K36Me2	SDG4/ASHR3, SDG8/ASHH2, SDG25/ATXR7	JMJ30	(8), (11), (21), (23),
H3K36Me3	SDG4/ASHR3, SDG8/ASHH2, SDG26/ASHH1	JMJ30	(8), (11), (21)
H3R17Me2	AtPRMT4a, AtPRMT4b	ND	(24)
H4R3Me2	AtPRMT1a, AtPRMT1b, AtPRMT5, AtPRMT10	ND	(25), (26), (27)

Histone Acetylation			
H3K9ac, H3K14ac H3K18ac, H3K23ac H4K5ac, H4K8ac H4K12ac, H4K16ac H4K20ac	HAG1/GCN5, HAG2, HAG3, MCC1, HAM1, HAM2, HAC1, HAC2, HAC4, HAC5, HAC12, HAF1, HAF2	HDA2, HDA5, HDA6/RPD3B, HDA7, HDA8, HDA9, HDA10, HDA14, HDA15, HDA17, HDA18, HDA19, HDT1/HD2A, HDT2/HD2B, HDT3, HDT4, SRT1, SRT2	(28)
Histone Ubiquitination			
H2Aub1	AtBMI1A, AtBMI1B	ND	(29)
H2Bub1	HUB1, HUB2	SUP32/UBP26	(30), (31), (32)

ND: no data. (1) Tamada, Yun, Woo, & Amasino, 2009, (2) Jiang, Gu, & He, 2009, (3) Jiang, Yang, He, & Amasino, 2007, (4) Liu et al., 2007, (5) Lu, Cui, & Cao, 2010, (6) Yang et al., 2012b, (7) Berr et al., 2010, (8) Cartagena et al., 2008, (9) Saleh et al., 2008, (10) Yang et al., 2012a, (11) Xu et al., 2008, (12) Ebbs & Bender, 2006, (13) Ebbs, Bartee, & Bender, 2005, (14) Thorstensen et al., 2006, (15) Saze, Shiraishi, Miura, & Kakutani, 2008, (16) Jacob et al., 2010, (17) Goodrich et al., 1997, (18) Grossniklaus, Vielle-Calzada, Hoeppner, & Gagliano, 1998, (19) Wang, Tyson, Jackson, & Yadegari, 2006, (20) Noh et al., 2004, (21) Lu et al., 2011, (22) Gan et al., 2014, (23) Berr et al., 2009, (24) Niu, Zhang, Pei, Liu, & Cao, 2008, (25) Pei et al., 2007, (26) Niu, Lu, Pei, Liu, & Cao, 2007, (27) Yan, Zhang, Niu, Yuan, & Cao, 2007, (28) Wang, Cao, Chen, & Liu, 2014, (29) Bratzel, Lopez-Torrejon, Koch, Del Pozo, & Calonje, 2010, (30) Fleury et al., 2007, (31) Liu, Koornneef, & Soppe, 2007, (32) Sridhar et al., 2007.

homologies to *Drosophila* SET domain-containing proteins: (1) Enhancer of Zeste [E(Z)] homologues, (2) ARABIDOPSIS TRITHORAX-LIKE (ATX) or TRITHORAX-RELATED (ATXR) PROTEIN, (3) ASH1-HOMOLOGUE (ASHH) or ASH1-RELATED (ASHR), and (4) SUPRESSOR OF VARIEGATION 3-9 HOMOLOG or RELATED (SUVH or SUVR) (Baumbusch et al., 2001). Arabidopsis histone lysine demethylases are classified into two subfamilies on the basis of their structure: Lysine Specific Demethylases (LSDs)-like family (LDL), and JmjC domain-containing protein family (JMJ) (Jiang, Yang, He, & Amasino, 2007; Lu et al., 2008).

In Arabidopsis, SDG2/ATXR3, SDG25/ATXR7, SDG27/ATX1, SDG30/ATX2, SDG4/ASHR3, and SDG26/ASHH1 can methylate H3K4 (Berr et al., 2010; Berr, Shafiq, Pinon, Dong, & Shen, 2015; Cartagena et al., 2008; Saleh et al., 2008; Tamada, Yun, Woo, & Amasino, 2009). SDG2/ATXR3 and SDG4/ASHR3 can function in both di- and tri-methylation. SDG25/ATXR7 can add methyl groups at all extents; mono-, di-, and tri-methylation. Demethylation of H3K4 involves proteins belonging to both subclasses of histone lysine demethylases, LDL (FLD, LDL1, and LDL2) and JMJ (JMJ14, JMJ15, and JMJ18) (Jiang, Gu, & He, 2009; Jiang, Yang, He, & Amasino, 2007; Liu et al., 2007; Lu, Cui, Zhang, Liu, & Cao, 2010; Yang et al., 2012a, 2012b). Analyses of the effects of mutations in these writers and erasers indicated that H3K4 methylation is involved in flowering, pollen tube formation and gametophyte development (Berr et al., 2015, 2010; Cartagena et al., 2008; Jiang et al., 2009, 2007; Liu et al., 2007; Lu et al., 2010; Saleh et al., 2008; Tamada et al., 2009; Yang et al., 2012a, 2012b).

In yeast and animals, H3K36Me3 is found in the 3′ end of the coding region of actively transcribed genes and is associated with transcriptional elongation (Pokholok et al., 2005; Wagner & Carpenter, 2012). In Arabidopsis, H3K36Me3 is enriched in the 5′ end of actively transcribed genes, while H3K36Me2 is enriched in the 3′ end and may be associated with transcriptional elongation (Roudier et al., 2011). H3K4 methyltransferases, SDG25/ATXR7, SDG4/ASHR3, and SDG26/ASHH1 also methylate H3K36 (Berr et al., 2010, 2009; Cartagena et al., 2008). The role of SDG8/ASHH2 in H3K4 methylation is unknown; however, it can catalyze H3K36 methylation (Xu et al., 2008). Studies of mutations in these histone methyltransferases have shown they cause defects in flowering and pollen tube formation (Berr et al., 2010, 2009; Cartagena et al., 2008; Xu et al. 2008). SDG8/ASHH2 and SDG4/ASHR3 are also involved in plant responses to mechanical stimuli and in root growth, respectively (Cazzonelli et al., 2014; Kumpf et al., 2014). JMJ30 has been shown to demethylate H3K36, which represses expression of the *FT* gene encoding florigen in response to ambient temperatures (Yan et al., 2014).

Generally, H3K9 methylation serves as a repressive epigenetic mark on constitutively silenced genes such as TEs and repeat sequences. As described in Section 2.4, SDG33/KYP/SUVH4 and IBM1 catalyze the methylation and demethylation, respectively, of H3K9Me2; this epigenetic mark is recognized by CMT DNA methyltransferases in a self-reinforcing loop of gene repression (Inagaki et al., 2010; Jackson et al., 2002; Johnson et al., 2007; Miura et al., 2009; Saze et al., 2008). Other proteins, namely, SDG9/SUVH5, SDG23/SUVH6 and SDG31/SUVR4, also function in H3K9 methylation (Ebbs & Bender, 2006; Ebbs, Bartee, & Bender, 2005; Thorstensen et al., 2006).

The POLYCOMB REPRESSIVE COMPLEX 2 (PRC2) catalyzes methylation of H3K27 in plants and animals (Simon & Kingston, 2013). The Arabidopsis genome has three homologues of Enhancer of Zeste [E(Z)] which is a catalytic subunit for the methyltransferase activity of PRC2: SDG1/CURLY LEAF (CLF), SDG5/MEDEA (MEA), and SDG10/SWINGER (SWN) (Goodrich et al., 1997; Grossniklaus, Vielle-Calzada, Hoeppner, & Gagliano, 1998; Wang, Tyson, Jackson, & Yadegari, 2006). These proteins catalyze di- and tri-methylation of H3K27, which is necessary for developmental transitions, including seed maturation, flowering and flower formation (Mozgova, Kohler, & Hennig, 2015; Xiao & Wagner, 2015). H3K27 tri-methylation also represses the expression of *FLOWERING LOCUS C* (*FLC*) during vernalization (Berry & Dean, 2015). Demethylation of H3K27Me2 and Me3 is catalyzed by EARLY FLOWERING 6 (JMJ11/ELF6), RELATIVE OF EARLY FLOWERING 6 (JMJ12/REF6), JMJ30, and JMJ32 (Gan et al., 2014; Lu et al., 2011; Noh et al., 2004). All of these demethylases function in the regulation of *FLC* expression; in particular, JMJ11/ELF6 resets *FLC* expression during gametogenesis (Crevillen et al., 2014). H3K27Me1 is found in heterochromatic chromosome regions and is catalyzed by SDG15/ATXR5 and SDG34/ATXR6 (Jacob et al., 2010). The enzymes that remove H3K27Me1 have not been identified to date.

Methylation of H4K20 and H3K79 has been identified in animals and yeast. In Arabidopsis, H4K20Me1 is detected throughout the genome, although the details of the regulation and function of this methylation are still largely unknown (Roudier et al., 2011). To date, H3K79 methylation has not been found in Arabidopsis; nor has any homolog of DOT1 that serves as an H3K79 methyltransferase in animals and yeast been identified (Zhang, Sridhar, Zhu, Kapoor, & Zhu, 2007).

Methylation of arginine in H3 and H4 is also involved in the regulation of gene expression (Ahmad & Cao, 2012; Liu et al., 2010). Arginine residues can be mono- or di-methylated; di-methylation occurs either symmetrically or asymmetrically (Ahmad & Cao, 2012). The Arabidopsis genome encodes homologues of protein arginine methyltransferases (PRMTs), some of which have been reported to act on histone arginine methylation. AtPRMT4a and AtPRMT4b catalyze asymmetrical di-methylation of H3R17, which affects *FLC* expression during flowering regulation (Niu, Zhang, Pei, Liu, & Cao, 2008). H4R3 is symmetrically di-methylated by AtPRMT5 and asymmetrically di-methylated by AtPRMT10, AtPRMT1a and AtPRMT1b (Niu, Lu, Pei, Liu, & Cao, 2007; Pei et al., 2007; Schmitz, Sung, & Amasino, 2008; Wang et al., 2007; Yan, Zhang, Niu, Yuan, & Cao, 2007). These methyltransferases also influence *FLC* expression. A mutation in AtRPMT5 has been shown to induce pleiotropic phenotypes during development (Pei et al., 2007).

3.2 Histone Acetylation

Histone acetylation is closely associated with active gene transcription (Roudier et al., 2011; Sequeira-Mendes et al., 2014). This modification influences transcription by neutralizing the positive charge in histone tails to decrease affinity for the DNA and/or "reader" proteins that regulate transcription (Lusser, Kolle, & Loidl, 2001). The prime acetylation targets are lysine residues 9, 14, 18, and 23 on histone H3 and lysine residues 5, 8, 12, 16, and 20 on histone H4 (Fuchs, Demidov, Houben, & Schubert, 2006). Histone acetylation on lysine residues is established by histone acetyltransferases (HATs) and erased by histone deacetylases (HDACs). Plant HATs fall into four categories: HATs of the GNAT superfamily (HAG), HATs of the MYST superfamily (HAM), HATs of the CREB-binding protein family (HAC), and HATs of the TATA-binding protein-associated factor family (HAF) (Lusser et al., 2001). Plant HDACs are sorted into three families: homologues of yeast Reduced Potassium Deficiency 3 (RPD3), HD-tuins (HDT), and homologues of yeast silent information regulator 2 (Sir2) (Lusser et al., 2001). Genetic analyses have shown that histone acetylation has various functions in development and environmental responses in plant species which are listed in a review by Wang, Cao, Chen, and Liu (2014).

3.3 Histone Ubiquitination

Ubiquitin is an important post-translational modification of proteins that marks them for degradation. Ubiquitination of histone H2A and H2B controls gene activity in eukaryotes (March & Farrona, 2018; Weake & Workman, 2008). In Arabidopsis, H2A K121 and H2B K143 can be mono-ubiquitinated (March & Farrona, 2018). The genomic distribution of H2B mono-ubiquitination is correlated with H3K4 and H3K36 methylation and acts as an active transcriptional mark (Roudier et al., 2011; Sequeira-Mendes et al., 2014). H2A mono-ubiquitination acts as a repression mark (March & Farrona, 2018; Weake & Workman, 2008); H2A mono-ubiquitination is catalyzed by the PRC1 complex, in which Bmi1, Ring1A and Ring1B subunits have E3 ubiquitin ligase activity (March & Farrona, 2018; Weake & Workman, 2008). Arabidopsis homologues of PRC1 subunits can catalyze H2A mono-ubiquitination *in vitro*, and double mutants of AtBMI1A and AtBMI1B show a decrease in H2A mono-ubiquitination levels (Bratzel, Lopez-Torrejon, Koch, Del Pozo, & Calonje, 2010). HUB1 and HUB2, and SUP32/UBP26 have been identified as the enzymes responsible for ubiquitination and de-ubiquitination activities, respectively, in H2B mono-ubiquitination (Fleury et al., 2007; Liu, Koornneef, & Soppe, 2007; Sridhar et al., 2007). Mutations in HUB1 and HUB2 cause defects in organ growth and seed dormancy (Fleury et al., 2007; Liu et al., 2007). Additionally, H2B de-ubiquitination by SUP32/UBP26 is required for generation of heterochromatic marks; H3K9Me2 and DNA methylation (Sridhar et al., 2007).

4. HISTONE VARIANTS

Histone proteins, except for H4, have a number of amino acid sequence variants (Talbert et al., 2012). Substitution of canonical histone proteins by the variants in the nucleosome can change chromatin properties, including gene expression (Talbert & Henikoff, 2017). In Arabidopsis, studies of histone variants of H2A and H3 have been reported.

Among H2A variants, H2A.Z is widely conserved and is found in euchromatic regions, especially at transcriptional start sites of genes (Zilberman, Coleman-Derr, Ballinger, & Henikoff, 2008). H2A.Z is also found in the gene body of genes responsive to environmental and developmental stimuli. Loss of this variant causes misregulation of these genes (Coleman-Derr & Zilberman, 2012). The inclusion of H2A.Z in chromatin is catalyzed by an ATP-dependent chromatin remodeling complex termed SWR1 (Mizuguchi et al., 2004). In Arabidopsis mutants of PHOTOPERIOD INDEPENDENT EARLY FLOWERING 1 (PIE1), the homolog of a catalytic subunit of SWR1, H2A.Z is lost from the genome (Deal, Topp, McKinney, & Meagher, 2007; Zilberman et al., 2008). The H2A variant, H2A.W, is found in heterochromatin and is involved in chromatin condensation in synergy with DNA methylation and H3K9Me2 (Yelagandula et al., 2014).

H3.1 and H3.3 are the most abundant H3 variants in euchromatic regions. Genome-wide profiling of their distribution revealed that H3.1 is found in silent regions and is associated with H3K9 and H3K27 methylation; H3.3 is found in actively transcribed genes, and is especially enriched in the 3′ end of genes (Stroud et al., 2012; Wollmann et al., 2012). Knockdown of genes encoding H3.3 causes developmental abnormalities and misregulation of expression of many genes as well as decrease of gene body methylation, indicating a requirement for H3.3 for proper gene expression via chromatin status (Wollmann et al., 2017).

5. PERSPECTIVES

Genetic and genomic studies, particularly in wild type and mutants in Arabidopsis, have revealed epigenetic mechanisms involving DNA methylation, histone modification, and histone variants. Some of these mechanisms are common to animals and yeast, while others appear to be unique to plants. Recently parts of epigenetic mechanisms were revealed in the other plant species including crops. In rice (*Oryza sativa*), one of major food crops, some of epigenetic regulators have different functions from Arabidopsis ones (Lanciano & Mirouze, 2017). The use of genome-wide analyses using deep sequencing technologies, such as bisulfite (BS)-seq and ChIP-seq, has enabled the investigation of epigenomic features in plant species other than Arabidopsis. Novel gene targeting techniques, such as CRIPR/Cas9, will make it possible to induce targeted mutations in such non-model plant species. The epigenetic mechanisms described in this chapter have been found in Arabidopsis and they might be differences in the epigenetic regulators in other plant species. Based on the findings obtained from Arabidopsis studies, however, new molecular breeding techniques will undoubtedly be developed and applied to commercially important plant species.

REFERENCES

Agius, F., Kapoor, A., & Zhu, J. K. (2006). Role of the *Arabidopsis* DNA glycosylase/lyase ROS1 in active DNA demethylation. *Proceedings of the National Academy of Sciences of the United States of America, 103*, 11796–11801.

Ahmad, A., & Cao, X. (2012). Plant PRMTs broaden the scope of arginine methylation. *Journal of Genetics and Genomics, 39*, 195–208.

Andreuzza, S., Li, J., Guitton, A. E., Faure, J. E., Casanova, S., Park, J. S., et al. (2010). DNA LIGASE I exerts a maternal effect on seed development in *Arabidopsis thaliana. Development, 137*, 73–81.

Avramova, Z. (2015). Transcriptional 'memory' of a stress: Transient chromatin and memory (epigenetic) marks at stress-response genes. *The Plant Journal, 83*, 149–159.

Baumbusch, L. O., Thorstensen, T., Krauss, V., Fischer, A., Naumann, K., Assalkhou, R., et al. (2001). The *Arabidopsis thaliana* genome contains at least 29 active genes encoding SET domain proteins that can be assigned to four evolutionarily conserved classes. *Nucleic Acids Research, 29*, 4319–4333.

Berr, A., McCallum, E. J., Menard, R., Meyer, D., Fuchs, J., Dong, A., et al. (2010). *Arabidopsis SET DOMAIN GROUP2* is required for H3K4 trimethylation and is crucial for both sporophyte and gametophyte development. *The Plant Cell, 22*, 3232–3248.

Berr, A., Shafiq, S., Pinon, V., Dong, A., & Shen, W. H. (2015). The trxG family histone methyltransferase SET DOMAIN GROUP 26 promotes flowering via a distinctive genetic pathway. *The Plant Journal, 81*, 316–328.

Berr, A., Xu, L., Gao, J., Cognat, V., Steinmetz, A., Dong, A., et al. (2009). *SET DOMAIN GROUP25* encodes a histone methyltransferase and is involved in *FLOWERING LOCUS C* activation and repression of flowering. *Plant Physiology, 151*, 1476–1485.

Berry, S., & Dean, C. (2015). Environmental perception and epigenetic memory: Mechanistic insight through *FLC. The Plant Journal, 83*, 133–148.

Bewick, A. J., Ji, L., Niederhuth, C. E., Willing, E. M., Hofmeister, B. T., Shi, X., et al. (2016). On the origin and evolutionary consequences of gene body DNA methylation. *Proceedings of the National Academy of Sciences of the United States of America, 113*, 9111–9116.

Bewick, A. J., & Schmitz, R. J. (2017). Gene body DNA methylation in plants. *Current Opinion in Plant Biology, 36*, 103–110.

Bird, A. (2002). DNA methylation patterns and epigenetic memory. *Genes & Development, 16*, 6–21.

Bratzel, F., Lopez-Torrejon, G., Koch, M., Del Pozo, J. C., & Calonje, M. (2010). Keeping cell identity in *Arabidopsis* requires PRC1 RING-finger homologs that catalyze H2A monoubiquitination. *Current Biology, 20*, 1853–1859.

Calarco, J. P., Borges, F., Donoghue, M. T., Van Ex, F., Jullien, P. E., Lopes, T., et al. (2012). Reprogramming of DNA methylation in pollen guides epigenetic inheritance via small RNA. *Cell, 151*, 194–205.

Cao, X., Aufsatz, W., Zilberman, D., Mette, M. F., Huang, M. S., Matzke, M., et al. (2003). Role of the DRM and CMT3 methyltransferases in RNA-directed DNA methylation. *Current Biology, 13*, 2212–2217.

Cao, X., & Jacobsen, S. E. (2002). Role of the *Arabidopsis DRM* methyltransferases in de novo DNA methylation and gene silencing. *Current Biology, 12*, 1138–1144.

Cartagena, J. A., Matsunaga, S., Seki, M., Kurihara, D., Yokoyama, M., Shinozaki, K., et al. (2008). The *Arabidopsis* SDG4 contributes to the regulation of pollen tube growth by methylation of histone H3 lysines 4 and 36 in mature pollen. *Developmental Biology, 315*, 355–368.

Cazzonelli, C. I., Nisar, N., Roberts, A. C., Murray, K. D., Borevitz, J. O., & Pogson, B. J. (2014). A chromatin modifying enzyme, SDG8, is involved in morphological, gene expression, and epigenetic responses to mechanical stimulation. *Frontiers in Plant Science, 5*, 533.

Chan, S. W., Henderson, I. R., & Jacobsen, S. E. (2005). Gardening the genome: DNA methylation in *Arabidopsis thaliana*. *Nature Reviews Genetics, 6*, 351–360.

Choi, Y., Gehring, M., Johnson, L., Hannon, M., Harada, J. J., Goldberg, R. B., et al. (2002). DEMETER, a DNA glycosylase domain protein, is required for endosperm gene imprinting and seed viability in arabidopsis. *Cell, 110*, 33–42.

Cokus, S. J., Feng, S., Zhang, X., Chen, Z., Merriman, B., Haudenschild, C. D., et al. (2008). Shotgun bisulphite sequencing of the Arabidopsis genome reveals DNA methylation patterning. *Nature, 452*, 215–219.

Coleman-Derr, D., & Zilberman, D. (2012). Deposition of histone variant H2A.Z within gene bodies regulates responsive genes. *PLoS Genetics, 8*, e1002988.

Crevillen, P., Yang, H., Cui, X., Greeff, C., Trick, M., Qiu, Q., et al. (2014). Epigenetic reprogramming that prevents transgenerational inheritance of the vernalized state. *Nature, 515*, 587–590.

Cubas, P., Vincent, C., & Coen, E. (1999). An epigenetic mutation responsible for natural variation in floral symmetry. *Nature, 401*, 157–161.

Cuerda-Gil, D., & Slotkin, R. K. (2016). Non-canonical RNA-directed DNA methylation. *Nature Plants, 2*, 16163.

Deal, R. B., Topp, C. N., McKinney, E. C., & Meagher, R. B. (2007). Repression of flowering in *Arabidopsis* requires activation of *FLOWERING LOCUS C* expression by the histone variant H2A.Z. *The Plant Cell, 19*, 74–83.

Dowen, R. H., Pelizzola, M., Schmitz, R. J., Lister, R., Dowen, J. M., Nery, J. R., et al. (2012). Widespread dynamic DNA methylation in response to biotic stress. *Proceedings of the National Academy of Sciences of the United States of America, 109*, E2183–E2191.

Du, J., Johnson, L. M., Groth, M., Feng, S., Hale, C. J., Li, S., et al. (2014). Mechanism of DNA methylation-directed histone methylation by KRYPTONITE. *Molecular Cell, 55*, 495–504.

Du, J., Zhong, X., Bernatavichute, Y. V., Stroud, H., Feng, S., Caro, E., et al. (2012). Dual binding of chromomethylase domains to H3K9me2-containing nucleosomes directs DNA methylation in plants. *Cell, 151*, 167–180.

Ebbs, M. L., Bartee, L., & Bender, J. (2005). H3 lysine 9 methylation is maintained on a transcribed inverted repeat by combined action of SUVH6 and SUVH4 methyltransferases. *Molecular and Cellular Biology, 25*, 10507–10515.

Ebbs, M. L., & Bender, J. (2006). Locus-specific control of DNA methylation by the *Arabidopsis* SUVH5 histone methyltransferase. *The Plant Cell, 18*, 1166–1176.

Feng, S., Jacobsen, S. E., & Reik, W. (2010). Epigenetic reprogramming in plant and animal development. *Science, 330*, 622–627.

Finnegan, E. J., Peacock, W. J., & Dennis, E. S. (1996). Reduced DNA methylation in *Arabidopsis thaliana* results in abnormal plant development. *Proceedings of the National Academy of Sciences of the United States of America, 93*, 8449–8454.

Fleury, D., Himanen, K., Cnops, G., Nelissen, H., Boccardi, T. M., Maere, S., et al. (2007). The *Arabidopsis thaliana* homolog of yeast *BRE1* has a function in cell cycle regulation during early leaf and root growth. *The Plant Cell, 19*, 417–432.

Fuchs, J., Demidov, D., Houben, A., & Schubert, I. (2006). Chromosomal histone modification patterns–from conservation to diversity. *Trends in Plant Science, 11*, 199–208.

Gan, E. S., Xu, Y., Wong, J. Y., Goh, J. G., Sun, B., Wee, W. Y., et al. (2014). Jumonji demethylases moderate precocious flowering at elevated temperature via regulation of *FLC* in *Arabidopsis*. *Nature Communications, 5*, 5098.

Gehring, M., Huh, J. H., Hsieh, T. F., Penterman, J., Choi, Y., Harada, J. J., et al. (2006). DEMETER DNA glycosylase establishes *MEDEA* polycomb gene self-imprinting by allele-specific demethylation. *Cell, 124*, 495–506.

Goll, M. G., & Bestor, T. H. (2005). Eukaryotic cytosine methyltransferases. *Annual Review of Biochemistry, 74*, 481–514.

Gong, Z., Morales-Ruiz, T., Ariza, R. R., Roldan-Arjona, T., David, L., & Zhu, J. K. (2002). *ROS1*, a repressor of transcriptional gene silencing in *Arabidopsis*, encodes a DNA glycosylase/lyase. *Cell, 111*, 803–814.

Goodrich, J., Puangsomlee, P., Martin, M., Long, D., Meyerowitz, E. M., & Coupland, G. (1997). A Polycomb-group gene regulates homeotic gene expression in *Arabidopsis*. *Nature, 386*, 44–51.

Grossniklaus, U., Vielle-Calzada, J. P., Hoeppner, M. A., & Gagliano, W. B. (1998). Maternal control of embryogenesis by *MEDEA*, a *polycomb* group gene in *Arabidopsis*. *Science, 280*, 446–450.

He, Y. F., Li, B. Z., Li, Z., Liu, P., Wang, Y., Tang, Q., et al. (2011). Tet-mediated formation of 5-carboxylcytosine and its excision by TDG in mammalian DNA. *Science, 333*, 1303–1307.

Henderson, I. R., Deleris, A., Wong, W., Zhong, X., Chin, H. G., Horwitz, G. A., et al. (2010). The *de novo* cytosine methyltransferase DRM2 requires intact UBA domains and a catalytically mutated paralog DRM3 during RNA-directed DNA methylation in *Arabidopsis thaliana*. *PLoS Genetics, 6*, e1001182.

Herr, A. J., Jensen, M. B., Dalmay, T., & Baulcombe, D. C. (2005). RNA polymerase IV directs silencing of endogenous DNA. *Science, 308*, 118–120.

Ibarra, C. A., Feng, X., Schoft, V. K., Hsieh, T. F., Uzawa, R., Rodrigues, J. A., et al. (2012). Active DNA demethylation in plant companion cells reinforces transposon methylation in gametes. *Science, 337*, 1360–1364.

Ikeda, Y., & Nishimura, T. (2015). The role of DNA methylation in transposable element silencing and genomic imprinting. In O. Pontes, & H. Jin (Eds.), *Nuclear functions in plant transcription, signaling and development* (pp. 13–29). New York, NY: Springer.

Inagaki, S., Miura-Kamio, A., Nakamura, Y., Lu, F., Cui, X., Cao, X., et al. (2010). Autocatalytic differentiation of epigenetic modifications within the Arabidopsis genome. *The EMBO Journal, 29*, 3496–3506.

Ingouff, M., Selles, B., Michaud, C., Vu, T. M., Berger, F., Schorn, A. J., et al. (2017). Live-cell analysis of DNA methylation during sexual reproduction in *Arabidopsis* reveals context and sex-specific dynamics controlled by noncanonical RdDM. *Genes & Development, 31*, 72–83.

Ishiga, Y., Ishiga, T., Ichinose, Y., & Mysore, K. (2017). *Pseudomonas syringae* flood-inoculation method in *Arabidopsis*. *Bio-protocol, 7*, e2106.

Ito, S., Shen, L., Dai, Q., Wu, S. C., Collins, L. B., Swenberg, J. A., et al. (2011). Tet proteins can convert 5-methylcytosine to 5-formylcytosine and 5-carboxylcytosine. *Science, 333*, 1300–1303.

Jackson, J. P., Lindroth, A. M., Cao, X., & Jacobsen, S. E. (2002). Control of CpNpG DNA methylation by the KRYPTONITE histone H3 methyltransferase. *Nature, 416*, 556–560.

Jacobsen, S. E., & Meyerowitz, E. M. (1997). Hypermethylated *SUPERMAN* epigenetic alleles in *Arabidopsis*. *Science, 277*, 1100–1103.

Jacob, Y., Stroud, H., Leblanc, C., Feng, S., Zhuo, L., Caro, E., et al. (2010). Regulation of heterochromatic DNA replication by histone H3 lysine 27 methyltransferases. *Nature, 466*, 987–991.

Jiang, D., Gu, X., & He, Y. (2009). Establishment of the winter-annual growth habit via *FRIGIDA*-mediated histone methylation at *FLOWERING LOCUS C* in *Arabidopsis*. *The Plant Cell, 21*, 1733–1746.

Jiang, D., Yang, W., He, Y., & Amasino, R. M. (2007). *Arabidopsis* relatives of the human lysine-specific Demethylase1 repress the expression of *FWA* and *FLOWERING LOCUS C* and thus promote the floral transition. *The Plant Cell, 19*, 2975–2987.

Johnson, L. M., Bostick, M., Zhang, X., Kraft, E., Henderson, I., Callis, J., et al. (2007). The SRA methyl-cytosine-binding domain links DNA and histone methylation. *Current Biology, 17*, 379–384.

Johnson, L. M., Du, J., Hale, C. J., Bischof, S., Feng, S., Chodavarapu, R. K., et al. (2014). SRA- and SET-domain-containing proteins link RNA polymerase V occupancy to DNA methylation. *Nature, 507*, 124–128.

Jullien, P. E., Mosquna, A., Ingouff, M., Sakata, T., Ohad, N., & Berger, F. (2008). Retinoblastoma and its binding partner MSI1 control imprinting in *Arabidopsis*. *PLoS Biology, 6*, e194.

Kanno, T., Huettel, B., Mette, M. F., Aufsatz, W., Jaligot, E., Daxinger, L., et al. (2005). Atypical RNA polymerase subunits required for RNA-directed DNA methylation. *Nature Genetics, 37*, 761–765.

Kawakatsu, T., Nery, J. R., Castanon, R., & Ecker, J. R. (2017). Dynamic DNA methylation reconfiguration during seed development and germination. *Genome Biology, 18*, 171.

Kim, J., Kim, J. H., Richards, E. J., Chung, K. M., & Woo, H. R. (2014). *Arabidopsis* VIM proteins regulate epigenetic silencing by modulating DNA methylation and histone modification in cooperation with MET1. *Molecular Plant, 7*, 1470–1485.

Kouzarides, T. (2007). Chromatin modifications and their function. *Cell, 128*, 693–705.

Kumpf, R., Thorstensen, T., Rahman, M. A., Heyman, J., Nenseth, H. Z., Lammens, T., et al. (2014). The ASH1-RELATED3 SET-domain protein controls cell division competence of the meristem and the quiescent center of the *Arabidopsis* primary root. *Plant Physiology, 166*, 632–643.

Lahmy, S., Pontier, D., Bies-Etheve, N., Laudié, M., Feng, S., Jobet, E., et al. (2016). Evidence for ARGONAUTE4-DNA interactions in RNA-directed DNA methylation in plants. *Genes & Development, 30*, 2565–2570.

Lämke, J., & Bäurle, I. (2017). Epigenetic and chromatin-based mechanisms in environmental stress adaptation and stress memory in plants. *Genome Biology, 18*, 124.

Lanciano, S., & Mirouze, M. (2017). DNA methylation in rice and relevance for breeding. *Epigenomes, 1*, 10.

Law, J. A., Du, J., Hale, C. J., Feng, S., Krajewski, K., Palanca, A. M., et al. (2013). Polymerase IV occupancy at RNA-directed DNA methylation sites requires SHH1. *Nature, 498*, 385–389.

Law, J. A., & Jacobsen, S. E. (2010). Establishing, maintaining and modifying DNA methylation patterns in plants and animals. *Nature Reviews Genetics, 11*, 204–220.

Li, Y., Duan, C. G., Zhu, X., Qian, W., & Zhu, J. K. (2015). A DNA ligase required for active DNA demethylation and genomic imprinting in *Arabidopsis*. *Cell Research, 25*, 757–760.

Lindroth, A. M., Shultis, D., Jasencakova, Z., Fuchs, J., Johnson, L., Schubert, D., et al. (2004). Dual histone H3 methylation marks at lysines 9 and 27 required for interaction with *CHROMOMETHYLASE3*. *The EMBO Journal, 23*, 4286–4296.

Lister, R., O'Malley, R. C., Tonti-Filippini, J., Gregory, B. D., Berry, C. C., Millar, A. H., et al. (2008). Highly integrated single-base resolution maps of the epigenome in Arabidopsis. *Cell, 133*, 523–536.

Lister, R., Pelizzola, M., Dowen, R. H., Hawkins, R. D., Hon, G., Tonti-Filippini, J., et al. (2009). Human DNA methylomes at base resolution show widespread epigenomic differences. *Nature, 462*, 315–322.

Liu, Y., Koornneef, M., & Soppe, W. J. (2007). The absence of histone H2B monoubiquitination in the *Arabidopsis hub1* (*rdo4*) mutant reveals a role for chromatin remodeling in seed dormancy. *The Plant Cell, 19*, 433–444.

Liu, C., Lu, F., Cui, X., & Cao, X. (2010). Histone methylation in higher plants. *Annual Review of Plant Biology, 61*, 395–420.

Liu, F., Quesada, V., Crevillen, P., Baurle, I., Swiezewski, S., & Dean, C. (2007). The *Arabidopsis* RNA-binding protein FCA requires a lysine-specific demethylase 1 homolog to downregulate *FLC. Molecular Cell, 28*, 398–407.

Liu, Z. W., Shao, C. R., Zhang, C. J., Zhou, J. X., Zhang, S. W., Li, L., et al. (2014). The SET domain proteins SUVH2 and SUVH9 are required for Pol V occupancy at RNA-directed DNA methylation loci. *PLoS Genetics, 10*, e1003948.

Li, J., Yang, Z., Yu, B., Liu, J., & Chen, X. (2005). Methylation protects miRNAs and siRNAs from a 3'-end uridylation activity in *Arabidopsis. Current Biology, 15*, 1501–1507.

Lu, F., Cui, X., Zhang, S., Liu, C., & Cao, X. (2010). JMJ14 is an H3K4 demethylase regulating flowering time in *Arabidopsis. Cell Research, 20*, 387–390.

Lu, S. X., Knowles, S. M., Webb, C. J., Celaya, R. B., Cha, C., Siu, J. P., et al. (2011). The Jumonji C domain-containing protein JMJ30 regulates period length in the Arabidopsis circadian clock. *Plant Physiology, 155*, 906–915.

Lu, F., Li, G., Cui, X., Liu, C., Wang, X. J., & Cao, X. (2008). Comparative analysis of JmjC domain-containing proteins reveals the potential histone demethylases in *Arabidopsis* and rice. *Journal of Integrative Plant Biology, 50*, 886–896.

Lusser, A., Kolle, D., & Loidl, P. (2001). Histone acetylation: Lessons from the plant kingdom. *Trends in Plant Science, 6*, 59–65.

Manning, K., Tor, M., Poole, M., Hong, Y., Thompson, A. J., King, G. J., et al. (2006). A naturally occurring epigenetic mutation in a gene encoding an SBP-box transcription factor inhibits tomato fruit ripening. *Nature Genetics, 38*, 948–952.

March, E., & Farrona, S. (2018). Plant deubiquitinases and their role in the control of gene expression through modification of histones. *Frontiers of Plant Science, 8*, 2274.

Mari-Ordonez, A., Marchais, A., Etcheverry, M., Martin, A., Colot, V., & Voinnet, O. (2013). Reconstructing *de novo* silencing of an active plant retrotransposon. *Nature Genetics, 45*, 1029–1039.

Miura, A., Nakamura, M., Inagaki, S., Kobayashi, A., Saze, H., & Kakutani, T. (2009). An *Arabidopsis* jmjC domain protein protects transcribed genes from DNA methylation at CHG sites. *The EMBO Journal, 28*, 1078–1086.

Mizuguchi, G., Shen, X., Landry, J., Wu, W. H., Sen, S., & Wu, C. (2004). ATP-driven exchange of histone H2AZ variant catalyzed by SWR1 chromatin remodeling complex. *Science, 303*, 343–348.

Morales-Ruiz, T., Ortega-Galisteo, A. P., Ponferrada-Marin, M. I., Martinez-Macias, M. I., Ariza, R. R., & Roldan-Arjona, T. (2006). *DEMETER* and *REPRESSOR OF SILENCING 1* encode 5-methylcytosine DNA glycosylases. *Proceedings of the National Academy of Sciences of the United States of America, 103*, 6853–6858.

Morel, J. B., Mourrain, P., Beclin, C., & Vaucheret, H. (2000). DNA methylation and chromatin structure affect transcriptional and post-transcriptional transgene silencing in *Arabidopsis. Current Biology, 10*, 1591–1594.

Mozgova, I., Kohler, C., & Hennig, L. (2015). Keeping the gate closed: Functions of the polycomb repressive complex PRC2 in development. *The Plant Journal, 83*, 121–132.

Ng, D. W., Wang, T., Chandrasekharan, M. B., Aramayo, R., Kertbundit, S., & Hall, T. C. (2007). Plant SET domain-containing proteins: Structure, function and regulation. *Biochimica et Biophysica Acta, 1769*, 316–329.

Niu, L., Lu, F., Pei, Y., Liu, C., & Cao, X. (2007). Regulation of flowering time by the protein arginine methyltransferase AtPRMT10. *EMBO Report, 8*, 1190–1195.

Niu, L., Zhang, Y., Pei, Y., Liu, C., & Cao, X. (2008). Redundant requirement for a pair of PROTEIN ARGININE METHYLTRANSFERASE4 homologs for the proper regulation of Arabidopsis flowering time. *Plant Physiology, 148*, 490–503.

Noh, B., Lee, S. H., Kim, H. J., Yi, G., Shin, E. A., Lee, M., et al. (2004). Divergent roles of a pair of homologous jumonji/zinc-finger-class transcription factor proteins in the regulation of Arabidopsis flowering time. *The Plant Cell, 16*, 2601–2613.

Nuthikattu, S., McCue, A. D., Panda, K., Fultz, D., DeFraia, C., Thomas, E. N., et al. (2013). The initiation of epigenetic silencing of active transposable elements is triggered by RDR6 and 21-22 nucleotide small interfering RNAs. *Plant Physiology, 162*, 116–131.

Onodera, Y., Haag, J. R., Ream, T., Costa Nunes, P., Pontes, O., & Pikaard, C. S. (2005). Plant nuclear RNA polymerase IV mediates siRNA and DNA methylation-dependent heterochromatin formation. *Cell, 120*, 613–622.

Park, K., Kim, M. Y., Vickers, M., Park, J. S., Hyun, Y., Okamoto, T., et al. (2016). DNA demethylation is initiated in the central cells of *Arabidopsis* and rice. *Proceedings of the National Academy of Sciences of the United States of America, 113*, 15138–15143.

Paszkowski, J., & Grossniklaus, U. (2011). Selected aspects of transgenerational epigenetic inheritance and resetting in plants. *Current Opinion in Plant Biology, 14*, 195–203.

Pavlopoulou, A., & Kossida, S. (2007). Plant cytosine-5 DNA methyltransferases: Structure, function, and molecular evolution. *Genomics, 90*, 530–541.

Pei, Y., Niu, L., Lu, F., Liu, C., Zhai, J., Kong, X., et al. (2007). Mutations in the Type II protein arginine methyltransferase AtPRMT5 result in pleiotropic developmental defects in Arabidopsis. *Plant Physiology, 144*, 1913–1923.

Penterman, J., Zilberman, D., Huh, J. H., Ballinger, T., Henikoff, S., & Fischer, R. L. (2007). DNA demethylation in the *Arabidopsis* genome. *Proceedings of the National Academy of Sciences of the United States of America, 104*, 6752–6757.

Pokholok, D. K., Harbison, C. T., Levine, S., Cole, M., Hannett, N. M., Lee, T. I., et al. (2005). Genome-wide map of nucleosome acetylation and methylation in yeast. *Cell, 122*, 517–527.

Pontier, D., Yahubyan, G., Vega, D., Bulski, A., Saez-Vasquez, J., Hakimi, M. A., et al. (2005). Reinforcement of silencing at transposons and highly repeated sequences requires the concerted action of two distinct RNA polymerases IV in *Arabidopsis. Genes & Development, 19*, 2030–2040.

Pumplin, N., & Voinnet, O. (2013). RNA silencing suppression by plant pathogens: Defence, counter-defence and counter-counter-defence. *Nature Reviews Microbiology, 11*, 745–760.

Reinders, J., Wulff, B. B., Mirouze, M., Mari-Ordonez, A., Dapp, M., Rozhon, W., et al. (2009). Compromised stability of DNA methylation and transposon immobilization in mosaic *Arabidopsis* epigenomes. *Genes & Development, 23*, 939–950.

Ronemus, M. J., Galbiati, M., Ticknor, C., Chen, J., & Dellaporta, S. L. (1996). Demethylation-induced developmental pleiotropy in *Arabidopsis. Science, 273*, 654–657.

Rossetto, D., Avvakumov, N., & Cote, J. (2012). Histone phosphorylation: A chromatin modification involved in diverse nuclear events. *Epigenetics, 7*, 1098–1108.

Roudier, F., Ahmed, I., Berard, C., Sarazin, A., Mary-Huard, T., Cortijo, S., et al. (2011). Integrative epigenomic mapping defines four main chromatin states in Arabidopsis. *The EMBO Journal, 30*, 1928–1938.

Saleh, A., Alvarez-Venegas, R., Yilmaz, M., Le, O., Hou, G., Sadder, M., et al. (2008). The highly similar *Arabidopsis* homologs of trithorax ATX1 and ATX2 encode proteins with divergent biochemical functions. *The Plant Cell, 20*, 568–579.

Saze, H., Shiraishi, A., Miura, A., & Kakutani, T. (2008). Control of genic DNA methylation by a jmjC domain-containing protein in *Arabidopsis thaliana. Science, 319*, 462–465.

Schmitz, R. J., Sung, S., & Amasino, R. M. (2008). Histone arginine methylation is required for vernalization-induced epigenetic silencing of *FLC* in winter-annual *Arabidopsis thaliana. Proceedings of the National Academy of Sciences of the United States of America, 105*, 411–416.

Schoft, V. K., Chumak, N., Choi, Y., Hannon, M., Garcia-Aguilar, M., Machlicova, A., et al. (2011). Function of the DEMETER DNA glycosylase in the *Arabidopsis thaliana* male gametophyte. *Proceedings of the National Academy of Sciences of the United States of America, 108*, 8042–8047.

Sequeira-Mendes, J., Araguez, I., Peiro, R., Mendez-Giraldez, R., Zhang, X., Jacobsen, S. E., et al. (2014). The functional topography of the *Arabidopsis* genome is organized in a reduced number of linear motifs of chromatin states. *The Plant Cell, 26*, 2351–2366.

Simon, J. A., & Kingston, R. E. (2013). Occupying chromatin: Polycomb mechanisms for getting to genomic targets, stopping transcriptional traffic, and staying put. *Molecular Cell, 49*, 808–824.

Slotkin, R. K., Vaughn, M., Borges, F., Tanurdzic, M., Becker, J. D., Feijo, J. A., et al. (2009). Epigenetic reprogramming and small RNA silencing of transposable elements in pollen. *Cell, 136*, 461–472.

Soppe, W. J., Jacobsen, S. E., Alonso-Blanco, C., Jackson, J. P., Kakutani, T., Koornneef, M., et al. (2000). The late flowering phenotype of *fwa* mutants is caused by gain-of-function epigenetic alleles of a homeodomain gene. *Molecular Cell, 6*, 791–802.

Sridhar, V. V., Kapoor, A., Zhang, K., Zhu, J., Zhou, T., Hasegawa, P. M., et al. (2007). Control of DNA methylation and heterochromatic silencing by histone H2B deubiquitination. *Nature, 447*, 735–738.

Stroud, H., Do, T., Du, J., Zhong, X., Feng, S., Johnson, L., et al. (2014). Non-CG methylation patterns shape the epigenetic landscape in *Arabidopsis*. *Nature Structural & Molecular Biology, 21*, 64–72.

Stroud, H., Otero, S., Desvoyes, B., Ramírez-Parra, E., Jacobsen, S. E., & Gutierrez, C. (2012). Genome-wide analysis of histone H3.1 and H3.3 variants in *Arabidopsis thaliana*. *Proceedings of the National Academy of Sciences of the United States of America, 109*, 5370–5375.

Talbert, P. B., Ahmad, K., Almouzni, G., Ausió, J., Berger, F., Bhalla, P. L., et al. (2012). A unified phylogeny-based nomenclature for histone variants. *Epigenetics & Chromatin, 5*, 7.

Talbert, P. B., & Henikoff, S. (2017). Histone variants on the move: Substrates for chromatin dynamics. *Nature Reviews Molecular Cell Biology, 18*, 115–126.

Tamada, Y., Yun, J. Y., Woo, S. C., & Amasino, R. M. (2009). *ARABIDOPSIS TRITHORAX-RELATED7* is required for methylation of lysine 4 of histone H3 and for transcriptional activation of *FLOWERING LOCUS C*. *The Plant Cell, 21*, 3257–3269.

Teixeira, F. K., Heredia, F., Sarazin, A., Roudier, F., Boccara, M., Ciaudo, C., et al. (2009). A role for RNAi in the selective correction of DNA methylation defects. *Science, 323*, 1600–1604.

Thorstensen, T., Fischer, A., Sandvik, S. V., Johnsen, S. S., Grini, P. E., Reuter, G., et al. (2006). The *Arabidopsis* SUVR4 protein is a nucleolar histone methyltransferase with preference for monomethylated H3K9. *Nucleic Acids Research, 34*, 5461–5470.

Wagner, E. J., & Carpenter, P. B. (2012). Understanding the language of Lys36 methylation at histone H3. *Nature Reviews. Molecular Cell Biology, 13*, 115–126.

Wang, Z., Cao, H., Chen, F., & Liu, Y. (2014). The roles of histone acetylation in seed performance and plant development. *Plant Physiology and Biochemistry, 84*, 125–133.

Wang, D., Tyson, M. D., Jackson, S. S., & Yadegari, R. (2006). Partially redundant functions of two SET-domain polycomb-group proteins in controlling initiation of seed development in *Arabidopsis*. *Proceedings of the National Academy of Sciences of the United States of America, 103*, 13244–13249.

Wang, X., Zhang, Y., Ma, Q., Zhang, Z., Xue, Y., Bao, S., et al. (2007). SKB1-mediated symmetric dimethylation of histone H4R3 controls flowering time in *Arabidopsis*. *The EMBO Journal, 26*, 1934–1941.

Weake, V. M., & Workman, J. L. (2008). Histone ubiquitination: Triggering gene activity. *Molecular Cell, 29*, 653–663.

Wierzbicki, A. T., Haag, J. R., & Pikaard, C. S. (2008). Noncoding transcription by RNA polymerase Pol IVb/Pol V mediates transcriptional silencing of overlapping and adjacent genes. *Cell, 135*, 635–648.

Wierzbicki, A. T., Ream, T. S., Haag, J. R., & Pikaard, C. S. (2009). RNA polymerase V transcription guides ARGONAUTE4 to chromatin. *Nature Genetics, 41*, 630–634.

Wollmann, H., Holec, S., Alden, K., Clarke, N. D., Jacques, P. E., & Berger, F. (2012). Dynamic deposition of histone variant H3.3 accompanies developmental remodeling of the *Arabidopsis transcriptome. PLoS Genetics, 8*, e1002658.

Wollmann, H., Stroud, H., Yelagandula, R., Tarutani, Y., Jiang, D., Jing, L., et al. (2017). The histone H3 variant H3.3 regulates gene body DNA methylation in *Arabidopsis thaliana. Genome Biology, 18*, 94.

Xiao, J., Lee, U. S., & Wagner, D. (2016). Tug of war: Adding and removing histone lysine methylation in *Arabidopsis. Current Opinion in Plant Biology, 34*, 41–53.

Xiao, J., & Wagner, D. (2015). Polycomb repression in the regulation of growth and development in *Arabidopsis. Current Opinion in Plant Biology, 23*, 15–24.

Xie, Z., Johansen, L. K., Gustafson, A. M., Kasschau, K. D., Lellis, A. D., Zilberman, D., et al. (2004). Genetic and functional diversification of small RNA pathways in plants. *PLoS Biology, 2*, E104.

Xu, L., Zhao, Z., Dong, A., Soubigou-Taconnat, L., Renou, J. P., Steinmetz, A., et al. (2008). Di- and tri- but not monomethylation on histone H3 lysine 36 marks active transcription of genes involved in flowering time regulation and other processes in *Arabidopsis thaliana. Molecular and Cellular Biology, 28*, 1348–1360.

Yang, H., Han, Z., Cao, Y., Fan, D., Li, H., Mo, H., et al. (2012a). A companion cell-dominant and developmentally regulated H3K4 demethylase controls flowering time in *Arabidopsis* via the repression of *FLC* expression. *PLoS Genetics, 8*, e1002664.

Yang, H., Mo, H., Fan, D., Cao, Y., Cui, S., & Ma, L. (2012b). Overexpression of a histone H3K4 demethylase, JMJ15, accelerates flowering time in *Arabidopsis. Plant Cell Reports, 31*, 1297–1308.

Yan, Y., Shen, L., Chen, Y., Bao, S., Thong, Z., & Yu, H. (2014). A MYB-domain protein EFM mediates flowering responses to environmental cues in *Arabidopsis. Developmental Cell, 30*, 437–448.

Yan, D., Zhang, Y., Niu, L., Yuan, Y., & Cao, X. (2007). Identification and characterization of two closely related histone H4 arginine 3 methyltransferases in *Arabidopsis thaliana. Biochemical Journal, 408*, 113–121.

Yao, Q., Song, C. X., He, C., Kumaran, D., & Dunn, J. J. (2012). Heterologous expression and purification of *Arabidopsis thaliana* VIM1 protein: *In vitro* evidence for its inability to recognize hydroxymethylcytosine, a rare base in *Arabidopsis* DNA. *Protein Expression and Purification, 83*, 104–111.

Yelagandula, R., Stroud, H., Holec, S., Zhou, K., Feng, S., Zhong, X., et al. (2014). The histone variant H2A.W defines heterochromatin and promotes chromatin condensation in *Arabidopsis. Cell, 158*, 98–109.

Yun, M., Wu, J., Workman, J. L., & Li, B. (2011). Readers of histone modifications. *Cell Research, 21*, 564–578.

Zemach, A., Kim, M. Y., Hsieh, P. H., Coleman-Derr, D., Eshed-Williams, L., Thao, K., et al. (2013). The *Arabidopsis* nucleosome remodeler DDM1 allows DNA methyltransferases to access H1-containing heterochromatin. *Cell, 153*, 193–205.

Zhang, K., Sridhar, V. V., Zhu, J., Kapoor, A., & Zhu, J. K. (2007). Distinctive core histone post-translational modification patterns in *Arabidopsis thaliana. PloS One, 2*, e1210.

Zheng, X., Zhu, J., Kapoor, A., & Zhu, J. K. (2007). Role of *Arabidopsis* AGO6 in siRNA accumulation, DNA methylation and transcriptional gene silencing. *The EMBO Journal, 26*, 1691–1701.

Zhong, X., Du, J., Hale, C. J., Gallego-Bartolome, J., Feng, S., Vashisht, A. A., et al. (2014). Molecular mechanism of action of plant DRM de novo DNA methyltransferases. *Cell, 157*, 1050–1060.

Zilberman, D., Cao, X., & Jacobsen, S. E. (2003). ARGONAUTE4 control of locus-specific siRNA accumulation and DNA and histone methylation. *Science, 299*, 716–719.

Zilberman, D., Coleman-Derr, D., Ballinger, T., & Henikoff, S. (2008). Histone H2A.Z and DNA methylation are mutually antagonistic chromatin marks. *Nature, 456*, 125–129.

CHAPTER THREE

Epigenetic Diversity and Application to Breeding

Sunil Kumar Kenchanmane Raju and Chad E. Niederhuth[1]
Department of Plant Biology, Michigan State University, East Lansing, MI, United States
[1]Corresponding author: e-mail address: niederhu@msu.edu

Contents

Advances in Botanical Research, Volume 88
ISSN 0065-2296
https://doi.org/10.1016/bs.abr.2018.08.001

Abstract

Natural phenotypic variation is the basis of crop improvement. Understanding the basis of this variation has long focused on the role of genetic diversity and its exploitation in breeding programs. Beyond the genetic level, the genome is regulated by various chromatin modifications, which can be referred to as the "epigenome". Differences in the epigenome between individuals can affect gene expression and be a contributing factor in phenotypic variation. In some instances, these differences may also be inherited independent of genetic variation. While much has been learned mechanistically, the study of epigenomic diversity and function in plants has remained limited to a handful of examples. Moving the study of epigenome from one of discovery to application will require addressing still unanswered questions of how extensive epigenomic variation is, its relationship to the genome, and ultimately its contribution to phenotypic variation.

1. INTRODUCTION

Crop improvement relies on a constant supply of variation (novel alleles at genes of interest), such that elite cultivars can be bred for favorable alleles that optimally perform under changing environments. Over the past century, plant breeders have been carrying out artificial selection on natural and/or induced variation for integration of desired traits. Availability of cheap genotyping platforms has led to new tools and approaches for exploiting DNA sequence variation and the ability to associate genotype with phenotypic trait variation (Morrell, Buckler, & Ross-Ibarra, 2011; Wallace, Larsson, & Buckler, 2014). However, plant genetic diversity has been eroded through domestication, inbreeding, selection, and intensive breeding for high-yielding cultivars (National Research Council. Committee on Genetic Vulnerability of Major 1972; Zamir, 2001). The narrow genetic base of important crop species (Hyten, Song et al., 2006) and rapid depletion of genetic variation in landraces and related species (Zamir, 2001) makes progress through traditional approaches increasingly difficult. Advances in breeding and agronomic practices have steadily increased crop yields over the past century (Rowntree, Suhre et al., 2013), but further improvement will require novel strategies to expand the source of heritable phenotypic variation (Giovannoni, 2016; Rincker, Nelson et al., 2014).

Although heritable phenotypic variation is primarily driven by genetic diversity, recent evidence indicates that factors like DNA methylation, small RNAs, and histone modifications affect gene expression and ultimately contribute towards heritable trait variation (Cortijo, Wardenaar et al., 2014; Johannes, Porcher et al., 2009; Reinders, Wulff et al., 2009). These factors are often referred to as the epigenome, and they can provide an additional layer of variation on top of the existing genetic variation and may themselves be a novel source of phenotypic variation. Here we will discuss what is currently known about the nature of epigenomic variation, with a focus on DNA methylation, and its potential use in plant breeding.

2. EPIGENETICS, EPIGENOMICS, AND DNA METHYLATION

2.1 Problems of Definition

Defining "epigenetics" and "epigenetic variation" can itself be challenging and has generated extensive debate (Deans & Maggert, 2015; Jablonka & Lamb, 2003). It was originally coined by the embryologist developmental biologist Conrad Waddington as a fusion of the words "epigenesis" and "genetics" and was seen as a way of explaining how cells containing the same genetic material differentiated. The definition was picked up by David Nanney, who used it to describe "auxiliary" processes involved in gene expression and importantly, the idea that these could be maintained across cell division (Nanney, 1958). Robin Holliday emphasized the maintenance of expression states and brought in the concept of heritability, in particular, the idea of inherited differences independent of variation in the DNA sequence (Holliday, 1994). Contemporary usage of "epigenetic" is more varied, sometimes being used to refer to chromatin modifications themselves. The varied usage of the term can lead to confusion. While different definitions may all have some utility, in the context of understanding epigenetic variation for application to plant breeding, the most useful definition is a variation of that put forward by Holliday which emphasizes heritability.

Multiple molecular mechanisms fall under the study of epigenetic variation. Prions, which transmit through protein folding, can even be considered a form of epigenetic inheritance. However, by far most of the focus is on chromatin marks such as DNA methylation and histone modifications and their effect on gene expression (Schmitz & Zhang, 2014). There are a number of histone protein variants and modifications made to these histones, with different variants and modifications associated with different chromatin states. For example, the histone modification H3K9me3 is typically found in constitutive heterochromatin, while H3K4me3 is associated with actively transcribed genes (Schmitz & Zhang, 2014). Even in the case of DNA methylation, specific patterns are associated with different chromatin and transcriptional states (Niederhuth & Schmitz, 2017). These chromatin marks, when studied genome-wide, are referred to as the epigenome. Epigenomic variants, such as differences in DNA methylation at a gene, are often referred to as "epialleles".

While epigenomics has made it more feasible to discover new epialleles, there is still the question of whether or not these differences are dependent upon underlying DNA sequence variation or independent of it. Based on this, Eric Richards proposed classifying epialleles into one of three classes (Fig. 1): obligate, facultative, and pure epialleles (Richards, 2006). Obligate epialleles are completely dependent upon the presence/absence of some genetic variant (Fig. 1A). For example, the insertion of a transposable element (TE) near a gene can lead to silencing of a gene through the spread of DNA methylation. For an obligate epiallele, removal of the TE insertion, say through recombination, will result in a loss of DNA methylation at the neighboring gene. In contrast, facultative epialleles are those that are induced by a genetic variant, but which persist after the inducing genetic variant has been removed (Fig. 1B). For instance, if after the TE is segregated away through recombination, DNA methylation persists at the neighboring gene. Finally, pure epialleles are those that are independent of any sequence variation (Fig. 2C). While this classification scheme is useful for thinking about epialleles and the nature of epigenetic variation, proving this for any given epiallele is very difficult. This is especially true in natural populations, where it's likely impossible to identify an initial cause. While obligate epialleles may not fit canonical notions of being epigenetic, they can still be important mechanistically as a source of phenotypic variation and of interest to breeders. Because of the difficulty in knowing the causal effect of epialleles and the potentially confusing usage of the word "epigenetic", we will instead use the terminology "epigenomic variation" or "epialleles" to avoid any connotation regarding the inheritance of these variants unless otherwise specifically discussed.

2.2 Mechanisms of DNA Methylation

Most epialleles with known phenotypes have been studied in the context of DNA methylation and this has been the primary focus of most population, quantitative, and evolutionary epigenomic studies. While histone modifications have been studied extensively in development and environmental responses, they are less studied at population or evolutionary scales in plants. A major limiting factor of this is likely the greater expense and difficulty of producing data on histone modifications, which typically relies on

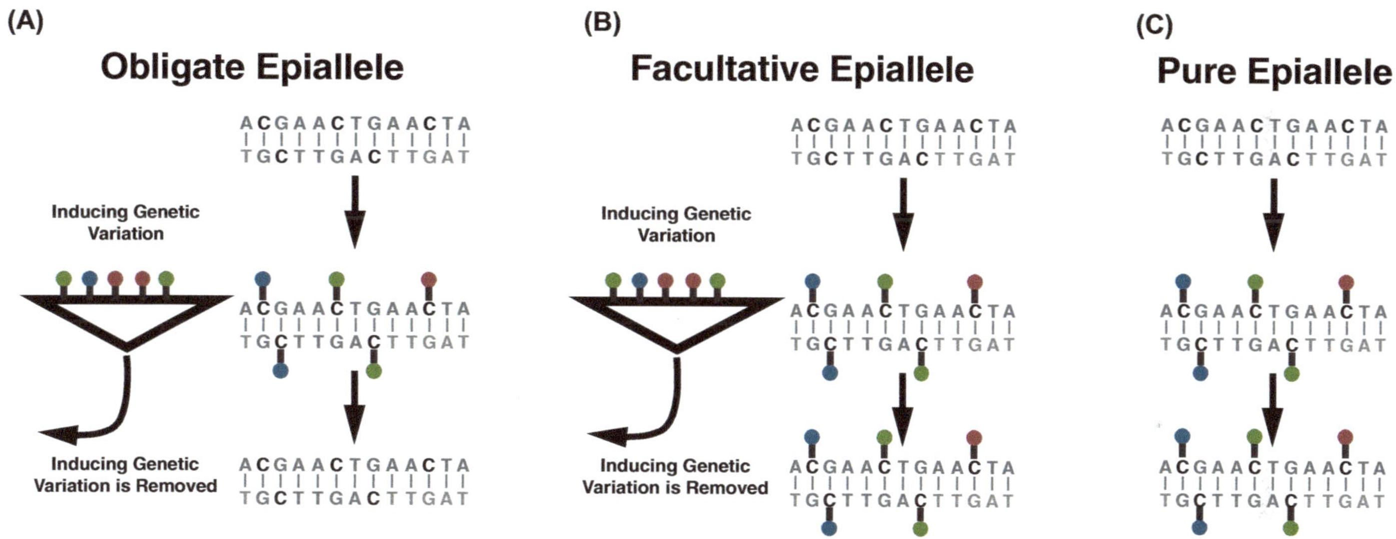

Figure 1 Epialleles classification scheme proposed by Richards (2006). (A) Obligate epialleles are dependent upon the presence/absence of a causative genetic variant. (B) Facilitative epialleles are induced by a genetic variant but are not dependent upon its presence/absence for the persistence of the epiallele in the next generation. (C) Pure epialleles occur and are transmitted independent of genetic variation.

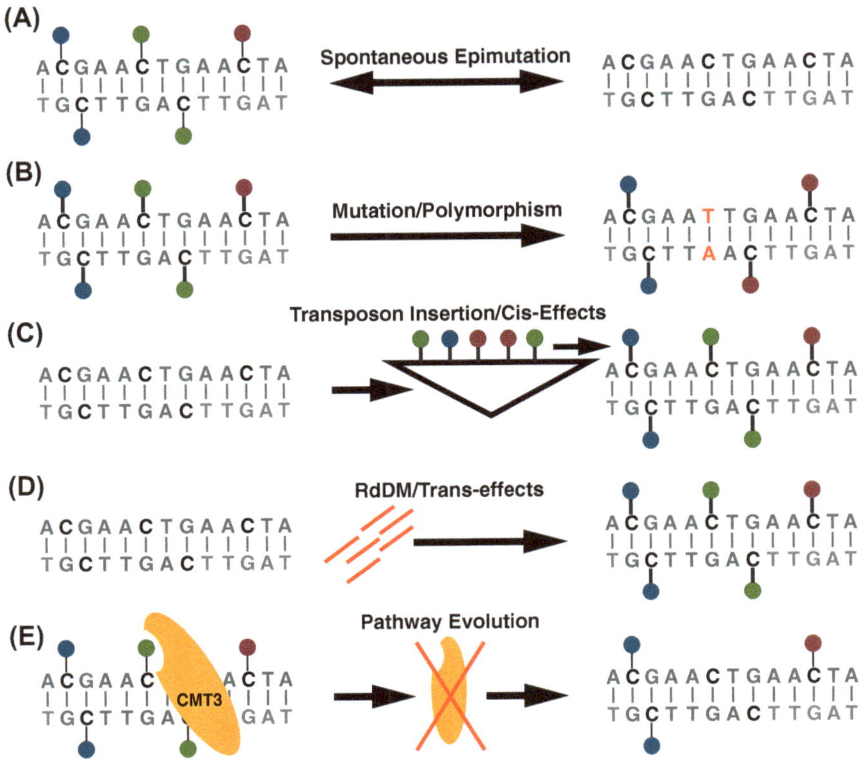

Figure 2 Potential sources of epigenomic variation. (A) Spontaneous epimutations can result in the loss or gain of a chromatin mark. (B) Mutation of the underlying sequence can create differences in chromatin states, such as the deamination of a methylated cytosine to thymine can eliminate a methylated site. (C) *Cis*-genetic variation can induce epigenomic variation at a region, such as the spreading of heterochromatin from a neighboring transposon insertion. (D) Epigenomic variation can result from *trans*-effects, such as silencing of a locus via small RNAs produced at another locus and RNA-directed DNA methylation (RdDM). (E) Genome-wide epigenomic differences might occur through the evolution of the pathways that establish and maintain the epigenome. For example, the loss of the CMT3 methyltransferase in some plant species.

chromatin immunoprecipitation (ChIP) followed by either quantitative PCR, hybridization to arrays, or sequencing (Schmitz & Zhang, 2011). The lack of such studies to date limits our understanding of epigenomic variation, which has prompted calls for programs to address these gaps (Lane, Niederhuth, Ji, & Schmitz, 2014). For this reason, we will focus primarily on examples and studies on DNA methylation, specifically methylation of cytosine bases.

DNA methylation in plants occurs primarily at cytosine bases, however recent results have shown evidence of adenine methylation in the model plant species *Arabidopsis thaliana* (Arabidopsis) and we still know very little about it (Liang, Shen et al., 2018). Cytosine methylation in plants is generally classified into three contexts based on composition of the neighboring three-prime bases: dinucleotide CpG (or CG), or trinucleotide CHG and CHH (H = A, T, or C) sites. In contrast, mammalian cytosine methylation is primarily limited to just the dinucleotide CG sites, except in specific cell types (Lister, Pelizzola et al., 2009). This classification scheme is mechanistically-based, with differing DNA methylation pathways having their own preferred targets. Different patterns of DNA methylation have been identified with different functional associations and for this reason it is important to consider methylation according to its context.

CG methylation (mCG) is found across kingdoms in plants, animals, and fungi and is maintained by the highly conserved METHYLTRANSFERASE 1 (MET1; known as DNMT1 in animals) (Law & Jacobsen, 2010). CG sites are symmetrical, with a corresponding CG site on the opposing strand. Due to the semi-conservative nature of DNA replication, new DNA molecules are a hybrid of an original methylated strand and a newly synthesized unmethylated strand. Methylated CG sites from the original strand serve as a template to guide MET1 in the methylation of the unmethylated strand (Finnegan, Peacock, & Dennis, 1996; Law & Jacobsen, 2010; Ronemus, Galbiati, Ticknor, Chen, & Dellaporta, 1996). The non-CG contexts, CHG and CHH, are unique to plants and arise from plant-specific methylation pathways. Methylation of CHG (mCHG) sites primarily involves a self-reinforcing loop between CHROMOMETHYLASE 3 (CMT3) and the histone modification H3K9me2 (Cao & Jacobsen, 2002; Du, Zhong et al., 2012; Law & Jacobsen, 2010). Both mCHG and H3K9me2 are enriched in regions of constitutive heterochromatin. As in the case of mCG, mCHG is symmetrical and results in stable propagation after replication. Another member of the CMT family, CMT2, also binds H3K9me2, where it methylates asymmetrical CHH sites (mCHHs), primarily in TE bodies (Zemach, Kim et al., 2013). More specifically, methylation by CMT2 seems to be biased towards CAA/CTA sites, as revealed through comparative studies and the use of mutants (Gouil & Baulcombe, 2016). RNA-directed DNA Methylation (RdDM) is another plant specific process, where the *de novo* methyltransferase DOMAINS REARRANGED METHYLTRANSFERASE 2 (DRM2) is targeted to specific sequences by short RNA molecules (Cao & Jacobsen, 2002). There are two variations of this pathway that are known. Pol IV-RdDM involves POLYMERASE IV (Pol IV) and the production of 24 nt short interfering RNAs (siRNAs) (Law & Jacobsen, 2010), while the RDR6-RdDM pathway involves 21–22 nt short RNAs produced by RNA-dependent RNA Polymerase 6 (RDR6) and seems to be involved in targeting sequences for *de novo* methylation (Panda & Slotkin, 2013). RdDM is often associated with mCHH in the literature, and vice versus, but in fact methylates in all three contexts (Law & Jacobsen, 2010).

2.3 Methods for Studying DNA Methylation

Multiple methods have been developed for the study of DNA methylation, but it is only with high-throughput array or sequencing based approaches that genome-wide surveys of epigenomic variation became practical (Lister & Ecker, 2009; Schmitz & Zhang, 2011). Broadly these methods can be classified into three categories: digestion-based, affinity purification, and treatment with sodium bisulfite. These differ in degree of resolution and both low- and high-throughput variants of each exist. Digestion-based methods are amongst the oldest and most widely used methods, making use of differing sensitivities of restriction enzymes to the presence or absence of DNA methylation (Schmitz & Zhang, 2011). They can be scaled across the entire genome without the need of genome sequence but are limited in resolution by the availability of target sites. Affinity purification-based approaches are used to enrich methylated regions of the genome; using either proteins with methylation binding domains (MBDs) or antibodies raised against methylated cytosines. These can then be analyzed through PCR, hybridization to arrays, or sequencing (Schmitz & Zhang, 2011). As they enrich for methylated sequences, they can cover more of the genome, but will miss unmethylated or potentially lowly methylated regions.

Sodium bisulfite treatment is currently the gold standard in studying DNA methylation. This approach takes advantage of the fact that when DNA is treated with sodium bisulfite (from here out "bisulfite"), unmethylated cytosines are deaminated to uracil and in subsequent PCR or sequencing, converted cytosines are read as thymines (Frommer, McDonald et al., 1992). Methylated cytosines, however, are protected and remain as cytosines. This allows one to distinguish methylated and unmethylated bases and when combined with whole-genome sequencing (whole-genome bisulfite sequencing or WGBS), allows for single base resolution of DNA methylation across the entire genome (Cokus, Feng et al., 2008; Lister, O'Malley et al., 2008). While bisulfite sequencing is currently the gold standard, new methods using the unique properties of nanopore sequencing technologies are being developed to call methylated bases without manipulation of the original DNA (Rand, Jain et al., 2017; Simpson et al., 2017). Such single-molecule long read technologies will be particularly advantageous, for example, allowing for phasing of DNA methylation data and a better understanding of how it is inherited.

3. KNOWN EPIALLELES

Likely the first epiallele discovered, although its molecular basis would not be known for another 253 years, was the *peloric* mutation of *Linaria vulgaris* (Cubas, Vincent, & Coen, 1999; Gustafsson, 1979). This mutation affects floral morphology and was first discovered by Magnus Ziörberg in 1742 while collecting plants in Roslagen archipelago in Sweden. The mutant flowers fascinated botanists, in particular Carl Linnaeus, who made an extensive study of it. The flowers challenged notions of species and so were given the name "Peloria", from the Greek-word for monster (Gustafsson, 1979). In 1999, Cubas et al. mapped the *peloric* mutation to a homolog of the *cycloidea* gene, *Lcyc*. Rather finding any differences in the DNA sequence, they instead found that in *peloric* mutants, *Lcyc* was methylated and silenced, while in the wild-type it was unmethylated (Cubas et al., 1999). So, this "monster" plant that so fascinated Carl Linnaeus was an epigenetic monster.

Since Linnaeus's time, an increasing number of epigenetic variants have been discovered (Table 1). Historically, these were primarily identified through unusual patterns of non-Mendelian inheritance in easily observed traits. For example, in 1915, Bateson and Pellew described the behavior of so-called "rogues" (it seems epigenetic variants have a way of upsetting scientists) that arose sporadically in *Pisum sativum* and which affected leaf morphology (Bateson & Pellew, 1915). These "rogues", whether selfed or crossed, would give rise exclusively to other rogues and this would be inherited in subsequent generations in violation of normal Mendelian inheritance. This was the first known example of a type of epigenetic inheritance known as paramutation, although it was not called that until subsequent discoveries in maize (Brink 1956, 1958). The exact mechanism of the rogue paramutation remains unknown, although one study found associations with differences in DNA methylation (Santo, Pereira, & Leitão, 2017).

Whole-genome approaches are increasing the rate of discovery of new epialleles, but few of these have been studied in any depth. In this section,

Table 1 Epialleles With Described Phenotypes. when Possible, Each Epiallele is Classified Either According to Richards (2006), as a Paramutation, or as Undetermined

Species	Epiallele	Phenotype	Classification	References
Linaria vulgaris	*peloric*	Flower development	Pure?	Cubas et al., 1999
Pisum sativum	*rogue*	Leaf morphology	Paramutation	Bateson & Pellew, 1915
Zea mays	*r1*	Anthocyanin content and pigmentation	Paramutation	Brink, 1956
	b1	Anthocyanin content and pigmentation	Paramutation	Coe, 1959
	p1	Flavonoid production and pigmentation	Paramutation	Das & Messing, 1994
	pl1	Anthocyanin content and pigmentation	Paramutation	Hollick et al., 1995
	lpa1	Low phytic acid	Paramutation	Shi et al., 2007, Pilu et al., 2008
Solanum lycopersicum	*sulfurea*	Chlorotic sectors	Paramutation	Hagemann, 1958
	Cnr	Fruit ripening and pericarp pigmentation	Pure?	Manning et al., 2006
	VTE3(1)	Vitamin E content	Facilitated?	Quadrana et al., 2014
Oryza sativa	*qWS8/ipa1-2D*	Plant Architecture and Yield	Obligate	Zhang et al., 2017
	Pigm	Rice Blast resistance	Obligate	Deng et al., 2017
Cucumis melo	*CmWIP1*	Flower sex determination	Obligate	Martin et al., 2009
Elaeis guineensis	*Bad Karma*	Flower development and yield	Pure—Tissue culture induced	Ong-Abdullah et al., 2015
Arabidopsis thaliana	*PAI*	Blue fluorescent phenotype and tryptophan production	Facilitated	Bender & Fink, 1995
	AtFOLT	Incompatibility	Facilitated	Durand et al., 2012
	HISN6	Incompatibility	Undetermined	Blevins et al., 2017
	QQS	Starch metabolism	Pure?	Silveira et al., 2013
Antirrhinum majus	*nivea*	Anthocyanin content and pigmentation	Paramutation	Harrison & Carpenter, 1973
Œnothera spp.	*cruciata*	Flower development	Paramutation	Renner, 1959

Those epialleles where there is some uncertainty in the classification are indicated by a question mark.

we will discuss epigenetic variants that have known phenotypes (Table 1). This list is by no means exhaustive but does tell us something about the mechanisms of epigenomic variation and its causes (Fig. 2) while demonstrating its effects on phenotype.

3.1 Zea mays

In 1956, Alexander Brink first described the inheritance of kernel anthocyanin content and pigmentation with the *red1* (*r*) locus in *Z. mays* (maize) (Brink, 1956). Like Bateson and Pellew's rogue peas, the *r* locus violated Mendelian inheritance. Brink coined the term "paramutation" to refer to this phenomenon where a "paramutagenic" allele from one parent was able to induce a heritable change in the other allele, referred to as the "paramutable" allele (Brink, 1958). In 1959, Ed Coe discovered another example of paramutation in maize called *booster1* (*b1*), which affects anthocyanin content and pigmentation of various parts of the maize plant (Coe, 1959). Other cases of paramutation in maize include *pericarp colour1* (*p1*) (Das & Messing, 1994), *purple plant1* (*pl1*) (Hollick, Patterson, Coe, Cone, & Chandler, 1995), and *low phytic acid1* (*lpa1*) (Pilu et al., 2008; Shi, Wang et al., 2007). As the names suggest, *p1* and *pl1* also affect pigmentation, either through the flavonoid or anthocyanin pathways (Das & Messing, 1994; Hollick et al., 1995), while *lpa1* affects phytic acid content in seeds.

Investigation of the molecular basis of paramutation in maize has implicated the importance of repetitive sequences. Across maize haplotypes, there is variation in gene copy number at the *r* locus and this is directly related to paramutagenic activity. Haplotypes with fewer gene copies are less paramutagenic, while those with more copies have increased activity (Kermicle, Eggleston, & Alleman, 1995; Panavas, Weir, & Walker, 1999; Walker & Panavas, 2001). Paramutagenic *r* alleles are also characterized by DNA methylation (Walker & Panavas, 2001). At the *b1* locus, paramutagenic activity is linked to tandem repeat copy number and fewer repeats are associated with weaker paramutagenic activity (Stam, Belele, Dorweiler, & Chandler, 2002). Interestingly, methylation of these repeats is also associated with paramutation (Stam et al., 2002). Genetic evidence for the role of DNA methylation in maize comes from mutants of *mediator of parmutation1* (*mop1*) (Dorweiler et al., 2000), which is an RNA-dependent RNA polymerase needed for RdDM and identified for altering the phenotypes of *b1* and pl1 (Alleman, Sidorenko et al., 2006).

3.2 Solanum lycopersicum

Contemporary to Brink and Coe, Rudolf Hagemann identified an instance of paramutation at the *SULFUREA* (*SULF*) locus in *S. lycopersicum* (tomato) (Hagemann, 1958). Homozygous *sulf* mutants are lethal in seedlings, but as a heterozygous the *sulf* mutation is paramutagenic and lead to silencing of the paramutable wild type allele, which can then be inherited (Hagemann, 1969). Phenotypically, this causes the development of chlorotic sectors throughout the plant. The gene underlying the *sulf* paramutation was identified as *SLTAB2*, an orthologue of Arabidopsis *ATAB2*, mutants of which share a similar phenotype. In the *sulf* plants *SLTAB2* has two differentially methylated regions (DMRs) in the promoter, one of which is found at the transcriptional start site, and there are also differences in small RNA production (Gouil, Novák, & Baulcombe, 2016). Unlike examples in maize, these regions of *SLTAB2* does not involve tandem repeats, although one of the DMRs is associated with TE fragments.

In 1990s, tomato fruit that failed to ripen and had a non-pigmented pericarp were identified in a commercial field (Thompson, Tor et al., 1999). The *Colorless non-ripening* (*Cnr*) locus maps to a *SQUAMOSA* promoter binding protein-like (SBP-box) gene called *LeSPL-CNR* and in *Cnr* mutants, expression of *LeSPL-CNR* is greatly reduced compared to wild-type (Manning, Tör et al., 2006). However, no sequence variation was identified in *Cnr* at this locus, rather in the mutants the *LeSPL-CNR* promoter is methylated. That *Cnr* is an epigenetic mutation is supported by observations of ripening in sectors of *Cnr* fruit, presumably that have reverted back to wild-type methylation levels (Manning, Tör et al., 2006). Maintenance and inheritance of *Cnr* has been shown to be dependent upon the tomato CMT3 methyltransferase (SlCMT3), as silencing of *SlCMT3* expression restored fruit ripening in *Cnr* plants (Chen, Kong et al., 2015).

Epigenomic variation can also affect quantitative trait loci, as in the case of the *VTE3* gene and vitamin E production in tomato fruit (Quadrana, Almeida et al., 2014). *VTE3* encodes a 2-methyl-6-phytylquinol methyltransferase involved in some of the last steps of vitamin E biosynthesis and *VTE3(1)* alleles are differentially expressed in introgression lines resulting from crosses of the M82 accession and wild tomato species. These differences in expression appear to be due to the production of small RNAs and higher methylation of *VTE3(1)* promoter regions resulting from the presence of a SINE retrotransposon in the M82 accession. Interestingly, when M82 is grown under field conditions, spontaneous reversion of DNA methylation at *VTE3(1)* can occur, leading to higher gene expression and higher vitamin E content.

3.3 Oryza sativa

The *qWS8/ipa1-2D* locus in rice is a QTL that affects plant architecture and yield (Zhang, Yu et al., 2017). Yongyou12 is high-yielding hybrid variety derived from crosses of the two subspecies *O. sativa* spp. *indica and O. sativa* spp. *japonica*. Yongyou12 has ideal plant architecture (IPA) consisting of traits such as production of more rice grains, reduced number of unproductive tillers, and more robust stems. Map-based cloning of the *qWS8/ipa1-2D* QTL, which contributes to IPA in Younyou12, found three tandem repeats that are associated with increased expression of the neighboring *IPA1* gene. These tandem repeats are associated with decreased DNA methylation levels, in particular mCHH, in the promoter regions of *IPA1* and an increase in open chromatin.

In another example, the *Pigm* locus derived from the Gumei 4 variety confers resistance to the fungal pathogen *Magnaporthe oryzae* (rice blast) and has been extensively used in breeding programs (Deng, Zhai et al., 2017). This locus contains several copies of nucleotide-binding leucine-rich repeat (NLR) receptors. One of these NLR receptors, *PigmR*, forms homodimers to confer resistance, while another, *PigmS*, inhibits homodimer formation through the creation of heterodimer. *PigmS* expression is largely restricted to panicles and pollen, while *PigmR* is expressed throughout the plant. The suppression of *PigmS* in other tissues is a result of two MITE TE sequences located upstream that are targeted by the RdDM pathway, demonstrating the role of the epigenome in fine-tuning traits such as disease resistance and yield.

3.4 Cucumis melo

Sex determination of flowers in *C. melo* (melon) is controlled by two loci, andromonoecious (*a*) and gynoecious (*g*) (Martin, Troadec et al., 2009). Plants homozygous for *g* produce either unisexual female flowers (*AAgg*) or hermaphrodites (*aagg*). The *g* locus maps to a TE insertion from the *hAT* TE family, subsequently named *Gyno-hAT*. Sex determination appears to result from spreading of DNA methylation from *Gyno-hAT* into a neighboring transcription factor, *CmWIP1*, resulting in its silencing. Methylation of *CmWIP1* appears to be dependent upon the presence of *Gyno-hAT*, as methylation was absent in recombinant plants where *Gyno-hAT* was not present. Although partial reversion of flowers did occur and was associated with spontaneous demethylation of *CmWIP1*, the dependence of DNA methylation on the presence of *Gyno-hAT* would classify it as an example of an obligate epiallele.

3.5 Elaeis guineensis

While it does not occur naturally, the *BAD KARMA* epiallele in *E. guineensis* (African oil palm) demonstrates the economic impact of epigenetic variation can have (Ong-Abdullah, Ordway et al., 2015). Micropropagation is used to propagate high-yielding African oil palm hybrids for planting in plantations, but also results in a high frequency of aberrant floral phenotypes, known as mantling, that greatly reduces yield and could not be identified until well after planting. An epigenome-wide association study finally identified the cause of mantling as a loss of methylation in a LINE retrotransposon (called *Karma*), located in the intron of the *DEFICIENS* gene. This loss of DNA methylation epiallele, named *BAD KARMA*, results in mis-splicing of *DEFICIENS* and the mantling phenotype. By identifying the causative epiallele of mantling it is now possible to design assays for screening clones before they are planted, with the potential to increase yields and reduce environmental impacts.

3.6 *Arabidopsis thaliana*

While it is not a crop species, the epigenome of *A. thaliana* has been studied more extensively than any other plant species, and a number of natural epigenetic variants have been studied in-depth: *PAI* (Bender & Fink, 1995), *AtFOLT* (Durand, Bouché, Perez Strand, Loudet, & Camilleri, 2012), *HISN6* (Blevins, Wang, Pflieger, Pontvianne, & Pikaard, 2017), and *QQS* (Silveira, Trontin et al., 2013). The classic example is the work of Judith Bender on the *PHOPHORRIBOSYLANTHRANILATE ISOMERASE* (*PAI*) genes, which is involved in tryptophan biosynthesis. These mutants were first identified as having a blue fluorescent phenotype in the Wassilewskija (Ws) ecotype (Bender & Fink, 1995). These plants were deficient in *PAI* enzyme and had both tryptophan and indole-3-acetic acid related phenotypes. The Ws ecotype has four copies of *PAI* in the genome, all four of which are methylated. Three of these are unlinked and reside on separate chromosomes, while the fourth copy, *PAI4*, is an inverted duplication next to *PAI1*. In contrast, the Columbia (Col) ecotype has only three copies, not having the *PAI4* duplication and all three genes are unmethylated (Bender & Fink, 1995). By introducing the *PAI4* duplication into the Col background, it was shown that this inverted repeat structure was able to trigger DNA methylation at the unlinked *PAI2* and *PAI3* loci (Luff, Pawlowski, & Bender, 1999). Even when the Ws *PAI1-PAI4* locus was segregated away, DNA methylation was maintained at Col-derived *PAI2* and *PAI3* for at least five generations (Luff et al., 1999). This is an excellent example of a facultative-epiallele, where genetic change (the *PAI4* inverted duplication) induces a heritable epigenetic change at loci, even in the absence of the inducing genetic variant.

The *FOLATE TRANSPORTER* (*AtFOLT*) (Durand et al., 2012) and *HISTIDINE BIOSYNTHESIS 6* (*HISN6*) (Blevins et al., 2017) epialleles both phenotypically result in genetic incompatibilities. Of these, *AtFOLT* was the first identified in recombinant inbred lines (RILs) of a cross between Col and the Shahdara (Sha) ecotypes due to non-independent segregation of a locus on Col chromosome 4 (K4) and a locus on Sha chromosome 5 (K5). Few plants homozygous for both the Col K4 and Sha K5 were isolated and these produced few seeds (Durand et al., 2012). Col has a single copy of the gene at K5, *AtFOLT1*, whereas Sha has two copies, *AtFOLT1* at K5 and *AtFOLT2* at K4. *AtFOLT1* in Sha is not expressed, so plants homozygous for Col K4 and Sha K5 do not have an expressed copy of *AtFOLT*, resulting in incompatibility. The lack of *AtFOLT1* expression in Sha does not appear to be due to sequence variation at *AtFOLT1* as identical sequences are observed in other ecotypes where it is expressed. Rather this sequence is methylated at the start of the gene and associated with small RNAs. The K4 locus, which contains *AtFOLT2* gene in Sha is structurally complex with two truncated and rearranged copies, including expressed sequences that match the methylated parts of the promoter at *AtFOLT1*. It was proposed that this could give rise to small RNAs resulting in silencing of *AtFOLT1*. Indeed, in some Col-Sha RILs, stable silencing of Col *AtFOLT1* was observed. Even in the absence of the inducing K4 locus, silencing of *AtFOLT1* is stable over multiple generations, demonstrating this to be another example of a facultative epiallele as in the *PAI* genes (Durand et al., 2012).

HISN6 is an essential gene involved in histidine biosynthesis and Arabidopsis has two copies, *HISN6A* and *HISN6B*, located on chromosomes five and 1 respectively (Blevins et al., 2017). In the Col ecotype, *HISN6A* is expressed while *HISN6B* is methylated and silence. In contrast, *HISN6A* is a non-functional mutant in the Cape Verde Islands (Cvi) ecotype, while *HISN6B* is the expressed copy. The silenced status of Col *HISN6B* is stably inherited, despite its ability to produce a functional protein if expressed, and incompatibility arises in the progeny of Col and Cvi crosses when plants are homozygous for Col *HISN6B* and Cvi *HISN6A*. This incompatibility could be alleviated, however, by reactivation of Col *HISN6B* using plants mutant for certain types of DNA methylation or histone modification (Blevins et al., 2017). *HISN6* does not appear to involve the sort of structural rearrangements that might give rise to small RNAs and silencing, as in the case of *PAI* or *AtFOLT*. So, while there is clearly an epiallele involved in silencing of a duplicate gene copy, the underlying cause of this silencing is unclear.

Unlike the previous examples, *Qua-Quine Starch* (*QQS*) is a single copy gene of *de novo* origin in Arabidopsis and is involved in starch metabolism in leaves (Silveira, Trontin et al., 2013). *QQS* is a small gene and is flanked by TE sequences on both sides, as well as tandem repeats in the promoter and 5′UTR, which are heavily methylated and produce small RNA. These methylated flanking sequences negatively impact *QQS* expression, as it is more highly expressed in mutants in DNA methylation pathways. Spontaneous hypomethylated epialleles were identified in the Col ecotype that resulted in increased expression of *QQS*. As no sequence variation was observed in these lines, it appears that the hypomethylated *QQS* lines may be an instance of pure epiallelic variation, independent of any inducing genetic change (Silveira, Trontin et al., 2013). Epialleles at *QQS* were also observed in other accessions (Fig. 3A), including those collected from the wild and in all examples studied, these appear to be independent of the methylation state of flanking TE sequences or other genetic variation at the *QQS* gene (Silveira, Trontin et al., 2013).

Aside from these examples of natural Arabidopsis epialleles, several have been found in laboratory settings arising from transgenic experiments or induced through other experimental means. Paramutation behavior has been observed in a number of independent cases of transgenic Arabidopsis (Gao & Zhao, 2013; Stokes & Richards, 2002; Xue, Ruprecht et al., 2012; Zheng, Yu et al., 2015). In one example, diploid plants carrying a transgene conferring resistance to hygromycin saw silencing of the transgene and subsequent paramutation-like behavior when outcrossed to tetraploid plants, but not to other diploids (Mittelsten Scheid, Afsar, & Paszkowski, 2003). Anti-sense lines targeting DNA methyltransferases resulted in overall decreases in DNA methylation, but oddly enough resulted in hypermethylation of the *SUPERMAN* gene, causing floral phenotypes (Jacobsen & Meyerowitz, 1997).

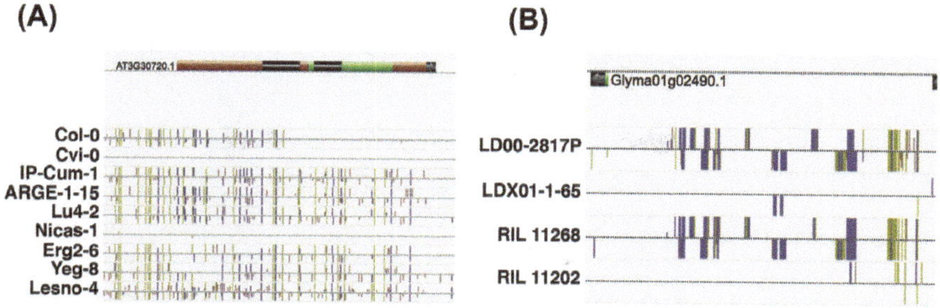

Figure 3 Browser screen shots displaying natural variation in DNA methylation. Gene models are displayed in the top track and DNA methylation for each sample below. Horizontal indicate methylation levels at individual cytosines for each sample. Gold bars represent mCG, blue represents mCHG, and pink represent mCHH. (A) DNA methylation at the QQS gene in *Arabidopsis thaliana* accessions from across the world. (B) Inheritance of DNA methylation in *Glycine max* recombinant inbred lines tracks the parental genotype. At this locus, RIL 11268 DNA methylation is inherited from parent LD00-2817P, while DNA methylation in RIL 11202 is inherited from the LDX01-1-65 parent. *(A) Data is from Kawakatsu, T., Huang, S-s. C., Jupe, F., Sasaki, E., Schmitz, R. J., Urich, M. A., et al. (2016). Epigenomic diversity in a global collection of* Arabidopsis thaliana *accessions.* Cell, 166*(2), 492–505. (B) Data is from Schmitz, R. J., Schultz, M. D., Urich, M. A., Nery, J. R., Pelizzola, M., Libiger, O., et al. (2013), Patterns of population epigenomic diversity.* Nature, 495*(7440), 193–198.*

This could be an indirect effect, due to misregulation of other genes which are themselves regulated by chromatin modifications (Rigal, Kevei, Pélissier, & Mathieu, 2012; Williams, Pignatta, Henikoff, & Gehring, 2015). It also demonstrates that perturbations to the pathways that establish and maintain chromatin modifications can induce novel epialleles, a strategy that has seen powerful application in the Arabidopsis epigenetic recombinant inbred line populations (Johannes, Porcher et al., 2009; Reinders, Wulff et al., 2009), which will be discussed later.

3.7 Other Species

Outside of Arabidopsis and crop species, only a limited number of epialleles linked to phenotypic variation have been identified. This includes the aforementioned *peloric* in *L. vulgaris* (Cubas et al., 1999), *nivea* in *Antirrhinum majus* (common snapdragon) (Harrison & Carpenter, 1973; Krebbers et al., 1987), and *cruciata* in *Œnothera spp.* (evening primrose) (Renner, 1959). Both *nivea* and *cruciata* are examples of paramutation. The *nivea* locus involves the excision of TEs in a *chalcone synthase* gene, affecting anthocyanin production and flower pigmentation, however, this does not appear to be necessary for paramutation in all cases (Almeida, Carpenter, Robbins, Martin, & Coen, 1989; Harrison & Carpenter, 1973; Krebbers et al., 1987). The *cruciata* locus, on the other hand affects floral morphology, resulting in sepaloid petals, however, this has not been studied in depth and to our knowledge the molecular basis of *cruciata* remains unknown (Renner, 1959).

4. EPIGENOMIC VARIATION WITHIN POPULATIONS

The past century saw a limited number of epialleles discovered through careful study of individual loci. Often these studies have been limited to specific lines, although a few, as in the case of Arabidopsis, have looked at these epialleles across populations (Durand et al., 2012; Melquist, Luff, & Bender, 1999; Silveira, Trontin et al., 2013). If epigenomic variation is to have any significant application in plant breeding, a much larger effort is needed to discover such variation and quantify its contribution to phenotype. Genome-wide array and sequencing-based approaches have now been applied to study epigenomic variation across diverse accessions in several species, which will be reviewed in this section.

4.1 *Arabidopsis thaliana*

Arabidopsis was the first plant species' methylome to be studied by array (Tran et al., 2005) and the first of any species studied by whole-genome bisulfite sequencing (Cokus, Feng et al., 2008; Lister, O'Malley et al., 2008). It has also been amongst the extensively studied at a population scale, with well over a thousand accessions now sequenced (Dubin, Zhang et al., 2015; Kawakatsu, Huang et al., 2016; Schmitz, Schultz et al., 2013; Vaughn, Tanurdžić et al., 2007). Vaughn et al. (2007) gave the first real glimpse of the epigenomic diversity of Arabidopsis by comparing the *Landsberg erecta* (Ler) and Col accessions using hybridization to tilling arrays after digestion of the DNA with the methylation-dependent restriction enzyme *MrcBC*. Around a third of genes were found to have detectable methylation, but these were polymorphic between the two ecotypes, with ~50% of the methylated genic tiles differing between the two accessions. Tracking of methylation in Col x Ler crosses in the F_1 and F_2 generations showed that methylation patterns typically followed the parental genotype, although some instability was observed. A subset of 18 loci were further investigated across 96 additional accessions and all were highly polymorphic with exception of only two loci.

Schmitz et al. (2013) used WGBS to sequence the methylomes of 152 accessions, which was then followed by sequencing of 1107 accessions by Kawakatsu et al. (2016). Most methylated cytosines are found in TE-rich regions of constitutive heterochromatin and stably methylated across accessions. However, these studies found hundreds of thousands of single methylation polymorphisms (SMPs; differences in methylation at a single cytosine) and tens of thousands of DMRs. Using the SMP data, a phylogeny of the 152 accessions was constructed and compared to those based on genetic data; showing that while there was a correlation of the CG SMP phylogeny to genetic based ones, there was little correlation of CHG or CHH SMP-based phylogenies (Schmitz, Schultz et al., 2013). This shows that methylation of individual non-CG sites can be highly variable, while mCG is likely more stable within populations. There is little evidence, however, of an individual SMP affecting gene expression, rather this is usually associated with differences in the methylation over a region. Schmitz et al. (2013) identified 40,269 CG-DMRs (CG context only) and 14,485 C-DMRs (all sequence contexts), while Kawakatsu et al. (2016) found 22,060 DMRs: 8917 CG-DMRs, 2208 CH-DMRs (non-CG contexts only), and 10,935 C-DMRs. The fewer DMRs identified in the latter is likely due to methodological differences in defining DMRs. The 22,060 DMRs identified in 1107 accessions cover 38% of the Arabidopsis genome, showing that epigenomic variation is extensive in this species (Kawakatsu, Huang et al., 2016).

As extensive as this variation is, how much of it might affect gene expression? CG-DMRs are enriched in gene-bodies and mostly represent variation in what is called "gene-body methylation" (gbM) (Kawakatsu, Huang et al., 2016; Schmitz, Schultz et al., 2013). GbM genes are methylated only in the CG context and have a rather distinct pattern of depletion of mCG around the transcriptional start site and termination site, with enrichment in the gene-body (Tran et al., 2005). While gbM is associated with constitutively expressed genes, variation in gbM seems to have a minimal to little impact on gene expression variation and its role in gene expression is questionable (Kawakatsu, Huang et al., 2016; Meng, Dubin et al., 2016; Teixeira & Colot, 2009).

The association of non-CG methylation with silencing, means that CH or C-DMRs are more likely to have an effect when found in association with genes. CH-DMRs are almost entirely found in TEs or unannotated regions, but a number are also found in genes (Kawakatsu, Huang et al., 2016). While TEs are enriched in C-DMRs (Schmitz, Schultz et al., 2013), the majority of those identified by Kawakatsu et al. (2016) can be found overlapping genes or overlapping both TEs and genes. Genes associated with C-DMRs tend to have lower expression in accessions where they are methylated and are typically enriched in genes with functions such as disease resistance and whose expression tends to be highly variable in differing environments (Kawakatsu, Huang et al., 2016; Schmitz, Schultz et al., 2013). Interestingly, these genes also show higher expression in pollen, a phenomenon that could be related to the re-establishment of methylated states in gametes (Schmitz, Schultz et al., 2013; Slotkin, Vaughn et al., 2009). Meng et al. (2016) tried to quantify the contribution of DNA methylation to gene expression in 135 Swedish lines and found that DNA methylation explained only a small proportion of the variation in gene expression. While it seems unlikely that most DMRs have a direct impact on gene expression, there are still many across different accessions that do have such an association and which could potentially affect phenotypes, especially in the wild.

Differences in methylation have been found to have geographic and environmental associations. Genes containing TE-like non-CG methylation (teM) are rarer than gbM genes and many are found in methylated in only a single accession (teM-singletons). These teM-singletons were often found enriched in accessions originating from ice age refugia around the Mediterranean (Kawakatsu, Huang et al., 2016). In a population of 150 Swedish accessions, temperature was observed to affect mCHH, with growth at higher temperatures associated with higher levels of mCHH. Most of the temperature dependent CHH-DMRs were associated with a subset of TEs concentrated in euchromatic regions and changes in these loci seem to be regulated by the RdDM pathway (Dubin, Zhang et al., 2015). In the same study, gbM was shown to have a latitudinal association. Northern accessions were hypermethylated in comparison to Southern accessions (Dubin, Zhang et al., 2015). This trend is observed even more broadly across the range of Arabidopsis, with the Northern (Swedish) accessions being hypermethylated in comparison to those from more Southern nations (Spain) (Kawakatsu, Huang et al., 2016). Dubin et al. (2015) did find a weak correlation between higher gene expression and higher gbM methylation, although whether the methylations differences were a cause or a result of the differences in gene expression is unclear.

There is a strong genetic basis for a lot of epigenomic variation. For example, almost none of the gbM variation in the Swedish lines was explained by environmental factors, rather genome-wide association studies (GWASs) showed that the variation mostly has a genetic basis, with the majority being explained by *trans*-acting SNPs. Northern Swedish lines have had a number of selective sweeps and the author conclude that the differences in gbM amongst Swedish lines likely reflect local adaptation at the genetic level (Dubin, Zhang et al., 2015). Schmitz et al. (2013) found that ~35% of C-DMRs were associated with a methylation QTL (mQTL) and many of these had a polygenic basis. While local variant, especially structural variation, explains a large portion of this (Kawakatsu, Huang et al., 2016; Schmitz, Schultz et al., 2013), there may be more overall distant mQTL (Schmitz, Schultz et al., 2013). Using GWAS, it was found that the mCHH methylation levels of many loci were associated with variants of CMT2 (Dubin, Zhang et al., 2015; Kawakatsu, Huang et al., 2016; Shen, De Jonge et al., 2014) and it has been suggested that CMT2 variants could be involved in adaptation to temperature stress (Shen, De Jonge et al., 2014). Other significant associations were found with SNPs near genes involved in RdDM, showing that variation in DNA methylation pathways can have global effects on the epigenome (Kawakatsu, Huang et al., 2016).

While much of the data from Arabidopsis shows a strong link between genetic and epigenomic variation, the study of natural populations alone might limit the ability to detect more subtle effects. These might be overcome with experimental populations. For example, DNA methylation has been studied in the Arabidopsis mutation accumulation lines, a mostly isogenic population of the Col ecotype that have been grown via selfing and single-seed descent for more than 30 generations (Becker, Hagmann et al., 2011; Schmitz, Schultz et al., 2011). Because genetic variation is largely controlled for, spontaneous (pure) epialleles can be more easily identified and the rate at which epimutations (changes in methylation at a single cytosine) occur can be calculated (Becker, Hagmann et al., 2011; Hofmeister, Lee, Rohr, Hall, & Schmitz, 2017; Schmitz, Schultz et al., 2011; van der Graaf, Wardenaar et al., 2015). Even more powerful are populations where widespread epigenetic variation has been induced. Mutants in DNA methylation pathway genes induce genome-wide losses of DNA methylation. These can then be crossed back to isogenic wild-type plants and the mutation segregated away, followed by selfing of non-mutant offspring. The subsequent generations are genetically isogenic, but are a mosaic of induced epialleles, resulting in creation of epigenetic recombinant inbred lines (epiRILs) (Johannes, Porcher et al., 2009; Reinders, Wulff et al., 2009). These epiRIL populations show a wide-range of phenotypic variation and it is possible to use the methylation differences to map epigenetic QTL (QTL^{epi}) (Cortijo, Wardenaar et al., 2014).

4.2 Zea mays

Maize has a significantly larger genome than Arabidopsis, largely owing to a larger TE and repeat content (Schnable, Ware et al., 2009). This makes genome-wide analyses of the epigenome both more technically challenging and expensive, however, this may also create more opportunity for epigenomic variation and epigenetic phenomena. Indeed the total amount of open chromatin in plant genomes scales with genome size and alters the nature of adaptive evolution (Mei, Stetter, Gates, Stitzer, & Ross-Ibarra, 2018), while also correlating with DNA methylation content (Alonso, Pérez, Bazaga, & Herrera, 2015; Niederhuth, Bewick et al., 2016; Takuno, Ran, & Gaut, 2016).

Using methylated DNA immunoprecipitation followed by hybridization to microarrays (meDIP-chip), ~700 DMRs were identified between the accessions B73 and Mo17 (Eichten, Swanson-Wagner et al., 2011). Examination of nine DMRs in regions of identity by descent (IBD) with little genetic variation between B73 and Mo17 showed potential evidence of pure epigenetic variation for eight of these, the ninth being associated with a recent transposon insertion. To study the inheritance of DMRs, 13 were tracked in near isogenic lines (NILs) from a B73 x Mo17 cross show that most (10/13) segregate with the parental genotype, however, 3/13 did not, providing evidence of *trans*-acting effects (Eichten, Swanson-Wagner et al., 2011). In another experiment, 415–804 DMRs per accession were identified in a comparison of 19 inbred lines from the maize nested association mapping (NAM) population to B73 (Eichten, Briskine et al., 2013). These DMRs were examined in an additional 31 maize accessions, showing that most ~51% associated with local SNPs, but the other ~49% did not. DMRs associated with TEs in the B73 genome were more likely to be associated with local SNPs than those that were not. Looking at 401 of these DMRs in a B73 × Mo17 RIL population allowed the authors to further examine the inheritance of these DMRs. While ~80% tracked the parental genotype, ~23% of these were unstable, showing that this inheritance is not perfect. Approximately 6% showed *trans*-effects and interestingly ~1% showed evidence of paramutation-like behavior.

The numbers of DMRs identified in the previous studies provide valuable insight into the nature of epigenomic variation in maize but are certainly underestimates due to the more limited resolution of meDIP-chip. Using WGBS, 9635 CG-DMRs and 6671 CHG-DMRs were identified between B73 and Mo17 alone (CHH-DMRs were not called) (Regulski, Lu et al., 2013) and other studies have found 5000–22,000 context-specific DMRs for any given pair-wise comparison between different accessions of maize (Li, Song et al., 2015). In this latter example, comparison of four different accessions to B73 produced 53,816 unique CG-DMRs, 50,899 unique CHG-DMRs, and 16,736 unique CHH-DMRs; far more diversity than what is found in Arabidopsis (Li, Song et al., 2015). As in other studies most DMRs are inherited with the parental genotype, however, a far greater number of DMRs were found to not follow rule. Regulski et al. (2013) showed that the methylation state of 1772 DMRs switched, with the majority becoming hypermethylated as opposed to hypomethylated. This switching was often associated with the production of small RNAs, indicating that this is most likely mediated in *trans* by RdDM.

Of course, the overarching question is what effect these epigenomic variants might have on gene expression and phenotype. In their meDIP-chip study, Eichten et al. (2013) found correlations between gene expression and DMRs for 111 genes and for 26% of these, there was an on/off relationship between unmethylated/methylated state. In Regulski et al. (2013), 1598 genes were found to be within 1 Kb of a DMR and differentially expressed between B73 and Mo17, with a negative correlation between methylation level and gene expression. Interestingly, they also found an association between alternative splicing and methylation of acceptor sites at intron–exon junctions and suggested that gbM might be important for exclusion of TE insertions into genes (Regulski, Lu et al., 2013). While Li et al. (2015) found that differential methylation was not associated with the majority of differential gene expression ~20% of genes that show on/off expression differences between maize accessions were also unmethylated/methylated in those accessions. Indeed, they find evidence of potential "cryptic" variation, where genes that are silenced by DNA methylation in one accession are unmethylated and expressed in another. Finally, analysis of maize epigenomes have demonstrated the effect of "spreading" and "non-spreading" TEs, where spreading TEs are characterized by the spread of heterochromatic states outside of the TE body and into neighboring regions, including genes (Eichten, Ellis et al., 2012). Such spreading TEs can be a source of novel obligate and facultative epialleles, as in the case of *CmWIP1* in melon (Martin, Troadec et al., 2009).

4.3 Oryza sativa

In rice, DNA methylation has been studied in the context of two subspecies of rice, *O. sativa* spp. *japonica* (accession: Nipponbare) and *O. sativa* spp. *indica* (accession: 93–11), and their hybrids (Chodavarapu, Feng et al., 2012). Nipponbare was more highly methylated in CG and CHG contexts, while 93–11 was more highly methylated in the CHH context. Globally, hybrids had intermediate methylation levels, but methylation levels matched that of the parent at individual chromosomes. Methylation differed at about ~7% of cytosines between the parents, but in the hybrids, only ~0.8% of cytosines were differentially methylated compared to the parent-of-origin. The authors clustered differentially methylated cytosines between the parents into DMRs, identifying 16,865 such DMRs, 8584 of higher methylation in Nipponbare and 5152 higher in 93–11. When applied to hybrids, many fewer DMRs were identified between the parents and the corresponding chromosome in the offspring: 340 for Nipponbare and 75 for 93–11. Approximately 29% of DMRs between the parents were associated with small RNAs, with regions of higher methylation in one parent having corresponding higher levels of small RNAs. A significant number of genes with allele-specific expression in the hybrids (~17–18%) were associated with differences in DNA methylation, indicating a potential role in imprinting.

Another study in rice looked at three varieties differing in tolerance to various stresses: IR64 (stress-sensitive), Nagina 22 (drought-tolerant), and Pokkali (salinity tolerant) (Garg, Narayana Chevala, Shankar, & Jain, 2015). 64,212 DMRs were identified between Nagina 22 and IR64, while 35,723 were identified between Pokkali and IR64. In the Nagina 22-IR64 comparison, DMRs were predominantly hypermethylated, reflecting higher global methylation levels. In Pokkali-IR64, the proportion of hyper and hypomethylated DMRs was similar. Approximately 57.9% genes in Nagina 22-IR64 and ~38.5% of genes in Pokkali-IR64 were associated with DMRs. Genes associated with DMRs had more variable expression, and amongst differentially expressed genes (DEGs) there was an enrichment of down-regulated genes associated with hypermethylated DMRs and up-regulated genes associated with hypomethylated DMRs. TEs were also enriched in DMR-associated DEGs compared to DEGs not associated with DMRs. The vast differences in the number of DMRs identified between this study (Zimmer, Jupe, & Walbot, 1988) and that of Chodavarapu et al. (2012) are primarily due to methodological differences in defining DMRs and emphasizes the challenges in defining epigenomic variation.

4.4 Glycine max

DNA methylation has been studied in RIL populations from crosses of the *G. max* (soybean) accessions LD00-2817P and LDX01-1-65 (Schmitz, He et al., 2013). Between the parents, methylation levels were higher in LD00-2817P. DMRs were called between the two parents and two RIL lines sequenced at high depth, identifying 3241 CG-DMRs and 1416 C-DMRs. In regions that could be called as homozygous for one parent or the other in the two RIL lines, ~82% of C-DMRs and ~94% of C-DMRs segregated with the parental genotype (Fig. 3B). However, 254 CG-DMRs and 122 C-DMRs did show the methylation state of the opposite parent. These could be due to *trans* effects or even potentially an example of paramutation. QTL mapping for C-DMRs was done using low-coverage sequencing of an additional 81 RILs, finding evidence of mQTL for 91% of C-DMRs. Most of these mapped locally, indicating the importance of *cis* factors in explaining methylation differences. Only 33 mQTL mapped to another chromosome, indicating a relatively small contribution of *trans* effects in the RIL populations.

4.5 Brachypodium distachyon

Eichten, Stuart, Srivastava, Lister, and Borevitz (2016) compared methylomes of seven accessions of *B. distachyon* (Brachypodium), an emerging model for C3 grasses. Pairwise comparisons were made between the accession Bd21 (reference genome) and each of the six other accessions, identifying 5588–9550 CG-DMRs, 4568–8396 CHG-DMRs, and 520–921 CHH DMRs. Most of these DMRs were rare, being found in only one or two accessions. As in other species, CG-DMRs were often enriched in genic regions, while non-CG DMRs were primarily found in TEs or intergenic regions. Interestingly, the density of DMRs (CG, CHG, or CHH) correlated with SNP density in the genome, showing that regions more variable genetically also were more variable in DNA methylation. This supports the idea of a link between genetic and epigenomic variation. The authors also looked at the relationship between variation in TE insertions/deletions between accessions and DNA methylation. Surprisingly, they found little evidence of an association between DMRs and novel TE insertions/deletions and suggest that this could be either due to younger TEs not being targeted for silencing or more insertions occurring in regions of heterochromatin where there is already little variation in DNA methylation.

5. EPIGENOMIC VARIATION BETWEEN SPECIES

While within-species epigenomic diversity is likely of more direct interest to breeders, between-species variation can be highly informative and the basis of new tool development. Furthermore, breeders are increasingly seeking novel diversity in crop wild relatives that can be introgressed through breeding or moved in through genome editing (Zamir, 2001). For example, comparative epigenomic approaches can help us understand changes in the regulatory landscape in polyploidy (both natural and synthetic), interspecific crosses, or chromosome addition lines (Dong, Yu et al., 2018; Edger, Smith et al., 2017). By looking at the epigenome over much longer evolutionary distances, deeper insights can be gained into how the genome and epigenome interact to shape each other's function and evolution and perhaps provide a novel source of epialleles in breeding programs.

5.1 Genome-Wide Variation

Considerable variation exists in DNA methylation between plant species (Ausin, Feng et al., 2016; Bewick, Ji et al., 2016; Bewick, Niederhuth et al., 2017; Niederhuth, Bewick et al., 2016; Seymour, Koenig, Hagmann, Becker, & Weigel, 2014; Takuno et al., 2016; Willing, Rawat et al., 2015), much more so than the variation that is found within a species (Niederhuth, Bewick et al., 2016). Genome-wide, DNA methylation in non-CG contexts tend to be more variable between species than in the CG context and while DNA methylation is typically concentrated in gene-poor heterochromatic regions, there are exceptions to this trend, particularly in the case of CHH methylation in the grasses (Niederhuth, Bewick et al., 2016). This variation is driven by a number of factors: phylogeny (Alonso et al., 2015; Niederhuth, Bewick et al., 2016), genome size (Alonso et al., 2015; Niederhuth, Bewick et al., 2016; Takuno et al., 2016), TE and repeat content (Niederhuth, Bewick et al., 2016; Seymour et al., 2014), polyploidy and hybridization (Edger, Smith et al., 2017; El Baidouri et al., 2018; Jiao, Yuan et al., 2018; Kim, El Baidouri et al., 2015; Lukens et al., 2006; Madlung et al., 2002; Parkin, Koh et al., 2014; Song, Zhang, Stelly, & Chen, 2017; Wang, Xie et al., 2018), and differences in the evolution of the underlying pathways that establish and maintain the epigenome (Bewick, Ji et al., 2016; Bewick, Niederhuth et al., 2017; Willing, Rawat et al., 2015; Zemach, Kim et al., 2013).

Genome-wide levels of methylation correlate with genome size and TE content, with the exception of mCHH, emphasizing the impact of TEs in shaping the epigenome (Alonso et al., 2015; Niederhuth, Bewick et al., 2016). Most species tend to show the same general trends in the patterns of methylation, with some noted examples. In some grass lineages, mCHH is depleted in gene-poor regions and enriched in gene-rich regions (Niederhuth, Bewick et al., 2016). Extensive variation is observed in various DNA methylation pathways, in particular the plant-specific CMTs (Bewick, Ji et al., 2016; Bewick, Niederhuth et al., 2017; Dubin, Zhang et al., 2015; Kawakatsu, Huang et al., 2016; Zemach, Kim et al., 2013). This can have profound effects on the epigenome, such as in *Eutrema salsugineum*, where the loss of CMT3 has resulted in a loss of most mCHG, except for in RdDM regions (Bewick, Ji et al., 2016). The evolutionary causes and consequences of such variation is unknown but could provide unique opportunities for epigenome engineering and epimutagensis. As a more extreme example plants lack the TEN ELEVEN TRANSLOCATION (TET) family of enzymes that are found in mammals and the conversion of 5-methylcytosine to 5-hydroxymethylcytosine, which it catalyzes (Erdmann, Souza, Clish, & Gehring, 2015). Transgenic expression of TET enzymes in Arabidopsis leads to epimutagensis through genome-wide reductions in DNA methylation (Hollwey, Watson, & Meyer, 2016; Ji et al., 2018) and has also been used for making targeted epimutations using modified CRISPR-CAS9 systems (Gallego-Bartolomé, Gardiner et al., 2018).

5.2 Genic Variation

Gene-body methylated genes, as previously mentioned, are methylated only in the CG context (Tran et al., 2005). Unlike other patterns of genic methylation, gbM tends to be conserved in orthologues between species and is associated with constitutively high gene expression, longer genes with more exons, and more slowly evolving genes (Seymour et al., 2014; Takuno et al., 2016; Takuno and Gaut 2012, 2013). The discovery that *E. salsugineum* and *Conringia planisiliqua* have both lost gbM (Bewick, Ji et al., 2016), along with both work in Arabidopsis (Bewick, Ji et al., 2016; Miura et al., 2009; Teixeira & Colot, 2009) and larger comparative studies (Bewick, Niederhuth et al., 2017), shows that gbM is dispensable over evolutionary time and may have no direct functional consequence on gene expression. Surprisingly, there is an association between the loss of gbM in evolution and the loss of CMT3 in plants (Bewick, Ji et al., 2016; Bewick, Niederhuth et al., 2017), which along with the results from Arabidopsis, suggests that gbM and its associations may be an indirect by-product of other pathways (Bewick, Ji et al., 2016; Bewick, Niederhuth et al., 2017; Teixeira & Colot, 2009).

There tends to be little conservation of heterochromatic methylation states associated with silencing of methylated regions between species (Seymour et al., 2014). This is also true for genes and their flanking regions, although there are some exceptions for closely related species (Niederhuth, Bewick et al., 2016; Seymour et al., 2014). Likely the major driving force is differences in transposable element content, which are targets for silencing and can be highly variable (Seymour et al., 2014). Indeed, genic non-CG methylation levels are correlated with TE and repeat content in or around the gene (Niederhuth, Bewick et al., 2016; Seymour et al., 2014). Non-CG genic methylation levels are also correlated with genome size (Niederhuth, Bewick et al., 2016; Takuno et al., 2016). Others have observed that larger genomes have more open chromatin space and differences in adaptive evolution (Mei et al., 2018). It's interesting to speculate that perhaps larger genomes allow more opportunity for epialleles and epigenetic phenomenon like paramutation.

Despite this lack of conservation, non-CG methylated genes tend to be enriched for certain molecular and biological functions (Niederhuth, Bewick et al., 2016). A variety of causes could explain this, for example, silencing of mitochondrial and chloroplast genes that have transferred to the nucleus or *trans*-silencing of gene families resulting from gene duplication or other structural variation. The role of gene duplication was previously seen with examples like *PAI* (Bender & Fink, 1995) and *AtFOLT* (Durand et al., 2012). The relationship between genic DNA methylation and gene duplication/polyploidy has been examined in a number of studies (El Baidouri et al., 2018; Jiao, Yuan et al., 2018; Kim, El Baidouri et al., 2015; Lukens et al., 2006; Madlung et al., 2002; Parkin, Koh et al., 2014; Schmitz, He et al., 2013; Song et al., 2017; Wang, Wang et al., 2011, Wang, Wang, Lee, Mansoor, & Paterson, 2013). Certain types of gene duplication are more likely to be targeted for silencing by DNA methylation (El Baidouri et al., 2018) and widespread epigenomic changes are observed in new polyploids, particularly in regulatory regions (Edger, Smith et al., 2017; Jiao, Yuan et al., 2018; Lukens et al., 2006).

6. APPLICATIONS TO PLANT BREEDING

There are a number of ways in which epigenomic variation might see application in plant breeding programs. The most direct is to simply breed and select for or against epialleles with desirable or undesirable phenotypes. The most important factor in such an application seems to be the behavior and stability of the epiallele. In most instances, stable inheritance of an epiallele over many generations will be desired. Most studies, particularly those in populations, have shown that the majority of epigenomic variants are stably inherited with the parental genotype (Eichten, Briskine et al., 2013; Eichten, Swanson-Wagner et al., 2011; Regulski, Lu et al., 2013; Schmitz, He et al., 2013; Vaughn, Tanurdžić et al., 2007). Whether or not such variants are caused by underlying genetic variation or simply linked to it is unknown for most cases, but for many purposes this might be irrelevant, as their stability means that they can be faithfully transmitted. This will be of particular importance for integrating epialleles from epimutagenized populations (Hollwey et al., 2016; Ji et al., 2018; Johannes, Porcher et al., 2009; Raju, Shao et al., 2018; Reinders, Wulff et al., 2009; Xu, Tanino, Horner, & Robinson, 2016, Xu, Tanino, & Robinson, 2016) or for the stability of traits created through epigenome editing (Gallego-Bartolomé, Gardiner et al., 2018). The creation of novel populations with induced epigenomic variation is now seeing application in crop species such as soybeans (Raju, Shao et al., 2018) and opens up the possibilities as a source of phenotypic variation.

Epigenomic variation can also itself be used as molecular markers, in addition to or apart from genetic variation (Cortijo, Wardenaar et al., 2014; Hu, Morota, Rosa, & Gianola, 2015; Meng, Dubin et al., 2016). Meng et al. (2016) noted that in their associations using SNPs and methylation with variation in gene expression, that methylation differences were often physically closer than the SNP to the gene in question and that such markers could be used in fine-mapping. The ability to use methylation markers in mapping has already been demonstrated in the epiRIL lines (Cortijo, Wardenaar et al., 2014) and in one example, used for genomic prediction of plant height (epigenomic prediction) instead of genetic markers (Hu et al., 2015). In some cases, especially for long-lived species, these could also be used to screen plants for unwanted phenotypes before they go to the field, as in the case of *BAD KARMA* in African Oil Palm (Ong-Abdullah, Ordway et al., 2015).

7. CURRENT CHALLENGES

Despite efforts to understand epigenomic variation, we still have a poor grasp of how extensive epigenomic variation is (beyond Arabidopsis) and its relative contribution to overall phenotypic variation. This first question will be addressed in part by improved sequencing methods and declining costs. It is not such a far leap from over a thousand Arabidopsis methylomes to population scale sequencing of crop species. Studies of other epigenomic modifications will be slightly more difficult, but could be achieved through concerted public efforts (Lane et al., 2014). Another limitation is less technical, more theoretical. Unlike population and quantitative genetics, which has over the decades developed metrics and mathematical models for describing genetic diversity and its behavior, much less work has been done for epigenomic modifications. Furthermore, while epigenomic variation can clearly generate phenotypic variation (Johannes, Porcher et al., 2009; Reinders, Wulff et al., 2009), it is itself a molecular phenotype of underlying genetic variation. With the production of large scale epigenomic data, new ways of using and understanding the data is necessary.

The second major question of what the relative contribution of epigenomic variation to overall phenotypic variation is, is more challenging. As has been seen in previously discussed examples, examples of spontaneous and pure epigenomic variation does exist, but the majority is linked either directly or indirectly to genetic variation. In some cases of obligate epialleles such as *CmWIP1* (Martin, Troadec et al., 2009), chromatin modifications are the direct cause of the phenotype, but are themselves dependent upon a genetic variant. In another sense, epigenomic variation can be an indirect cause of phenotypic variation. As one of the primary functions of the epigenome is to control the spread of transposable elements, epigenomic instability could result in bursts of transposition, leading to genetic and phenotypic variation (Thieme et al., 2017). Additionally, DNA methylation is mutagenic, as methylated cytosines can be spontaneously deaminated to thymine and C to T transitions occurs more frequently than any other transition (Holliday & Grigg, 1993). Quantitative epigenetic studies, such as those in the Arabidopsis epiRIL lines (Aller, Jagd, Kliebenstein, & Burow, 2018; Cortijo, Wardenaar et al., 2014; Johannes, Porcher et al., 2009; Reinders, Wulff et al., 2009) or which might result from other manipulated populations such as through TET-mediated epimutagensis

(Hollwey et al., 2016; Ji et al., 2018) are necessary. However, these insights must ultimately be combined with ways of studying natural populations to ultimately answer this question. As one example Aller et al. (2018), recently did a meta-analysis variation in glucosinolate levels in Arabidopsis RIL and epiRIL populations, concluding that genetic variation contributed much more than epigenetic variation.

8. CONCLUSION

Epigenetic phenomenon has had a long history in genetics since the identification of Bateson and Pellew's *rogue* peas. For many years, the identification of novel epigenomic variation was largely limited to a few examples which could be identified through careful genetic analysis. Since the development of whole-genome approaches to studying chromatin modifications, there has been a rapid expansion of our knowledge of the epigenome and the identification of epialleles. This is now making it possible to apply epigenomics to plant breeding and crop improvement. However, open questions still remain that will require innovative approaches to make this a reality.

REFERENCES

Alleman, M., Sidorenko, L., McGinnis, K., Seshadri, V., Dorweiler, J. E., White, J., et al. (2006). An RNA-dependent RNA polymerase is required for paramutation in maize. *Nature, 442*, 295.

Aller, E. S. T., Jagd, L. M., Kliebenstein, D. J., & Burow, M. (2018). Comparison of the relative potential for epigenetic and genetic variation to contribute to trait stability. *G3: Genes|Genomes|Genetics, 8*.

Almeida, J., Carpenter, R., Robbins, T. P., Martin, C., & Coen, E. S. (1989). Genetic interactions underlying flower color patterns in Antirrhinum majus. *Genes & Development, 3*(11), 1758–1767.

Alonso, C., Pérez, R., Bazaga, P., & Herrera, C. M. (2015). Global DNA cytosine methylation as an evolving trait: Phylogenetic signal and correlated evolution with genome size in angiosperms. *Frontiers in Genetics, 6*(4).

Ausin, I., Feng, S., Yu, C., Liu, W., Kuo, H. Y., Jacobsen, E. L., et al. (2016). DNA methylome of the 20-gigabase Norway spruce genome. *Proceedings of the National Academy of Sciences, 113*(50), E8106–E8113.

Bateson, W., & Pellew, C. (1915). On the genetics of "Rogues" among culinary peas (*Pisum sativum*). *Journal of Genetics, 5*(1), 13–36.

Becker, C., Hagmann, J., Muller, J., Koenig, D., Stegle, O., Borgwardt, K., et al. (2011). Spontaneous epigenetic variation in the *Arabidopsis thaliana* methylome. *Nature, 480*(7376), 245–249.

Bender, J., & Fink, G. R. (1995). Epigenetic control of an endogenous gene family is revealed by a novel blue fluorescent mutant of Arabidopsis. *Cell, 83*(5), 725–734.

Bewick, A. J., Ji, L., Niederhuth, C. E., Willing, E.-M., Hofmeister, B. T., Shi, X., et al. (2016). On the origin and evolutionary consequences of gene body DNA methylation. *Proceedings of the National Academy of Sciences, 113*(32), 9111–9116.
Bewick, A. J., Niederhuth, C. E., Ji, L., Rohr, N. A., Griffin, P. T., Leebens-Mack, J., et al. (2017). The evolution of CHROMOMETHYLASES and gene body DNA methylation in plants. *Genome Biology, 18*(1), 65.
Blevins, T., Wang, J., Pflieger, D., Pontvianne, F., & Pikaard, C. S. (2017). Hybrid incompatibility caused by an epiallele. *Proceedings of the National Academy of Sciences, 114*(14), 3702–3707.
Brink, R. A. (1956). A genetic change associated with the R locus in maize which is directed and potentially reversible. *Genetics, 41*(6), 872–889.
Brink, R. A. (1958). Paramutation at the R Locus in maize. *Cold Spring Harbor Symposia on Quantitative Biology, 23*, 379–391.
Cao, X., & Jacobsen, S. E. (2002). Locus-specific control of asymmetric and CpNpG methylation by the DRM and CMT3 methyltransferase genes. *Proceedings of the National Academy of Sciences, 99*(suppl. 4), 16491–16498.
Chen, W., Kong, J., Qin, C., Yu, S., Tan, J., Chen, Y-r., et al. (2015). Requirement of CHROMOMETHYLASE3 for somatic inheritance of the spontaneous tomato epimutation Colourless non-ripening. *Scientific Reports, 5*, 9192.
Chodavarapu, R. K., Feng, S., Ding, B., Simon, S. A., Lopez, D., Jia, Y., et al. (2012). Transcriptome and methylome interactions in rice hybrids. *Proceedings of the National Academy of Sciences, 109*(30), 12040–12045.
Coe, E. H. (1959). A regular and continuing conversion-type phenomenon at the B locus in maize. *Proceedings of the National Academy of Sciences, 45*(6), 828–832.
Cokus, S. J., Feng, S., Zhang, X., Chen, Z., Merriman, B., Haudenschild, C. D., et al. (2008). Shotgun bisulphite sequencing of the Arabidopsis genome reveals DNA methylation patterning. *Nature, 452*, 215.
Cortijo, S., Wardenaar, R., Colomé-Tatché, M., Gilly, A., Etcheverry, M., Labadie, K., et al. (2014). Mapping the epigenetic basis of complex traits. *Science, 343*(6175), 1145–1148.
Cubas, P., Vincent, C., & Coen, E. (1999). An epigenetic mutation responsible for natural variation in floral symmetry. *Nature, 401*, 157.
Das, O. P., & Messing, J. (1994). Variegated phenotype and developmental methylation changes of a maize allele originating from epimutation. *Genetics, 136*(3), 1121–1141.
Deans, C., & Maggert, K. A. (2015). What do you mean, "epigenetic"? *Genetics, 199*(4), 887–896.
Deng, Y., Zhai, K., Xie, Z., Yang, D., Zhu, X., Liu, J., et al. (2017). Epigenetic regulation of antagonistic receptors confers rice blast resistance with yield balance. *Science, 355*(6328), 962–965.
Dong, Z., Yu, J., Li, H., Huang, W., Xu, L., Zhao, Y., et al. (2018). Transcriptional and epigenetic adaptation of maize chromosomes in Oat-Maize addition lines. *Nucleic Acids Research*. gky209-gky209.
Dorweiler, J. E., Carey, C. C., Kubo, K. M., Hollick, J. B., Kermicle, J. L., & Chandler, V. L. (2000). Mediator of paramutation1 is required for establishment and maintenance of paramutation at multiple maize loci. *The Plant Cell Online, 12*(11), 2101–2118.
Dubin, M. J., Zhang, P., Meng, D., Remigereau, M.-S., Osborne, E. J., Paolo Casale, F., et al. (2015). DNA methylation in Arabidopsis has a genetic basis and shows evidence of local adaptation. *eLife, 4*, e05255.
Durand, S., Bouché, N., Perez Strand, E., Loudet, O., & Camilleri, C. (2012). Rapid establishment of genetic incompatibility through natural epigenetic variation. *Current Biology, 22*(4), 326–331.

Du, J., Zhong, X., Bernatavichute, Y. V., Stroud, H., Feng, S., Caro, E., et al. (2012). Dual binding of chromomethylase domains to H3K9me2-Containing nucleosomes directs DNA methylation in plants. *Cell, 151*(1), 167–180.

Edger, P. P., Smith, R., McKain, M. R., Cooley, A. M., Vallejo-Marin, M., Yuan, Y., et al. (2017). Subgenome dominance in an interspecific hybrid, synthetic allopolyploid, and a 140-year-old naturally established neo-allopolyploid monkeyflower. *The Plant Cell Online, 29*(9), 2150–2167.

Eichten, S. R., Briskine, R., Song, J., Li, Q., Swanson-Wagner, R., Hermanson, P. J., et al. (2013). Epigenetic and genetic influences on DNA methylation variation in maize populations. *The Plant Cell Online, 25*(8), 2783–2797.

Eichten, S. R., Ellis, N. A., Makarevitch, I., Yeh, C.-T., Gent, J. I., Guo, L., et al. (2012). Spreading of heterochromatin is limited to specific families of maize retrotransposons. *PLoS Genetics, 8*(12), e1003127.

Eichten, S. R., Stuart, T., Srivastava, A., Lister, R., & Borevitz, J. O. (2016). DNA methylation profiles of diverse Brachypodium distachyon align with underlying genetic diversity. *Genome Research, 26*(11), 1520–1531.

Eichten, S. R., Swanson-Wagner, R. A., Schnable, J. C., Waters, A. J., Hermanson, P. J., Liu, S., et al. (2011). Heritable epigenetic variation among maize inbreds. *PLoS Genetics,* 7(11), e1002372.

El Baidouri, M., Kim, K. D., Abernathy, B., Li, Y.-H., Qiu, L.-J., & Jackson, S. A. (2018). Genic C-Methylation in soybean is associated with gene paralogs relocated to transposable element-rich pericentromeres. *Molecular Plant, 11*(3), 485–495.

Erdmann, R. M., Souza, A. L., Clish, C. B., & Gehring, M. (2015). 5-Hydroxymethylcytosine is not present in appreciable quantities in Arabidopsis DNA. *G3: Genes|Genomes|Genetics, 5*(1), 1–8.

Finnegan, E. J., Peacock, W. J., & Dennis, E. S. (1996). Reduced DNA methylation in *Arabidopsis thaliana* results in abnormal plant development. *Proceedings of the National Academy of Sciences, 93*(16), 8449–8454.

Frommer, M., McDonald, L. E., Millar, D. S., Collis, C. M., Watt, F., Grigg, G. W., et al. (1992). A genomic sequencing protocol that yields a positive display of 5-methylcytosine residues in individual DNA strands. *Proceedings of the National Academy of Sciences, 89*(5), 1827–1831.

Gallego-Bartolomé, J., Gardiner, J., Liu, W., Papikian, A., Ghoshal, B., Kuo, H. Y., et al. (2018). Targeted DNA demethylation of the Arabidopsis genome using the human TET1 catalytic domain. *Proceedings of the National Academy of Sciences.*

Gao, Y., & Zhao, Y. (2013). Epigenetic suppression of T-DNA insertion mutants in Arabidopsis. *Molecular Plant, 6*(2), 539–545.

Garg, R., Narayana Chevala, V. V. S., Shankar, R., & Jain, M. (2015). Divergent DNA methylation patterns associated with gene expression in rice cultivars with contrasting drought and salinity stress response. *Scientific Reports, 5*, 14922.

Giovannoni, J. (2016). Harnessing epigenome modifications for better crops. *Journal of Experimental Botany, 67*(9), 2535–2537.

Gouil, Q., & Baulcombe, D. C. (2016). DNA methylation signatures of the plant chromomethyltransferases. *PLoS Genetics, 12*(12), e1006526.

Gouil, Q., Novák, O., & Baulcombe, D. C. (2016). SLTAB2 is the paramutated SULFUREA locus in tomato. *Journal of Experimental Botany, 67*(9), 2655–2664.

Gustafsson, Å. (1979). Linnaeus' Peloria: The history of a monster. *TAG. Theoretical and Applied Genetics. Theoretische und angewandte Genetik, 54*(6), 241–248.

Hagemann, R. (1958). Somatische konversion bei *Lycopersicon esculentum* mill. *Molecular Genetics and Genomics : Molecular & General Genetics, 89*(4), 587–613.

Hagemann, R. (1969). Somatische Konversion (Paramutation) am sulfurea Locus von *Lycopersicon esculentum* Mill. *TAG. Theoretical and Applied Genetics. Theoretische und angewandte Genetik, 39*(7), 295–305.

Harrison, B. J., & Carpenter, R. (1973). A comparison of the instabilities at the Nivea and Pallida loci in *Antirrhinum majus. Heredity, 31*, 309.

Hofmeister, B. T., Lee, K., Rohr, N. A., Hall, D. W., & Schmitz, R. J. (2017). Stable inheritance of DNA methylation allows creation of epigenotype maps and the study of epiallele inheritance patterns in the absence of genetic variation. *Genome Biology, 18*(1), 155.

Hollick, J. B., Patterson, G. I., Coe, E. H., Cone, K. C., & Chandler, V. L. (1995). Allelic interactions heritably alter the activity of a metastable maize pl allele. *Genetics, 141*(2), 709–719.

Holliday, R. (1994). Epigenetics: An overview. *Developmental Genetics, 15*(6), 453–457.

Holliday, R., & Grigg, G. W. (1993). DNA methylation and mutation. *Mutation Research: Fundamental and Molecular Mechanisms of Mutagenesis, 285*(1), 61–67.

Hollwey, E., Watson, M., & Meyer, P. (2016). Expression of the C-Terminal domain of mammalian TET3 DNA dioxygenase in *Arabidopsis thaliana* induces induces heritable methylation changes at rDNA loci. *Advances in Bioscience and Biotechnology, 07*(05), 243–250.

Hu, Y., Morota, G., Rosa, G. J. M., & Gianola, D. (2015). Prediction of plant height in *Arabidopsis thaliana* using DNA methylation data. *Genetics, 201*(2), 779–793.

Hyten, D. L., Song, Q., Zhu, Y., Choi, I.-Y., Nelson, R. L., Costa, J. M., et al. (2006). Impacts of genetic bottlenecks on soybean genome diversity. *Proceedings of the National Academy of Sciences, 103*(45), 16666–16671.

Jablonka, E., & Lamb, M. (2003). *The changing concept of epigenetics.*

Jacobsen, S. E., & Meyerowitz, E. M. (1997). Hypermethylated SUPERMAN epigenetic alleles in Arabidopsis. *Science, 277*(5329), 1100–1103.

Jiao, W., Yuan, J., Jiang, S., Liu, Y., Wang, L., Liu, M., et al. (2018). Asymmetrical changes of gene expression, small RNAs and chromatin in two resynthesized wheat allotetraploids. *The Plant Journal, 93*(5), 828–842.

Ji, L., Jordan, W. T., Shi, X., Hu, L., He, C., & Schmitz, R. J. (2018). TET-mediated epimutagenesis of the *Arabidopsis thaliana* methylome. *Nature Communications, 9*(1), 895.

Johannes, F., Porcher, E., Teixeira, F. K., Saliba-Colombani, V., Simon, M., Agier, N., et al. (2009). Assessing the impact of transgenerational epigenetic variation on complex traits. *PLoS Genetics, 5*(6), e1000530.

Kawakatsu, T., Huang, S-s. C., Jupe, F., Sasaki, E., Schmitz, R. J., Urich, M. A., et al. (2016). Epigenomic diversity in a global collection of *Arabidopsis thaliana* accessions. *Cell, 166*(2), 492–505.

Kermicle, J. L., Eggleston, W. B., & Alleman, M. (1995). Organization of paramutagenicity in R-stippled maize. *Genetics, 141*(1), 361–372.

Kim, K. D., El Baidouri, M., Abernathy, B., Iwata-Otsubo, A., Chavarro, C., Gonzales, M., et al. (2015). A comparative epigenomic analysis of polyploidy-derived genes in soybean and common bean. *Plant Physiology, 168*(4), 1433–1447.

Krebbers, E., Hehl, R., Piotrowiak, R., Lonnig, W. E., Sommer, H., & Saedler, H. (1987). Molecular analysis of paramutant plants of *Antirrhinum majus* and the involvement of transposable elements. *Molecular Genetics and Genomics, 209*(3), 499–507.

Lane, A. K., Niederhuth, C. E., Ji, L., & Schmitz, R. J. (2014). pENCODE: A plant encyclopedia of DNA elements. *Annual Review of Genetics, 48*(1), 49–70.

Law, J. A., & Jacobsen, S. E. (2010). Establishing, maintaining and modifying DNA methylation patterns in plants and animals. *Nature Reviews. Genetics, 11*, 204.

Liang, Z., Shen, L., Cui, X., Bao, S., Geng, Y., Yu, G., et al. (2018). DNA N6-Adenine methylation in *Arabidopsis thaliana. Developmental Cell.*

Li, Q., Song, J., West, P. T., Zynda, G., Eichten, S. R., Vaughn, M. W., et al. (2015). Examining the causes and consequences of context-specific differential DNA methylation in maize. *Plant Physiology, 168*(4), 1262–1274.

Lister, R., & Ecker, J. R. (2009). Finding the fifth base: Genome-wide sequencing of cytosine methylation. *Genome Research, 19*(6), 959–966.

Lister, R., O'Malley, R. C., Tonti-Filippini, J., Gregory, B. D., Berry, C. C., Millar, A. H., et al. (2008). Highly integrated single-base resolution maps of the epigenome in Arabidopsis. *Cell, 133*(3), 523–536.

Lister, R., Pelizzola, M., Dowen, R. H., Hawkins, R. D., Hon, G., Tonti-Filippini, J., et al. (2009). Human DNA methylomes at base resolution show widespread epigenomic differences. *Nature, 462*, 315.

Luff, B., Pawlowski, L., & Bender, J. (1999). An inverted repeat triggers cytosine methylation of identical sequences in Arabidopsis. *Molecular Cell, 3*(4), 505–511.

Lukens, L. N., Pires, J. C., Leon, E., Vogelzang, R., Oslach, L., & Osborn, T. (2006). Patterns of sequence loss and cytosine methylation within a population of newly resynthesized Brassica napus allopolyploids. *Plant Physiology, 140*(1), 336–348.

Madlung, A., Masuelli, R. W., Watson, B., Reynolds, S. H., Davison, J., & Comai, L. (2002). Remodeling of DNA methylation and phenotypic and transcriptional changes in synthetic Arabidopsis allotetraploids. *Plant Physiology, 129*(2), 733–746.

Manning, K., Tör, M., Poole, M., Hong, Y., Thompson, A. J., King, G. J., et al. (2006). A naturally occurring epigenetic mutation in a gene encoding an SBP-box transcription factor inhibits tomato fruit ripening. *Nature Genetics, 38*, 948.

Martin, A., Troadec, C., Boualem, A., Rajab, M., Fernandez, R., Morin, H., et al. (2009). A transposon-induced epigenetic change leads to sex determination in melon. *Nature, 461*, 1135.

Mei, W., Stetter, M. G., Gates, D. J., Stitzer, M. C., & Ross-Ibarra, J. (2018). Adaptation in plant genomes: Bigger is different. *American Journal of Botany, 105*(1), 16–19.

Melquist, S., Luff, B., & Bender, J. (1999). Arabidopsis PAI gene arrangements, cytosine methylation and expression. *Genetics, 153*(1), 401–413.

Meng, D., Dubin, M., Zhang, P., Osborne, E. J., Stegle, O., Clark, R. M., et al. (2016). Limited contribution of DNA methylation variation to expression regulation in *Arabidopsis thaliana. PLoS Genetics, 12*(7), e1006141.

Mittelsten Scheid, O., Afsar, K., & Paszkowski, J. (2003). Formation of stable epialleles and their paramutation-like interaction in tetraploid *Arabidopsis thaliana. Nature Genetics, 34*, 450.

Miura, A., Nakamura, M., Inagaki, S., Kobayashi, A., Saze, H., & Kakutani, T. (2009). An Arabidopsis jmjC domain protein protects transcribed genes from DNA methylation at CHG sites. *The EMBO Journal, 28*(8), 1078–1086.

Morrell, P. L., Buckler, E. S., & Ross-Ibarra, J. (2011). Crop genomics: Advances and applications. *Nature Reviews. Genetics, 13*(2), 85–96.

Nanney, D. L. (1958). Epigenetic control systems. *Proceedings of the National Academy of Sciences, 44*(7), 712–717.

National Research Council Committee on Genetic Vulnerability of Major C. (1972). *Genetic vulnerability of major crops*. Washington: National Academy of Sciences.

Niederhuth, C. E., Bewick, A. J., Ji, L., Alabady, M. S., Kim, K. D., Li, Q., et al. (2016). Widespread natural variation of DNA methylation within angiosperms. *Genome Biology, 17*(1), 194.

Niederhuth, C. E., & Schmitz, R. J. (2017). Putting DNA methylation in context: from genomes to gene expression in plants. *Biochim Biophys Acta, 1860*, 149–156.

Ong-Abdullah, M., Ordway, J. M., Jiang, N., Ooi, S. E., Kok, S. Y., Sarpan, N., et al. (2015). Loss of Karma transposon methylation underlies the mantled somaclonal variant of oil palm. *Nature, 525*(7570), 533–537.

Panavas, T., Weir, J., & Walker, E. L. (1999). The structure and paramutagenicity of the *r-marbled* haplotype of *Zea mays*. *Genetics, 153*(2), 979–991.

Panda, K., & Slotkin, R. K. (2013). Proposed mechanism for the initiation of transposable element silencing by the RDR6-directed DNA methylation pathway. *Plant Signaling & Behavior, 8*(8).

Parkin, I. A., Koh, C., Tang, H., Robinson, S. J., Kagale, S., Clarke, W. E., et al. (2014). Transcriptome and methylome profiling reveals relics of genome dominance in the mesopolyploid *Brassica oleracea*. *Genome Biology, 15*(6), R77.

Pilu, R., Panzeri, D., Cassani, E., Badone, F. C., Landoni, M., & Nielsen, E. (2008). A paramutation phenomenon is involved in the genetics of maize low phytic acid1-241 (lpa1-241) trait. *Heredity, 102*, 236.

Quadrana, L., Almeida, J., Asís, R., Duffy, T., Dominguez, P. G., Bermúdez, L., et al. (2014). Natural occurring epialleles determine vitamin E accumulation in tomato fruits. *Nature Communications, 5*, 4027.

Raju, S. K. K., Shao, M. R., Sanchez, R., Xu, Y. Z., Sandhu, A., Graef, G., et al. (2018). An epigenetic breeding system in soybean for increased yield and stability. *Plant Biotechnologyogy Journal.*

Rand, A. C., Jain, M., Eizenga, J. M., Musselman-Brown, A., Olsen, H. E., Akeson, M., et al. (2017). Mapping DNA methylation with high-throughput nanopore sequencing. *Nature Methods, 14*(4), 411–413.

Regulski, M., Lu, Z., Kendall, J., Donoghue, M. T. A., Reinders, J., Llaca, V., et al. (2013). The maize methylome influences mRNA splice sites and reveals widespread paramutation-like switches guided by small RNA. *Genome Research, 23*(10), 1651–1662.

Reinders, J., Wulff, B. B., Mirouze, M., Mari-Ordonez, A., Dapp, M., Rozhon, W., et al. (2009). Compromised stability of DNA methylation and transposon immobilization in mosaic Arabidopsis epigenomes. *Genes & Development, 23*(8), 939–950.

Renner, O. (1959). Somatic conversion in the heredity of the Cruciata character in Œnothera. *Heredity, 13*, 283.

Richards, E. J. (2006). Inherited epigenetic variation — revisiting soft inheritance. *Nature Reviews. Genetics,* 7, 395.

Rigal, M., Kevei, Z., Pélissier, T., & Mathieu, O. (2012). DNA methylation in an intron of the IBM1 histone demethylase gene stabilizes chromatin modification patterns. *The EMBO Journal, 31*(13), 2981–2993.

Rincker, K., Nelson, R., Specht, J., Sleper, D., Cary, T., Cianzio, S. R., et al. (2014). Genetic improvement of U.S. Soybean in maturity groups II, III, and IV. *Crop Science, 54*(4), 1419–1432.

Ronemus, M. J., Galbiati, M., Ticknor, C., Chen, J., & Dellaporta, S. L. (1996). Demethylation-induced developmental pleiotropy in Arabidopsis. *Science, 273*(5275), 654–657.

Rowntree, S. C., Suhre, J. J., Weidenbenner, N. H., Wilson, E. W., Davis, V. M., Naeve, S. L., et al. (2013). Genetic gain × management interactions in soybean: I. Planting date. *Crop Science, 53*(3).

Santo, T., Pereira, R., & Leitão, J. (2017). The pea (*Pisum sativum* L.) rogue paramutation is accompanied by alterations in the methylation pattern of specific genomic sequences. *Epigenomes, 1*(1), 6.

Schmitz, R. J., He, Y., Valdés-López, O., Khan, S. M., Joshi, T., Urich, M. A., et al. (2013). Epigenome-wide inheritance of cytosine methylation variants in a recombinant inbred population. *Genome Research, 23*(10), 1663–1674.

Schmitz, R. J., Schultz, M. D., Lewsey, M. G., O'Malley, R. C., Urich, M. A., Libiger, O., et al. (2011). Transgenerational epigenetic instability is a source of novel methylation variants. *Science, 334*(6054), 369–373.

Schmitz, R. J., Schultz, M. D., Urich, M. A., Nery, J. R., Pelizzola, M., Libiger, O., et al. (2013). Patterns of population epigenomic diversity. *Nature, 495*(7440), 193–198.

Schmitz, R. J., & Zhang, X. (2011). High-throughput approaches for plant epigenomic studies. *Current Opinion in Plant Biology, 14*(2), 130–136.

Schmitz, R. J., & Zhang, X. (2014). 'Chapter Ten - Decoding the Epigenomes of Herbaceous Plants.'. In Andrew H. Paterson (Ed.), *Advances in Botanical Research* (Academic Press).

Schnable, P. S., Ware, D., Fulton, R. S., Stein, J. C., Wei, F., Pasternak, S., et al. (2009). The B73 maize genome: Complexity, diversity, and dynamics. *Science, 326*(5956), 1112.

Seymour, D. K., Koenig, D., Hagmann, J., Becker, C., & Weigel, D. (2014). Evolution of DNA methylation patterns in the Brassicaceae is driven by differences in genome organization. *PLoS Genetics, 10*(11), e1004785.

Shen, X., De Jonge, J., Forsberg, S. K. G., Pettersson, M. E., Sheng, Z., Hennig, L., et al. (2014). Natural CMT2 variation is associated with genome-wide methylation changes and temperature seasonality. *PLoS Genetics, 10*(12), e1004842.

Shi, J., Wang, H., Schellin, K., Li, B., Faller, M., Stoop, J. M., et al. (2007). Embryo-specific silencing of a transporter reduces phytic acid content of maize and soybean seeds. *Nature Biotechnology, 25*, 930.

Silveira, A. B., Trontin, C., Cortijo, S., Barau, J., Del Bem, L. E. V., Loudet, O., et al. (2013). Extensive natural epigenetic variation at a de novo originated gene. *PLoS Genetics, 9*(4), e1003437.

Simpson, J. T., Workman, R. E., Zuzarte, P. C., David, M., Dursi, L. J., & Timp, W. (2017). Detecting DNA cytosine methylation using nanopore sequencing. *Nature Methods, 14*(4), 407–410.

Slotkin, R. K., Vaughn, M., Borges, F., Tanurdžić, M., Becker, J. D., Feijó, J. A., et al. (2009). Epigenetic reprogramming and small RNA silencing of transposable elements in pollen. *Cell, 136*(3), 461–472.

Song, Q., Zhang, T., Stelly, D. M., & Chen, Z. J. (2017). Epigenomic and functional analyses reveal roles of epialleles in the loss of photoperiod sensitivity during domestication of allotetraploid cottons. *Genome Biology, 18*(1), 99.

Stam, M., Belele, C., Dorweiler, J. E., & Chandler, V. L. (2002). Differential chromatin structure within a tandem array 100 kb upstream of the maize b1 locus is associated with paramutation. *Genes & Development, 16*(15), 1906–1918.

Stokes, T. L., & Richards, E. J. (2002). Induced instability of two Arabidopsis constitutive pathogen-response alleles. *Proceedings of the National Academy of Sciences, 99*(11), 7792–7796.

Takuno, S., & Gaut, B. S. (2012). Body-methylated genes in *Arabidopsis thaliana* are functionally important and evolve slowly. *Molecular Biology and Evolution, 29*(1), 219–227.

Takuno, S., & Gaut, B. S. (2013). Gene body methylation is conserved between plant orthologs and is of evolutionary consequence. *Proceedings of the National Academy of Sciences, 110*(5), 1797–1802.

Takuno, S., Ran, J.-H., & Gaut, B. S. (2016). Evolutionary patterns of genic DNA methylation vary across land plants. *Nature Plants, 2*, 15222.

Teixeira, F. K., & Colot, V. (2009). Gene body DNA methylation in plants: A means to an end or an end to a means? *The EMBO Journal, 28*(8), 997–998.

Thieme, M., Lanciano, S., Balzergue, S., Daccord, N., Mirouze, M., & Bucher, E. (2017). Inhibition of RNA polymerase II allows controlled mobilisation of retrotransposons for plant breeding. *Genome Biology, 18*(1), 134.

Thompson, A. J., Tor, M., Barry, C. S., Vrebalov, J., Orfila, C., Jarvis, M. C., et al. (1999). Molecular and genetic characterization of a novel pleiotropic tomato-ripening mutant. *Plant Physiology, 120*(2), 383–390.

Tran, R. K., Henikoff, J. G., Zilberman, D., Ditt, R. F., Jacobsen, S. E., & Henikoff, S. (2005). DNA methylation profiling identifies CG methylation clusters in Arabidopsis genes. *Current Biology, 15*(2), 154–159.

van der Graaf, A., Wardenaar, R., Neumann, D. A., Taudt, A., Shaw, R. G., Jansen, R. C., et al. (2015). Rate, spectrum, and evolutionary dynamics of spontaneous epimutations. *Proceedings of the National Academy of Sciences, 112*(21), 6676–6681.

Vaughn, M. W., Tanurdžić, M., Lippman, Z., Jiang, H., Carrasquillo, R., Rabinowicz, P. D., et al. (2007). Epigenetic natural variation in *Arabidopsis thaliana*. *PLoS Biology, 5*(7), e174.

Walker, E. L., & Panavas, T. (2001). Structural features and methylation patterns associated with paramutation at the *r1* locus of *Zea mays*. *Genetics, 159*(3), 1201–1215.

Wallace, J. G., Larsson, S. J., & Buckler, E. S. (2014). Entering the second century of maize quantitative genetics. *Heredity, 112*(1), 30–38.

Wang, Y., Wang, X., Lee, T. H., Mansoor, S., & Paterson, A. H. (2013). Gene body methylation shows distinct patterns associated with different gene origins and duplication modes and has a heterogeneous relationship with gene expression in *Oryza sativa* (rice). *New Phytologist, 198*(1), 274–283.

Wang, Y., Wang, X., Tang, H., Tan, X., Ficklin, S. P., Feltus, F. A., et al. (2011). Modes of gene duplication contribute differently to genetic novelty and redundancy, but show parallels across divergent angiosperms. *PLoS One, 6*(12), e28150.

Wang, L., Xie, J., Hu, J., Lan, B., You, C., Li, F., et al. (2018). Comparative epigenomics reveals evolution of duplicated genes in potato and tomato. *The Plant Journal, 93*(3), 460–471.

Williams, B. P., Pignatta, D., Henikoff, S., & Gehring, M. (2015). Methylation-sensitive expression of a DNA demethylase gene serves as an epigenetic rheostat. *PLoS Genetics, 11*(3), e1005142.

Willing, E.-M., Rawat, V., Mandáková, T., Maumus, F., James, G. V., Nordström, K. J. V., et al. (2015). Genome expansion of Arabis alpina linked with retrotransposition and reduced symmetric DNA methylation. *Nature Plants, 1*, 14023.

Xue, W., Ruprecht, C., Street, N., Hematy, K., Chang, C., Frommer, W. B., et al. (2012). Paramutation-like interaction of T-DNA loci in Arabidopsis. *PLoS One,* 7(12), e51651.

Xu, J., Tanino, K. K., Horner, K. N., & Robinson, S. J. (2016). Quantitative trait variation is revealed in a novel hypomethylated population of woodland strawberry (*Fragaria vesca*). *BMC Plant Biology, 16*(1), 240.

Xu, J., Tanino, K. K., & Robinson, S. J. (2016). Stable epigenetic variants selected from an induced hypomethylated fragaria vesca population. *Frontiers in Plant Science,* 7(1768).

Zamir, D. (2001). Improving plant breeding with exotic genetic libraries. *Nature Reviews. Genetics, 2*, 983.

Zemach, A., Kim, M. Y., Hsieh, P.-H., Coleman-Derr, D., Eshed-Williams, L., Thao, K., et al. (2013). The Arabidopsis nucleosome remodeler DDM1 allows DNA methyltransferases to access H1-Containing heterochromatin. *Cell, 153*(1), 193–205.

Zhang, L., Yu, H., Ma, B., Liu, G., Wang, J., Wang, J., et al. (2017). A natural tandem array alleviates epigenetic repression of IPA1 and leads to superior yielding rice. *Nature Communications, 8*, 14789.

Zheng, Z., Yu, H., Miki, D., Jin, D., Zhang, Q., Ren, Z., et al. (2015). Involvement of multiple gene-silencing pathways in a paramutation-like phenomenon in Arabidopsis. *Cell Reports, 11*(8), 1160–1167.

Zimmer, E. A., Jupe, E. R., & Walbot, V. (1988). Ribosomal gene structure, variation and inheritance in maize and its ancestors. *Genetics, 120*(4), 1125–1136.

CHAPTER FOUR

EpiRILs: Lessons From Arabidopsis

Marco Catoni and Sandra Cortijo[1]
The Sainsbury Laboratory, University of Cambridge, Cambridge, United Kingdom
[1]Corresponding author: e-mail address: sandra.cortijo@slcu.cam.ac.uk

Contents

Abstract

In recent times, epigenetic marks have emerged as important players involved in the regulation of gene expression and transposable element silencing in many organisms. In plants, many epigenetic changes, mainly at the level of DNA methylation, are transgenerational stable and contribute to formation of epialleles, affecting developmental and agronomical traits. In this scenario, it becomes critical to differentiate the genetic from the epigenetic contribution to plant phenotypes. In Arabidopsis, epigenetic Recombinant Inbred Lines (epiRILs), obtained by an initial cross of isogenic parents with different DNA methylation profiles, provide a powerful tool to investigate the role and significance of epigenetic alteration in identical or almost identical genetic backgrounds. Such populations have greatly increased our knowledge in mechanisms involved in epialleles formation and stability, as well as in the consequences of DNA methylation changes in genomic stability, transposable elements activation and phenotypic traits.

Advances in Botanical Research, Volume 88
ISSN 0065-2296
https://doi.org/10.1016/bs.abr.2018.08.002

1. INTRODUCTION

While it is known that DNA is the support of heredity, it is more and more recognised that heritable phenotypic variation can be also caused by epigenetic changes, and not only by change in the DNA sequence. Methylation of cytosines (DNA methylation) is an epigenetic mark conserved across many species and plays an important role in regulating gene expression (He, Chen, & Zhu, 2011). Widespread perturbation in DNA methylation has been shown to lead to heritable phenotypic changes in plants (He et al., 2011; Seymour & Becker, 2017). Moreover, in plants changes in DNA methylation can be transmitted through generations because, contrary to what occurs in mammals, there is no clear evidence of global DNA methylation resetting at each generation (Heard & Martienssen, 2014). DNA methylation in plants occurs at cytosines and can be observed in all three contexts CG, CHG and CHH. DNA methylation in genes is almost only observed in the CG context, and seems to be associated with gene activation. On the other hand, dense DNA methylation at Transposable Elements (TEs) is observed in all three contexts and is associated with transcriptional repression (Gehring & Henikoff, 2007).

To date only a handful of natural epialleles (see definitions) have been described in eukaryotes (Table 1). These epialleles are characterised by a gain or loss of DNA methylation, often associated with a change in gene expression, as well as strong phenotypes. Such changes of DNA methylation at epialleles are usually observed in repeated sequences or TEs that are either in close proximity of genes or inside genes. While these epialleles are stable over a generation, some are metastable, which means that some level of instability as well as gradual reversions have been observed. This is in striking contrast to genetic alleles (see definitions), as changes in the DNA sequence are more stable than changes in DNA methylation for several orders of magnitude (Becker et al., 2011).

The small number of natural epialleles described so far could be explained by the fact they were either identified thanks to a strong phenotypes (Bender & Fink, 1995; Cubas, Vincent, & Coen, 1999; Manning et al., 2006; Martin et al., 2009; Melquist, Luff, & Bender, 1999; Miura et al., 2009; Zhang et al., 2012; Zhang, Sun, Cao, & Song, 2015), to an allelic incompatibility between accessions (Agorio et al., 2017; Blevins, Wang, Pflieger, Pontvianne, & Pikaard, 2017; Durand, Bouche, Perez Strand, Loudet, & Camilleri, 2012) or by chance (Silveira et al., 2013). We can

Table 1 Non-exhaustive List of Known Epialleles in Plants

Species	Epimutation Name	DNA Methylation	Gene	Phenotype	References
Linaria vulgaris	*Peloric*	DNA methylation gain in promoter region	*CYCLOIDEA*	Floral symmetry affected	Cubas et al., 1999
Solanum lycopersicum	*Cnr*	DNA methylation gain in promoter region	*LeSPL-CNR*	Loss of fruit maturation	Manning et al., 2006
Oryza sativa	*Epi-d1*	DNA methylation gain in promoter region	*DWARF1*	Dwarf plant	Miura et al., 2009
Oryza sativa	*Epi-df*	DNA methylation loss in 5' of the gene	*FIE1*	Dwarf plant	Zhang et al., 2012
Oryza sativa	*Epi-rav6*	DNA methylation loss at promoter	*RAV6*	Larger lamina inclination and smaller grain size	Zhang et al., 2015
Cucumis melo		DNA methylation gain in promoter region	*CmWIP1*	Male flowers transitioned to femelle flowers	Martin et al., 2009
Arabidopsis thaliana		DNA methylation gain for all 4 genes of this family	*PAI*	Tryptophane deficiency	Bender and Fink, 1995; Melquist et al., 1999
Arabidopsis thaliana		DNA methylation gain in Nok-1 and Est-1 accessions	*TAD3*	Allelic incompatibility	Agorio et al., 2017
Arabidopsis thaliana		DNA methylation gain in 5' of the gene	*ATFOLT1*	Allelic Incompatibility between Col-0 and Sha	Durand et al., 2012
Arabidopsis thaliana		DNA methylation gain in promoter	HISN6B	Allelic incompatibility between Col-0 and Cvi-0	Blevins et al., 2017
Arabidopsis thaliana		DNA methylation loss at promoter and 5' UTR of the gene	QQS	Expression level negatively regulated by DNA methylation	Silveira et al., 2013

suppose that many epialleles have not been discovered yet as they might be associated with mild phenotypes, or phenotypes only visible under certain circumstances (e.g environmental stress). Moreover, while identifying alleles underlying a certain phenotype is nowadays straight-forward, identifying DNA methylation changes at an unknown position associated with a phenotype is still challenging and requires more sophisticated analysis. In the context of this chapter, we will use a broad definition of epialleles: any stably transmitted change in methylated profiles, with or without phenotypic consequences.

Perturbing DNA methylation, by mutating genes involved in DNA methylation deposition or maintenance, is a way to increase the chance of detecting epialleles. Hence epialleles have been detected in *Arabidopsis thaliana* mutants characterised with a global loss of DNA methylation. Some of these epialleles are characterised by DNA hypomethylation, such as *fwa*, associated with a late flowering phenotype (Kakutani, 1997; Kakutani, Jeddeloh, Flowers, Munakata, & Richards, 1996; Lippman et al., 2004; Ronemus, Galbiati, Ticknor, Chen, & Dellaporta, 1996; Soppe et al., 2000) and *sqn*, associated with an increased expression (Catoni et al., 2017; Habu et al., 2006). Others are characterised by DNA hypermethylation, such as *sup*, associated with an excess of stamens (Jacobsen & Meyerowitz, 1997; Jacobsen, Sakai, Finnegan, Cao, & Meyerowitz, 2000), *ag*, associated with an absence of carpels (Jacobsen et al., 2000) and *bns*, associated with a dwarf phenotype (Saze & Kakutani, 2007). These epialleles are stably maintained after removal of the inducible mutation, with a certain degree of metastability, as also observed for natural epialleles. Except for QQS (Silveira et al., 2013), until now, none of these induced epialleles have been naturally observed in *A. thaliana*.

In order to identify alleles with milder or quantitative phenotypes (in contrast to strong qualitative phenotypes), recombinant inbred lines (RILs, see definitions) are commonly used (Mackay, 2001). These populations are used to identify loci at which the segregation of parental alleles are associated with phenotypic changes. Such an approach could also allow the detection of epialleles associated with mild or quantitative phenotypes. However, as this will be described in more detail in this chapter, alleles as well as epialleles are segregating in RIL populations, making it difficult to separate epigenetic from genetic impact on phenotypes (Johannes, Colot, & Jansen, 2008). In order to specifically identify epialleles associated with phenotypic changes, epigenetic RIL (epiRILs, see definitions) have been generated in *A. thaliana* (Johannes et al., 2009; Reinders et al., 2009).

In short these populations have been created in order to maximise DNA methylation changes, while reducing (if not completely removing) DNA sequence differences.

In this chapter we will be discussing the many aspects in which epiRIL populations have been of a great use and how the acquired knowledge could be translated in crops in the future.

1.1 Definitions

Allele: Genetic variants of a gene. Different alleles can result in different phenotypic traits.

Epiallele: Epigenetic variants of a gene. The genetic sequence of the epialleles is identical, but the level of DNA methylation, or other epigenetic marks, are different. Epialleles can result in differences in gene expression, which can potentially lead to differences in phenotypic traits.

RILs (Recombinant Inbred Lines): Set of homozygous lines that incorporate a combination of genomic regions derived from the cross of two parent lines. Each RIL is developed by self-pollination and single seed descent propagation of a segregating F2 plant obtained from the initial cross. Inbreeding continues for at least six/eight generations, determining the fixation in homozygous form of most of alleles and epialleles. RILs are often used for mapping QTLs.

EpiRILs (Epigenetic Recombinant Inbred Lines): Similarly to RILs, epiRILs are a set of fixed homozygous lines, descending from a F2 population. However, contrary to RILs, the parents used to generate epiRIL population have identical (or almost identical) genomic sequence but different DNA methylation profiles. EpiRILs are thus maximising epialleles segregation, while reducing (if not removing completely) allelic segregation. EpiRILs can be used for mapping epiQTLs.

QTLs (Quantitative Trait Locus): A QTL is a locus of the genome at which genetic variation correlates with variation of a quantitative trait.

EpiQTLs (Epigenetic Quantitative Trait Locus): An epiQTL is a locus of the genome at which variation in DNA methylation correlates with variation of a quantitative trait.

Additive alleles: Different alleles of a gene that combine in a way that the phenotype or expression level of the heterozygous is equal to the sum of each allele.

Dominant alleles: The dominant allele dictates the phenotype or expression level of the heterozygous, when paired with a recessive allele.

Transgressive transcripts: In the context of a hybrid, locus expression level that is not explained neither as additive, nor than as dominant allelic effect.

2. DEFINITION AND DESCRIPTION OF THE EpiRIL POPULATIONS

The study of epiallele stability and phenotypic consequences can be performed by taking advantage of natural variation in *A. thaliana*. DNA methylation at the level of genes has been shown to be highly polymorphic between *A. thaliana* accessions, making possible to follow epialleles segregation and their stability in F2 populations (Vaughn et al., 2007). Natural accessions not only differ in their levels of DNA methylation, but also in their genetic sequences, and genetic polymorphism can be used to identify the parent of origin for genomic regions in F1 and F2 populations (Greaves et al., 2012; Shen et al., 2012; Vaughn et al., 2007; Zhang, Shiu, Cal, & Borevitz, 2008).

However, the presence of genetic and epigenetic variation across *A. thaliana* natural accessions often impairs proper quantification of the epigenetic contribution to phenotypic differences. Indeed, several studies in plants (but also in mammals) reported many examples of DNA methylation variations associated to either local (*cis*) or distant (*trans*) changes in DNA sequence (Eichten et al., 2011; Gibbs et al., 2010; Hellman & Chess, 2010; Zhang et al., 2010). On the other hand, mutation rate of methylated cytosines is higher than non methylated cytosines (Xia, Han, & Zhao, 2012), suggesting that DNA methylation and DNA sequence polymorphisms can be linked and also influence each others.

Therefore, a classification of epialleles has been proposed depending on their link with DNA sequence polymorphism (Richards, 2006): (i) obligatory epialleles, for which a *cis* or *trans* genetic polymorphism influences the DNA methylation status; (ii) facilitated epialleles, which can be linked to or caused by a genetic polymorphism, but that are not fully dependent on it; and (iii) pure epialleles, that are not affected by any genetic changes.

Two epiRIL populations have been independently created in *A. thaliana* (Fig. 1), to maximise DNA methylation variation and minimise (if not abolishing) DNA sequence polymorphisms, in order to discriminate epialleles that are not influenced by DNA sequence polymorphisms (Johannes et al., 2008). These epiRILs have been generated by crossing an epigenetic mutant, *met1-3* (Reinders et al., 2009) or *ddm1-2* (Johannes et al., 2009), with its corresponding wild-type (Columbia-0 accession). The two parents thus have the same genome, except for the mutated gene, but they have very

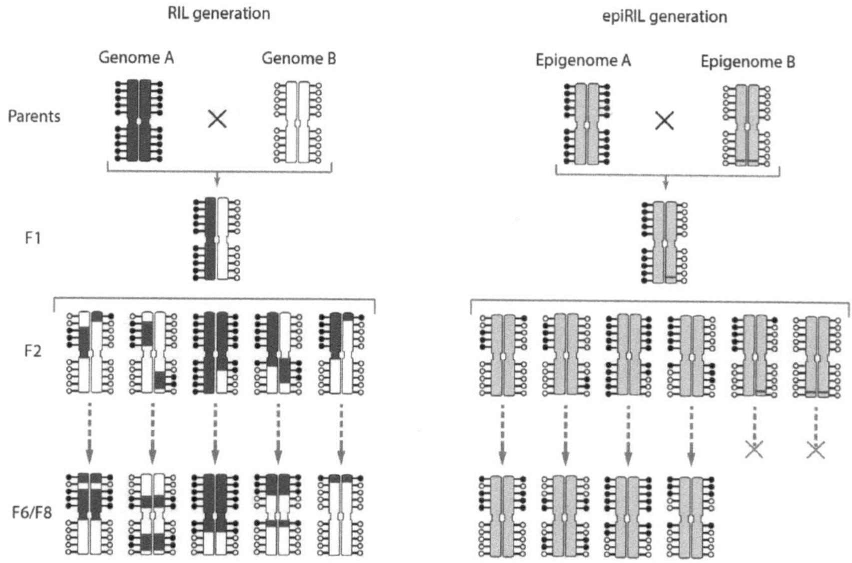

Figure 1 ***Allelic and epiallelic segregation in RIL and epiRIL populations.*** RIL populations (left) are usually created by crossing two distinct Arabidopsis accessions that are different in their genomes (depicted with different chromosome colours) and epigenomes (depicted as full or empty dots beside chromosomes). Alleles and epialleles are thus segregating in F2 population derived by this cross, and fixed in homozygous form by self-pollination and single seed-descend. By contrast, epiRILs (right) are created by crossing parents that have identical (or almost identical) genomic sequence but different DNA methylation profiles. This is obtained in Arabidopsis by mutation of *MET1* or *DDM1* genes (represented by a red horizontal line on chromosome sequence), coding for factors involved in DNA methylation maintenance. During the generation of epiRILs, only F2 plants with a MET1 or DDM1 wild-type allele are carried out, to avoid new events of genome wide hypomethylation. EpiRILs are thus maximising epialleles segregation, while reducing (if not removing completely) allelic segregation.

contrasting DNA methylation profiles. Each epiRIL within the population essentially contains a mosaic epigenome derived from either wild-type and *ddm1-2* or wild-type and *met1-3*.

Although both *met1-3* and *ddm1-2* mutants are hypomethylated, their effects on genome wide DNA methylation are different, and these differences are conserved in the epigenetic perturbations segregating in two epiRILs populations. The DNA methyltransferase MET1 maintains CG methylation in *A. thaliana* and the *met1-3* null mutant is characterised by a virtual complete erasure of CG methylation and indirect loss of plant-specific non-CG methylation (Saze, Mittelsten Scheid, & Paszkowski, 2003). On the other hand, *DDM1* encodes an ATPase chromatin remodeler primarily involved in allowing DNA methyltransferases to access heterochromatin (Zemach et al., 2013). *ddm1-2* mutation mainly affects DNA methylation in all cytosine contexts (CG, CHG and CHH) at heterochromatic TEs (Kakutani, Jeddeloh, & Richards, 1995; Lippman et al., 2004; Vongs, Kakutani, Martienssen, & Richards, 1993). Consequently, the epialleles generated in *met1*-derived epiRILs are equally distributed in euchromatic and heterochromatic areas, including gene bodies that are exclusively CG methylated (Bewick et al., 2016; Catoni et al., 2017), while epialleles in *ddm1*-derived epiRILs are mostly involving TE loci (Cortijo et al., 2014).

The *met1-3* mutant used to create the *met1*-epiRIL population also shows very severe phenotypic defects, including reduced fertility (Mathieu, Reinders, Caikovski, Smathajitt, & Paszkowski, 2007). Hence, a high level of mortality (29%) has also been observed while propagating 100 individuals of the *met1*-epiRIL population over generations (Reinders et al., 2009). On the contrary *ddm1*-derived epiRILs have been generated starting from the *ddm1-2* mutant, which displays only minor developmental defects. This strategy allowed the production of a large population of 505 different *ddm1*-derived epiRILs, with no evidence of selection against deleterious phenotypes (Colome-Tatche et al., 2012).

The crossing scheme of the two populations also differs. In both cases, the mutant (*met1-3* or *ddm1-2*) has been crossed with a wild-type plant and only F2 plants segregating the wild-type allele have been used to generate the epiRIL populations. The *met1*-epiRILs originate directly from the F2 individuals resulting from this cross, while the *ddm1*-epiRILs descend from a second back cross of the F1 with the wild-type. Thus, DNA methylation changes segregate with a 1:1 ratio in the *met1*-epiRILs and with a 1:3 (mut/WT) ratio in the *ddm1*-epiRILs.

3. DNA METHYLATION TRANSGENERATIONAL STABILITY AND ITS PHENOTYPIC CONSEQUENCES

3.1 Understanding of the Stability of DNA Methylation Perturbations

Contrary to mammals, in plants there is no evidence of a consistent global resetting of DNA methylation during development, making the transmission of epialleles over generations more probable. Indeed, it has been shown that the loss of DNA methylation induced by the *ddm1-2* mutation can be stably inherited over many generations once the DDM1 wild-type allele is re-introduced (Kakutani, Munakata, Richards, & Hirochika, 1999). The analysis of the transmission of *ddm1-2* and *met1-3* induced hypomethylation at six TEs, after a cross with wild-type, showed that the hypomethylation is transmitted at some loci and reversed to a wild-type methylation state at other loci (Lippman et al., 2004). Methylated regions have been divided into two categories: (i) those that can form two distinct epialleles that are maintained over generations once in a WT background, and (ii) those that revert to the WT epigenetic state (remethylatable). EpiRILs are a great tool to study the mechanisms and consequences of DNA hypomethylation stability or reversion over generations. Indeed, the analysis of the DNA hypomethylation stability at multiple loci in three *ddm1*-epiRILs and three *met1*-epiRILs confirmed the presence of stable and remethylatable *ddm1*- and *met1*-induced hypomethylation (Reinders et al., 2009; Teixeira et al., 2009). Transgressive DNA methylation patterns have also been observed in these populations. Using bisulfite sequencing it was shown that remethylation to a level similar to wild-type was observed occurring at many loci in all cytosine contexts. This remethylation requires sRNA and factors involved in RNA-directed DNA methylation and is progressive over generations in the *ddm1*-epiRILs (Teixeira et al., 2009), while remethylation has been observed directly occurring in the F1 in the case of *met1*-induced hypomethylation (Rigal et al., 2016). Further analysis of *cis* factors influencing remethylation in the *met1*-epiRILs as well as in the F2, containing the wild-type allele of MET1, originating from a backcross between *met1-3* and wild-type, showed that remethylation is associated with repetitiveness and relative scarcity of CpGs. In contrast, stable epialleles are associated with low copy number and high CpG content (Catoni et al., 2017). The link between these *cis* factors and the level of epigenetic stability was confirmed in rice (Catoni et al., 2017), and also observed generally associated to the susceptibility of transgenes to be epigenetically silenced (Sidorenko et al., 2017). This observation shows how epiRILs in *A. thaliana* could be of great help to identify general rules associated to epiallele stability in different plant species or even for synthetic or heterologous DNA sequences (like transgenes).

3.2 From Epialleles to Epigenomic Recombination Maps

The identification of epialleles in epiRILs has been used advantageously to identify the parental origin of genomic regions along the genome, exclusively using DNA methylation information (Colome-Tatche et al., 2012; Reinders et al., 2009). Parental origin was identified for three *met1*-epiRILs using whole-genome methylation analysis (Reinders et al., 2009). Genome-wide DNA methylation data for 123 *ddm1*-epiRILs were also used in order to construct a recombination map derived from 126 epialleles covering 81.9% of the total genome (Colome-Tatche et al., 2012). The genetic length of this map is comparable to those obtained from classical *Arabidopsis* crosses, suggesting that the hypomethylated loci segregating in the *ddm1*-epiRILs do not affect the global meiotic recombination rates. However, it has been seen on a local scale that recombination rates are reduced within repeat-rich pericentromeric regions and increased in chromosome arms (Colome-Tatche et al., 2012). This remodelling of recombination hotspots, without changing the global rate, was also independently observed using *met1*-epiRILs (Mirouze et al., 2012). A later study shown that this remodelling of local recombination requires genes involved in the redistribution of interfering crossovers (Yelina et al., 2015).

Interestingly, the creation of epigenomic recombination maps using epialleles has also been done using mutation accumulation (MA) lines in *A. thaliana* (Hofmeister, Lee, Rohr, Hall, & Schmitz, 2017). MA lines are self-pollinated single-seed descent lines originating from a single founder, such that the lines are nearly genetically identical. MA lines display DNA methylation variation, and more than half of the differentially methylated regions identified in MA lines were stably transmitted in the progeny of a cross between two of them (Hofmeister et al., 2017). The creation of epigenomic recombination maps using stable DNA methylation variation is thus not restricted to epiRILs and will be of great interest to identify epialleles underlying phenotypic variation.

3.3 Epialleles and Phenotypic Consequences

Knowing that a proportion of DNA methylation perturbations are transmitted through generations in *met1* and *ddm1*-epiRILs, one important question is to define if these can have phenotypic consequences. The two epiRIL populations have been extensively phenotyped for qualitative as well as quantitative traits such as flowering time, biomass or response to biotic and abiotic stresses. Two types of phenotypic variation have been observed. A first type of recessive variation has been observed sporadically occurring in only one epiRIL line and thus arose specifically during the creation of that line (Fig. 2). These specific phenotypic changes are unlikely to be transmitted from the parents used in the creation of the epiRIL populations, and it was shown that TE mobilization impairing gene functions were the cause of such specific phenotypes in the *met1*-epiRIL population

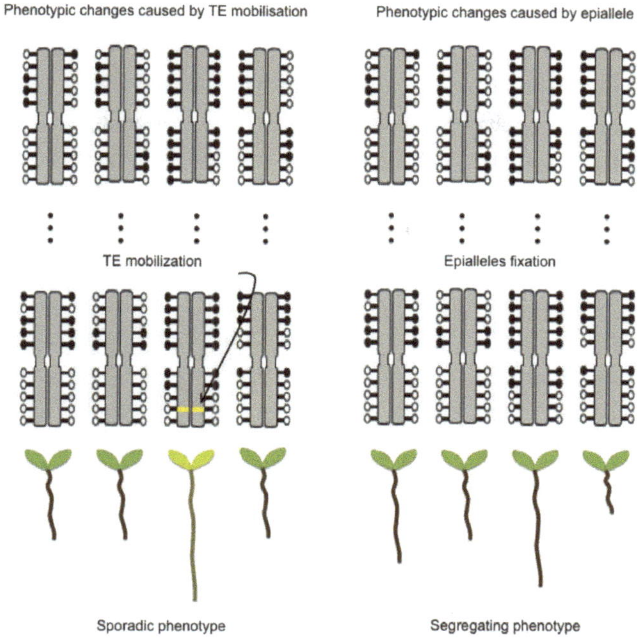

Figure 2 ***Origin of phenotypic changes observed in epiRILs.*** Phenotypic changes occurring in epiRILs are of two types. The first type (left) is sporadic and recessive and occurring specifically in one line, probably caused by TE mobilization or other genetic mutation. These phenotypes are unlikely to be transmitted from the parents used in the creation of the epiRIL populations. The second type of phenotypic changes (right) appears on a significant proportion of epiRIL lines. These traits are potentially inherited from the parents and likely caused by epialleles segregating in the epiRIL populations.

(Mirouze et al., 2009). The second type of phenotypic change is affecting a significant proportion of the epiRIL lines and is thus potentially inherited from the parents. We will discuss more in detail this second type of phenotypic change, as they are more likely to be caused by epialleles segregating in the epiRIL populations.

One strong phenotype observed in the two epiRIL populations is delayed flowering time, which has been shown to be associated with the hypomethylated epiallele at the *FWA* locus (Johannes et al., 2009; Reinders et al., 2009). However, continuous variation for flowering time is still observed in the *ddm1*-epiRIL population after removing individuals for which late flowering is caused by this *fwa* epiallele (Johannes et al., 2009). This suggests that DNA methylation changes at other loci are also involved in the segregation of this trait in the *ddm1*-epiRIL population.

A large proportion of the *met1*-epiRIL population is also characterised with retarded growth (85% of *met1*-epiRILs) as well as delayed germination under elevated salinity (60% of *met1*-epiRILs). Moreover, 34% and 4% of *met1*-epiRILs showed respectively increased resistance or susceptibility to the biotrophic bacterial pathogen *Pseudomonas syringae* pv. tomato (Pst) (Reinders et al., 2009).

Given the high number of lines in the *ddm1*-epiRIL population (505 lines), many quantitative traits have been measured in this population and their heritability estimated (i.e. the degree of variation in the phenotypic trait in the population due to genetic, and here epigenetic, variation between individuals). A continuous variation and high heritability have been observed for several traits such as flowering time, plant height, primary root length, fruit number, total biomass and others (Cortijo et al., 2014; Johannes et al., 2009; Roux et al., 2011). Many traits such as flowering time, plant height, fruit size, dry biomass and rosette diameter have also been measured in common garden experiments, alongside natural accessions of *A. thaliana* (Roux et al., 2011). It was found that phenotypic variation in the *ddm1*-epiRIL population displays a level of trait heritability similar to the natural Arabidopsis accessions grown in parallel. Phenotypic plasticity, which is the ability of one genotype to produce multiple phenotypes in response to the environment, has also been measured for flowering time, plant height, fruit number, total biomass and root:shoot ratio in response to drought and nutrient addition (Zhang et al., 2012). A high heritability was observed for these traits in the absence and presence of environmental perturbations, but also for their plasticity (Zhang, Fischer, Colot, & Bossdorf, 2013). Theoretical predictions indicate that these heritability values are consistent with a small number of parentally derived quantitative trait loci (QTL, see definitions). These results suggest that phenotypic variability in the *ddm1*-epiRILs can be caused by the segregation of epialleles, or by DNA sequence polymorphisms caused by mobilization of transposable elements, reactivated by DNA hypomethylation.

In order to identify the loci underlying heritable phenotypic variability in the *ddm1*-epiRIL population, and to define their genetic or epigenetic origin, epigenetic quantitative trait loci (epiQTL, see definitions) have been mapped in *ddm1*-epiRILs for flowering time and primary root length (Cortijo et al., 2014). This was done taking advantage of a genetic map generated using differentially methylated regions in 123 *ddm1*-epiRILs, and covering 81.9% of the total genome (Colome-Tatche et al., 2012). Several epiQTLs were detected on chromosomes 1, 4 and 5 for flowering time, and on chromosomes 1, 2 and 4 for primary root length (Fig. 3). These QTLs could be associated to epigenetic polymorphisms, but also caused by TE mobilization. In order to discriminate between these two possibilities, association between DNA methylation status and primary root length was confirmed for the markers at the peak of the three epiQTLs in an independent F3 population. Moreover, new TE mobilisations detected at these epiQTLs in the epiRIL population are not present in this F3 population. These results strongly suggest that changes in DNA methylation are causing the epiQTLs detected for primary root length (Cortijo et al., 2014).

The next step will be to identify the epialleles underlying these epiQTLs. However, as for mapping alleles underlying QTLs, this operation is challenging and will require more time and work. A first step would be to generate a fine mapping population in order to reduce the size of

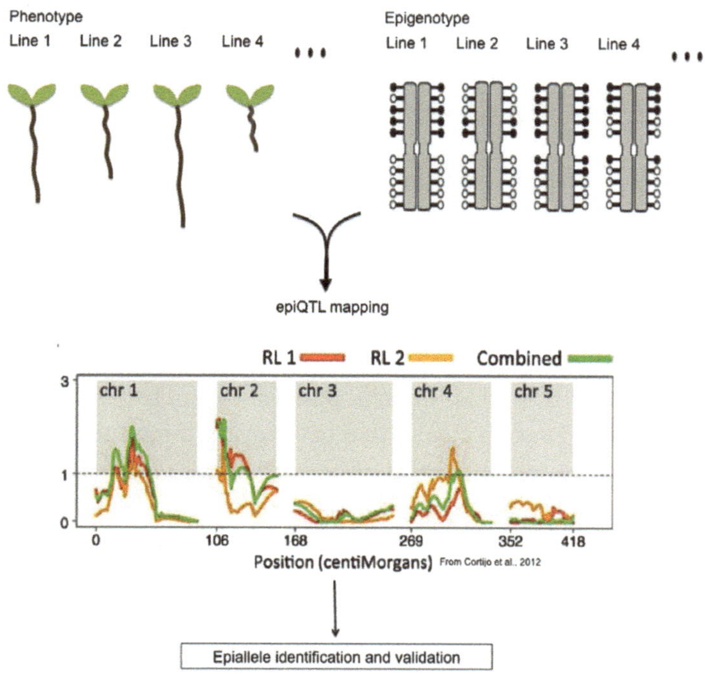

Figure 3 ***Principle of epiQTL mapping in epiRILs for root length, followed by epiallele identification and validation.*** In order to identify epiQTLs for a quantitative trait, every line of the population is phenotyped (top left) and epigenotyped (top right). EpiQTLs are then identified by measuring the co-segregation of phenotype and epigenotype. Several QTLs were identified on chromosomes 1, 2 and 4 for root length in the *ddm1*-derived epiRILs (middle). The next step is to identify epialleles underlying epiQTLs and to validate them by changing their DNA methylation level (bottom).

QTLs and thus the number of potential epialleles (Loudet, Gaudon, Trubuil, & Daniel-Vedele, 2005). Once potential epialleles will be detected, manipulating their DNA methylation status will be required in order to confirm the link between DNA methylation and phenotypic variability at this locus. Targeted DNA methylation is still challenging but could be now achieved using a deactivated Cas9 fused with a DNA methyltransferase (Vojta et al., 2016), by VIGS (Bond & Baulcombe, 2015) or by using RNA hairpins to trigger RdDM (Mette, Aufsatz, van der Winden, Matzke, & Matzke, 2000).

However, the complete characterization of epialleles responsible for the identified epiQTL associated to traits of interest is not necessarily a requirement in order to use this knowledge to improve plants. Methods such as marker-assisted selection could be used to introgress the desired trait in the cultivar of interest, taking advantage of markers associated to the identified epiQTL (Kumar et al., 2017). The DNA methylation status of these markers, rather than the DNA sequence polymorphisms, would have to be used during the selection process. Assays based on DNA digestion with enzymes sensitive to DNA methylation, as for example McrBC (Teixeira et al., 2009), associated to qPCR, would provide a cheap and high throughput approach to perform such selection based on the markers epigenetic status.

4. USING EpiRILs TO UNDERSTAND TE MOBILIZATION

Transposable elements (TEs) are a heterogeneous group of mobile DNA elements integrated in the genome of virtually all organisms, with the ability to move from their original position to a new genomic location. TEs can be classified in two main classes based on their transposition strategy: (i) Class I TEs (or retrotransposons), which transpose with a copy-and-paste mechanism through reverse transcription of a RNA intermediate and (ii) Class II TEs, transposing with a cut-and-paste mechanism mediated by a transposase (Wicker et al., 2007). Although initially considered as selfish genes and assimilated to "junk DNA" (Doolittle & Sapienza, 1980), the importance of the contribution of TEs to gene and genome structure and evolution is currently recognised across the entire tree of life (Hurst & Werren, 2001; Rebollo, Romanish, & Mager, 2012), including plants (Lisch, 2013). Consequently, transcriptional silencing of TEs ensures genetic stability, and is controlled in plants by a network of self-reinforcing epigenetic pathways, marking TEs with repressive marks at the level of DNA (cytosine methylation) and chromatin (histone repressive marks). Therefore, epigenetics mutants often show release of TE expression, and have been used to reveal and study real time TE mobilization (Ito et al., 2011; Miura et al., 2001; Tsukahara et al., 2009). In this context, epiRILs represent a valuable alternative to homozygous *met1*, *ddm1* and other epigenetic mutants in studying TE mobilization for several reasons. First, epiRILs are in the wild-type genetic background and are therefore genetically and phenotypically more stable compared to the mutant from which they derived (Reinders et al., 2009). Moreover, the epiallele segregation and homozygous fixation that occurred through many inbred generations contributed to "dilute" the epialleles with deleterious effects, reducing the amount of developmental defects that are normally displayed in the homozygous mutant. For example, the Arabidopsis *met1-3* mutation is semi-lethal with transgenerational decrease of fitness, and homozygous mutant plants can be maintained viable for a maximum of four generations (Mathieu et al., 2007). Although not as severe as for *met1-3* mutants, *ddm1-1* and *ddm1-2* homozygous mutants accumulate strong phenotypic defects through generations (Kakutani et al., 1996). Stochastic bursts of several TEs independently occur in different *ddm1* inbred lines, and are contributing to at least some of the developmental phenotypes observed in *ddm1* (Miura et al., 2001; Tsukahara et al., 2009).

By contrast, *met1* and *ddm1*-derived epiRILs have been maintained for more than eight generations without noticing a significant decrease in fertility (Johannes et al., 2009; Reinders et al., 2009), providing a much more reliable platform to study transposition events. Indeed, the mobilization of the Class II DNA transposon CACTA1 (Reinders et al., 2009) and the Class I retrotransposon EVADE (EVD) (Mirouze et al., 2009) were reported in *met1*-derived epiRILs, while not detected in the *met1-3* mutant. Similarly, many transposons have been found active in *ddm1*-epiRILs, indicating that *ddm1-2* mutation is necessary to release TE silencing, and that TEs can remain active after re-introduction of the DDM1 wilt-type allele (Cortijo et al., 2014; Gilly et al., 2014). In *ddm1*-epiRILs the fraction of the demethylated genome was initially diluted through one *ddm1* backcross of the F1 with the wild-type, reducing in average to 25% the fraction of hypomethylated genome inherited from the *ddm1-2* mutant parent, and contributing to stabilize epiRILs phenotypes at late generations.

Therefore, both *met1* and *ddm1* derived epiRIL populations demonstrated a longer transgenerational viability and stability compared to the mutant parents from which they are derived. The advantage of this condition is that the plethora of epiallelic effects and multiple TE activation observed in the homozygous mutants can be isolated in independent epiRILs, making it possible to study the activation and *de novo* silencing of independent TEs in real time experiments. For example, the transgenerational dynamic evolution of EVD mobilization was studied in inbred epiRILs (Mari-Ordonez et al., 2013). The EVD burst and its *de novo* silencing was reconstructed in a *met1*-epiRIL, observing that efficient silencing is associated to a change in small RNA composition, and consistently occurs approximately at the 14th generation after EVD activation, when its copy number in the genome reaches a threshold of 40 copies (Mari-Ordonez et al., 2013).

Although the first events of real time transposition were discovered in maize more than 50 years ago (Mc Clintock, 1950), the impact of TE mobilization on genome stability and the biology of complex organisms is still poorly investigated, and essentially extrapolated from comparative genomics and phylogenetic studies. This limitation is the direct consequence of the rarity of TE mobilization events so far observed in nature, likely due to the epigenetic silencing normally associated to repeated DNA sequences.

The most evident effect of TE mobilization is the recessive mutation of genes with a new TE insertion occurring in their coding region, in many cases producing a visible phenotype. Nonetheless, phylogenetic studies produced evidence of several TE-induced non-destructive effects on gene expression responsible for agricultural important traits in crops (Lisch, 2013). It is however unclear if these non-destructive effects derived from positively selected exceptional aberrant transposition events or are the result of transposition strategies of different TE families. In this scenario, epiRILs offer the opportunity to identify and characterize new active TEs, and to study the impact of their real time mobilization across generations in a limited number of plant lines. Therefore, the study of epiRILs may contribute to elucidate the role of TE on genetic and biology in higher plants, and more generally in eukaryotic multicellular organisms.

5. HETEROSIS

Heterosis, or hybrid vigour, is a phenomenon describing the improved phenotype of a hybrid offspring compared to the average of both parents, first recorded by Charles Darwin in 1876 (Darwin, 1876). In agriculture, heterosis has been adopted as a routine strategy for plant breeding, leading to improved biomass, yield or resistance to biotic and abiotic stimuli in hybrids (Baranwal, Mikkilineni, Zehr, Tyagi, & Kapoor, 2012). Despite such an extensive use in agriculture, the underlying mechanisms of heterosis are still poorly understood.

Traditionally, it is generally accepted that heterosis directly correlates with the level of genetic distance between the two parents (Birchler, Yao, Chudalayandi, Vaiman, & Veitia, 2010). However, more recent experiments performed in Arabidopsis have shown that hybrids generated from accessions with very similar genome can also display a high level of hybrid vigour (Groszmann, Greaves, Fujimoto, Peacock, & Dennis, 2013; Schneeberger et al., 2011), suggesting that epigenetic differences could also contribute to heterosis (Fig. 4).

Indeed, changes in small RNA level and DNA methylation have been associated to hybrid vigour in both interspecific (i.e. between species) or intraspecific (i.e. between accessions) hybrids systems studied in

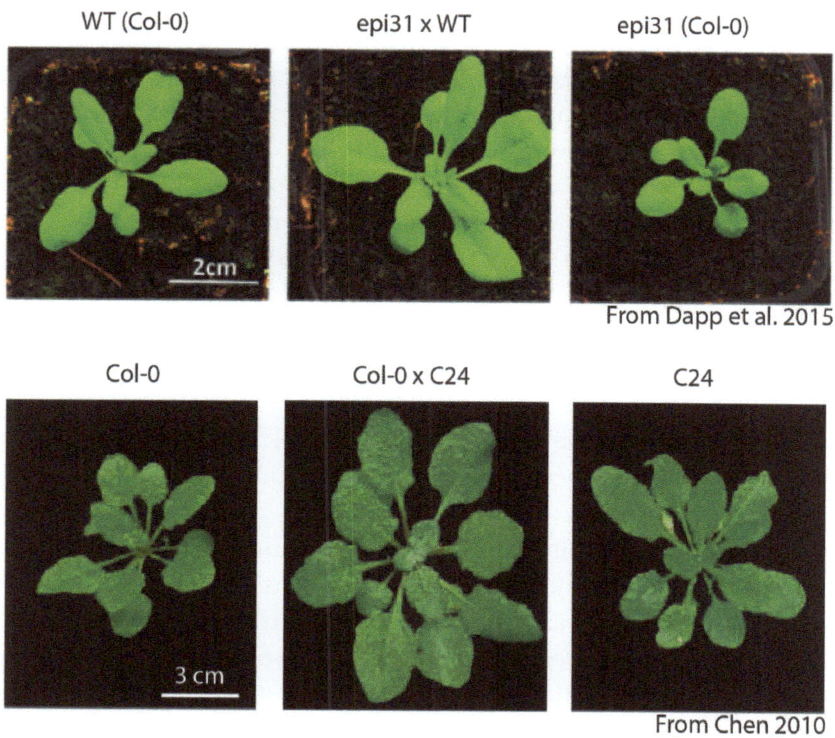

Figure 4 ***Comparison of epi-hybrid and intraspecies hybrid in* Arabidopsis thaliana.** Examples on enhanced vigour in an epi-hybrid, compared with its two parents, epi31 and wild-type Col-0 (top), and in an intraspecies hybrid compared to its two parent accessions, Col-0 and C24 (bottom). In both cased, the epi-hybrid and the intraspecies hybrid are bigger than their parents, indicating a heterotic effect.

Arabidopsis (Greaves et al., 2012; Groszmann et al., 2011; Shen et al., 2012) and other plant species, including rice (Chodavarapu et al., 2012; He, He, & Deng, 2013), maize (Barber et al., 2012; He, Chen, et al., 2013), wheat (Kenan-Eichler et al., 2011) and tomato (Shivaprasad, Dunn, Santos, Bassett, & Baulcombe, 2012). However, the coexistence of genetic and epigenetic differences in hybrids makes it intrinsically difficult to quantify the epigenetic contribution to heterosis.

In contrast, epiRILs are isogenic to wild-type but differ at localized hypomethylated chromosomal areas. Interestingly, some lines from both *met1*-derived and *ddm1*-derived epiRIL populations displayed increased biomass or higher resistance to a pathogen if compared to wild-type Columbia-0 accession (Johannes et al., 2009; Reinders et al., 2009), similar to what is observed in heterotic hybrids. These results suggest that epigenetic variation by itself might be involved in the generation of hybrid vigour.

In a recent work, heterosis for growth-related traits was investigated in epigenetic hybrids generated by pollinating *met1*-derived epiRIL plants with pollen from their isogenic wild-type line (Col-0) (Dapp et al., 2015). In the case of one *met1*-derived epiRIL (epi31), a consistent and reproducible increase in rosette size was observed in F1 plants compared to both parental lines. Remarkably, epi31 displayed a clear parent-of-origin effect on hybrid vigour, as also observed in certain crosses between Arabidopsis accessions (Barth, Busimi, Friedrich Utz, & Melchinger, 2003; Meyer, Torjek, Becher, & Altmann, 2004). Although the authors could not associate any change in gene expression with the hybrid vigour observed, several additive, dominant and transgressive (see definitions) transcripts have been identify in the F1 hybrids (Dapp et al., 2015), supporting the existence of multiple scenarios for DNA methylation-mediated gene regulation in epi-hybrids.

More recently, the contribution of differences in parental methylation to heterosis was quantified measuring six different traits in a larger panel of over 500 *A. thaliana* epi-hybrids obtained starting from *ddm1*-derived epiRILs (Lauss et al., 2018). Several positive and negative heterotic effects were documented, and specific differentially methylated regions in parental genomes were associated with heterotic phenotypes observed in nineteen epi-hybrids (Lauss et al., 2018).

In conclusion, there is growing evidence supporting the epigenetic contribution to heterosis. In this context, epiRILs may be the optimal tool to isolate and characterize epigenetic determinants of hybrid vigour, for example by mapping epiQTLs associated to different favourable traits. In addition, altering the epigenetic landscape of parents can potentially increase the heterotic effect of hybrids, and could be used as a tool to increase plant productivity.

6. CHALLENGES WITH CROPS

The investigation of the epigenetic landscape in Arabidopsis epiRILs critically contributed to reveal general plant epigenetic proprieties and mechanisms. Such findings include the mapping of epiQTLs (Cortijo et al., 2014), the discovery of genetic proprieties that predict epialleles, common in Arabidopsis and rice (Catoni et al., 2017), and a model for origin and evolutionary consequences of gene body DNA methylation in Angiosperms (Bewick et al., 2016). However, despite a general conservation of most epigenetic factors and proprieties across plants, epiRILs are so far only available for *A. thaliana*. Creating epiRILs in crops could improve our understanding of the source of epiallelic creation and also help detecting epialleles with potential agronomic advantages.

The introduction in crops of a level of epigenetic variation similar to that observed in Arabidopsis epiRILs might be of great interest for agriculture. Especially when considering that crops have larger genomes containing a much higher number of transposons and repetitive DNA, suggesting an elevated potential for the generation of epialleles. Consistently, rice, maize and tomato mutants in components of epigenetic regulation display strong developmental phenotypes and partial or complete infertility (Gouil & Baulcombe, 2016; Hu et al., 2014; Li et al., 2014). Remarkably, developmental phenotypes described in crop epigenetic mutants do not correlate with extensive genome hypomethylation as observed in Arabidopsis (Mathieu et al., 2007), suggesting that in most plants small perturbations of the methylome have stronger deleterious phenotypic effects than in Arabidopsis.

Taking this into account, the generation of crop epiRILs may be impaired by the inability of producing viable hypomethylated mutants required for the initial cross. However, alternative strategies should be considered to induce stable epiallele formation without affecting plant viability (Fig. 5).

One possibility to reduce genome methylation is the use of hypomethylated partial loss-of-function epigenetic mutants with mitigated deleterious developmental phenotypes. In Arabidopsis, while the null *met1-3* allele

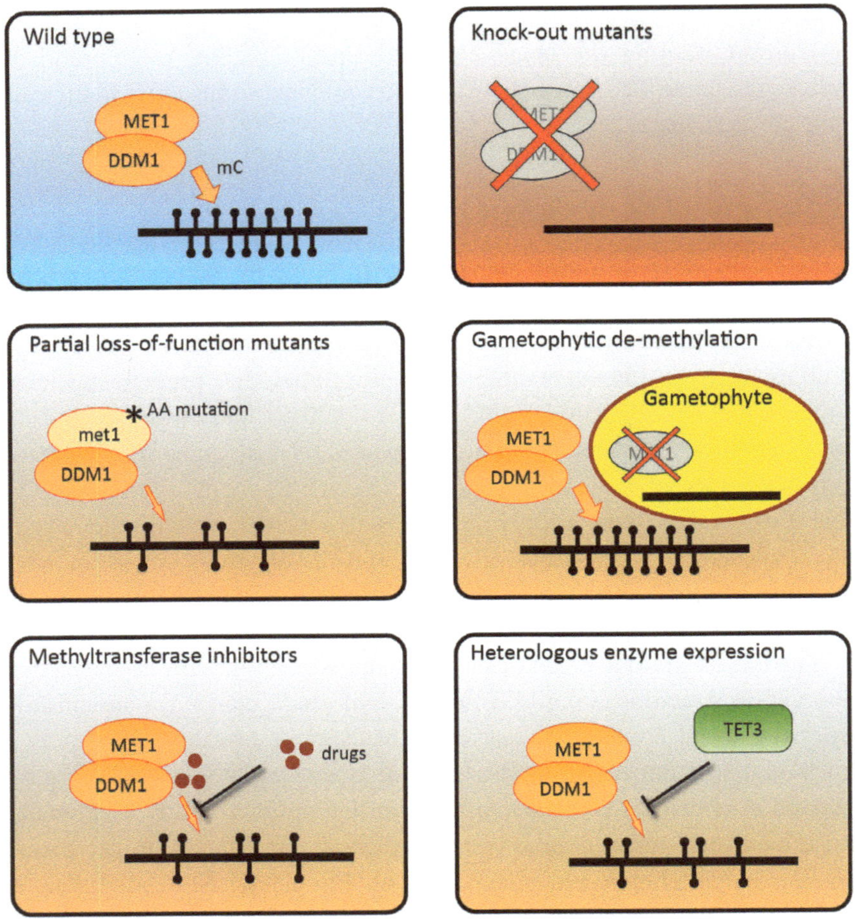

Figure 5 ***Different approaches to induce global DNA demethylation in order to create epiRIL populations.*** In wild-type, DNA maintenance mechanisms ensure conservation of epigenetic marks (i.e. DNA methylation, represented as black dots). In *met1* or *ddm1* knock-out mutants, DNA methylation is strongly impaired and normally associated to strong developmental phenotype. Alternative strategies to reduce DNA methylation limiting the impact on plant fitness include the use of partial loss-of-function mutations with partial de-methylation; the self-propagation of heterozygous knock-out mutants, resulting in gametophyte hypomethylation; the application of drugs interfering with methyltransferase activity; and the ectopic overexpression of TET methylcytosine dioxygenases.

causes complete loss of CpG methylation and is semi-lethal (Mathieu et al., 2007), the partially functional MET1 protein produced in the *met1-1* allele can retain CpG methylation in approximately one quarter of the genome, causing only minor developmental defects and allowing transgenerational conservation of the *met1-1* mutation in the homozygous form (Kankel et al., 2003). In addition, mobilization of TEs has also been observed in the *met1-1* mutant background (Griffiths, Catoni, Iwasaki, & Paszkowski, 2018) as well as the formation of epialleles that are stably maintained for several generations after transgenic complementation with a wild-type MET1 allele (Catoni et al., 2017). This suggests that the use of partial loss-of-function mutants might replace null alleles in epiRIL construction, if a viable knock-out mutant cannot be obtained. However, the production of partial-loss of function mutants for a chosen gene may be difficult to achieve in plants, and is normally associated to fortuitous screening starting from random mutagenized populations. Nonetheless, DNA editing strategies, such as CRISPR/CAS9 (Cong et al., 2013) and TALEN (Miller et al., 2011) have been successfully extended to plants, allowing an unprecedented high level of accuracy in targeting chromosomal sequences to induce mutations (Malzahn, Lowder, & Qi, 2017). Using these approaches, the effect of well know partial loss-of-function mutations observed in Arabidopsis might be more easily obtained in the species of interest by targeting a similar mutation in the corresponding homologous genes.

Alternatively, passive DNA hypomethylation has been proposed to occur during gametogenesis in heterozygous *met1* mutant. The haploid male and female gametophytes undergo two and three post-meiotic divisions, respectively. Therefore, genomic DNA is duplicated in gametophytes with the *met1* mutant allele, in absence of the MET1 methylation maintenance system, leading to the passive reduction to 50% and 75% of the genome methylation respectively in male and female gametes (Saze et al., 2003). This hypothesis was confirmed by later studies, observing also a genome-wide demethylation and the formation of stable epialleles in heterozygous inbred *met1* mutant lines, similar to what was observed in epiRILs (Catoni et al., 2017; Stroud, Greenberg, Feng, Bernatavichute, & Jacobsen, 2013). Therefore, genome-wide hypomethylation in crop plants may be simply achieved by inbreeding the usually more fertile heterozygous *met1* mutant, without the necessity of a viable homozygous mutant allele.

One alternative to the generation of epigenetic mutants is the use of drugs interfering with epigenetic pathways. Inhibitors of DNA methylases, such as 5-Azacytidine and Zebularine, have been successfully used to induce DNA demethylation in plants (Griffin, Niederhuth, & Schmitz, 2016; Pecinka & Liu, 2014), including crops (Sano, Kamada, Youssefian, Katsumi, & Wabiko, 1990; Santos et al., 2002; Zhu et al., 2018). Although most of hypomethylation and transcriptional changes induced by these drugs are only transient (Baubec, Pecinka, Rozhon, & Mittelsten Scheid, 2009), transgenerational effects have been observed in rice treated with 5-Azacytidine (Sano et al., 1990). Recently, simultaneous application of Zebularine and the RNA polymerase II inhibitor α-amanitin on Arabidopsis wild-type seedlings was sufficient to mobilize the heat-responsive Class I retrotransposon ONSEN, demonstrating that drug application can efficiently release transposon transcriptional silencing (Thieme et al., 2017).

Finally, another very valuable alternative in order to reduce DNA methylation in plant is the heterologous expression of enzymes promoting DNA hypomethylation. For example, the human Ten-eleven translocation (TET) methylcytosine dioxygenases are an enzyme family catalysing the conversion of 5 mC in 5-hydroxymethylcytosine (5 hmC), and are involved in active DNA demethylation in embryonic stem cells (Tahiliani et al., 2009). The transgenic expression of TET3 catalytic subunit in Arabidopsis was enough to decrease DNA methylation at ribosomal repeats (Hollwey, Watson, & Meyer, 2016). In addition, the transgenic expression of the same TET3 gene in Tomato induced hypomethylation and ectopic expression of the CEN1.1 gene in leaves, promoting vegetative growth (Hollwey, Out, Watson, Heidmann, & Meyer, 2017). In a more recent work, ectopic overexpression of a different TET gene in Arabidopsis induced widespread DNA demethylation and phenotypic variations, mimicking the effects of *met1* mutation (Ji et al., 2018). In addition, a Cas9-based targeted demethylation system using the TET1 catalytic subunit was recently generated and was shown to be able to target demethylation and activate gene expression when directed to known switchable epialleles in Arabidopsis (Gallego-Bartolomé et al., 2018).

The combination of these approaches could thus potentially be used in order to promote global or specific changes in DNA methylation profiles and be the first step to create epiRILs in crops.

7. CONCLUSION

Arabidopsis epiRIL populations have allowed major advances in understanding the genetic determinant controlling DNA methylation stability as well as mechanisms involved in the transgenerational transmission of epigenetic information. Several studies used epiRILs to highlight the phenotypic consequences of epiallele segregation and the epigenetic contribution to quantitative traits. While epiRILs have been initially created with the intention of minimising DNA polymorphisms, the TE reactivation induced by the global loss of DNA methylation has been used advantageously in order to better understand how TE mobilization is controlled, and to study the transgenerational effect of TE activation. EpiRILs have also helped to better understand the importance of DNA methylation on heterosis, commonly used in crops to improve yield.

The next step to extend the epigenetic potential to improve agricultural traits will be the creation of epiRILs in crops. This step is challenged by the amount of developmental defects associated to genome wide hypomethylation observed in epigenetic mutants. Nonetheless, the better understanding of the epigenetic contribution to phenotypes, and the use of more sophisticated genome editing strategies might be critical to successfully obtain crop epiRILs in the near future.

ACKNOWLEDGEMENTS

S.C. is supported by the Gatsby foundation, M.C. is supported by European Research Council (EVOBREED) [322621]. We thank Dr. Jayne Griffiths (Sainsbury Laboratory—Cambridge, UK) for critical reading of the manuscript draft.

REFERENCES

Agorio, A., Durand, S., Fiume, E., Brousse, C., Gy, I., Simon, M., et al. (2017). An arabidopsis natural epiallele maintained by a feed-forward silencing loop between histone and DNA. *PLoS Genetics, 13*(1), e1006551. https://doi.org/10.1371/journal.pgen.1006551.

Baranwal, V. K., Mikkilineni, V., Zehr, U. B., Tyagi, A. K., & Kapoor, S. (2012). Heterosis: Emerging ideas about hybrid vigour. *Journal of Experimental Botany, 63*(18), 6309–6314. https://doi.org/10.1093/jxb/ers291.

Barber, W. T., Zhang, W., Win, H., Varala, K. K., Dorweiler, J. E., Hudson, M. E., et al. (2012). Repeat associated small RNAs vary among parents and following hybridization in maize. *Proceedings of the National Academy of Sciences of the United States of America, 109*(26), 10444–10449. https://doi.org/10.1073/pnas.1202073109.

Barth, S., Busimi, A. K., Friedrich Utz, H., & Melchinger, A. E. (2003). Heterosis for biomass yield and related traits in five hybrids of *Arabidopsis thaliana* L. Heynh. *Heredity, 91*(1), 36–42. https://doi.org/10.1038/sj.hdy.6800276.

Baubec, T., Pecinka, A., Rozhon, W., & Mittelsten Scheid, O. (2009). Effective, homogeneous and transient interference with cytosine methylation in plant genomic DNA by zebularine. *The Plant Journal: for Cell and Molecular Biology, 57*(3), 542–554. https://doi.org/10.1111/j.1365-313X.2008.03699.x.

Becker, C., Hagmann, J., Muller, J., Koenig, D., Stegle, O., Borgwardt, K., et al. (2011). Spontaneous epigenetic variation in the *Arabidopsis thaliana* methylome. *Nature, 480*(7376), 245–249. https://doi.org/10.1038/nature10555.

Bender, J., & Fink, G. R. (1995). Epigenetic control of an endogenous gene family is revealed by a novel blue fluorescent mutant of Arabidopsis. *Cell, 83*(5), 725–734.

Bewick, A. J., Ji, L., Niederhuth, C. E., Willing, E. M., Hofmeister, B. T., Shi, X., et al. (2016). On the origin and evolutionary consequences of gene body DNA methylation. *Proceedings of the National Academy of Sciences of the United States of America, 113*(32), 9111–9116. https://doi.org/10.1073/pnas.1604666113.

Birchler, J. A., Yao, H., Chudalayandi, S., Vaiman, D., & Veitia, R. A. (2010). Heterosis. *The Plant Cell Online, 22*(7), 2105–2112. https://doi.org/10.1105/tpc.110.076133.

Blevins, T., Wang, J., Pflieger, D., Pontvianne, F., & Pikaard, C. S. (2017). Hybrid incompatibility caused by an epiallele. *Proceedings of the National Academy of Sciences of the United States of America, 114*(14), 3702–3707. https://doi.org/10.1073/pnas.1700368114.

Bond, D. M., & Baulcombe, D. C. (2015). Epigenetic transitions leading to heritable, RNA-mediated de novo silencing in *Arabidopsis thaliana. Proceedings of the National Academy of Sciences of the United States of America, 112*(3), 917–922. https://doi.org/10.1073/pnas.1413053112.

Catoni, M., Griffiths, J., Becker, C., Zabet, N. R., Bayon, C., Dapp, M., et al. (2017). DNA sequence properties that predict susceptibility to epiallelic switching. *The EMBO Journal, 36*(5), 617–628. https://doi.org/10.15252/embj.201695602.

Chodavarapu, R. K., Feng, S., Ding, B., Simon, S. A., Lopez, D., Jia, Y., et al. (2012). Transcriptome and methylome interactions in rice hybrids. *Proceedings of the National Academy of Sciences of the United States of America, 109*(30), 12040–12045. https://doi.org/10.1073/pnas.1209297109.

Colome-Tatche, M., Cortijo, S., Wardenaar, R., Morgado, L., Lahouze, B., Sarazin, A., et al. (2012). Features of the Arabidopsis recombination landscape resulting from the combined loss of sequence variation and DNA methylation. *Proceedings of the National Academy of Sciences of the United States of America, 109*(40), 16240–16245. https://doi.org/10.1073/pnas.1212955109.

Cong, L., Ran, F. A., Cox, D., Lin, S., Barretto, R., Habib, N., et al. (2013). Multiplex genome engineering using CRISPR/Cas systems. *Science, 339*(6121), 819–823. https://doi.org/10.1126/science.1231143.

Cortijo, S., Wardenaar, R., Colome-Tatche, M., Gilly, A., Etcheverry, M., Labadie, K., et al. (2014). Mapping the epigenetic basis of complex traits. *Science, 343*(6175), 1145–1148. https://doi.org/10.1126/science.1248127.

Cubas, P., Vincent, C., & Coen, E. (1999). An epigenetic mutation responsible for natural variation in floral symmetry. *Nature, 401*(6749), 157–161. https://doi.org/10.1038/43657.

Dapp, M., Reinders, J., Bediee, A., Balsera, C., Bucher, E., Theiler, G., et al. (2015). Heterosis and inbreeding depression of epigenetic Arabidopsis hybrids. *Nature Plants, 1*, 15092. https://doi.org/10.1038/nplants.2015.92.

Darwin, C. R. (1876). *The effects of cross and self fertilisation in the vegetable kingdom*. London: John Murray.

Doolittle, W. F., & Sapienza, C. (1980). Selfish genes, the phenotype paradigm and genome evolution. *Nature, 284*(5757), 601–603.

Durand, S., Bouche, N., Perez Strand, E., Loudet, O., & Camilleri, C. (2012). Rapid establishment of genetic incompatibility through natural epigenetic variation. *Current Biology: CB, 22*(4), 326–331. https://doi.org/10.1016/j.cub.2011.12.054.

Eichten, S. R., Swanson-Wagner, R. A., Schnable, J. C., Waters, A. J., Hermanson, P. J., Liu, S., et al. (2011). Heritable epigenetic variation among maize inbreds. *Plos Genetics, 7*(11), e1002372. https://doi.org/10.1371/journal.pgen.1002372.

Gallego-Bartolomé, J., Gardiner, J., Liu, W., Papikian, A., Ghoshal, B., Kuo, H. Y., et al. (2018). Targeted DNA demethylation of the *Arabidopsis* genome using the human TET1 catalytic domain. *Proceedings of the National Academy of Sciences of the United States of America, 115*(9), E2125–E2134. https://doi.org/10.1073/pnas.1716945115.

Gehring, M., & Henikoff, S. (2007). DNA methylation dynamics in plant genomes. *Biochimica et Biophysica Acta, 1769*(5–6), 276–286. https://doi.org/10.1016/j.bbaexp.2007.01.009.

Gibbs, J. R., van der Brug, M. P., Hernandez, D. G., Traynor, B. J., Nalls, M. A., Lai, S. L., et al. (2010). Abundant quantitative trait loci exist for DNA methylation and gene expression in human brain. *Plos Genetics, 6*(5), e1000952. https://doi.org/10.1371/journal.pgen.1000952.

Gilly, A., Etcheverry, M., Madoui, M. A., Guy, J., Quadrana, L., Alberti, A., et al. (2014). TE-tracker: Systematic identification of transposition events through whole-genome resequencing. *BMC Bioinformatics, 15*, 377. https://doi.org/10.1186/s12859-014-0377-z.

Gouil, Q., & Baulcombe, D. C. (2016). DNA methylation signatures of the plant chromomethyltransferases. *Plos Genetics, 12*(12), e1006526. https://doi.org/10.1371/journal.pgen.1006526.

Greaves, I. K., Groszmann, M., Ying, H., Taylor, J. M., Peacock, W. J., & Dennis, E. S. (2012). Trans chromosomal methylation in Arabidopsis hybrids. *Proceedings of the National Academy of Sciences of the United States of America, 109*(9), 3570–3575. https://doi.org/10.1073/pnas.1201043109.

Griffin, P. T., Niederhuth, C. E., & Schmitz, R. J. (2016). A comparative analysis of 5-azacytidine- and zebularine-induced DNA demethylation. *G3 (Bethesda), 6*(9), 2773–2780. https://doi.org/10.1534/g3.116.030262.

Griffiths, J., Catoni, M., Iwasaki, M., & Paszkowski, J. (2018). Sequence-independent identification of active LTR retrotransposons in arabidopsis. *Molecular Plant, 11*(3), 508–511. https://doi.org/10.1016/j.molp.2017.10.012.

Groszmann, M., Greaves, I. K., Albertyn, Z. I., Scofield, G. N., Peacock, W. J., & Dennis, E. S. (2011). Changes in 24-nt siRNA levels in Arabidopsis hybrids suggest an epigenetic contribution to hybrid vigor. *Proceedings of the National Academy of Sciences of the United States of America, 108*(6), 2617–2622. https://doi.org/10.1073/pnas.1019217108.

Groszmann, M., Greaves, I. K., Fujimoto, R., Peacock, W. J., & Dennis, E. S. (2013). The role of epigenetics in hybrid vigour. *Trends in Genetics, 29*(12), 684–690. https://doi.org/10.1016/j.tig.2013.07.004.

Habu, Y., Mathieu, O., Tariq, M., Probst, A. V., Smathajitt, C., Zhu, T., et al. (2006). Epigenetic regulation of transcription in intermediate heterochromatin. *EMBO Reports, 7*(12), 1279–1284. https://doi.org/10.1038/sj.embor.7400835.

Heard, E., & Martienssen, R. A. (2014). Transgenerational epigenetic inheritance: Myths and mechanisms. *Cell, 157*(1), 95–109. https://doi.org/10.1016/j.cell.2014.02.045.

He, G., Chen, B., Wang, X., Li, X., Li, J., He, H., et al. (2013). Conservation and divergence of transcriptomic and epigenomic variation in maize hybrids. *Genome Biology, 14*(6), R57. https://doi.org/10.1186/gb-2013-14-6-r57.

He, X. J., Chen, T., & Zhu, J. K. (2011). Regulation and function of DNA methylation in plants and animals. *Cell Research, 21*(3), 442–465. https://doi.org/10.1038/cr.2011.23.

He, G., He, H., & Deng, X. W. (2013). Epigenetic variations in plant hybrids and their potential roles in heterosis. *J Genet Genomics, 40*(5), 205–210. https://doi.org/10.1016/j.jgg.2013.03.011.

Hellman, A., & Chess, A. (2010). Extensive sequence-influenced DNA methylation polymorphism in the human genome. *Epigenetics & Chromatin, 3*(1), 11. https://doi.org/10.1186/1756-8935-3-11.

Hofmeister, B. T., Lee, K., Rohr, N. A., Hall, D. W., & Schmitz, R. J. (2017). Stable inheritance of DNA methylation allows creation of epigenotype maps and the study of epiallele inheritance patterns in the absence of genetic variation. *Genome Biology, 18*(1), 155. https://doi.org/10.1186/s13059-017-1288-x.

Hollwey, E., Out, S., Watson, M. R., Heidmann, I., & Meyer, P. (2017). TET3-mediated demethylation in tomato activates expression of a CETS gene that stimulates vegetative growth. *Plant Direct, 1*(4), e00022. https://doi.org/10.1002/pld3.22.

Hollwey, E., Watson, M., & Meyer, P. (2016). Expression of the C-Terminal domain of mammalian *TET3* DNA dioxygenase in *Arabidopsis thaliana* induces heritable methylation changes at *rDNA* loci. *Advances in Bioscience and Biotechnology, 07*(05), 8. https://doi.org/10.4236/abb.2016.75023.

Hu, L., Li, N., Xu, C., Zhong, S., Lin, X., Yang, J., et al. (2014). Mutation of a major CG methylase in rice causes genome-wide hypomethylation, dysregulated genome expression, and seedling lethality. *Proceedings of the National Academy of Sciences of the United States of America, 111*(29), 10642–10647. https://doi.org/10.1073/pnas.1410761111.

Hurst, G. D., & Werren, J. H. (2001). The role of selfish genetic elements in eukaryotic evolution. *Nature Reviews. Genetics, 2*(8), 597–606. https://doi.org/10.1038/35084545.

Ito, H., Gaubert, H., Bucher, E., Mirouze, M., Vaillant, I., & Paszkowski, J. (2011). An siRNA pathway prevents transgenerational retrotransposition in plants subjected to stress. *Nature, 472*(7341), 115–119. https://doi.org/10.1038/nature09861.

Jacobsen, S. E., & Meyerowitz, E. M. (1997). Hypermethylated SUPERMAN epigenetic alleles in arabidopsis. *Science, 277*(5329), 1100–1103.

Jacobsen, S. E., Sakai, H., Finnegan, E. J., Cao, X., & Meyerowitz, E. M. (2000). Ectopic hypermethylation of flower-specific genes in Arabidopsis. *Current Biology: CB, 10*(4), 179–186.

Ji, L., Jordan, W. T., Shi, X., Hu, L., He, C., & Schmitz, R. J. (2018). TET-mediated epimutagenesis of the *Arabidopsis thaliana* methylome. *Nature Communications, 9*(1), 895. https://doi.org/10.1038/s41467-018-03289-7.

Johannes, F., Colot, V., & Jansen, R. C. (2008). Epigenome dynamics: A quantitative genetics perspective. *Nature Reviews. Genetics, 9*(11), 883–890. https://doi.org/10.1038/nrg2467.

Johannes, F., Porcher, E., Teixeira, F. K., Saliba-Colombani, V., Simon, M., Agier, N., et al. (2009). Assessing the impact of transgenerational epigenetic variation on complex traits. *Plos Genetics, 5*(6), e1000530. https://doi.org/10.1371/journal.pgen.1000530.

Kakutani, T. (1997). Genetic characterization of late-flowering traits induced by DNA hypomethylation mutation in *Arabidopsis thaliana. The Plant Journal: for Cell and Molecular Biology, 12*(6), 1447–1451.

Kakutani, T., Jeddeloh, J. A., Flowers, S. K., Munakata, K., & Richards, E. J. (1996). Developmental abnormalities and epimutations associated with DNA hypomethylation mutations. *Proceedings of the National Academy of Sciences of the United States of America, 93*(22), 12406–12411.

Kakutani, T., Jeddeloh, J. A., & Richards, E. J. (1995). Characterization of an *Arabidopsis thaliana* DNA hypomethylation mutant. *Nucleic Acids Research, 23*(1), 130–137.

Kakutani, T., Munakata, K., Richards, E. J., & Hirochika, H. (1999). Meiotically and mitotically stable inheritance of DNA hypomethylation induced by ddm1 mutation of *Arabidopsis thaliana. Genetics, 151*(2), 831–838.

Kankel, M. W., Ramsey, D. E., Stokes, T. L., Flowers, S. K., Haag, J. R., Jeddeloh, J. A., et al. (2003). Arabidopsis MET1 cytosine methyltransferase mutants. *Genetics, 163*(3), 1109–1122.

Kenan-Eichler, M., Leshkowitz, D., Tal, L., Noor, E., Melamed-Bessudo, C., Feldman, M., et al. (2011). Wheat hybridization and polyploidization results in deregulation of small RNAs. *Genetics, 188*(2), 263–272. https://doi.org/10.1534/genetics.111.128348.

Kumar, J., Gupta, D. S., Gupta, S., Dubey, S., Gupta, P., & Kumar, S. (2017). Quantitative trait loci from identification to exploitation for crop improvement. *Plant Cell Reports, 36*(8), 1187–1213. https://doi.org/10.1007/s00299-017-2127-y.

Lauss, K., Wardenaar, R., Oka, R., van Hulten, M. H. A., Guryev, V., Keurentjes, J. J. B., et al. (2018). Parental DNA methylation states are associated with heterosis in epigenetic hybrids. *Plant Physiology, 176*(2), 1627–1645. https://doi.org/10.1104/pp.17.01054.

Li, Q., Eichten, S. R., Hermanson, P. J., Zaunbrecher, V. M., Song, J., Wendt, J., et al. (2014). Genetic perturbation of the maize methylome. *The Plant Cell Online, 26*(12), 4602–4616. https://doi.org/10.1105/tpc.114.133140.

Lippman, Z., Gendrel, A. V., Black, M., Vaughn, M. W., Dedhia, N., McCombie, W. R., et al. (2004). Role of transposable elements in heterochromatin and epigenetic control. *Nature, 430*(6998), 471–476. https://doi.org/10.1038/nature02651.

Lisch, D. (2013). How important are transposons for plant evolution? *Nature Reviews. Genetics, 14*(1), 49–61. https://doi.org/10.1038/nrg3374.

Loudet, O., Gaudon, V., Trubuil, A., & Daniel-Vedele, F. (2005). Quantitative trait loci controlling root growth and architecture in *Arabidopsis thaliana* confirmed by heterogeneous inbred family. *TAG. Theoretical and Applied Genetics. Theoretische und angewandte Genetik, 110*(4), 742–753. https://doi.org/10.1007/s00122-004-1900-9.

Mackay, T. F. (2001). The genetic architecture of quantitative traits. *Annual Review of Genetics, 35*, 303–339. https://doi.org/10.1146/annurev.genet.35.102401.090633.

Malzahn, A., Lowder, L., & Qi, Y. (2017). Plant genome editing with TALEN and CRISPR. *Cell & Bioscience, 7*, 21. https://doi.org/10.1186/s13578-017-0148-4.

Manning, K., Tor, M., Poole, M., Hong, Y., Thompson, A. J., King, G. J., et al. (2006). A naturally occurring epigenetic mutation in a gene encoding an SBP-box transcription factor inhibits tomato fruit ripening. *Nature Genetics, 38*(8), 948–952. https://doi.org/10.1038/ng1841.

Mari-Ordonez, A., Marchais, A., Etcheverry, M., Martin, A., Colot, V., & Voinnet, O. (2013). Reconstructing de novo silencing of an active plant retrotransposon. *Nature Genetics, 45*(9), 1029–1039. https://doi.org/10.1038/ng.2703.

Martin, A., Troadec, C., Boualem, A., Rajab, M., Fernandez, R., Morin, H., et al. (2009). A transposon-induced epigenetic change leads to sex determination in melon. *Nature, 461*(7267), 1135–1138. https://doi.org/10.1038/nature08498.

Mathieu, O., Reinders, J., Caikovski, M., Smathajitt, C., & Paszkowski, J. (2007). Transgenerational stability of the Arabidopsis epigenome is coordinated by CG methylation. *Cell, 130*(5), 851–862. https://doi.org/10.1016/j.cell.2007.07.007.

Mc Clintock, B. (1950). The origin and behavior of mutable loci in maize. *Proceedings of the National Academy of Sciences of the United States of America, 36*(6), 344–355.

Melquist, S., Luff, B., & Bender, J. (1999). Arabidopsis PAI gene arrangements, cytosine methylation and expression. *Genetics, 153*(1), 401–413.

Mette, M. F., Aufsatz, W., van der Winden, J., Matzke, M. A., & Matzke, A. J. (2000). Transcriptional silencing and promoter methylation triggered by double-stranded RNA. *The EMBO Journal, 19*(19), 5194–5201. https://doi.org/10.1093/emboj/19.19.5194.

Meyer, R. C., Torjek, O., Becher, M., & Altmann, T. (2004). Heterosis of biomass production in Arabidopsis. Establishment during early development. *Plant Physiology, 134*(4), 1813–1823. https://doi.org/10.1104/pp.103.033001.

Miller, J. C., Tan, S., Qiao, G., Barlow, K. A., Wang, J., Xia, D. F., et al. (2011). A TALE nuclease architecture for efficient genome editing. *Nature Biotechnology, 29*(2), 143–148. https://doi.org/10.1038/nbt.1755.

Mirouze, M., Lieberman-Lazarovich, M., Aversano, R., Bucher, E., Nicolet, J., Reinders, J., et al. (2012). Loss of DNA methylation affects the recombination landscape in Arabidopsis. *Proceedings of the National Academy of Sciences of the United States of America, 109*(15), 5880–5885. https://doi.org/10.1073/pnas.1120841109.

Mirouze, M., Reinders, J., Bucher, E., Nishimura, T., Schneeberger, K., Ossowski, S., et al. (2009). Selective epigenetic control of retrotransposition in Arabidopsis. *Nature, 461*(7262), 427–430. https://doi.org/10.1038/nature08328.

Miura, K., Agetsuma, M., Kitano, H., Yoshimura, A., Matsuoka, M., Jacobsen, S. E., et al. (2009). A metastable DWARF1 epigenetic mutant affecting plant stature in rice. *Proceedings of the National Academy of Sciences of the United States of America, 106*(27), 11218–11223. https://doi.org/10.1073/pnas.0901942106.

Miura, A., Yonebayashi, S., Watanabe, K., Toyama, T., Shimada, H., & Kakutani, T. (2001). Mobilization of transposons by a mutation abolishing full DNA methylation in Arabidopsis. *Nature, 411*(6834), 212–214. https://doi.org/10.1038/35075612.

Pecinka, A., & Liu, C. H. (2014). Drugs for plant chromosome and chromatin research. *Cytogenetic and Genome Research, 143*(1–3), 51–59. https://doi.org/10.1159/000360774.

Rebollo, R., Romanish, M. T., & Mager, D. L. (2012). Transposable elements: An abundant and natural source of regulatory sequences for host genes. *Annual Review of Genetics, 46*, 21–42. https://doi.org/10.1146/annurev-genet-110711-155621.

Reinders, J., Wulff, B. B., Mirouze, M., Mari-Ordonez, A., Dapp, M., Rozhon, W., et al. (2009). Compromised stability of DNA methylation and transposon immobilization in mosaic Arabidopsis epigenomes. *Genes & Development, 23*(8), 939–950. https://doi.org/10.1101/gad.524609.

Richards, E. J. (2006). Inherited epigenetic variation–revisiting soft inheritance. *Nature Reviews. Genetics*, 7(5), 395–401. https://doi.org/10.1038/nrg1834.

Rigal, M., Becker, C., Pelissier, T., Pogorelcnik, R., Devos, J., Ikeda, Y., et al. (2016). Epigenome confrontation triggers immediate reprogramming of DNA methylation and transposon silencing in *Arabidopsis thaliana* F1 epihybrids. *Proceedings of the National Academy of Sciences of the United States of America, 113*(14), E2083–E2092. https://doi.org/10.1073/pnas.1600672113.

Ronemus, M. J., Galbiati, M., Ticknor, C., Chen, J., & Dellaporta, S. L. (1996). Demethylation-induced developmental pleiotropy in Arabidopsis. *Science, 273*(5275), 654–657.

Roux, F., Colome-Tatche, M., Edelist, C., Wardenaar, R., Guerche, P., Hospital, F., et al. (2011). Genome-wide epigenetic perturbation jump-starts patterns of heritable variation found in nature. *Genetics, 188*(4), 1015–1017. https://doi.org/10.1534/genetics.111.128744.

Sano, H., Kamada, I., Youssefian, S., Katsumi, M., & Wabiko, H. (1990). A single treatment of rice seedlings with 5-azacytidine induces heritable dwarfism and undermethylation of genomic DNA. *Molecular and General Genetics MGG, 220*(3), 441–447. https://doi.org/10.1007/bf00391751.

Santos, A. P., Abranches, R., Stoger, E., Beven, A., Viegas, W., & Shaw, P. J. (2002). The architecture of interphase chromosomes and gene positioning are altered by changes in DNA methylation and histone acetylation. *Journal of Cell Science, 115*(Pt 23), 4597–4605.

Saze, H., & Kakutani, T. (2007). Heritable epigenetic mutation of a transposon-flanked Arabidopsis gene due to lack of the chromatin-remodeling factor DDM1. *The EMBO Journal, 26*(15), 3641–3652. https://doi.org/10.1038/sj.emboj.7601788.

Saze, H., Mittelsten Scheid, O., & Paszkowski, J. (2003). Maintenance of CpG methylation is essential for epigenetic inheritance during plant gametogenesis. *Nature Genetics, 34*(1), 65–69. https://doi.org/10.1038/ng1138.

Schneeberger, K., Ossowski, S., Ott, F., Klein, J. D., Wang, X., Lanz, C., et al. (2011). Reference-guided assembly of four diverse *Arabidopsis thaliana* genomes. *Proceedings of the National Academy of Sciences of the United States of America, 108*(25), 10249–10254. https://doi.org/10.1073/pnas.1107739108.

Seymour, D. K., & Becker, C. (2017). The causes and consequences of DNA methylome variation in plants. *Current Opinion in Plant Biology, 36*, 56–63. https://doi.org/10.1016/j.pbi.2017.01.005.

Shen, H., He, H., Li, J., Chen, W., Wang, X., Guo, L., et al. (2012). Genome-wide analysis of DNA methylation and gene expression changes in two Arabidopsis ecotypes and their reciprocal hybrids. *The Plant Cell Online, 24*(3), 875–892. https://doi.org/10.1105/tpc.111.094870.

Shivaprasad, P. V., Dunn, R. M., Santos, B. A., Bassett, A., & Baulcombe, D. C. (2012). Extraordinary transgressive phenotypes of hybrid tomato are influenced by epigenetics and small silencing RNAs. *The EMBO Journal, 31*(2), 257–266. https://doi.org/10.1038/emboj.2011.458.

Sidorenko, L. V., Lee, T. F., Woosley, A., Moskal, W. A., Bevan, S. A., Merlo, P. A. O., et al. (2017). GC-rich coding sequences reduce transposon-like, small RNA-mediated transgene silencing. *Nature Plants, 3*(11), 875–884. https://doi.org/10.1038/s41477-017-0040-6.

Silveira, A. B., Trontin, C., Cortijo, S., Barau, J., Del Bem, L. E., Loudet, O., et al. (2013). Extensive natural epigenetic variation at a de novo originated gene. *Plos Genetics, 9*(4), e1003437. https://doi.org/10.1371/journal.pgen.1003437.

Soppe, W. J., Jacobsen, S. E., Alonso-Blanco, C., Jackson, J. P., Kakutani, T., Koornneef, M., et al. (2000). The late flowering phenotype of fwa mutants is caused by gain-of-function epigenetic alleles of a homeodomain gene. *Molecular Cell, 6*(4), 791–802.

Stroud, H., Greenberg, M. V., Feng, S., Bernatavichute, Y. V., & Jacobsen, S. E. (2013). Comprehensive analysis of silencing mutants reveals complex regulation of the Arabidopsis methylome. *Cell, 152*(1–2), 352–364. https://doi.org/10.1016/j.cell.2012.10.054.

Tahiliani, M., Koh, K. P., Shen, Y., Pastor, W. A., Bandukwala, H., Brudno, Y., et al. (2009). Conversion of 5-methylcytosine to 5-hydroxymethylcytosine in mammalian DNA by MLL partner TET1. *Science, 324*(5929), 930–935. https://doi.org/10.1126/science.1170116.

Teixeira, F. K., Heredia, F., Sarazin, A., Roudier, F., Boccara, M., Ciaudo, C., et al. (2009). A role for RNAi in the selective correction of DNA methylation defects. *Science, 323*(5921), 1600–1604. https://doi.org/10.1126/science.1165313.

Thieme, M., Lanciano, S., Balzergue, S., Daccord, N., Mirouze, M., & Bucher, E. (2017). Inhibition of RNA polymerase II allows controlled mobilisation of retrotransposons for plant breeding. *Genome Biology, 18*(1), 134. https://doi.org/10.1186/s13059-017-1265-4.

Tsukahara, S., Kobayashi, A., Kawabe, A., Mathieu, O., Miura, A., & Kakutani, T. (2009). Bursts of retrotransposition reproduced in Arabidopsis. *Nature, 461*(7262), 423–426. https://doi.org/10.1038/nature08351.

Vaughn, M. W., Tanurdzic, M., Lippman, Z., Jiang, H., Carrasquillo, R., Rabinowicz, P. D., et al. (2007). Epigenetic natural variation in *Arabidopsis thaliana. Plos Biology, 5*(7), e174. https://doi.org/10.1371/journal.pbio.0050174.

Vojta, A., Dobrinic, P., Tadic, V., Bockor, L., Korac, P., Julg, B., et al. (2016). Repurposing the CRISPR-Cas9 system for targeted DNA methylation. *Nucleic Acids Research, 44*(12), 5615–5628. https://doi.org/10.1093/nar/gkw159.

Vongs, A., Kakutani, T., Martienssen, R. A., & Richards, E. J. (1993). *Arabidopsis thaliana* DNA methylation mutants. *Science, 260*(5116), 1926–1928.

Wicker, T., Sabot, F., Hua-Van, A., Bennetzen, J. L., Capy, P., Chalhoub, B., et al. (2007). A unified classification system for eukaryotic transposable elements. *Nature Reviews. Genetics, 8*(12), 973–982. https://doi.org/10.1038/nrg2165.

Xia, J., Han, L., & Zhao, Z. (2012). Investigating the relationship of DNA methylation with mutation rate and allele frequency in the human genome. *BMC Genomics, 13*(Suppl 8), S7. https://doi.org/10.1186/1471-2164-13-s8-s7.

Yelina, N. E., Lambing, C., Hardcastle, T. J., Zhao, X., Santos, B., & Henderson, I. R. (2015). DNA methylation epigenetically silences crossover hot spots and controls chromosomal domains of meiotic recombination in Arabidopsis. *Genes & Development, 29*(20), 2183–2202. https://doi.org/10.1101/gad.270876.115.

Zemach, A., Kim, M. Y., Hsieh, P. H., Coleman-Derr, D., Eshed-Williams, L., Thao, K., et al. (2013). The Arabidopsis nucleosome remodeler DDM1 allows DNA methyltransferases to access H1-containing heterochromatin. *Cell, 153*, 193–205.

Zhang, D., Cheng, L., Badner, J. A., Chen, C., Chen, Q., Luo, W., et al. (2010). Genetic control of individual differences in gene-specific methylation in human brain. *The American Journal of Human Genetics, 86*(3), 411–419. https://doi.org/10.1016/j.ajhg.2010.02.005.

Zhang, L., Cheng, Z., Qin, R., Qiu, Y., Wang, J. L., Cui, X., et al. (2012). Identification and characterization of an epi-allele of FIE1 reveals a regulatory linkage between two epigenetic marks in rice. *The Plant Cell Online, 24*(11), 4407–4421. https://doi.org/10.1105/tpc.112.102269.

Zhang, Y. Y., Fischer, M., Colot, V., & Bossdorf, O. (2013). Epigenetic variation creates potential for evolution of plant phenotypic plasticity. *The New Phytologist, 197*(1), 314–322. https://doi.org/10.1111/nph.12010.

Zhang, X., Shiu, S. H., Cal, A., & Borevitz, J. O. (2008). Global analysis of genetic, epigenetic and transcriptional polymorphisms in *Arabidopsis thaliana* using whole genome tiling arrays. *Plos Genetics, 4*(3), e1000032. https://doi.org/10.1371/journal.pgen.1000032.

Zhang, X., Sun, J., Cao, X., & Song, X. (2015). Epigenetic mutation of RAV6 affects leaf angle and seed size in rice. *Plant Physiology, 169*(3), 2118–2128. https://doi.org/10.1104/pp.15.00836.

Zhu, J., Fang, L., Yu, J., Zhao, Y., Chen, F., & Xia, G. (2018). 5-Azacytidine treatment and TaPBF-D over-expression increases glutenin accumulation within the wheat grain by hypomethylating the Glu-1 promoters. *TAG. Theoretical and Applied Genetics. Theoretische und angewandte Genetik, 131*(3), 735–746. https://doi.org/10.1007/s00122-017-3032-z.

CHAPTER FIVE

Sexual and Non-sexual Reproduction: Inheritance and Stability of Epigenetic Variations and Consequences for Breeding Application

Yu-Hung Hung[*,§], **Fang Liu**[§,¶], **Xiang-Qian Zhang**[§,||], **Wenyan Xiao**[#] **and Tzung-Fu Hsieh**[*,§,1]

[*]Department of Plant and Microbial Biology, North Carolina State University, Raleigh, NC, United States
[§]Plants for Human Health Institute, North Carolina State University, North Carolina Research Campus, Kannapolis, NC, United States
[¶]State Key Laboratory for Conservation and Utilization of Subtropical Agro-bioresources and College of Agriculture, Guangxi University, Nanning, China
[||]Guangdong Engineering Research Center of Grassland Science, College of Forestry and Landscape Architecture, South China Agricultural University, Guangzhou, China
[#]Department of Biology, St. Louis University, St. Louis, MO, United States
[1]Corresponding author: e-mail address: thsieh3@ncsu.edu

Contents

Advances in Botanical Research, Volume 88
ISSN 0065-2296
https://doi.org/10.1016/bs.abr.2018.09.002

Abstract

Although plant breeding has traditionally relied on exploiting genetic diversity, epigenetic variations are attracting new attentions in breeding design. Epigenetic modifications play a critical role in regulating gene expression during development and in response to environmental stimulation, and thereby contribute to phenotypic variation. When epigenetic modifications persist during reproduction and are stably transmitted to the next generation, transgenerational inheritance of epigenetic variations has a potential to produce heritable phenotypic diversity. In addition to naturally occurring epialleles, genetic variation due to organization of duplicated genes or transposon polymorphisms are a potential source for inducing epialleles. Furthermore, stress-induced activation and mobilization of transposable elements represent another promising avenue to create new genetic and epigenetic diversity that can be sexually or asexually propagated. Finally, methods for artificially creating epigenetic diversity in experimental models have been developed and applied to select crop species in some cases. With the advance in epigenome profiling techniques, dissection of epigenetic-based complex traits, and the development of molecular tools for locus-specific epigenome editing, the effects of epigenetics for underlying phenotypic traits can be unequivocally elucidated.

1. INTRODUCTION

Agriculture is the cornerstone of the whole human society. We depend on the cultivation and breeding of plants that provide food, animal feed, fibre, medicine, fuel, and other resources to sustain and improve our lives. With decreasing arable land, an exploding population and global climate change, agriculture is facing unprecedented challenges in the near future. Feeding the world in this millennium will require a step change in the efficiency of crop production, and this can only come from a deeper understanding of plant reproductive development and a more powerful strategy for crop improvement.

Human have been practicing plant breeding since the dawn of agriculture. Plant scientists and breeders have been trying to develop novel research, new technologies and modern breeding methods to enhance agronomic traits such as increased stress tolerance, bigger seed size, increased yield, superior nutritional quality. Variation in phenotypic traits is determined by genetic and epigenetic factors. In the past decades, technological advances have provided great opportunities for scientists to study epigenetic variations in plants leading to significant progress in understanding epigenetic mechanisms, variations, inheritance and stability in plants. This information opens a new avenue in advancing modern breeding. In this chapter, we provide readers a review of current knowledge of epigenetic variation and its promising application for crop improvement.

2. GENETIC DIVERSITY AND EPIGENETIC VARIATION

For thousands of years, people have been using, developing and relying on plant breeding for agriculture production to feed and improve the human society. Essentially, traditional plant breeding is the art and science of genetic improvement of agricultural plant species. Traditional plant breeding relies on exploring the genetic diversity presents within a crop species to transfer desirable traits into preferred cultivars. Thus, it is generally believed that phenotypic diversity is primarily driven by genetic variation. With the advance of molecular techniques, genomic and marker-assisted selection strategies have significantly improved the speed and precision of selecting multiple desirable traits simultaneously by monitoring molecular markers. However, in the past decade, there is an ever-increasing appreciation that epigenetic diversity can provide a new source of phenotypic variation. Epigenetics refers to the study of heritable change in gene activity that is not due to changes in DNA sequence. Consequently, changes in epigenetics can cause phenotypic variations if the underlying gene is directly or indirectly involved in a phenotypic trait.

Epigenetic changes can result from an array of chemical modifications to the DNA or chromatin proteins, and when occuring near a coding gene, can affect chromatin structure, accessibility to transcriptional machinery, and expression states. Although many epigenetic marks have been associated with transcriptionally active or repressed genes, they often reflect dynamics in transcriptional programs due to development switch or in response to environmental challenge. By contrast, heritable epigenetic marks can persist through mitosis or meiosis and can be stably transmitted to the next generation, resulting in transgenerational epigenetic inheritance.

3. MOLECULAR MECHANISMS OF EPIGENETICS IN PLANT

Traditional plant breeding strategies of hybridization, backcrossing and selection have successfully generated crops with desirable traits associated with yield, stress tolerance, disease resistance and nutritional quality by harnessing the desirable genetic variations. However, genetic variation is not the only factor that contributes to phenotypic variations among organisms. In addition to genetic variation, epigenetic variations underpin the phenotypic plasticity of an organism. Epigenetic marks, such as DNA methylation and histone modification, can alter the chromatin structure and DNA accessibility thereby regulating gene expression and resulting in phenotypic variation. In this section, we briefly summarize the molecular mechanisms of epigenetic regulations in flowering plants.

3.1 Histone Modification

Histone modification is one of the epigenetic regulations that greatly affect chromatin structure. There are several kinds of reversible chemical modification on histone tails, including acetylation, methylation, deacetylation, and phosphorylation. These modifications can recruit different chromatin reader complexes to change chromatin structure and affect gene expression. Histone H3 lysine methylation is the most well-studied histone modification in plants. Several lysine residues on histone H3 tail can be methylated. These methylations can be associated with either active or repressed gene expression states, depending on the position and the number of methyl groups added to the lysine residue. For example, trimethylation of histone H3 lysine 4 and lysine 36 are generally linked to active gene expression, while dimethylation of lysine 9 (H3K9me2) and trimethylation of lysine 27 (H3K27me3) are associated with gene repression (Bannister & Kouzarides, 2011; Berger, 2007; Margueron & Reinberg, 2011). The H3K27me2/3 is deposited by an evolutionarily conserved Polycomb group (PcG) protein complex called POLYCOMB REPRESSIVE COMPLEX 2 (PRC2) (Margueron & Reinberg, 2011). Plants need to undergo several developmental transitions during their life cycles. PcGs are chromatin modifiers that control many of these important developmental processes by regulating gene expression (Bemer & Grossniklaus, 2012). In Arabidopsis, the

FIS-MEA-PRC2 has been shown to be critical during early seed development. FIS-MEA-PRC2 repress central cell proliferation prior to fertilization. After fertilization, FIS-MEA-PRC2 regulates the transition from syncytial development to cellularization. Seeds deficient in FIS-MEA-PRC2 function fail to undergo proper endosperm cellularization, leading to embryo arrest and seed abortion (Chaudhury et al., 1997, 2001; Hehenberger, Kradolfer, & Kohler, 2012; Ohad et al., 1996).

3.2 DNA Methylation

DNA methylation is another important epigenetic mechanism that regulates gene expression. The majority of DNA methylation at cytosine (5-methylcytosine) is found in repetitive sequences and transposable elements (TEs), which are enriched in heterochromatic regions including centromeres, pericentromeres, and telomeres. DNA methylation is crucial for the repression of TE expression and transposition, and is essential for genome stability. In plants, DNA methylation is found in three sequence contexts: CG, CHG, and CHH (where H represents A, C or T). The maintenance of DNA methylation is necessary to preserve DNA methylation after every cellular DNA replication cycle. Without the maintenance DNA methyltransferases, the replication machinery would produce unmethylated daughter strands and, over time, lead to passive demethylation. The symmetry of CG and CHG sequences allows the methyltransferases to copy the methylation pattern from the template strand to the new unmethylated daughter strand. The maintenance of CG methylation is catalysed by METHYLTRANSFERASE 1 (MET1). The CHG methylation is maintained by the CHROMOMETHYLASE 3 (CMT3) through a reinforcing loop between CMT3 and dimethylation of histone H3 lysine 9 (H3K9m2) (Law & Jacobsen, 2010; Lindroth et al., 2001). Due to its asymmetric nature, methylation of CHH sites must be constantly established *de novo* after each round of DNA replication. The CHH methylation is catalysed by DOMAINS REARRANGED METHYLTRANSFERASE 2 (DRM2) through plant-specific RNA-directed DNA methylation (RdDM) pathway (Henderson & Jacobsen, 2007). In addition to DRM2, CHH methylation is also maintained by a homolog of CMT3, the CHROMOMETHYLASE 2 (CMT2), independent of RdDM. DRM2-mediated CHH methylation mainly takes place in more euchromatic regions, including short TEs and the edge of long TEs in pericentromeric regions. On the other hand, CHH methylation catalysed by CMT2 is found in more heterochromatic regions, such as the bodies of long TEs. CMT2-mediated CHH methylation is facilitated by a nucleosome remodeler DECREASE IN DNA METHYLATION 1 (DDM1) which remodels heterochromatic nucleosomes to allow CMT2 access to the heterochromatic TEs (Law & Jacobsen, 2010; Zemach et al., 2013).

Loss of DNA methylation can occur either passively or actively. Passive DNA demethylation occurs when there is a loss of DNA methyltransferase activity, which leads to failure in the maintenance of DNA methylation. Active DNA demethylation requires an enzyme that recognizes and enzymatically remove methylated cytosines. In Arabidopsis, active DNA demethylation is accomplished by a family of novel DNA glycosylases: REPRESSOR OF SILENCING 1 (ROS1), DEMETER (DME), DME-LIKE 2 (DML2), DML3. These glycosylases work with base excision repair pathway to remove 5-methylcytosines and replace it with unmethylated cytosines (Agius, Kapoor, & Zhu, 2006; Gehring et al., 2006). In Arabidopsis, ROS1 is the main DNA demethylase in vegetative tissues. During plant development, DNA methylation pattern is a dynamic balance between methylation and demethylation. RdDM preferentially targets small TEs and repetitive sequences to silence and prevent their transposition. However, spreading of methylation to nearby genes might alter their expression and create undesirable outcomes. ROS1 counteracts DNA methylation established by RdDM pathway and many genomic loci are regulated by both ROS1 and RdDM (Tang, Lang, Zhang, & Zhu, 2016). Genetic screens identified many RdDM components as *ros1* repressors, supporting the antagonistic relationship between ROS1 and RdDM (Matzke & Mosher, 2014; Zhang & Zhu, 2012). Intriguing evidence shows that a methylation monitoring sequence in the ROS1 promoter region can adjust the transcriptional output of ROS1 by sensing the strength of RdDM (Lei et al., 2015; Williams, Pignatta, Henikoff, & Gehring, 2015). These studies highlight the delicate balance between DNA methylation and demethylation to maintain the DNA methylation homeostasis during plant development (Hsieh, 2016; Tang et al., 2016). Unlike the other members of this DNA glycosylase/lyase family that are expressed in vegetative tissues, DME is primarily expressed in reproductive organs, particularly the central cell in ovule and the vegetative cell in pollen. In ovule, the expression of DME is detected in diploid polar nuclei in the central cell, and quickly disappears after fertilization (Choi et al., 2002). DME-mediated DNA demethylation in the central cell regulates gene imprinting and is critical for seed viability (Choi et al., 2002; Gehring et al., 2006). RNA Polymerase IV and V (Pol IV and Pol V) are plant-specific homologs of RNA Polymerase II with specialized function in small RNA production that is required for

RdDM (Huang et al., 2009; Lahmy et al., 2009; Ream et al., 2009). Pol IV is responsible for the biogenesis of most 24-nucleotide siRNAs, which guide methylation in the canonical RdDM pathway (Ream et al., 2009). Pol IV transcribes single-stranded RNAs (ssRNAs) at its target loci, which are primarily transposons and other repeat sequences. The ssRNA is copied into double-stranded RNA (dsRNA) by RNA-DEPENDENT RNA POLYMERASE (RDR2). DICER-LIKE 3 (DCL3) processes dsRNA to 24-nt small interfering RNA (siRNA). The 24-nt siRNAs are then loaded onto ARGONAUTE 4 (AGO4). AGO4-bound siRNAs are believed to base-pair with Pol V transcripts and recruit DRM2, a DNA methyltransferase, to catalyse *de novo* DNA methylation at both CG and non-CG contexts at the homologous genomic sites (Matzke & Mosher, 2014).

4. SEXUAL REPRODUCTION IN FLOWERING PLANT

The angiosperms (flowering plants) are the most diverse group of land plants, with over 250,000 species (Christenhusz & Byng, 2016). Double fertilization is a unique feature distinguishing angiosperms from another clade of seed-plant gymnosperms. This feature is believed to have provided angiosperms an evolutionary advantage to become the dominant form of land plant, although the evolutionary advantage of double fertilization is not fully understood.

Unlike animals that define all cell lineages during embryogenesis, plants are able to establish novel cell fates through a small population of undifferentiated stem cells, producing vegetative tissues and organs during vegetative growth, and reproductive organs during sexual reproduction (Berger & Twell, 2011). In flowering plants, sexual reproduction can be divided into three phases: sporogenesis, gametogenesis and seed development. Sporogenesis is initiated in both male and female reproductive tissues. In the stamen primordium, one sub-epidermal nucellar cell gives rise to a microspore or pollen mother cell (PMC). The PMC then undergoes meiosis to form four haploid microspores before proceeding to gametogenesis (Fig. 1) (Berger & Twell, 2011). Female gametophyte development in Arabidopsis follows the *polygonum*-type pattern, which is the most prevalent pattern among flowering plants (Drews & Koltunow, 2011). In the Arabidopsis ovule primordium, a sub-epidermal nucellus cell differentiates into a megaspore mother cell (MMC), the MMC then undergoes meiosis to form four haploid megaspores. Only one megaspore survives to become functional

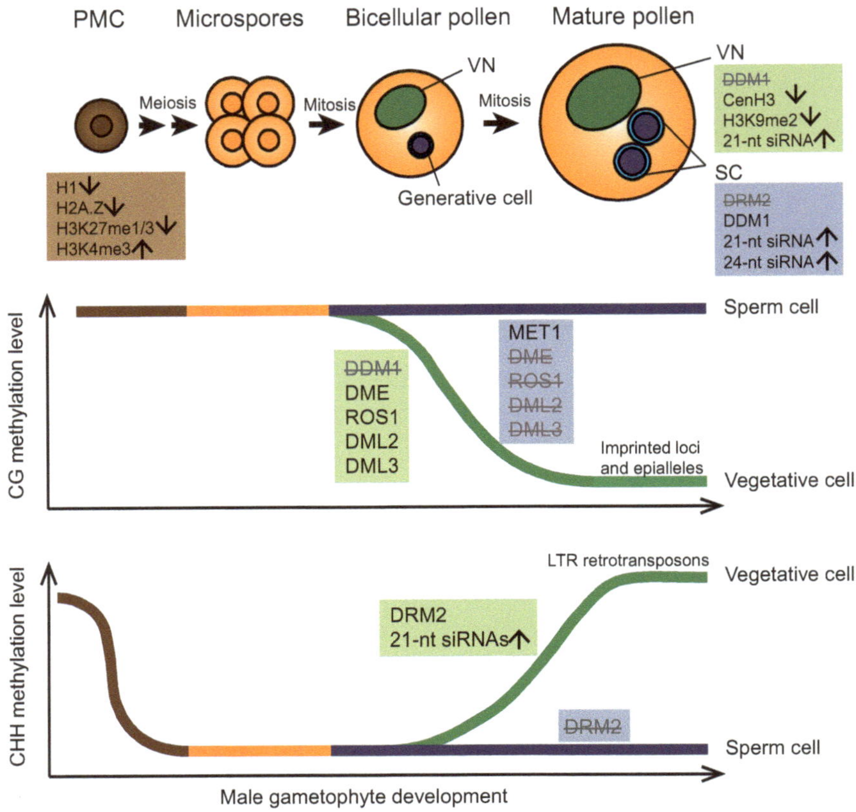

Figure 1 ***Chromatin reprogramming during male gametogenesis.*** During PMC specification, large-scale chromatin reorganization occurs including reduction of linker histone H1 and H2A.Z, reduction of repressive chromatin marks (H3K27me1 and H3K27me3), and upregulation of permissive chromatin marks (H3K4me3). After two rounds of meiosis, four haploid microspores are produced. Each microspore undergoes an asymmetrical cell division to generate a bicellular pollen consisting of a vegetative cell encasing a generative cell. The generative cell undergoes one more mitosis to create two sperm cells. The VN lose CG methylation due to activity of DME-family DNA demethylase. The 24-nt siRNAs are produced from DME target sites in VN, and are required for *de novo* CHH methylation in sperm cell. In the absence of DME and presence of MET1, sperm cells retain CG methylation. The CHH methylation is lost from heterochromatin in microspores and sperm cells, but is restored in the VN. The vegetative cell nuclei (VN) undergo loss of CenH3 and chromatin decondensation. Downregulation of DDM1 in VN leads to de-repression of TEs. Activation of TEs results in production of 21-nt siRNAs that accumulate in both VN and sperm cells. PMC, pollen mother cell; VN, vegetative cell nuclei; SC, sperm cell; DDM1, DECREASED DNA METHYLATION 1; CenH3, centromere-specific histone H3 variant; DRM2, DOMAINS REARRANGED METHYLTRANSFERASE 2; MET1, DNA METHYLTRANSFERASE 1; DME, DEMETER; ROS1, REPRESSOR OF SILENCING1; DML2, DEMETER-LIKE 2; DML3, DEMETER-LIKE 3.

megaspore (FM) and proceeds to gametogenesis, while the other three degenerate (Fig. 2).

During gametogenesis, the haploid FM in Arabidopsis is subjected to three rounds of mitosis to produce a syncytial female gametophyte (embryo sac) with eight nuclei. Following nuclei migration and cellularization, the seven cells are differentiated in the mature female gametophyte. The haploid egg cell and two synergid cells (called the egg apparatus) are located at the micropylar pole. The homodiploid central cell containing the two fused nuclei can be found in the central part of the embryo sac. The three antipodal cells are located at the chalazal pole opposite to the egg apparatus (Fig. 2) (Drews & Koltunow, 2011). By contrast, each microspore undergoes an asymmetrical cell division to generate a larger vegetative cell and a smaller generative cell. The generative cell is encased by the vegetative cell, and then undergoes once more mitosis to create two sperm cells. As a result, the mature pollen is a tricellular male gametophyte with a unique cell within cell structure (Fig. 1) (Berger & Twell, 2011). Seed development in angiosperm is triggered by the process of double fertilization that leads to differentiation and development of three genetically distinct components: embryo, endosperm and seed coat. The sperm cells are immotile. The major task of the vegetative cell is to deliver the two sperm cells to the female gametophyte. Upon landing on the tip of a stigma, the pollen grain rehydrates and receives signals from female sporophytic tissue that stimulate pollen germination and pollen tube growth (Higashiyama & Takeuchi, 2015). The vegetative cell elongates to form the pollen tube and acts as a vehicle to transport the two sperm cells to the mature female gametophyte. After arriving at the boundary region between egg cell and central cell, the sperm cell is activated by female gamete-secreted protein, and fuses with the female gamete (Sprunck et al., 2012). One sperm cell fertilizes the egg cell to become diploid zygote which initiates a new sporophytic generation with the development of the embryo. The other sperm cell fuses with the homodiploid central cell to generate a triploid embryo-nourishing tissue, the endosperm (Bleckmann, Alter, & Dresselhaus, 2014).

5. CHROMATIN REPROGRAMMING AND DYNAMIC DNA METHYLATION DURING MALE GAMETOGENESIS

Morphological and cytological evidences observed in many plant species (Armstrong & Jones, 2003) and in Arabidopsis (She & Baroux, 2015) during pollen mother cell specification and differentiation have suggested

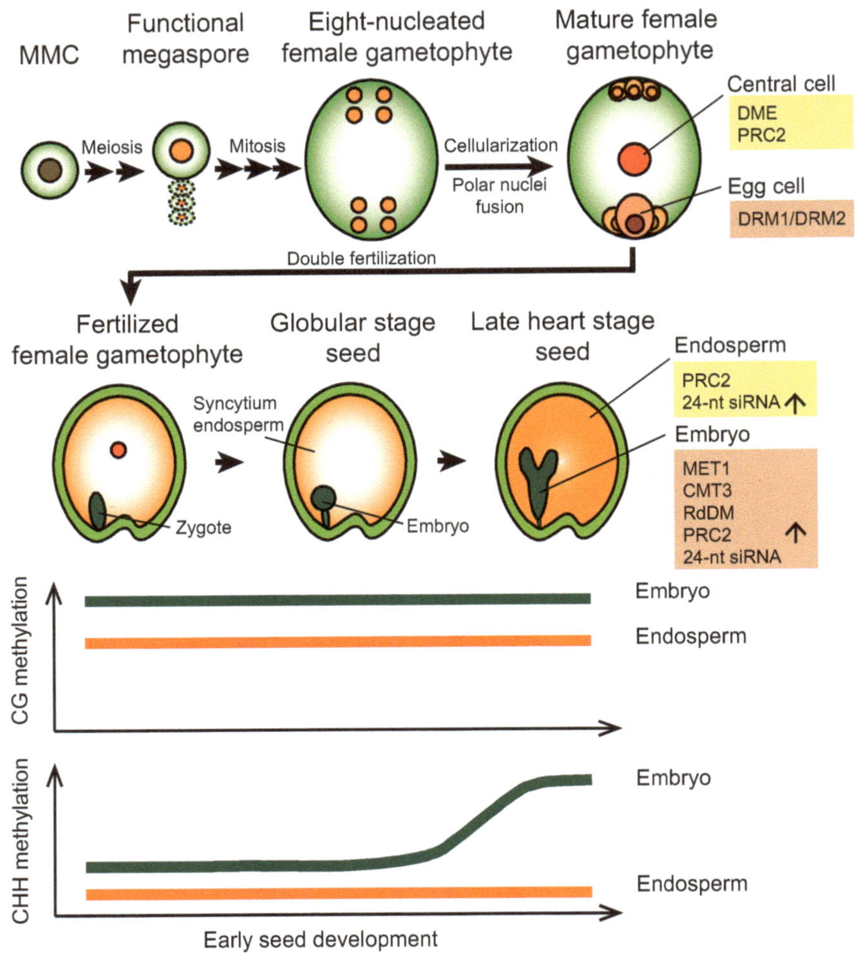

Figure 2 ***Epigenetic reprogramming during female gametogenesis and early seed development.*** The Arabidopsis megaspore mother cell (MMC) undergoes two rounds of meiosis to form four haploid megaspores. Three of the megaspores degenerates, and only one functional megaspore (FM) proceeds to three rounds of mitosis and becomes eight-nucleated female gametophyte. After cellularization and polar nuclei fusion, the mature female gametophyte contains an egg cell, an homodiploid central cell, two synergid cells, and three antipodal cells. Demethylation by DME in central cell establishes gene imprinting, promotes PRC2 complex, and likely induces 24-nt siRNAs expression for TE silencing in gamete and zygote. Upon double fertilization, one sperm cell fertilizes egg cell and the other sperm cell fuses with central cell, giving rise to diploid embryo and triploid endosperm. The DME-mediated DNA demethylation in central cell, together with passive DNA demethylation result in genome-wide CG hypomethylation in Arabidopsis endosperm. The CHH methylation remains low in endosperm, while CHH methylation is gradually restored during embryo development. CMT3, CHROMOMETHYLASE 3; DRM1/2, DOMAINS REARRANGED METHYLTRANSFERASE 1/2; MET1, DNA METHYLTRANSFERASE 1; DME, DEMETER; PRC2, POLYCOMB REPRESSIVE COMPLEX 2; RdDM, RNA directed DNA methylation.

presence of large-scale chromatin reorganization, including cellular and nuclear enlargement, decondensation of chromatin, reduction of histone H1 and H2A.Z expression in the PMC nucleus. This is accompanied by a concomitant decrease in H3K27me1 and H3K27me3 signal intensities (both mark repressive chromatins), and an increased in H3K4me3 permissive chromatin signal in the PMC (She & Baroux, 2015), and is thought to be a rewiring of transcriptional program to facilitate meiosis progression (Chen et al., 2010; Yang, Lu, Wang, & Ma, 2011; Zhou & Pawlowski, 2014).

After meiosis, the haploid microspore undergoes an asymmetric mitotic division to differentiate a larger vegetative cell and a smaller generative cell, the male germline. The vegetative cell nuclei undergo an extensive loss of centromere-specific histone H3 variant (CenH3), decondensation of the centromeric heterochromatin, and a prominent activation of transposable elements (Probst, Fransz, Paszkowski, & Scheid, 2003; Schoft et al., 2009; Slotkin et al., 2009). Consequently, activation of TEs in the vegetative cell leads to the production of a class of 21-nt siRNAs called epigenetically activated small RNAs, or easiRNAs, that accumulate in the sperm cells (Slotkin et al., 2009).

Cell-type specific whole genome methylation profiling of flow cytometer-sorted male gametes revealed that microspores, sperm cells, and vegetative cells maintain high CG and CHG methylations throughout development whereas a substantial CHH methylation is lost from the pericentromeric heterochromatin in microspores and sperm cells, but the CHH methylation is restored in the vegetative cells (Calarco et al., 2012) (Fig. 1). In addition to high CHH methylation, local CG hypomethylation is obvious in the vegetative cell nuclei (VN) due to active DNA demethylation by the DME/ROS1 family of DNA demethylases (Calarco et al., 2012; Ibarra et al., 2012). By comparing CG methylation between wildtype and *dme/+*, CG sites demethylated by DME in vegetative nuclei show preferential CHH hypomethylation in *dme+/*-sperm, indicating that DME activity in the vegetative cell is required for normal methylation of a subset of sperm TEs, possibly to reinforce DNA methylation in the sperms and subsequently in the zygotes (Ibarra et al., 2012). Upon fertilization, the CHH-hypomethylated paternal genomes are quickly remethylated in the embryo and the endosperm as both tissues have relatively high levels of CHH methylation (Hsieh et al., 2009; Ibarra et al., 2012). It is speculated that this remethylation is likely guided by the maternal 24-nt siRNAs as the previous report showed that most of the 24-nt siRNAs in early developing seeds are maternally derived (Calarco et al., 2012; Lu, Zhang, Baulcombe, &

Chen, 2012; Mosher & Melnyk, 2010). Thus, one purpose of stripping the CHH methylation of the paternal genomes might be to allow the maternal plants to gain control to distinguish self from non-self pollen (Creasey et al., 2014), and might have functional implications in cross-species hybridization or hybrid vigour.

6. CHROMATIN REPROGRAMMING DURING ARABIDOPSIS FEMALE GAMETOGENESIS AND ACTIVE DNA DEMETHYLATION IN THE CENTRAL CELL

Cell–cell communication through a specialized siRNA pathway has been shown to play an important role during female gametogenesis. Mutations in *AGO9*, *RDR6*, and *SUPPRESSOR OF GENE SILENCING 3* (*SDG3*) required for the biogenesis of trans-acting mobile siRNAs (ta-siRNAs) lead to supernumerary functional megaspore formation in the ovule, indicating that megaspore mother cell specification and differentiation in Arabidopsis is regulated by an AGO9-RDR6-SDG3 dependent siRNA silencing pathway (Olmedo-Monfil et al., 2010). Similarly, mutations in the maize AGO104 that expressed specifically in somatic cells surrounding the female meiocyte affect meiosis and produce unreduced gametes (Singh et al., 2011). Furthermore, perturbing the Arabidopsis *AGO5* expression surrounding the MMC impairs the initiation of megagametogenesis (Tucker et al., 2012). These results indicate that mobile siRNAs are important during early stages of gametogenesis in plants.

Specification of MMC in Arabidopsis is also associated with visible morphological and cytological changes, including nuclear enlargement, chromatin decondensation, depletion of linker histone H1, reduction in heterochromatin content, suggesting a common scheme of chromatin reorganization similar to the that occurs during male gametogenesis, likely to promote meiosis competent transcriptional program in the MMC (Armstrong & Jones, 2003; Schneitz, Hülskamp, & Pruitt, 1995; She et al., 2013; Sniezko, 2006).

7. EPIGENETIC REPROGRAMMING OF THE ARABIDOPSIS CENTRAL CELL FOR ENDOSPERM IMPRINTING

In the mature Arabidopsis female gametophyte, mutations in the components of the PRC2 Polycomb group complex (*MEA*, *FIS2*, *FIE*, and

MSI1) produce autonomous endosperm in the absence of fertilization, and all four genes are expressed in the central cell (Grossniklaus, Vielle-Calzada, Hoeppner, & Gagliano, 1998; Guitton, 2004; Kiyosue et al., 1999; Köhler et al., 2003; Luo, Bilodeau, Dennis, Peacock, & Chaudhury, 2000; Ohad et al., 1999). Thus, one primary function of this central cell PRC2 complex is to repress central cell proliferation until fertilization. Mutations in these PRC2 genes also cause parent-of-origin effects on seed viability, partly due to the imprinted expression of *FIS2* and *MEA* genes in the endosperm (Gehring et al., 2006; Jullien, Katz, Oliva, Ohad, & Berger, 2006; Kinoshita, Yadegari, Harada, Goldberg, & Fischer, 1999; Vielle-Calzada et al., 1999). Genomic imprinting is the differential expression of two alleles of a gene dependent on their parent of origins, and DNA methylation plays an important role in regulating gene imprinting in Arabidopsis. For example, maternal allele-specific expression of *MEA*, *FIS2*, and *FWA* depends on their activation in the central cell by DME-mediated active DNA demethylation whereas MET1 is required for silencing the expression of *FIS2* and *FWA* paternal alleles (Gehring et al., 2006; Jullien et al., 2006; Kinoshita et al., 2004; Tiwari et al., 2008; Xiao et al., 2003). In addition to establishing imprinted expression of many genes, DME also targets gene-poor heterochromatin and intergenic regions for demethylation, suggesting that establishment of gene imprinting is likely not the primary function of DME in the central cell (Ibarra et al., 2012). Instead, it was proposed that DME demethylation in the gamete companion cells (i.e., the vegetative and the central cells) activates TE expression and generates mobile siRNAs to reinforce TE silencing in the gametes (Ibarra et al., 2012; Slotkin et al., 2009).

During fertilization, in addition to DME-induced global methylation difference between the two maternal genomes (the central cell and the egg cell), endosperm and embryo also inherit different chromatin and transcriptional states from the gametes, including enlarged nucleus, de-condensed chromatin, and distinct heterochromatin organization in the endosperm compared to embryo (Baroux, Pecinka, Fuchs, Schubert, & Grossniklaus, 2007; Pillot et al., 2010). Since the two sperm cells are thought to be functionally identical, these differences in chromatin states are likely maternally derived (Faure et al., 2003; Russell, 1991). The chromatin state difference between the egg and the central cell is likely due to DME activity. Demethylation of the central cell by DME activates PRC2 complex that in turn repressed other genes, a process critical for seed viability in Arabidopsis (Gehring, 2013). Failure to activate the PRC2 complex in the central cell due to mutations either in DME or in the components of the

PRC2 complex results in seed failure, impaired gene imprinting in the endosperm, and de-repression of other genes regulated by this PRC2 complex (Gehring, 2013; Hsieh et al., 2011; Jullien et al., 2006). In addition, PRC2 is also responsible for maintaining the silencing of the maternal allele of the *VIM5* gene, a paternally expressed imprinted gene importance for the maintenance of CG methylation in the endosperm. For reason not yet fully understood, the expression of *VIM5* paternal allele is delayed during early endosperm development, causing a slight but global loss of CG methylation due to passive demethylation (Hsieh et al., 2009, 2011; Ibarra et al., 2012). Thus, genome-wide CG hypomethylation in the Arabidopsis endosperm is caused by active DNA demethylation in the central cell and passive DNA demethylation during early endosperm development (Fig. 2).

Based on GFP reporters and laser capture microdissection coupled with microarray hybridization assays, it was previously thought that maintenance DNA methyltransferases are barely expressed during Arabidopsis female gametogenesis whereas maintenance and *de novo* methyltransferases are expresses strongly in the embryo (Belmonte et al., 2013; Jullien et al., 2008; Jullien, Susaki, Yelagandula, Higashiyama, & Berger, 2012). This led to the speculation that CG methylation in the central cell is relatively low, which facilitates DME demethylation and contributes to the global CG hypomethylation observed in the endosperm (Hsieh et al., 2009; Jullien et al., 2008). However, multiple recent genome-scale methylation studies have shown that CG methylation is robustly maintained during gametogenesis and persists throughout embryogenesis (Bouyer et al., 2017; Hsieh et al., 2016; Kawakatsu, Nery, Castanon, & Ecker, 2017; Park et al., 2016). By contrast, methylation of CHH, and to a lesser degree CHG, increases during embryo development, presumably due to an increase in RdDM activity (Bouyer et al., 2017; Jullien et al., 2012; Kawakatsu et al., 2017; Lin et al., 2017).

The stability of epigenetic inheritance across multiple generations is one of the most important considerations for crop improvement. Although heritable changes of chromatin properties can involve a broad range of chromatin factors, DNA methylation and histone modifications are the most studied and better understood at the molecular levels. Among them, mechanisms of DNA methylation have been largely elucidated. Furthermore, earlier research on the transgene silencing had uncovered DNA methyltransferase *MET1* and the SNF2/SWI2 chromatin remodeler *DDM1* as prominent regulators for transgene silencing (Eun et al., 2012). Although RdDM can methylate cytosines in all sequence contexts, CG methylation is generally maintained by *MET1* whereas *DRM2* and *CMT3* are important for non-CG methylation and loss of non-CG methylation in the *drm2/cmt3* mutants induce developmental abnormalities (Cao & Jacobsen, 2002; Chan et al., 2006). However, the abnormalities can be rapidly complemented when *DRM2* and *CMT3* activities were reintroduced (Chan et al., 2006). By contrast, loss of CG methylation in *met1* null mutants trigger genome-wide aberrant, compensatory *de novo* non-CG methylation that acts in a stochastic manner, highlighting the importance of CG methylation in maintaining the epigenetic landscape during development. Interestingly, when *MET1* activity is reintroduced into the *met1* mutants, remethylation in certain loci is very inefficient (Mathieu, Reinders, Čaikovski, Smathajitt, & Paszkowski, 2007). Further study found that remethylation is specific to the subset of heavily methylated repeats that are targeted by the RNA interference (RNAi) machinery, revealing the existence of an efficient mechanism that protects against transgenerational loss of DNA methylation in Arabidopsis (Teixeira et al., 2009).

In addition to the maintenance of CG methylation, recent studies have elucidated the intricate interplay between DNA demethylation and RdDM in maintaining the methylation homeostasis for normal development and for stable transgenerational epigenetic inheritance. A recent study reported that a large portion of ROS1 targets are previously unrecognized RdDM targets that escaped prior identification due to strong ROS1 demethylation, indicating that the antagonistic relationship between RdDM and ROS1 demethylation has been significantly underappreciated and many unmethylated genomic regions are in fact actively targeted by *de novo* methylation and the demethylation pathways (Hsieh et al., 2016; Tang et al., 2016). Since one major function of the ROS1 demethylation is to counter the spread of RdDM, a DNA methylation monitoring *cis*-element within the transcriptional control region of ROS1 promoter has been identified (Lei et al., 2015; Williams et al., 2015). Methylation of this *cis* element is required for ROS1 transcription and ROS1 expression is greatly reduced in various *rddm* mutants. Thus, this methylation sensing *cis* element functions as an "epigenetic rheostat" that adjusts ROS1 transcriptional output based on the strength of local RdDM. Significantly, when this methylation sensing *cis* element is experimentally rendered "constitutively" methylated independent of the physiological RdDM strength, the homeostasis of DNA methylation is disrupted and a widespread methylation loss is inherited to the progenies and phenotypic abnormalities accumulate over multiple generations (Williams & Gehring, 2017). These results uncovered the importance of an actively maintained equilibrium between methylation and demethylation activities that is required for stable transgenerational epigenetic inheritance.

8. INHERITANCE AND STABILITY OF EPIGENETIC VARIATION

The epigenetic variations must have two characteristics to be potentially useful in plant breeding: the effects on traits, and heritability. Inheritance of epigenome across multiple generations in plants relies on the faithful replication of DNA CG methylation. The stability of transgenerational DNA methylation inheritance has been evaluated in a mutation accumulation population in Arabidopsis *thaliana* (Becker et al., 2011; Schmitz et al., 2011). A single founder plant was used to propagate a population by single-seed descent for 30 generations (Shaw, Byers, & Darmo, 2000). Genome and methylome sequencing of parental plant and offspring after 30 generations revealed the rate of spontaneous genetic mutation and epimutation, respectively (Becker et al., 2011; Ossowski et al., 2010; Schmitz et al., 2011). Although only about 20 SNPs per individual plants accumulate over 30 generations, methylation change was found at thousands of single cytosines. Although the DNA methylation stability appears to be less stable at single cytosines compared with genetic sequence, consistent methylation changes in larger regions (DMRs) over 30 generations, however, appear to be as rare as genetic mutations. Changes in DNA methylation was observed more in transposable elements than in genic sequence, which is consistent with the methylation difference between natural Arabidopsis strains (Vaughn et al., 2007).

9. NON-SEXUAL REPRODUCTION IN PLANTS

In addition to sexual reproduction, plants can also produce offspring by non-sexual reproduction. Both sexual and non-sexual reproduction have their specific advantages for natural plant populations and for agriculture applications. Sexual reproduction generates offspring with genetic and phenotypic variation for adaptation to changing environment, which enables selection of desirable traits in breeding programs. On the other hand, non-sexual reproduction produces genetically identical progenies that provide significant value to agriculture for its ability to fix complex desirable genotypes.

Flowering plants can produce seeds without fertilization, in a process known as apomixis. Unlike sexual seed production, apomixis can occur by various mechanisms and produce viable seeds, despite bypassing the critical process of sexual reproduction. The critical developmental components of apomixis include: female gametogenesis without meiosis (apomeiosis), development of embryo independent of fertilization (parthenogenesis), and formation of viable endosperm either with or without fertilization (Koltunow & Grossniklaus, 2003). Apomixis is documented in at least 30 families of angiosperms, but is rare among crop species (Carman, 1997). The ability to produce genetically identical progeny by seeds makes apomixis a useful agronomic trait for plant breeding. Therefore, understanding the molecular mechanism and utilization of apomixis in crop plant have great potential in improving crop production (Calzada, Crane, & Stelly, 1996; Hand & Koltunow, 2014; Koltunow, Bicknell, & Chaudhury, 1995; Spillane, Curtis, & Grossniklaus, 2004). The other form of non-sexual reproduction in plants is vegetative propagation. Vegetative propagation uses plant structures (propagules) other than seeds to generate a new plant. Natural propagules are usually specialized fleshy organs, such as runners, rhizomes, bulbs, tubers, suckers, bulbils (Begon, Harper, & Townsend, 1986; Frankel & Galun, 2012; Harada & Iwasa, 1994; Pate & Dixon, 1982).

Several strategies of artificial non-sexual reproduction are frequently used in horticultural and agricultural practices, including cutting, layering, grafting and plant tissue culture (Forbes & Watson, 1992; Frankel & Galun, 2012). Asexual reproduction methods used in agriculture are usually based on natural propagules mentioned above or artificial propagules such as leaf, stem or root cuttings (Swingle, 1952). The most important advantage of non-sexual propagation is that it allows the production of genetically identical plants which retain all the quantitative and qualitative characteristics of the parents. Once a breeding program identifies an elite plant with desirable traits, the selected plant can be propagated asexually without losing superiority. Thus, it is especially important and commonly used in commercial-scale horticultural and agricultural practices (Davies, Davis, & Kester, 1994; Soh et al., 2011).

Plant tissue culture is also an important method of non-sexual propagation. Propagation by tissue culture relies on the totipotency of many plant cells, the ability to regenerate a whole new plant. Plant tissue culture is grown in sterile containers to ensure the regeneration of plants. Living plant materials from natural environment are usually contaminated with microorganisms such as bacteria and fungi. Excised plant fragments (explants) collected from source plant are carefully sterilized with appropriate chemical reagents. Sterile explants are then placed on suitable sterile solid culture medium, or sometimes into sterile liquid medium when cell suspension culture is needed. Different compositions of culture medium have been used for tissue culture of various plant species, but MS medium is the most commonly used with minor modification in different applications (Murashige & Skoog, 1962). In general, plant tissue culture medium consists of all essential major and minor plant nutrient elements, vitamins, carbohydrates, other organic nutrients and plant growth regulators. The two main classes of growth regulators in tissue culture are auxins and cytokinins. The content of growth regulators is crucial for the morphological development of the plant tissue culture. After shoot and root induction, the mature platelets can be transferred to soil for growth in growth chamber or greenhouse as normal plants.

Plant tissue culture technique is widely used in research, forestry, horticulture and agriculture practices. In breeding programs, the cultivars have to reach high homozygosity before being released for commercial cultivation to avoid further segregation in the field. The induction of haploid and double haploid plants developed through tissue culture is of great interests in plant breeding since it drastically reduces the time and effort for breeders to select for desirable traits. Cell suspension culture is also successfully used to produce valuable compounds, such as secondary metabolites and recombinant proteins (Hellwig, Drossard, Twyman, & Fischer, 2004; Park & Wi, 2016; Shahzad et al., 2017; Varma, 2010). Tissue culture can also be used as a strategy for germplasm conservation purpose or breeding programs for species with recalcitrant seeds or those that are vegetatively propagated (Shahzad et al., 2017). Successful tissue culture is also a critical process for genetic transformation in many plant species. Plant scientists have successfully used genetic engineering and genetic transformation to transfer desired genes into plants and crops (Prado et al., 2014). Similar transgenic approaches enable plant scientists to study gene function and regulation during plant development. The transgenic technology also offers enormous potential to improve crops for better yield and quality by transferring novel genes that confer desirable traits between different species or even genus. Several transgenic crops have shown greatly improved characteristics, such as herbicide tolerance, insect resistance, or drought tolerance (Prado et al., 2014).

10. EPIGENETIC TRANSMISSION DURING NON-SEXUAL REPRODUCTION

Epigenetic modifications can be induced by environmental changes and potentially contribute to phenotypic plasticity (Lira-Medeiros et al., 2010; Pecinka et al., 2010; Tittel-Elmer et al., 2010). Although some epigenetic modifications (especially DNA methylation) and the resulting phenotypic variations are heritable in the subsequent generations, most of the environmentally induced epigenetic changes are reset in the epigenome reprogramming during gametogenesis, thus are not transmitted to the sexually reproduced offspring (Feng, Jacobsen, & Reik, 2010; Kawashima & Berger, 2014; Paszkowski & Grossniklaus, 2011). By contrast, plants reproduced from asexual propagation do not undergo gametogenesis, the environmentally induced epigenetic modifications might persist in clonal plants. One study demonstrated that the clonal offspring of white clover (*Trifolium repens*) showed different biomass if the parental plants experienced different intensities and durations of drought stress (González et al., 2016). The transgenerational effects were not observed in the clonal offsprings of the parents experienced the same treatment if the parents were treated with DNA methyltransferase inhibitor, suggesting that such transgenerational effect was mediated by DNA methylation (González et al., 2016). Another study showed that although the genetic diversity is limited among the asexually propagated white poplar (*Populus alba* L.) populations of Sardinia island, the methylation sensitive amplified polymorphism (MSAP) analysis revealed that these genetically identical poplar trees populations have a diverse DNA methylation status (Guarino, Cicatelli, Brundu, Heinze, & Castiglione, 2015). Moreover, the ramets of the same clonal population showed different DNA methylation status related to their geographical position, suggesting that such epigenetic diversity might provide the poplar trees with adaptive capability of growing in different natural environment on the island, and facilitate the success of their vegetative reproduction strategy (Guarino et al., 2015). These results support the idea that epigenetic modification might be an important mechanism contributing to the environmental adaptive capability of asexually propagated plants, as it can be more effectively transferred to clonal offsprings (Latzel, Rendina González, & Rosenthal, 2016). Another recent study compared the methylomes and transcriptomes of Arabidopsis plants regenerated from roots and leaves through induced asexual propagation (Wibowo et al., 2018). The results showed that plant somatic cells do not completely reset their cell type-specific epigenomes during asexual reproduction through the induction of embryonic developmental program. The progenies of the regenerated plants retain epigenetic marks as well as transcriptional and phenotypical signatures of their parental tissues for multiple generations (Wibowo et al., 2018). These observations suggest a possibility of creating new phenotypic variation through incomplete resetting of cell type-specific epigenetic marks during asexual reproduction.

11. EPIGENETIC VARIATION IN NON-SEXUAL REPRODUCTION

Plant breeding by exploiting natural genetic variation has been practiced to improve yields and resistance to biotic and abiotic stress since the dawn of agriculture. Plant breeding usually involved several crossing and selection cycles to produce seeds that give rise to cultivars with best and consistent quality. Thus, the development of conventional plant breeding has several limitations, such as long time and great effort requirement, limited gene pool for sufficient trait improvement in some species. The modern genetic engineering techniques provide greater potential and flexibility to create plants with superior yield traits, quality and stress resistance by introducing novel genes that even from different species or genera, such as genes from bacteria to resist insect pests (Hilder & Boulter, 1999) or genes that synthesize beta-carotene in rice grain (Beyer et al., 2002).

Tissue culture has been used for decades to generate genetic engineering crops such as rice, corn and many other plants. Undifferentiated plant cell mass (callus) is induced from explant on medium and then subjected to genetic transformation. Transformed cells with novel genes are then selected and proliferated on the medium containing selective reagents. The transgenic plants are eventually regenerated by hormone induction. Tissue culture represents an unusual and extreme environment that exerts several stresses to the plant tissues. Previous studies have shown that plants regenerated from tissue culture have significant different epigenomic profile comparing to non-regenerated plants, including a loss of DNA methylation in TEs (Stroud, Greenberg, Feng, Bernatavichute, & Jacobsen, 2013; Tanurdzic et al., 2008). In addition, farmers are forced to use clonal propagation such as grafting to maintain the best quality of elite varieties for many crops, such as banana and many other fruit and nut trees. Studies have shown that the epigenetic information such as mobile siRNA can direct DNA methylation changes across the grafts (Lewsey et al., 2016; Melnyk, Molnar, &

Baulcombe, 2011). These epigenetic variations induced by clonal propagation can be harnessed for crop improvement. Therefore, it is important to have a better understanding on the epigenetics variation in clonally propagated species.

The study of "*bad karma*" in oil palm is an excellent example of epigenetic information being used for agricultural improvement in asexually propagated crops. The African Oil palm (*Elaeis guineensis*) is one of the most efficient oil-bearing crops, which contributes to nearly one-half of the world's supply of edible vegetable oil and is also a promising source for biofuel. Due to its high value and growing demand, as well as the concern of its tropical habitats sustainability, oil palm has been under extensive breeding program and study to improve yield. Most commercial oil palms are the hybrids of two species, resulting in up to 30% of yield enhancement (Ong-Abdullah et al., 2015). The highest-yielding elite hybrid is then selected and propagated by cell culture to ensure consistent quality. However, despite their clonal origin, some cloned oil palm trees develop "mantled" fruits that produce almost no oil, drastically affecting oil yield. The oil palm fruit mantled abnormality is only observable after the plants grow to maturity— about six years after planting in the tropical plantation. This results in millions of dollars wasted on these mantled plants, and on the precious rainforest suffered from these crop failures. DNA methylation analysis revealed a differential DNA methylation region at a retrotransposon, *Karma*, within a B-class MADS-box *EgDEF1* gene (also named *MANTLED*), which is involved in sexual organ determination (Ong-Abdullah et al., 2015). Loss of DNA methylation at a splice site in this gene (termed *bad karma* epiallele) is found in all mantled clones and is associated with alternative splicing and premature termination of the EGDEF1 protein. Besides, loss of small RNA is also found in mantled clones. Although the reason for "bad karma" derived from non-sexually propagated clones is not fully understood, the growers can now use a simple epigenetic test to identify cloned oil palm nurslings with bad karma and discard them before transferring to plantations, saving millions of dollars and precious tropical land.

12. POSSIBLE DRAWBACKS OF NONSEXUAL PROPAGATION CONCERNING EPIGENETIC REGULATION

Somatic embryogenesis (SE) is the process in which embryos develop from somatic cells in the absence of fertilization. This process can occur spontaneously *in vivo* or can be induced *in vitro* by treating explants with plant growth regulators. Recently, epigenetic mechanism has been recognized as a significant molecular basis of SE, as it coordinates regulation of gene expression in cellular dedifferentiation, re-differentiation and somatic embryogenesis (Hemberger, Dean, & Reik, 2009; Nic-Can et al., 2013). The epigenetic regulatory mechanisms are known to be sensitive to environmental stress that is presented in *in vitro* SE process. The disturbance of the epigenetic mechanisms might result in phenotypic variations arising within clonal progeny regenerated from the single mother plant, known as somaclonal variation (SV) (Larkin & Scowcroft, 1981). One classical example of such epigenetic somaclonal variation is the previously described *mantled* oil palm, in which the altered DNA methylation resulted in abnormal floral orang development (Jaligot, Beulé, Baurens, Billotte, & Rival, 2004; Jaligot, Beulé, & Rival, 2002; Jaligot, Rival, Beulé, Dussert, & Verdeil, 2000; Matthes, Singh, Cheah, & Karp, 2001; Ong-Abdullah et al., 2015). In addition, transposable elements reactivation has been observed as a source of genetic and epigenetic somaclonal variations (Grandbastien, Spielmann, & Caboche, 1989; Hirochika, 1993; Hirochika, Sugimoto, Otsuki, Tsugawa, & Kanda, 1996). Under normal developmental condition, TEs are repressed through chromatin condensation by histone modification and DNA methylation (Cao et al., 2003; Lippman et al., 2004; Lisch, 2009; Mirouze & Vitte, 2014; Slotkin & Martienssen, 2007). Under the stress environment presented in tissue culture process, the epigenetic repressive mechanism was alleviated and such alleviation of epigenetic repression has been correlated with TE remobilization (Ding et al., 2007; Komatsu, Shimamoto, & Kyozuka, 2003; La et al., 2011; Liu et al., 2004). Production of clonal plants via tissue culture is often used to propagate elite cultivar with stable and homogeneous superior phenotype. In this regard, somaclonal variation is considered a drawback of the commercial development of nonsexual reproduction via SE. Therefore, elucidating the epigenetic mechanisms underpinning the origin of SV is an important future research direction to limit and control the incidence of SV.

13. NATURALLY OCCURRING EPIGENETIC VARIATION

Although the extent to which epigenetic variations affect phenotypic diversity in plants is not entirely clear, many studies have indicated that, like genetic variations, heritable epigenetic variations can arise spontaneously and contribute to phenotypic variation (Becker & Weigel, 2012). Epialleles or epimutations are the epigenetic variations that lead to heritable traits. However, it remains a challenge to determine the cause of an epiallele without comprehensive genome analysis because epigenetic variation can result from genetic variations acting either *in cis* or *in trans* (Becker & Weigel, 2012; Paszkowski & Grossniklaus, 2011). Nonetheless, a growing number of phenotypes attributed to epigenetic variation have been discovered (Becker & Weigel, 2012; He et al., 2018; Quadrana et al., 2014). Since the variation and inheritance mechanisms of DNA methylation are known in the greatest detail, it is not surprising that most epialleles discovered so far are associated with DNA methylation variation.

The first natural epiallele was found in a mutant of *Linaria vulgaris* (toadflax). Flowers of peloric *L. vulgaris* mutants have radial symmetry whereas wild-type flowers are bilaterally symmetrical. Genetic analysis shows that the peloric phenotype is caused by alternation in *Lcyc*, a homolog of *cycloidea* gene that controls flower symmetry in *Antirrhinum* (snapdragon). No RNA accumulation of *Lcyc* was detected in the peloric mutants. However, no changes in DNA sequence specific to the peloric mutants were found in *Lcyc*. Instead, increased DNA methylation was found in *Lcyc* locus in peloric mutants compared to wild-type. Furthermore, the peloric phenotype reverts back to semi-peloric or almost wild-type flowers occasionally, accompanied with reduced DNA methylation and restoration of *Lcyc* expression. Notably, the revertant phenotype correlates with the level of DNA methylation: the near-wild-type flowers has more profound demethylation whereas the peloric and semi-peloric flowers are more hypermethylated at *Lcyc*. These evidence support a direct relationship between the DNA methylation and gene expression of *Lcyc* leading to the peloric phenotype, and confirm that an epiallele can lead to a continuous variation (Pilar, Vincent, & Coen, 1999).

Another example of naturally occurring epiallele was found in the *Colorless non-ripening* (*Cnr*) locus in tomato (Manning et al., 2006). The *Cnr* mutation inhibits fruit ripening in tomato and produces colourless, mealy fruits. Genetic mapping and expression data indicated that the colourless non-ripening phenotype is caused by a mutant allele at *LeSPL-CNR* gene, encoding an SBP-box transcription factor. The *LeSPL-CNR* gene is expressed at a much lower level in *Cnr* mutant fruits compared to the control wild-type fruits. However, there is no nucleotide sequence difference between the *Cnr* mutant and wild-type fruit. Instead, the non-ripening phenotype is associated with significantly higher DNA methylation in a contiguous region upstream of *LeSPL-CNR*. The *Cnr* mutant was found in the Liberto cultivar. Interestingly, bisulfite sequencing of fruit and leaf tissues shows that the wild-type Liberto is hypermethylated in the region upstream of *LeSPL-CNR*, although not as high as its corresponding *Cnr* mutants. On the contrary, another wild-type cultivar Alisa Craig has a much lower DNA methylation level in the same region, while there is no obvious DNA sequence difference between the two genetic backgrounds (Manning et al., 2006). The higher DNA methylation level might explain the *Cnr* mutant plants is more likely to emerge from the Liberto cultivar than the Alisa Craig. Although the event that initiated the *Cnr* mutation is not yet known, genetic and methylome analysis indicate that a *COPIA*-like retrotransposon upstream of *LeSPL-CNR* promoter is highly methylated (Manning et al., 2006). The study in Arabidopsis has shown that DNA methylation in TEs can spread to the nearby genes (Ahmed, Sarazin, Bowler, Colot, & Quesneville, 2011). It is likely the *Cnr* epiallele was triggered by the presence of TE that directs RdDM to the *LeSPL-CNR* locus. The revertant of *Cnr* is very rare, only three were observed from more than 3000 plants, indicating the *Cnr* epimutation is very stable (Manning et al., 2006).

Epialleles of many Arabidopsis genes have been described, including *FWA*, *SUPERMAN*, *PAI*, *FOLT*, *BONSAI* and *QUA-QUINE STARCH* (*QQS*) (Bender, 2004; Durand, Bouché, Perez Strand, Loudet, & Camilleri, 2012; Fujimoto et al., 2008; Jacobsen & Meyerowitz, 1997; Saze & Kakutani, 2007; Silveira et al., 2013). Among them, several epimutations are associated with DNA methylation mediated by TE or other repeat sequences (Weigel & Colot, 2012). For example, the Arabidopsis *flowering wageningen* (*fwa*) is a well- characterized epiallele with *SHORT INTERSPERSED ELEMENTs* (*SINEs*) insertion in the 5′ regulatory region. During vegetative development, the DNA hypermethylation at the *SINE* repeats leads to transcriptional silencing of *FWA*. The *fwa* epimutant loses DNA methylation at *SINE* repeats and de-represses *FWA* expression, resulting in delayed flowering (Kinoshita et al., 2006; Soppe et al., 2000).

Gene duplication and their arrangement can also induce epigenetic variation. In Arabidopsis, the *PAI* tryptophan biosynthesis gene family of Columbia (Col) and many other ecotypes includes three unlinked genes (*PAI1*, *PAI2* and *PAI3*). However, Wassilewskija (WS) ecotype carries an additional copy of *PAI* gene at one of the three *PAI* loci. In *WS*, the *PAI4* gene presents as an almost perfectly inverted repeat of *PAI1* gene (referred to as the *PAI1-PAI4* locus). All four *PAI* loci are heavily methylated in WS ecotype. Deletion of *PAI1-PAI4 locus* reduces DNA methylation at *PAI2* and *PAI3* loci, indicating that the hypermethylation is triggered by the *PAI1-PAI4* locus (Judith Bender & Fink, 1995). Several recent studies uncovered more natural epialleles that cause hybrid incompatibility among Arabidopsis accessions (Agorio et al., 2017; Blevins, Wang, Pflieger, Pontvianne, & Pikaard, 201, p.20; Durand et al., 2012). For example, a previously identified case of hybrid incompatibility between accession Col-0 and Cvi-0 involves the two copies of a duplicated histidine biosynthesis gene, *HISN6A* and *HISN6B* (Bikard et al., 2009). In Arabidopsis accession Cvi-0, *HISN6A* is mutated, so *HISN6B* is essential. By contrast, *HISN6A* is essential in Col-0 because *HISN6B* is hypermethylated in its promoter region and is a stably silenced epiallele. Therefore, the hybrid progeny between the two accessions that are homozygous for Cvi-0 *HISN6A* and Col-0 *HISN6B* will die. Inheritance of *HISN6B* silent epiallele is dependent on MET1-, CMT3- and HD6-mediated DNA methylation, but independent of RdDM. Reactivating *HISN6B* by erasing its DNA methylation can complement *hisn6a* mutant and restore hybrid compatibility. This study shows that naturally occurring epigenetic variation could contribute to gene flow limitation between populations (Blevins et al., 2017).

A more recent study discovered the retrotransposon, *NMR19-4*, as a novel epiallele that controls leaf senescence by regulating the expression of *PHEOPHYTIN PHEOPHORBIDE HYDROLASE* (*PPH*) in Arabidopsis (He et al., 2018). *NMR19* is highly methylated in some Arabidopsis accessions such as C24 but not in others including Col-0. Genetic analysis shows that DNA methylation difference at *NMR19-4* is independent of genetic variation. Maintenance of the methylation at *NMR19-4* does not depend on RdDM, but the methylated *NMR19-4* allele might lose its DNA methylation without chromatin remodeler DDM1. The DNA methylation at *NMR19-4* negatively correlates with the expression of *PPH*, an enzyme critical for chlorophylls breakdown during leaf senescence. Moreover, further analysis indicates that the DNA methylation status of *NMR19-4* correlates with local climate, suggesting that *NMR19-4* might be an environmental adaptation associated epiallele (He et al., 2018).

Some epigenetic variations resulted in agronomically important traits. In rice, a spontaneous epimutant, *Epi-d1*, shows a metastable dwarf phenotype. The phenotype in epi-d1 mutant is due to the metastable expression of *DWARF1* (*D1*) gene. The silencing of *D1* gene is associated with repressive histone marks in genic region and DNA hypermethylation in *D1* promoter region (Miura et al., 2009).

Epigenetic variation has also been reported to involve in sexual determination in melon. Expression of a transcription factor, *CmWIP1*, leads to carpel abortion, resulting in the development of unisexual male flowers (Martin et al., 2009). In a melon sex determination mutant, a transposon inserted into upstream region of *CmWIP1* gene where the initiation and spreading of DNA methylation from the transposon to the promoter leads to its silencing. When the TE is removed by recombination, the promoter is demethylated and the *CmWIP1* is reactivated. These results indicate that the epigenetic variation induced by TE insertion is the major cause of *CmWIP1* silencing (Martin et al., 2009).

Another example of naturally occurring epiallele related to an agronomic trait is found in tomato fruits. A study shows that a major tomato vitamin E (VTE) content quantitative trait loci (QTL9-2-6) is associated with differential DNA methylation of a *SINE* retrotransposon in the promoter region of *VTE3(1)* gene (encoding an enzyme in VTE biosynthesis pathway) (Quadrana et al., 2014). The promoter DNA methylation can be spontaneously reverted. DNA methylation status of different epiallele leads to differential *VTE3* expression and vitamin E content in tomato fruits (Quadrana et al., 2014). These naturally occurring epialleles provide direct evidence that epigenetic variation plays a role in the determination of agronomic traits.

From the studies discussed so far, we can conclude that the plant DNA methylomes do not vary randomly between generations, nor are they all completely stable. Many epigenetic variations described above are stably inherited across many generations. The *Cnr* epiallele in tomato is very stable. In the 3000 colourless mutant plants examined, only three fruits on three separate plants showed revertant sectors (Manning et al., 2006). In wild-type *Arabidopsis thaliana, FWA* is stably silenced by DNA methylation in vegetative tissue, but in *fwa* epimutant, the demethylation at the *SINE*-related sequence upstream *FWA* resulting in ectopic *FWA* expression and late flowering phenotype (Kinoshita et al., 2006; Soppe et al., 2000). The unmethylated *fwa* epiallele and the phenotype are stably inherited across multiple generations. Interestingly, the *SINE*-related sequence structures vary among different Arabidopsis species, and the DNA methylation patterns in this region correlated with *FWA* expression, suggesting that the flowering time variation among different Arabidopsis species might be due to the presence of these *fwa* epialleles (Fujimoto et al., 2011).

14. EXPERIMENTALLY INDUCED EPIGENETIC VARIATIONS

14.1 Phenotypic Variance Induced in epiRILs

A twist in the traditional recombinant inbreeding strategy is used to create epigenetic inbred lines (epiRILs) of Arabidopsis— population of plants with identical genomes, but differ in their epigenomes. The epiRILs provide a useful tool to understand the function of DNA methylation in different loci, and also enable the scientists to track the inheritance of methylome in plants. The epiRILs exploit the mutant of the first two genes known to be involved in DNA methylation in Arabidopsis, *DDM1* and *MET1*, in which mutants show up to 70% to 90% reduction in CG methylation, respectively (Finnegan, Peacock, & Dennis, 1996; Ronemus, Galbiati, Ticknor, Chen, & Dellaporta, 1996; Vongs, Kakutani, Martienssen, & Richards, 1993). The epiRIL population was established by crossing two parent plants with the same genetic background, but one parent carried homozygous *ddm1* or *met1* mutation, while the other parent carried homozygous wild-type gene (Johannes et al., 2009; Reinders et al., 2009). The progenies inherited their parents' methylomes: with one set of chromosome from the wild-type parent with normal methylation and the other set from the mutant parent with depleted DNA methylation. Then, the female progeny was back-crossed to wild-type male to generate the segregating population, which includes individuals that were homozygous for *DDM1* allele. In the *met1*-derived epiRILs case, the F1 hybrids were selfed to generate F2 population containing individuals with homozygous wild-type *MET1* allele. These individuals were then selected and carried through multiple rounds of self-pollination, resulting in a population of inbred lines that have mosaic epigenome comprising genome segments from wild-type and methylation-depleted chromosomes, but are otherwise genetically identical. After eight or nine generations, about 30% of the differentially methylated sequences are faithfully inherited from the parents. The others demethylated segments regained wild-type methylation pattern over generations, likely through RNA directed DNA methylation at repetitive sequences (Teixeira et al., 2009).

These epiRIL populations display widespread phenotypic variance for traits such as stress response, flowering time and plant height. These lines could serve as populations for mapping of quantitative traits to methylation variation. However, some of the variability might arise from the genetic variation induced by TE remobilization observed in these epiRILs, which need to be controlled during QTL mapping (Richards, 2009). The widespread phenotypic variance present in the epiRILs suggests that the number of naturally occurring epiallele is probably underestimated. Further study confirmed that the phenotypic variation in these epiRILs is associated with DNA methylation and some traits are heritable over several generations (Cortijo et al., 2014). The epiRIL experiment and the known natural epialleles also indicate that altering the methylation machinery could establish heritable epigenetic variation resulting in phenotypic diversity for trait selection.

14.2 Induce Epigenetic Variations by Modulating DNA Methylation Regulatory Enzyme Expression

As discussed above, mutations in epigenetic pathways provide powerful tools to induce epigenetic variation, modulating the expression of DNA methylation enzyme genes can also be explored to induce epigenetic variation. For example, if MET1 function involves interaction with other epigenetic factors, increase in MET1 concentration by overexpression could perturb the stoichiometric relationship with interacting partners. Interestingly, overexpression of either catalytically active or inactive version of *MET1* stochastically generates novel and heritable epialleles at transposable elements, coding, or non-coding sequences (Brocklehurst et al., 2018). Furthermore, these transformants share some common phenotypes and genes with altered expression that can be inherited in the absence of the causal transgenes, indicating the heritability of induced epigenetic effects. Surprisingly, a majority of genes with altered expression in *MET1* over-expression lines show an increase in expression, including the majority of genes encoding transposable elements. DNA methylation analysis on selected genes with increased expression revealed DNA hypomethylation in these loci (Brocklehurst et al., 2018). In another recent independent study, it was demonstrated that expressing the catalytic domain of human DNA demethylase *TEN-ELEVEN TRANSLOCATION1* (TET1) protein in Arabidopsis could lead to widespread CG hypomethylation which is inheritable in the subsequent generations and induce phenotypic variation (Ji et al., 2018). Taken together, these studies suggest that manipulating the expression level of epigenetic regulator genes can be an effective means to create epigenetic variation. On the other hand, these results also implied that fluctuation in the expression levels of epigenetic regulators during development could be a primary factor dictating the stability of epigenetic inheritance.

Likewise, the two Arabidopsis epiRIL studies also suggested that *ddm1*- and *met1*- induced epigenetic variations can be stably inherited, and the heritable variations for several complex traits, such as flowering time and plant height, were inherited in the progeny (Johannes et al., 2009; Reinders et al., 2009). However, the change in DNA methylation pattern might not be the only attribute to the observed phenotypic diversity. Loss of DNA methylation might enable TE remobilization. For example, several *met1* epiRIL lines showed retrotransposon *EVD* mobilization (Reinders et al., 2009). Therefore, it is highly possible that some traits observed in the epiRILs are due to TE mobilization induced genetic variation.

14.3 MSH1 as a Potential Global Epigenome Regulator

Although disruption of the plant methylation machinery could create heritable epigenetic variation, the resulting developmental behaviour is not necessarily agronomically advantageous for breeding purpose. A few years ago, researchers discovered an intriguing plant nuclear-encoded gene called *MutS HOMOLOG1* (*MSH1*) (Ou et al., 2012). MSH1 is localized to mitochondria and plastid, and is involved in organelle genome stability (Abdelnoor et al., 2003). Silencing of *MSH1* gene tricks the plants into believing that they are suffering from stress environment, leading to variable phenotypes such as dwarfed growth, enhanced branching or tillering, leaf variegation, delayed flowering and enhanced abiotic stress tolerance (Xu et al., 2011, 2012). This developmental reprogramming is accompanied by altered genome methylation. These phenotypes are observed in several species, including tomato, sorghum, soybean, and Arabidopsis, by suppressing *MSH1* with RNA interference (RNAi) (de la Rosa Santamaria et al., 2014; Raju et al., 2017; Virdi et al., 2015; Yang et al., 2015). Curiously, in all cases, the observed phenotype persisted in the progeny, independent of the RNAi transgene segregation (de la Rosa Santamaria et al., 2014). Surprisingly, crossing these modified plants to their isogenic non-modified wild-type plants produces offsprings with heritable enhanced growth vigour that persists for five generations. Methylome analysis shows that plants derived from crosses between *msh1* and WT plant have an overall increased DNA methylation compared to WT control plants (Virdi et al., 2015). The methylome analysis of enhanced growth epi-lines shows differential methylation in large chromosomal segments, suggesting a remodelling of epigenetic landscape in these epi-lines. The boosted growth vigour is transmittable by grafting and is repressed by DNA methylation inhibitor, suggesting that mobile signal and DNA methylation are involved in the *MSH1* alteration-induced growth phenotype (Yang et al., 2015).

Studies in various species show that repressing *MSH1* expression can lead to epigenome remodelling that creates phenotypic variations for better fitness in changing environment (de la Rosa Santamaria et al., 2014; Virdi et al., 2015; Xu et al., 2011; Yang et al., 2015). Although the molecular mechanism of *MSH1* repression-mediated growth vigour remains unclear, these studies suggest that manipulating a potential epigenome regulator such as *MSH1* could be a potential strategy to generate plants with enhanced growth vigour for the future breeding program (de la Rosa Santamaria et al., 2014).

15. INDUCED RETROTRANSPOSITION FOR PLANT BREEDING APPLICATION

Plant genomes are littered with transposable elements (TEs). The most abundant class of TEs, retrotransposons, can propagate through a "copy and paste" mechanism using RNA as a template for reverse transcription. Consequently, one major biological role of epigenetic regulation is to prevent the transcription and transposition of TEs to maintain genome integrity. However, many studies have reported that activation and mobilization of TEs can be observed in tissue culture (Fukai et al., 2008; Grandbastien et al., 1989; Hirochika et al., 1996), indicating that cellular dedifferentiation process during tissue culture disrupts epigenetic suppression (Stroud, Ding, et al., 2013). For example, retrotransposition of *Tnt1*, a *copia*-like LTR-retrotransposon, was reproducibly induced in tobacco tissue culture (Hirochika, 1993). Likewise, retrotransposition of *Tos17* has also been found to occur in prolonged rice tissue culture (Hirochika et al., 1996), and has been adopted as a gene-tagging strategy for large-scale insertional mutagenesis (Miyao et al., 2003).

Studies in Arabidopsis have revealed that activation and mobilization of retrotransposons are tightly regulated by epigenetic mechanisms. The Arabidopsis *copia*-type retrotransposon *Evade*, was found to be transcriptionally activated during the inbreeding of *met1*-WT epiRILs, and retrotransposed on rare occasions. Interestingly, *Evade* transposition is significantly enhanced in *met1* plants that are also deficient in RdDM or KYP activities (Mirouze et al., 2009), suggesting a possibility that simultanuous loss of CG and non-CG methylation might be the reason behind *Evade* mobilization (Mirouze et al., 2009; Zemach et al., 2013).

Abiotic stress such as prolonged heat treatment can transiently override epigenetic regulation, and destabilize silencing of transcription (Pecinka et al., 2010; Tittel-Elmer et al., 2010). The Arabidopsis *copia*-like retrotransposon *ONSEN* becomes transcriptionally activated and accumulates its extrachromosomal cDNA copies. However, retrotransposition does not normally occur unless the heat-stressed plants are also deficient in siRNA biogenesis (Cavrak et al., 2014; Ito et al., 2011), or the heat stress occurs during tissue culture (Masuta et al., 2016), indicating additional layer of epigenetic regulation prohibiting active retrotransposion upon environmental stress. Such layer of epigenetic regulation limits the development of controlled retrotransposion as a mean to create novel genetic and epigenetic diversity for breeding purpose without the risk of genome-wide epigenetic perturbation. However, a recent study demonstrated that combined inhibition of both DNA methylation and Pol II activity by chemical inhibitors leads to a robust stress-dependent mobilization of the *ONSEN* retrotransposon in Arabidopsis seedlings, providing a possibility to induce retrotransposition for crop improvement purpose (Thieme et al., 2017).

16. MAPPING FOR EPIGENETIC QUANTITATIVE TRAIT LOCI (EPI-QTL)

Many important agronomic traits such as yield, plant height, nutrient content are complex quantitative traits that are controlled by multiple loci. Natural genetic variation has been exploited to obtain desirable traits for crop improvement (Alonso-Blanco & Koornneef, 2000). Using segregating populations for genetic mapping has successfully identified causal loci underlying many important quantitative traits. Most QTL mapping uses RILs as mapping population. These approaches use sequence polymorphism between the parental lines to mark the parent-of-origin genomic fragments in the progeny. Screening a large population of lines enables associating the particular genomic loci with a specific quantitative trait (Alonso-Blanco & Koornneef, 2000; Weigel, 2012).

In addition to DNA sequence variation, epigenetic variation can also contribute to phenotypic divergence. However, using classical RIL populations can only detect QTL contributed by genetic variation. Creation of epi-RILs that have almost identical DNA sequence but polymorphic in DNA methylation profile has become a powerful tool for identifying epigenetic variation associated with complex traits by population study (Cortijo et al., 2014). Similar to the RILs being used to identify causal genetic loci associated with complex trait variation, epiRILs are being used to dissect the epigenetic loci explaining trait variation via epi-QTL mapping. In a groundbreaking study, this was achieved by using the hundreds of differentially methylated regions (DMRs) identified among the epiRILs as molecular markers to map flowering time and primary root length epi-QTLs, providing a proof-of-concept example for mapping the epigenetic-based complex traits (Cortijo et al., 2014).

17. EPIGENETICS BASED SELECTION FOR CROP IMPROVEMENT

Examples are accumulating that epigenetic influence can direct the artificial selection during crop breeding process. Energy use efficiency (EUE) is an important factor in determining seed yield in oilseed crop canola (*Brassica napus*). EUE is calculated as the ratio of NAD(P)H versus cellular respiration, and higher EUE is correlated with higher seed yield. An isogenic double haploid canola line was used as starting material to recursively select for plants with the lowest and highest cellular respiration (Hauben et al., 2009). About four rounds of selection generates lines with distinct respiration and EUE. In field trials, lines with improved EUE showed on average up to 8% and 20% higher seed yield compared with control plants under normal and moderate drought stress conditions, respectively. Since these lines were derived from an isogenic line, the distinct EUE and seed yield were not attributed to genetic variations, as verified in amplified fragment-length polymorphism (AFLP) analysis results. Indeed, methylation-sensitive AFLP analysis revealed differentially methylated fragments in gene coding sequence specific for each selected line compared with the parental plants. Moreover, these "EUE epigenetic component" can be stably inherited for at least eight generations. Besides, altered global DNA methylation and histone modification were also observed in these selected lines. Although the mechanism underlying the fixation of the epigenetic component as well as the relationship between the genes and the differential methylated patterns remain unclear, this report suggests that it is feasible to select for plants with altered phenotype caused by emerging epigenetic variations during breeding process (Hauben et al., 2009).

18. CONCLUDING REMARKS

In the past several decades, researchers have uncovered numerous examples of epialleles that lead to inheritable morphological variations (Summarized in Table 1). The Arabidopsis epiRIL populations showed extensive phenotypic variations (Johannes et al., 2009; Reinders et al., 2009). Various species with *msh1*mutation induced epigenome alteration also showed

Table 1 Summary of Epigenetic Diversity

Trigger of Epigenetic Variation	Target	Epigenetic Variation	Species	Phenotype/Results
Naturally occurring	*Lcyc*	DNA methylation and silencing of Lcyc	*Linaria vulgaris*	Peloric flower
Naturally occurring	*Cnr*	DNA methylation and silencing of CNR	*Solanum lycopersicum*	Fruit ripening defect
Naturally occurring	*DWARF1 (D1)*	DNA methylation and silencing of D1	*Oryza sativa*	Dwarf
Naturally occurring	*QQS*	DNA methylation and silencing of QQS	*Arabidopsis thaliana*	No specific phenotype
TE polymorphism	*FWA*	SINE DNA methylation and silencing of FWA	*Arabidopsis*	Late flowering
EMS mutagenesis	*SUPERMAN*	DNA methylation and silencing of SUP	*Arabidopsis thaliana*	Floral organ defects
ddm1	*BONSAI*	Spreading of DNA methylation from LINE to BSN	*Arabidopsis thaliana*	Stunted growth
TE polymorphism	*CmWIP1*	Spreading of DNA methylation to CmWIP1	*Cucumis melo*	Promote female flower development
TE polymorphism	*VTE3*	DNA methylation and silencing of VTE3	*Solanum*	Reduced vitamine E content in fruit
PAI1-PAI4 inverted repeat	PAI genes	DNA methylaion and silencing of all PAI genes	*Arabidopsis thaliana*	No specific phenotype
FOLT duplication	*FOLT1*	DNA methylaion and silencing of all FOLT1	*Arabidopsis thaliana*	Reduced fertility
Gene duplication	*HISN6B*	DNA methylation and silencing of HISN6B	*Arabidopsis thaliana*	Hybrid incompatibility between certain genotypes
TE polymorphism	*NMR19*	DNA methylation and silencing of PPH	*Arabidopsis thaliana*	Leaf senescence

(*Continued*)

Table 1 Summary of Epigenetic Diversity—cont'd

Trigger of Epigenetic Variation	Target	Epigenetic Variation	Species	Phenotype/Results
met1-derived epiRIL	Global	Mosaic epigenome	*Arabidopsis thaliana*	Widespread phenotypic varaition
ddm1-derived epiRIL	Global	Mosaic epigenome	*Arabidopsis thaliana*	Widespread phenotypic varaition
msh1-mediated	Global	Mosaic epigenome	*Arabidopsis thaliana*	Enhanced growth
msh1-mediated	Global	Mosaic epigenome	*Sorghum bicolor*	Enhanced growth
msh1-mediated	Global	Mosaic epigenome	*Solanum lycopersicum*	Enhanced growth
MET1-mediated	Global	Stochastic and heritable epialleles	*Arabidopsis thaliana*	Shoot and root phenotype
TET1-mediated	Global	Stochastic and heritable epialleles	*Arabidopsis thaliana*	Flowering time
Tissue culture	Random	Retrotransposition of Tnt1	*Tobacco*	Retrotransposition
Tissue culture	Random	Retrotransposition of Tos17	*Oryza sativa*	Retrotransposition
Tissue culture	Random	Retrotransposition of ONSEN	*Arabidopsis thaliana*	Retrotransposition
Tissue culture	EgDEF1	DNA methylation and silencing of EgDEF1	*Elaeis guineensis*	Mentled fruit
met1-derived epiRIL	*Evade*	Activation and transposition of Evade	*Arabidopsis thaliana*	Retrotransposition
ddm1; drd or ddm1; drm2	*Evade*	Activation and transposition of Evade	*Arabidopsis thaliana*	Retrotransposition
TET1cd	Locus-specific	DNA methylation alteration	*Arabidopsis*	Targeted demethylation
ZFN/TALEN/ CRISPR-Cas system	Locus-specific	Epigenetic modification alteration	*Mammalia cell*	Targeted demethylation
Breeding selection	Energy use efficiency (EUE)	Heritable DNA methylation change	*Brassica napus*	High EUE

growth vigour (de la Rosa Santamaria et al., 2014; Virdi et al., 2015; Xu et al., 2011). These studies demonstrate that epigenetic variations could provide novel opportunities for crop improvement.

Although the effect of DNA methylation on gene transcription is well established, the extent to which it contributes to the phenotypic diversity among the population of a given species remains elusive in part due to lack of experimental tools for locus-specific assessment of methylation effects. Recent advances in the development of site-specific genome and epigenome editing systems provide powerful tools to create precise epigenetic variations in various plant species. Various enzymes such as transcription activator-like effectors nucleases (TALENs), zinc fingers nucleases (ZFNs), and CRISPR-Cas9 system have been developed for site-specific epigenome editing (Amabile et al., 2016; Liu et al., 2016; Park, Keung, & Khalil, 2016). The engineered DNA binding domains can be fused to DNA methyltransferase or DNA demethylase domain to introduce locus-specific DNA methylation or DNA demethylation, respectively. For instance, a study reports that a fusion between a zinc finger and the catalytic domain of the human DNA demethylase *TET1* (*TET1cd*) can efficiently target *FWA* promoter for demethylation and cause a heritable late-flowering phenotype (Gallego-Bartolomé et al., 2018).

Although many epialleles have been uncovered, there are seemly a lot more possible epialleles and epiQTLs that can lead to desirable phenotypic variations in crop genomes. Site-specific epigenome editing system could be a powerful tool to identify and test these epigenetic variations from epiRIL or populations with chemical- or epigenome regulator- induced epigenome alterations. Taken together, these biotechnologies and the information obtained by these approaches will no doubt provide novel avenues for crop improvement in the near future.

ACKNOWLEDGEMENT

This work is supported by the Hatch Project (No. 02413) of National Institute of Food and Agriculture, U.S.D.A to T.-F.H., the National Science Foundation Grant MCB-1715115 to T.-F.H. and W.Y.X., and by the State of NC appropriations through the University of North Carolina General Administration and the NC Agricultural Research Service Office of NCSU to T.-F.H.

REFERENCES

Abdelnoor, R. V., Yule, R., Elo, A., Christensen, A. C., Meyer-Gauen, G., & Mackenzie, S. A. (2003). Substoichiometric shifting in the plant mitochondrial genome is influenced by a gene homologous to MutS. *Proceedings of the National Academy of Sciences, 100*(10), 5968–5973. https://doi.org/10.1073/pnas.1037651100.

Agius, F., Kapoor, A., & Zhu, J.-K. (2006). Role of the Arabidopsis DNA glycosylase/lyase ROS1 in active DNA demethylation. *Proceedings of the National Academy of Sciences, 103*(31), 11796–11801.

Agorio, A., Durand, S., Fiume, E., Brousse, C., Gy, I., Simon, M., et al. (2017). An arabidopsis natural epiallele maintained by a feed-forward silencing loop between histone and DNA. *PLoS Genetics, 13*(1), e1006551. https://doi.org/10.1371/journal.pgen.1006551.

Ahmed, I., Sarazin, A., Bowler, C., Colot, V., & Quesneville, H. (2011). Genome-wide evidence for local DNA methylation spreading from small RNA-targeted sequences in Arabidopsis. *Nucleic Acids Research, 39*(16), 6919–6931. https://doi.org/10.1093/nar/gkr324.

Alonso-Blanco, C., & Koornneef, M. (2000). Natural variation, an underexploited resource of genetic variation for plant genetics. *Trends in Plant Science, 5*, 22–29.

Amabile, A., Migliara, A., Capasso, P., Biffi, M., Cittaro, D., Naldini, L., et al. (2016). Inheritable silencing of endogenous genes by hit-and-run targeted epigenetic editing. *Cell, 167*(1). https://doi.org/10.1016/j.cell.2016.09.006, 219–232.e14.

Armstrong, S. J., & Jones, G. H. (2003). Meiotic cytology and chromosome behaviour in wild-type *Arabidopsis thaliana*. *Journal of Experimental Botany, 54*(380), 1–10.

Bannister, A. J., & Kouzarides, T. (2011). Regulation of chromatin by histone modifications. *Cell Research, 21*(3), 381.

Baroux, C., Pecinka, A., Fuchs, J., Schubert, I., & Grossniklaus, U. (2007). The triploid endosperm genome of Arabidopsis adopts a peculiar, parental-dosage-dependent chromatin organization. *The Plant Cell, 19*(6), 1782–1794.

Becker, C., Hagmann, J., Müller, J., Koenig, D., Stegle, O., Borgwardt, K., et al. (2011). Spontaneous epigenetic variation in the *Arabidopsis thaliana* methylome. *Nature, 480*, 245.

Becker, C., & Weigel, D. (2012). Epigenetic variation: Origin and transgenerational inheritance. *Current Opinion in Plant Biology, 15*(5), 562–567. https://doi.org/10.1016/j.pbi.2012.08.004.

Begon, M., Harper, J. L., & Townsend, C. R. (1986). *Ecology. Individuals, populations and communities*. Blackwell Scientific Publications.

Belmonte, M. F., Kirkbride, R. C., Stone, S. L., Pelletier, J. M., Bui, A. Q., Yeung, E. C., et al. (2013). Comprehensive developmental profiles of gene activity in regions and subregions of the Arabidopsis seed. *Proceedings of the National Academy of Sciences, 110*(5), E435–E444. https://doi.org/10.1073/pnas.1222061110.

Bemer, M., & Grossniklaus, U. (2012). Dynamic regulation of Polycomb group activity during plant development. *Current Opinion in Plant Biology, 15*(5), 523–529. https://doi.org/10.1016/j.pbi.2012.09.006.

Bender, J. (2004). DNA methylation of the endogenous PAI genes in arabidopsis. *Cold Spring Harbor Symposia on Quantitative Biology, 69*, 10.

Bender, J., & Fink, G. R. (1995). Epigenetic control of an endogenous gene family is revealed by a novel blue fluorescent mutant of arabidopsis. *Cell, 83*(5), 725–734. https://doi.org/10.1016/0092-8674(95)90185-X.

Berger, S. L. (2007). The complex language of chromatin regulation during transcription. *Nature, 447*(7143), 407.

Berger, F., & Twell, D. (2011). Germline specification and function in plants. *Annual Review of Plant Biology, 62*(1), 461–484. https://doi.org/10.1146/annurev-arplant-042110-103824.

Beyer, P., Al-Babili, S., Ye, X., Lucca, P., Schaub, P., Welsch, R., et al. (2002). Golden rice: Introducing the β-carotene biosynthesis pathway into rice endosperm by genetic engineering to defeat vitamin a deficiency. *Journal of Nutrition, 132*(3), 506S–510S. https://doi.org/10.1093/jn/132.3.506S.

Bikard, D., Patel, D., Le Metté, C., Giorgi, V., Camilleri, C., Bennett, M. J., et al. (2009). Divergent evolution of duplicate genes leads to genetic incompatibilities within A. thaliana. *Science, 323*(5914), 623–626.

Bleckmann, A., Alter, S., & Dresselhaus, T. (2014). The beginning of a seed: Regulatory mechanisms of double fertilization. *Frontiers in Plant Science, 5*. https://doi.org/10.3389/fpls.2014.00452.

Blevins, T., Wang, J., Pflieger, D., Pontvianne, F., & Pikaard, C. S. (2017). Hybrid incompatibility caused by an epiallele. *Proceedings of the National Academy of Sciences, 114*, 3702–3707. https://doi.org/10.1073/pnas.1700368114, 201700368.

Bouyer, D., Kramdi, A., Kassam, M., Heese, M., Schnittger, A., Roudier, F., et al. (2017). DNA methylation dynamics during early plant life. *Genome Biology, 18*(1), 179.

Brocklehurst, S., Watson, M., Carr, I. M., Out, S., Heidmann, I., & Meyer, P. (2018). Induction of epigenetic variation in Arabidopsis by over-expression of DNA METHYLTRANSFERASE1 (MET1). *PLoS One, 13*(2), e0192170. https://doi.org/10.1371/journal.pone.0192170.

Calarco, J. P., Borges, F., Donoghue, M. T., Van Ex, F., Jullien, P. E., Lopes, T., et al. (2012). Reprogramming of DNA methylation in pollen guides epigenetic inheritance via small RNA. *Cell, 151*(1), 194–205.

Calzada, J.-P. V., Crane, C. F., & Stelly, D. M. (1996). Apomixis–The asexual revolution. *Science, 274*(5291), 1322. https://doi.org/10.1126/science.274.5291.1322.

Cao, X., Aufsatz, W., Zilberman, D., Mette, M. F., Huang, M. S., Matzke, M., et al. (2003). Role of the DRM and CMT3 methyltransferases in RNA-directed DNA methylation. *Current Biology, 13*(24), 2212–2217.

Cao, X., & Jacobsen, S. E. (2002). Locus-specific control of asymmetric and CpNpG methylation by the DRM and CMT3 methyltransferase genes. *Proceedings of the National Academy of Sciences, 99*(Suppl 4), 16491–16498. https://doi.org/10.1073/pnas.162371599.

Carman, J. G. (1997). Asynchronous expression of duplicate genes in angiosperms may cause apomixis, bispory, tetraspory, and polyembryony. *Biological Journal of the Linnean Society, 61*(1), 51–94. https://doi.org/10.1111/j.1095-8312.1997.tb01778.x.

Cavrak, V. V., Lettner, N., Jamge, S., Kosarewicz, A., Bayer, L. M., & Scheid, O. M. (2014). How a retrotransposon exploits the plant's heat stress response for its activation. *PLoS Genetics, 10*(1), e1004115.

Chan, S. W.-L., Henderson, I. R., Zhang, X., Shah, G., Chien, J. S.-C., & Jacobsen, S. E. (2006). RNAi, DRD1, and histone methylation actively target developmentally important non-CG DNA methylation in Arabidopsis. *PLoS Genetics, 2*(6), e83.

Chaudhury, A. M., Koltunow, A., Payne, T., Luo, M., Tucker, M. R., Dennis, E. S., et al. (2001). Control of early seed development. *Annual Review of Cell and Developmental Biology, 17*(1), 677–699.

Chaudhury, A. M., Ming, L., Miller, C., Craig, S., Dennis, E. S., & Peacock, W. J. (1997). Fertilization-independent seed development in *Arabidopsis thaliana*. *Proceedings of the National Academy of Sciences, 94*(8), 4223–4228.

Chen, C., Farmer, A. D., Langley, R. J., Mudge, J., Crow, J. A., May, G. D., et al. (2010). Meiosis-specific gene discovery in plants: RNA-seq applied to isolated arabidopsis male meiocytes. *BMC Plant Biology, 10*(1), 280.

Choi, Y., Gehring, M., Johnson, L., Hannon, M., Harada, J. J., Goldberg, R. B., et al. (2002). DEMETER, a DNA glycosylase domain protein, is required for endosperm gene imprinting and seed viability in arabidopsis. *Cell, 110*(1), 33–42. https://doi.org/10.1016/S0092-8674(02)00807-3.

Christenhusz, M. J. M., & Byng, J. W. (2016). The number of known plants species in the world and its annual increase. *Phytotaxa, 261*(3), 201. https://doi.org/10.11646/phytotaxa.261.3.1.

Cortijo, S., Wardenaar, R., Colomé-Tatché, M., Gilly, A., Etcheverry, M., Labadie, K., et al. (2014). Mapping the epigenetic basis of complex traits. *Science, 343*(6175), 1145. https://doi.org/10.1126/science.1248127.

Creasey, K. M., Zhai, J., Borges, F., Van Ex, F., Regulski, M., Meyers, B. C., et al. (2014). miRNAs trigger widespread epigenetically activated siRNAs from transposons in Arabidopsis. *Nature, 508*(7496), 411.

Davies, F. T., Davis, T. D., & Kester, D. E. (1994). Commercial importance of adventitious rooting to horticulture. In *Biology of adventitious root formation* (pp. 53–59). Springer.

de la Rosa Santamaria, R., Shao, M.-R., Wang, G., Nino-Liu, D. O., Kundariya, H., Wamboldt, Y., et al. (2014). MSH1-Induced non-genetic variation provides a source of phenotypic diversity in sorghum bicolor. *PLoS One, 9*(10), e108407. https://doi.org/10.1371/journal.pone.0108407.

Ding, Y., Wang, X., Su, L., Zhai, J., Cao, S., Zhang, D., et al. (2007). SDG714, a histone H3K9 methyltransferase, is involved in Tos17 DNA methylation and transposition in rice. *The Plant Cell, 19*(1), 9–22.

Drews, G. N., & Koltunow, A. M. (2011). The female gametophyte. *The Arabidopsis Book, 9*, e0155. https://doi.org/10.1199/tab.0155.

Durand, S., Bouché, N., Perez Strand, E., Loudet, O., & Camilleri, C. (2012). Rapid establishment of genetic incompatibility through natural epigenetic variation. *Current Biology, 22*(4), 326–331. https://doi.org/10.1016/j.cub.2011.12.054.

Eun, C., Lorkovic, Z., Sasaki, T., Naumann, U., Matzke, A., & Matzke, M. (2012). Use of forward genetic screens to identify genes required for RNA-directed DNA methylation in *Arabidopsis thaliana*. In *Presented at the Cold Spring Harbor symposia on quantitative biology* (Vol. 77, pp. 195–204). Cold Spring Harbor Laboratory Press.

Faure, J., Rusche, M. L., Thomas, A., Keim, P., Dumas, C., Mogensen, H. L., et al. (2003). Double fertilization in maize: The two male gametes from a pollen grain have the ability to fuse with egg cells. *The Plant Journal, 33*(6), 1051–1062.

Feng, S., Jacobsen, S. E., & Reik, W. (2010). Epigenetic reprogramming in plant and animal development. *Science, 330*(6004), 622–627.

Finnegan, E. J., Peacock, W. J., & Dennis, E. S. (1996). Reduced DNA methylation in Arabidopsis thaliana results in abnormal plant development. *Proceedings of the National Academy, 6*.

Forbes, J. C., & Watson, D. (1992). *Plants in agriculture*. Cambridge University Press.

Frankel, R., & Galun, E. (2012). *Pollination mechanisms, reproduction and plant breeding* (Vol. 2). Springer Science & Business Media.

Fujimoto, R., Kinoshita, Y., Kawabe, A., Kinoshita, T., Takashima, K., Nordborg, M., et al. (2008). Evolution and control of imprinted FWA genes in the genus arabidopsis. *PLoS Genetics, 4*(4), e1000048. https://doi.org/10.1371/journal.pgen.1000048.

Fujimoto, R., Sasaki, T., Kudoh, H., Taylor, J. M., Kakutani, T., & Dennis, E. S. (2011). Epigenetic variation in the FWA gene within the genus Arabidopsis: Epigenetic variation in the FWA gene. *The Plant Journal, 66*(5), 831–843. https://doi.org/10.1111/j.1365-313X.2011.04549.x.

Fukai, E., Dobrowolska, A. D., Madsen, L. H., Madsen, E. B., Umehara, Y., Kouchi, H., et al. (2008). Transposition of a 600 thousand-year-old LTR retrotransposon in the model legume Lotus japonicus. *Plant Molecular Biology, 68*(6), 653–663.

Gallego-Bartolomé, J., Gardiner, J., Liu, W., Papikian, A., Ghoshal, B., Kuo, H. Y., et al. (2018). Targeted DNA demethylation of the Arabidopsis genome using the human TET1 catalytic domain. *Proceedings of the National Academy of Sciences, 115*(9), E2125–E2134.

Gehring, M. (2013). Genomic imprinting: Insights from plants. *Annual Review of Genetics, 47*(1), 187–208. https://doi.org/10.1146/annurev-genet-110711-155527.

Gehring, M., Huh, J. H., Hsieh, T.-F., Penterman, J., Choi, Y., Harada, J. J., et al. (2006). DEMETER DNA glycosylase establishes MEDEA polycomb gene self-imprinting by allele-specific demethylation. *Cell, 124*(3), 495–506. https://doi.org/10.1016/j.cell.2005.12.034.

González, A. P. R., Chrtek, J., Dobrev, P. I., Dumalasová, V., Fehrer, J., Mráz, P., et al. (2016). Stress-induced memory alters growth of clonal offspring of white clover (Trifolium repens). *American Journal of Botany, 103*(9), 1567–1574.

Grandbastien, M.-A., Spielmann, A., & Caboche, M. (1989). Tnt1, a mobile retroviral-like transposable element of tobacco isolated by plant cell genetics. *Nature, 337*(6205), 376.

Grossniklaus, U., Vielle-Calzada, J.-P., Hoeppner, M. A., & Gagliano, W. B. (1998). Maternal control of embryogenesis by MEDEA, a polycomb group gene in Arabidopsis. *Science, 280*(5362), 446–450.

Guarino, F., Cicatelli, A., Brundu, G., Heinze, B., & Castiglione, S. (2015). Epigenetic diversity of clonal white poplar (*Populus alba* L.) populations: Could methylation support the success of vegetative reproduction strategy? *PLos One, 10*(7), e0131480.

Guitton, A.-E. (2004). Identification of new members of Fertilisation Independent Seed Polycomb Group pathway involved in the control of seed development in *Arabidopsis thaliana. Development, 131*(12), 2971–2981. https://doi.org/10.1242/dev.01168.

Hand, M. L., & Koltunow, A. M. G. (2014). The genetic control of apomixis: Asexual seed formation. *Genetics, 197*(2), 441–450. https://doi.org/10.1534/genetics.114.163105.

Harada, Y., & Iwasa, Y. (1994). Lattice population dynamics for plants with dispersing seeds and Vegetative propagation. *Researches on Population Ecology, 36*(2), 237–249. https://doi.org/10.1007/BF02514940.

Hauben, M., Haesendonckx, B., Standaert, E., Van Der Kelen, K., Azmi, A., Akpo, H., et al. (2009). Energy use efficiency is characterized by an epigenetic component that can be directed through artificial selection to increase yield. *Proceedings of the National Academy of Sciences, 106*(47), 20109–20114. https://doi.org/10.1073/pnas.0908755106.

Hehenberger, E., Kradolfer, D., & Kohler, C. (2012). Endosperm cellularization defines an important developmental transition for embryo development. *Development, 139*(11), 2031–2039. https://doi.org/10.1242/dev.077057.

Hellwig, S., Drossard, J., Twyman, R. M., & Fischer, R. (2004). Plant cell cultures for the production of recombinant proteins. *Nature Biotechnology, 22*(11), 1415–1422. https://doi.org/10.1038/nbt1027.

Hemberger, M., Dean, W., & Reik, W. (2009). Epigenetic dynamics of stem cells and cell lineage commitment: Digging Waddington's canal. *Nature Reviews Molecular Cell Biology, 10*(8), 526.

Henderson, I. R., & Jacobsen, S. E. (2007). Epigenetic inheritance in plants. *Nature, 447*(7143), 418–424. https://doi.org/10.1038/nature05917.

He, L., Wu, W., Zinta, G., Yang, L., Wang, D., Liu, R., et al. (2018). A naturally occurring epiallele associates with leaf senescence and local climate adaptation in Arabidopsis accessions. *Nature Communications, 9*(1). https://doi.org/10.1038/s41467-018-02839-3.

Higashiyama, T., & Takeuchi, H. (2015). The mechanism and key molecules involved in pollen tube guidance. *Annual Review of Plant Biology, 66*(1), 393–413. https://doi.org/10.1146/annurev-arplant-043014-115635.

Hilder, V. A., & Boulter, D. (1999). Genetic engineering of crop plants for insect resistance — a critical review. *Crop Protection, 18*(3), 177–191. https://doi.org/10.1016/S0261-2194(99)00028-9.

Hirochika, H. (1993). Activation of tobacco retrotransposons during tissue culture. *The EMBO Journal, 12*(6), 2521–2528.

Hirochika, H., Sugimoto, K., Otsuki, Y., Tsugawa, H., & Kanda, M. (1996). Retrotransposons of rice involved in mutations induced by tissue culture. *Proceedings of the National Academy of Sciences, 93*(15), 7783–7788.

Hsieh, T.-F. (2016). Epigenetics: A tug of war for DNA methylation. *Nature Plants, 2*(11), 16171.

Hsieh, P.-H., He, S., Buttress, T., Gao, H., Couchman, M., Fischer, R. L., et al. (2016). Arabidopsis male sexual lineage exhibits more robust maintenance of CG methylation than somatic tissues. *Proceedings of the National Academy of Sciences, 113*(52), 15132–15137.

Hsieh, T.-F., Ibarra, C. A., Silva, P., Zemach, A., Eshed-Williams, L., Fischer, R. L., et al. (2009). Genome-wide demethylation of Arabidopsis endosperm. *Science, 324*(5933), 1451–1454.

Hsieh, T.-F., Shin, J., Uzawa, R., Silva, P., Cohen, S., Bauer, M. J., et al. (2011). Regulation of imprinted gene expression in Arabidopsis endosperm. *Proceedings of the National Academy of Sciences, 108*(5), 1755–1762.

Huang, L., Jones, A. M. E., Searle, I., Patel, K., Vogler, H., Hubner, N. C., et al. (2009). An atypical RNA polymerase involved in RNA silencing shares small subunits with RNA polymerase II. *Nature Structural & Molecular Biology, 16*(1), 91–93. https://doi.org/10.1038/nsmb.1539.

Ibarra, C. A., Feng, X., Schoft, V. K., Hsieh, T.-F., Uzawa, R., Rodrigues, J. A., et al. (2012). Active DNA demethylation in plant companion cells reinforces transposon methylation in gametes. *Science, 337*(6100), 1360–1364.

Ito, H., Gaubert, H., Bucher, E., Mirouze, M., Vaillant, I., & Paszkowski, J. (2011). An siRNA pathway prevents transgenerational retrotransposition in plants subjected to stress. *Nature, 472*(7341), 115.

Jacobsen, S. E., & Meyerowitz, E. M. (1997). Hypermethylated SUPERMAN epigenetic alleles in arabidopsis. *Science, New Series, 277*(5329), 1100–1103.

Jaligot, E., Beulé, T., Baurens, F.-C., Billotte, N., & Rival, A. (2004). Search for methylation-sensitive amplification polymorphisms associated with the" mantled" variant phenotype in oil palm (Elaeis guineensis Jacq.). *Genome, 47*(1), 224–228.

Jaligot, E., Beulé, T., & Rival, A. (2002). Methylation-sensitive RFLPs: Characterisation of two oil palm markers showing somaclonal variation-associated polymorphism. *TAG. Theoretical and Applied Genetics. Theoretische und angewandte Genetik, 104*(8), 1263–1269.

Jaligot, E., Rival, A., Beulé, T., Dussert, S., & Verdeil, J.-L. (2000). Somaclonal variation in oil palm (*Elaeis guineensis* Jacq.): The DNA methylation hypothesis. *Plant Cell Reports, 19*(7), 684–690.

Ji, L., Jordan, W. T., Shi, X., Hu, L., He, C., & Schmitz, R. J. (2018). TET-mediated epimutagenesis of the *Arabidopsis thaliana* methylome. *Nature Communications, 9*(1), 895.

Johannes, F., Porcher, E., Teixeira, F. K., Saliba-Colombani, V., Simon, M., Agier, N., et al. (2009). Assessing the impact of transgenerational epigenetic variation on complex traits. *PLoS Genetics, 5*(6), e1000530. https://doi.org/10.1371/journal.pgen.1000530.

Jullien, P. E., Katz, A., Oliva, M., Ohad, N., & Berger, F. (2006). Polycomb group complexes self-regulate imprinting of the polycomb group gene MEDEA in arabidopsis. *Current Biology, 16*(5), 486–492. https://doi.org/10.1016/j.cub.2006.01.020.

Jullien, P. E., Mosquna, A., Ingouff, M., Sakata, T., Ohad, N., & Berger, F. (2008). Retinoblastoma and its binding partner MSI1 control imprinting in arabidopsis. *PLoS Biology, 6*(8), e194. https://doi.org/10.1371/journal.pbio.0060194.

Jullien, P. E., Susaki, D., Yelagandula, R., Higashiyama, T., & Berger, F. (2012). DNA methylation dynamics during sexual reproduction in *Arabidopsis thaliana. Current Biology, 22*(19), 1825–1830. https://doi.org/10.1016/j.cub.2012.07.061.

Kawakatsu, T., Nery, J. R., Castanon, R., & Ecker, J. R. (2017). Dynamic DNA methylation reconfiguration during seed development and germination. *Genome Biology, 18*(1), 171.

Kawashima, T., & Berger, F. (2014). Epigenetic reprogramming in plant sexual reproduction. *Nature Reviews Genetics, 15*(9), 613–624. https://doi.org/10.1038/nrg3685.

Kinoshita, T., Miura, A., Choi, Y., Kinoshita, Y., Cao, X., Jacobsen, S. E., et al. (2004). One-way control of FWA imprinting in Arabidopsis endosperm by DNA methylation.pdf. *Science, 303*.

Kinoshita, Y., Saze, H., Kinoshita, T., Miura, A., Soppe, W. J. J., Koornneef, M., et al. (2006). Control of FWA gene silencing in *Arabidopsis thaliana* by SINE-related direct repeats: FWA gene silencing in A. Thaliana. *The Plant Journal, 49*(1), 38–45. https://doi.org/10.1111/j.1365-313X.2006.02936.x.

Kinoshita, T., Yadegari, R., Harada, J. J., Goldberg, R. B., & Fischer, R. L. (1999). Imprinting of the MEDEA polycomb gene in the Arabidopsis endosperm. *The Plant Cell Online, 11*(10), 1945–1952.

Kiyosue, T., Ohad, N., Yadegari, R., Hannon, M., Dinneny, J., Wells, D., et al. (1999). Control of fertilization-independent endosperm development by the MEDEA polycomb gene in Arabidopsis. *Proceedings of the National Academy of Sciences, 96*(7), 4186–4191.

Köhler, C., Hennig, L., Bouveret, R., Gheyselinck, J., Grossniklaus, U., & Gruissem, W. (2003). Arabidopsis MSI1 is a component of the MEA/FIE Polycomb group complex and required for seed development. *The EMBO Journal, 22*(18), 4804–4814.

Koltunow, A. M., Bicknell, R. A., & Chaudhury, A. M. (1995). Apomixis: Molecular strategies for the generation of genetically identical seeds without. *Fertilization, 108*, 8.

Koltunow, A. M., & Grossniklaus, U. (2003). APOMIXIS : A developmental perspective. *Annual Review of Plant Biology, 54*(1), 547–574. https://doi.org/10.1146/annurev.arplant.54.110901.160842.

Komatsu, M., Shimamoto, K., & Kyozuka, J. (2003). Two-step regulation and continuous retrotransposition of the rice LINE-type retrotransposon Karma. *The Plant Cell, 15*(8), 1934–1944.

La, H., Ding, B., Mishra, G. P., Zhou, B., Yang, H., del Rosario Bellizzi, M., et al. (2011). A 5-methylcytosine DNA glycosylase/lyase demethylates the retrotransposon Tos17 and promotes its transposition in rice. *Proceedings of the National Academy of Sciences*, 201112704.

Lahmy, S., Pontier, D., Cavel, E., Vega, D., El-Shami, M., Kanno, T., et al. (2009). PolV (PolIVb) function in RNA-directed DNA methylation requires the conserved active site and an additional plant-specific subunit. *Proceedings of the National Academy of Sciences, 106*(3), 941–946.

Larkin, P. J., & Scowcroft, W. R. (1981). Somaclonal variation—a novel source of variability from cell cultures for plant improvement. *TAG. Theoretical and Applied Genetics. Theoretische und angewandte Genetik, 60*(4), 197–214.

Latzel, V., Rendina González, A. P., & Rosenthal, J. (2016). Epigenetic memory as a basis for intelligent behavior in clonal plants. *Frontiers in Plant Science*, 7, 1354.

Law, J. A., & Jacobsen, S. E. (2010). Establishing, maintaining and modifying DNA methylation patterns in plants and animals. *Nature Reviews Genetics, 11*(3), 204–220. https://doi.org/10.1038/nrg2719.

Lei, M., Zhang, H., Julian, R., Tang, K., Xie, S., & Zhu, J.-K. (2015). Regulatory link between DNA methylation and active demethylation in Arabidopsis. *Proceedings of the National Academy of Sciences, 112*(11), 3553–3557. https://doi.org/10.1073/pnas.1502279112.

Lewsey, M. G., Hardcastle, T. J., Melnyk, C. W., Molnar, A., Valli, A., Urich, M. A., et al. (2016). Mobile small RNAs regulate genome-wide DNA methylation. *Proceedings of the National Academy of Sciences, 113*(6), E801–E810. https://doi.org/10.1073/pnas.1515072113.

Lindroth, A. M., Cao, X., Jackson, J. P., Zilberman, D., McCallum, C. M., Henikoff, S., et al. (2001). Requirement of CHROMOMETHYLASE3 for maintenance of CpXpG methylation. *Science, 292*(5524), 2077. https://doi.org/10.1126/science.1059745.

Lin, J.-Y., Le, B. H., Chen, M., Henry, K. F., Hur, J., Hsieh, T.-F., et al. (2017). Similarity between soybean and Arabidopsis seed methylomes and loss of non-CG methylation does not affect seed development. *Proceedings of the National Academy of Sciences, 114*(45), E9730–E9739.

Lippman, Z., Gendrel, A.-V., Black, M., Vaughn, M. W., Dedhia, N., McCombie, W. R., et al. (2004). Role of transposable elements in heterochromatin and epigenetic control. *Nature, 430*(6998), 471.

Lira-Medeiros, C. F., Parisod, C., Fernandes, R. A., Mata, C. S., Cardoso, M. A., & Ferreira, P. C. G. (2010). Epigenetic variation in mangrove plants occurring in contrasting natural environment. *PLoS One, 5*(4), e10326.

Lisch, D. (2009). Epigenetic regulation of transposable elements in plants. *Annual Review of Plant Biology, 60*(1), 43–66. https://doi.org/10.1146/annurev.arplant.59.032607.092744.

Liu, Z., Han, F., Tan, M., Shan, X., Dong, Y., Wang, X., et al. (2004). Activation of a rice endogenous retrotransposon Tos17 in tissue culture is accompanied by cytosine demethylation and causes heritable alteration in methylation pattern of flanking genomic regions. *TAG. Theoretical and Applied Genetics. Theoretische und angewandte Genetik, 109*(1), 200–209.

Liu, X. S., Wu, H., Ji, X., Stelzer, Y., Wu, X., Czauderna, S., et al. (2016). Editing DNA methylation in the mammalian genome. *Cell, 167*(1). https://doi.org/10.1016/j.cell.2016.08.056, 233–247.e17.

Luo, M., Bilodeau, P., Dennis, E. S., Peacock, W. J., & Chaudhury, A. (2000). Expression and parent-of-origin effects for FIS2, MEA, and FIE in the endosperm and embryo of developing Arabidopsis seeds. *Proceedings of the National Academy of Sciences, 97*(19), 10637–10642.

Lu, J., Zhang, C., Baulcombe, D. C., & Chen, Z. J. (2012). Maternal siRNAs as regulators of parental genome imbalance and gene expression in endosperm of Arabidopsis seeds. *Proceedings of the National Academy of Sciences, 109*(14), 5529–5534. https://doi.org/10.1073/pnas.1203094109.

Manning, K., Tör, M., Poole, M., Hong, Y., Thompson, A. J., King, G. J., et al. (2006). A naturally occurring epigenetic mutation in a gene encoding an SBP-box transcription factor inhibits tomato fruit ripening. *Nature Genetics, 38*(8), 948–952. https://doi.org/10.1038/ng1841.

Margueron, R., & Reinberg, D. (2011). The Polycomb complex PRC2 and its mark in life. *Nature, 469*(7330), 343–349. https://doi.org/10.1038/nature09784.

Martin, A., Troadec, C., Boualem, A., Rajab, M., Fernandez, R., Morin, H., et al. (2009). A transposon-induced epigenetic change leads to sex determination in melon. *Nature, 461*, 1135.

Masuta, Y., Nozawa, K., Takagi, H., Yaegashi, H., Tanaka, K., Ito, T., et al. (2016). Inducible transposition of a heat-activated retrotransposon in tissue culture. *Plant & Cell Physiology, 58*(2), 375–384.

Mathieu, O., Reinders, J., Čaikovski, M., Smathajitt, C., & Paszkowski, J. (2007). Transgenerational stability of the arabidopsis epigenome is coordinated by CG methylation. *Cell, 130*(5), 851–862. https://doi.org/10.1016/j.cell.2007.07.007.

Matthes, M., Singh, R., Cheah, S.-C., & Karp, A. (2001). Variation in oil palm (Elaeis guineensis Jacq.) tissue culture-derived regenerants revealed by AFLPs with methylation-sensitive enzymes. *TAG. Theoretical and Applied Genetics. Theoretische und angewandte Genetik, 102*(6–7), 971–979.

Matzke, M. A., & Mosher, R. A. (2014). RNA-directed DNA methylation: An epigenetic pathway of increasing complexity. *Nature Reviews Genetics, 15*. https://doi.org/10.1038/nrg3683.

Melnyk, C. W., Molnar, A., & Baulcombe, D. C. (2011). Intercellular and systemic movement of RNA silencing signals: Intercellular and systemic movement of RNA silencing signals. *The EMBO Journal, 30*(17), 3553–3563. https://doi.org/10.1038/emboj.2011.274.

Mirouze, M., Reinders, J., Bucher, E., Nishimura, T., Schneeberger, K., Ossowski, S., et al. (2009). Selective epigenetic control of retrotransposition in Arabidopsis. *Nature, 461*(7262), 427.

Mirouze, M., & Vitte, C. (2014). Transposable elements, a treasure trove to decipher epigenetic variation: Insights from arabidopsis and crop epigenomes. *Journal of Experimental Botany, 65*(10), 2801–2812.

Miura, K., Agetsuma, M., Kitano, H., Yoshimura, A., Matsuoka, M., Jacobsen, S. E., et al. (2009). A metastable DWARF1 epigenetic mutant affecting plant stature in rice. *Proceedings of the National Academy of Sciences, 106*(27), 11218–11223.

Miyao, A., Tanaka, K., Murata, K., Sawaki, H., Takeda, S., Abe, K., et al. (2003). Target site specificity of the Tos17 retrotransposon shows a preference for insertion within genes and against insertion in retrotransposon-rich regions of the genome. *The Plant Cell, 15*(8), 1771–1780.

Mosher, R. A., & Melnyk, C. W. (2010). siRNAs and DNA methylation: seedy epigenetics. *Trends in Plant Science, 15*(4), 204–210. https://doi.org/10.1016/j.tplants.2010.01.002.

Murashige, T., & Skoog, F. (1962). A revised medium for rapid growth and bio assays with tobacco tissue cultures. *Physiologia Plantarum, 15*(3), 473–497. https://doi.org/10.1111/j.1399-3054.1962.tb08052.x.

Nic-Can, G. I., López-Torres, A., Barredo-Pool, F., Wrobel, K., Loyola-Vargas, V. M., Rojas-Herrera, R., et al. (2013). New insights into somatic embryogenesis: LEAFY COTYLEDON1, BABY BOOM1 and WUSCHEL-RELATED HOMEOBOX4 are epigenetically regulated in *Coffea canephora*. *PLoS One, 8*(8), e72160.

Ohad, N., Margossian, L., Hsu, Y.-C., Williams, C., Repetti, P., & Fischer, R. L. (1996). A mutation that allows endosperm development without fertilization. *Proceedings of the National Academy of Sciences, 93*(11), 5319–5324.

Ohad, N., Yadegari, R., Margossian, L., Hannon, M., Michaeli, D., Harada, J. J., et al. (1999). Mutations in FIE, a WD polycomb group gene, allow endosperm development without fertilization. *The Plant Cell, 11*(3), 407–415.

Olmedo-Monfil, V., Durán-Figueroa, N., Arteaga-Vázquez, M., Demesa-Arévalo, E., Autran, D., Grimanelli, D., et al. (2010). Control of female gamete formation by a small RNA pathway in Arabidopsis. *Nature, 464*(7288), 628.

Ong-Abdullah, M., Ordway, J. M., Jiang, N., Ooi, S.-E., Kok, S.-Y., Sarpan, N., et al. (2015). Loss of Karma transposon methylation underlies the mantled somaclonal variant of oil palm. *Nature, 525*(7570), 533–537. https://doi.org/10.1038/nature15365.

Ossowski, S., Schneeberger, K., Lucas-Lledó, J. I., Warthmann, N., Clark, R. M., Shaw, R. G., et al. (2010). The rate and molecular spectrum of spontaneous mutations in *Arabidopsis thaliana*. *Science, 327*(5961), 92. https://doi.org/10.1126/science.1180677.

Ou, X., Zhang, Y., Xu, C., Lin, X., Zang, Q., Zhuang, T., et al. (2012). Transgenerational inheritance of modified DNA methylation patterns and enhanced tolerance induced by heavy metal stress in rice (*Oryza sativa* L.). *PLoS One,* 7(9), e41143. https://doi.org/10.1371/journal.pone.0041143.

Park, M., Keung, A. J., & Khalil, A. S. (2016). The epigenome: The next substrate for engineering. *Genome Biology, 17*(1). https://doi.org/10.1186/s13059-016-1046-5.

Park, K., Kim, M. Y., Vickers, M., Park, J.-S., Hyun, Y., Okamoto, T., et al. (2016). DNA demethylation is initiated in the central cells of Arabidopsis and rice. *Proceedings of the National Academy of Sciences*, 201619047. https://doi.org/10.1073/pnas.1619047114.

Park, K. Y., & Wi, S. J. (2016). Potential of plants to produce recombinant protein products. *Journal of Plant Biology, 59*(6), 559–568. https://doi.org/10.1007/s12374-016-0482-9.

Paszkowski, J., & Grossniklaus, U. (2011). Selected aspects of transgenerational epigenetic inheritance and resetting in plants. *Current Opinion in Plant Biology, 14*(2), 195–203.

Pate, J. S., & Dixon, K. W. (1982). *Tuberous, cormous and bulbous plants. Biology of an adaptive strategy in Western Australia.*

Pecinka, A., Dinh, H. Q., Baubec, T., Rosa, M., Lettner, N., & Scheid, O. M. (2010). Epigenetic regulation of repetitive elements is attenuated by prolonged heat stress in Arabidopsis. *The Plant Cell, 22*(9), 3118–3129.

Pilar, C., Vincent, C., & Coen, E. (1999). An epigenetic mutation responsible for natural variation in floral symmetry. *Nature, 401*(6749), 157–161. https://doi.org/10.1038/43657.

Pillot, M., Baroux, C., Vazquez, M. A., Autran, D., Leblanc, O., Vielle-Calzada, J. P., et al. (2010). Embryo and endosperm inherit distinct chromatin and transcriptional states from the female gametes in arabidopsis. *The Plant Cell, 22*(2), 307–320. https://doi.org/10.1105/tpc.109.071647.

Prado, J. R., Segers, G., Voelker, T., Carson, D., Dobert, R., Phillips, J., et al. (2014). Genetically engineered crops: From idea to product. *Annual Review of Plant Biology, 65*(1), 769–790. https://doi.org/10.1146/annurev-arplant-050213-040039.

Probst, A. V., Fransz, P. F., Paszkowski, J., & Scheid, O. M. (2003). Two means of transcriptional reactivation within heterochromatin. *The Plant Journal, 33*(4), 743–749.

Quadrana, L., Almeida, J., Asís, R., Duffy, T., Dominguez, P. G., Bermúdez, L., et al. (2014). Natural occurring epialleles determine vitamin E accumulation in tomato fruits. *Nature Communications, 5*. https://doi.org/10.1038/ncomms5027.

Raju, S. K. K., Shao, M.-R., Sanchez, R., Xu, Y.-Z., Sandhu, A., Graef, G., et al. (2017). An epigenetic breeding system in soybean for increased yield and stability. *BioRxiv*, 232819.

Ream, T. S., Haag, J. R., Wierzbicki, A. T., Nicora, C. D., Norbeck, A. D., Zhu, J.-K., et al. (2009). Subunit compositions of the RNA-silencing enzymes Pol IV and Pol V reveal their origins as specialized forms of RNA polymerase II. *Molecular Cell, 33*(2), 192–203. https://doi.org/10.1016/j.molcel.2008.12.015.

Reinders, J., Wulff, B. B. H., Mirouze, M., Mari-Ordonez, A., Dapp, M., Rozhon, W., et al. (2009). Compromised stability of DNA methylation and transposon immobilization in mosaic Arabidopsis epigenomes. *Genes & Development, 23*(8), 939–950. https://doi.org/10.1101/gad.524609.

Richards, E. J. (2009). Quantitative epigenetics: DNA sequence variation need not apply. *Genes & Development, 23*(14), 1601–1605. https://doi.org/10.1101/gad.1824909.

Ronemus, M. J., Galbiati, M., Ticknor, C., Chen, J., & Dellaporta, S. L. (1996). Demethylation-induced developmental pleiotropy in arabidopsis. *Science, 273*(5275), 654. https://doi.org/10.1126/science.273.5275.654.

Russell, S. D. (1991). Isolation and characterization of sperm cells in flowering plants. *Annual Review of Plant Biology, 42*(1), 189–204.

Saze, H., & Kakutani, T. (2007). Heritable epigenetic mutation of a transposon-flanked Arabidopsis gene due to lack of the chromatin-remodeling factor DDM1. *The EMBO Journal, 26*(15), 3641–3652. https://doi.org/10.1038/sj.emboj.7601788.

Schmitz, R. J., Schultz, M. D., Lewsey, M. G., O'Malley, R. C., Urich, M. A., Libiger, O., et al. (2011). Transgenerational epigenetic instability is a source of novel methylation variants. *Science, 334*(6054), 369. https://doi.org/10.1126/science.1212959.

Schneitz, K., Hülskamp, M., & Pruitt, R. E. (1995). Wild-type ovule development in *Arabidopsis thaliana*: A light microscope study of cleared whole-mount tissue. *The Plant Journal, 7*(5), 731–749.

Schoft, V. K., Chumak, N., Mosiolek, M., Slusarz, L., Komnenovic, V., Brownfield, L., et al. (2009). Induction of RNA-directed DNA methylation upon decondensation of constitutive heterochromatin. *EMBO Reports, 10*(9), 1015–1021. https://doi.org/10.1038/embor.2009.152.

Shahzad, A., Parveen, S., Sharma, S., Shaheen, A., Saeed, T., Yadav, V., et al. (2017). Plant tissue culture: Applications in plant improvement and conservation. In M. Z. Abdin,

U. Kiran, Kamaluddin, & A. Ali (Eds.), *Plant biotechnology: Principles and applications* (pp. 37–72). Singapore: Springer Singapore. https://doi.org/10.1007/978-981-10-2961-5_2.

Shaw, R. G., Byers, D. L., & Darmo, E. (2000). Spontaneous mutational effects on reproductive traits of *Arabidopsis thaliana*. *Genetics, 155*(1), 369–378.

She, W., & Baroux, C. (2015). Chromatin dynamics in pollen mother cells underpin a common scenario at the somatic-to-reproductive fate transition of both the male and female lineages in Arabidopsis. *Frontiers in Plant Science, 6*, 294.

She, W., Grimanelli, D., Rutowicz, K., Whitehead, M. W., Puzio, M., Kotlinski, M., et al. (2013). Chromatin reprogramming during the somatic-to-reproductive cell fate transition in plants. *Development, 140*, 4008–4019. dev-095034.

Silveira, A. B., Trontin, C., Cortijo, S., Barau, J., Del Bem, L. E. V., Loudet, O., et al. (2013). Extensive natural epigenetic variation at a de novo originated gene. *PLoS Genetics, 9*(4), e1003437. https://doi.org/10.1371/journal.pgen.1003437.

Singh, M., Goel, S., Meeley, R. B., Dantec, C., Parrinello, H., Michaud, C., et al. (2011). Production of viable gametes without meiosis in maize deficient for an ARGONAUTE protein. *The Plant Cell, 23*, 443–458. https://doi.org/10.1105/tpc.110.079020.

Slotkin, R. K., & Martienssen, R. (2007). Transposable elements and the epigenetic regulation of the genome. *Nature Reviews Genetics, 8*(4), 272–285. https://doi.org/10.1038/nrg2072.

Slotkin, R. K., Vaughn, M., Borges, F., Tanurdžić, M., Becker, J. D., Feijó, J. A., et al. (2009). Epigenetic reprogramming and small RNA silencing of transposable elements in pollen. *Cell, 136*(3), 461–472. https://doi.org/10.1016/j.cell.2008.12.038.

Sniezko, R. (2006). Meiosis in plants. *Plant Cell Biology*, 227–258.

Soh, A., Wong, G., Tan, C., Chew, P., Chong, S., Ho, Y., et al. (2011). Commercial-scale propagation and planting of elite oil palm clones: Research and development towards realization. *Journal of Oil Palm Research, 23*(1), 935–952.

Soppe, W. J., Jacobsen, S. E., Alonso-Blanco, C., Jackson, J. P., Kakutani, T., Koornneef, M., et al. (2000). The late flowering phenotype of fwa mutants is caused by gain-of-function epigenetic alleles of a homeodomain gene. *Molecular Cell, 6*(4), 791–802.

Spillane, C., Curtis, M. D., & Grossniklaus, U. (2004). Apomixis technology development—virgin births in farmers' fields? *Nature Biotechnology, 22*(6), 687–691. https://doi.org/10.1038/nbt976.

Sprunck, S., Rademacher, S., Vogler, F., Gheyselinck, J., Grossniklaus, U., & Dresselhaus, T. (2012). Egg cell-secreted EC1 triggers sperm cell activation during double fertilization. *Science, 338*(6110), 1093–1097. https://doi.org/10.1126/science.1223944.

Stroud, H., Ding, B., Simon, S. A., Feng, S., Bellizzi, M., Pellegrini, M., et al. (2013). Plants regenerated from tissue culture contain stable epigenome changes in rice. *ELife, 2*. https://doi.org/10.7554/eLife.00354.

Stroud, H., Greenberg, M. V. C., Feng, S., Bernatavichute, Y. V., & Jacobsen, S. E. (2013). Comprehensive analysis of silencing mutants reveals complex regulation of the arabidopsis methylome. *Cell, 152*(1–2), 352–364. https://doi.org/10.1016/j.cell.2012.10.054.

Swingle, C. F. (1952). Regeneration and vegetative propagation. II. *The Botanical Review, 18*(1), 1–13. https://doi.org/10.1007/BF02861764.

Tang, K., Lang, Z., Zhang, H., & Zhu, J.-K. (2016). The DNA demethylase ROS1 targets genomic regions with distinct chromatin modifications. *Nature Plants, 2*(11), 16169. https://doi.org/10.1038/nplants.2016.169.

Tanurdzic, M., Vaughn, M. W., Jiang, H., Lee, T.-J., Slotkin, R. K., Sosinski, B., et al. (2008). Epigenomic consequences of immortalized plant cell suspension culture. *PLoS Biology, 6*(12), e302. https://doi.org/10.1371/journal.pbio.0060302.

Teixeira, F. K., Heredia, F., Sarazin, A., Roudier, F., Boccara, M., Ciaudo, C., et al. (2009). A role for RNAi in the selective correction of DNA methylation defects. *Science, 323*(5921), 1600–1604.

Thieme, M., Lanciano, S., Balzergue, S., Daccord, N., Mirouze, M., & Bucher, E. (2017). Inhibition of RNA polymerase II allows controlled mobilisation of retrotransposons for plant breeding. *Genome Biology, 18*(1), 134.

Tittel-Elmer, M., Bucher, E., Broger, L., Mathieu, O., Paszkowski, J., & Vaillant, I. (2010). Stress-induced activation of heterochromatic transcription. *PLoS Genetics, 6*(10), e1001175.

Tiwari, S., Schulz, R., Ikeda, Y., Dytham, L., Bravo, J., Mathers, L., et al. (2008). MATERNALLY EXPRESSED PAB C-terminal, a novel imprinted gene in arabidopsis, encodes the conserved C-terminal domain of polyadenylate binding proteins. *The Plant Cell Online, 20*(9), 2387–2398. https://doi.org/10.1105/tpc.108.061929.

Tucker, M. R., Okada, T., Hu, Y., Scholefield, A., Taylor, J. M., & Koltunow, A. M. G. (2012). Somatic small RNA pathways promote the mitotic events of megagametogenesis during female reproductive development in Arabidopsis. *Development, 139*(8), 1399–1404. https://doi.org/10.1242/dev.075390.

Varma, V. (2010). Advancements in the production of secondary metabolites. *Journal of Natural Products, 3*.

Vaughn, M. W., Tanurdžić, M., Lippman, Z., Jiang, H., Carrasquillo, R., Rabinowicz, P. D., et al. (2007). Epigenetic natural variation in *Arabidopsis thaliana*. *PLoS Biology, 5*(7), e174. https://doi.org/10.1371/journal.pbio.0050174.

Vielle-Calzada, J.-P., Thomas, J., Spillane, C., Coluccio, A., Hoeppner, M. A., & Grossniklaus, U. (1999). Maintenance of genomic imprinting at the Arabidopsis medea locus requires zygotic DDM1 activity. *Genes & Development, 13*(22), 2971–2982.

Virdi, K. S., Laurie, J. D., Xu, Y.-Z., Yu, J., Shao, M.-R., Sanchez, R., et al. (2015). Arabidopsis MSH1 mutation alters the epigenome and produces heritable changes in plant growth. *Nature Communications, 6*(1). https://doi.org/10.1038/ncomms7386.

Vongs, A., Kakutani, T., Martienssen, R., & Richards, E. (1993). *Arabidopsis thaliana* DNA methylation mutants. *Science, 260*(5116), 1926. https://doi.org/10.1126/science.8316832.

Weigel, D. (2012). Natural variation in arabidopsis: From molecular genetics to ecological genomics. *Plant Physiology, 158*(1), 2–22.

Weigel, D., & Colot, V. (2012). Epialleles in plant evolution. *Genome Biology, 13*(10), 249. https://doi.org/10.1186/gb-2012-13-10-249.

Wibowo, A., Becker, C., Durr, J., Price, J., Staepen, S., Hilton, S., et al. (2018). Incomplete reprogramming of cell-specific epigenetic marks during asexual reproduction leads to heritable phenotypic variation in plants. *BioRxiv*, 267955. https://doi.org/10.1101/267955.

Williams, B. P., & Gehring, M. (2017). Stable transgenerational epigenetic inheritance requires a DNA methylation-sensing circuit. *Nature Communications, 8*(1). https://doi.org/10.1038/s41467-017-02219-3.

Williams, B. P., Pignatta, D., Henikoff, S., & Gehring, M. (2015). Methylation-sensitive expression of a DNA demethylase gene serves as an epigenetic rheostat. *PLoS Genetics, 11*(3), e1005142. https://doi.org/10.1371/journal.pgen.1005142.

Xiao, W., Gehring, M., Choi, Y., Margossian, L., Pu, H., Harada, J. J., et al. (2003). Imprinting of the MEA Polycomb gene is controlled by antagonism between MET1 methyltransferase and DME glycosylase. *Developmental Cell, 5*(6), 891–901.

Xu, Y.-Z., Arrieta-Montiel, M. P., Virdi, K. S., de Paula, W. B., Widhalm, J. R., Basset, G. J., et al. (2011). MutS HOMOLOG1 is a nucleoid protein that alters mitochondrial and plastid properties and plant response to high light. *The Plant Cell, 23*(9), 3428–3441.

Xu, Y.-Z., De la Rosa Santamaria, R., Virdi, K. S., Arrieta-Montiel, M. P., Razvi, F., Li, S., et al. (2012). The chloroplast triggers developmental reprogramming when MUTS HOMOLOG1 is suppressed in plants. *Plant Physiology, 159*(2), 710–720.

Yang, X., Kundariya, H., Xu, Y.-Z., Sandhu, A., Yu, J., Hutton, S. F., et al. (2015). MutS HOMOLOG1-derived epigenetic breeding potential in tomato. *Plant Physiology, 168*(1), 222–232. https://doi.org/10.1104/pp.15.00075.

Yang, H., Lu, P., Wang, Y., & Ma, H. (2011). The transcriptome landscape of arabidopsis male meiocytes from high-throughput sequencing: The complexity and evolution of the meiotic process. *The Plant Journal, 65*(4), 503–516.

Zemach, A., Kim, M. Y., Hsieh, P.-H., Coleman-Derr, D., Eshed-Williams, L., Thao, K., et al. (2013). The arabidopsis nucleosome remodeler DDM1 allows DNA methyltransferases to access H1-containing heterochromatin. *Cell, 153*(1), 193–205. https://doi.org/10.1016/j.cell.2013.02.033.

Zhang, H., & Zhu, J.-K. (2012). Active DNA demethylation in plants and animals. *Cold Spring Harbor Symposia on Quantitative Biology,* 77(0), 161–173. https://doi.org/10.1101/sqb.2012.77.014936.

Zhou, A., & Pawlowski, W. P. (2014). Regulation of meiotic gene expression in plants. *Frontiers in Plant Science, 5*, 413.

CHAPTER SIX

Transposable Elements as Tool for Crop Improvement

Michael Thieme* **and Etienne Bucher**[§,1]
*Institute of Plant and Microbial Biology, University of Zürich, Zollikerstrasse, Zürich, Switzerland
[§]IRHS (Institut de Recherche en Horticulture et Semences), UMR 1345, INRA, Agrocampus-Ouest, Université d'Angers, SFR 4207 QuaSaV, Beaucouzé, France
[1]Corresponding author: E-mail: etienne.bucher@inra.fr

Contents

Advances in Botanical Research, Volume 88
ISSN 0065-2296
https://doi.org/10.1016/bs.abr.2018.09.001

Abstract

Transposable elements (TEs) are major contributors to the size of plant genomes. For a long time, TEs had a very negative connotation being primarily considered as "parasites" or "junk DNA". In recent years, this view has started to change: Even though TEs can damage genes present in the host genome, they also contribute to stress response and played a key role in crop domestication. Here, we discuss the molecular mechanisms involved in taming TEs, how TEs link the genome and epigenome to the environment, their contribution to gene structure and expression and finally discuss how TE could be used for crop breeding.

1. GENERAL INTRODUCTION

Prior to the groundbreaking discovery of transposable elements (TEs) by Barbara McClintock (McClintock, 1950) in the mid of the 20th century, the genetic information encoding for a certain phenotype had been assumed to be organized in a unidimensional and static way. Challenging this generally accepted theory, McClintock's observation of unmappable "mutable" loci underlying the highly diverse variegated kernel color of maize, initially did not attract a great deal of attention. Today it is known that the presence of potentially mobile genetic elements, like the *Ac/Ds* system discovered by McClintock, is the rule rather than the exception. Indeed, TEs have been detected in all organisms tested so far and in some extreme cases like maize and barley they make out more than 80% of the genome (Schnable et al., 2009; Wendel, Jackson, Meyers, & Wing, 2016).

In stark contrast to their previous image as "parasitic elements" or "junk DNA" (Ohno, 1972), increasing knowledge in the field of TE-research has rather revealed a multitude of cases illustrating their indispensable function during development, response to environmental triggers and as drivers of evolution. Apart from a multitude of examples found in plants, prominent findings from other kingdoms of life including humans are completing the latest image of TEs as vital basic modules of life (Chuong, Elde, & Feschotte, 2017).

Currently, especially breeders in the organic sector are facing a challenging situation. They have to develop new crop varieties with the potential to build the basis for food security of a growing global population under rapidly changing environmental conditions (Fischer, Byerlee, & Edmeades, 2014). Besides knowledge gained through recent significant advances in genomics (Abberton et al., 2016) especially new insights into the field of epigenetics and epigenetic memory has become a focus of attention of plant breeders (Gallusci et al., 2017). Situated at the interface of genetics and epigenetics, also the awareness of TEs as a valuable genetic resource that could be used to cope with future challenges in agriculture has increased significantly (Paszkowski, 2015). However, as described above, it was so far not possible to access this "(epi)genetic treasure" (Mirouze & Vitte, 2014) without using genetic engineering. In order to be able to harness these elements for plant breeding it is key to gain further knowledge about the mechanisms of their regulation.

2. TRANSPOSABLE ELEMENTS IN PLANT GENOMES

2.1 Life Cycle and Classification

Depending on the mechanism of transposition, TEs are assigned to two major classes: Class I TEs are evolutionary closely related to retroviruses and also referred to as retrotransposons (retroTEs). These elements fully depend on transcription by the host RNA polymerases II (Pol II) - or in special cases Pol III - and amplify themselves via an RNA intermediate resulting in a "copy and paste" transposition. In contrast, class II elements, also known as DNA-transposons, transpose without producing an RNA-intermediate, usually resulting in a "cut and paste" movement. Because of the increasing taxonomical complexity of known TEs, a regularly updated classification system is required (Piegu, Bire, Arensburger, & Bigot, 2015). However, in general, both main classes of TEs can be further subdivided according to their distinct structural and enzymatic characteristics.

For retroTEs, the main criterion for differentiation is the presence or absence of two long terminal repeats (LTRs) with the same orientation at the 5′ and 3′ end of the transposing unit. Plant LTR-retroTEs can range in size from less than 1 kb to up to 22 kb (*Ogre* element in pea) (Neumann, Požárková, & Macas, 2003). The LTRs, 85 bp to 5 kb in length (Zhao, Ferguson, & Jiang, 2016), contain both regulatory regions and the transcriptional start site (TSS) needed for transcription of the TE by the host Pol II (Fig. 1) (Grandbastien, 2015). Transcripts originating from the 5′ LTR play two important roles in the retroTE life cycle. On one hand, they code for the retroTE-replication machinery or polyprotein (POL) consisting of the aspartic proteinase (AP), the reverse transcriptase (RT), the ribonuclease RNaseH (RH), the integrase (INT) and for the structural GAG capsid proteins that form a virus-like particle (VLP) (Grandbastien, 2015). On the other hand, they serve as the template for reverse transcription resulting in

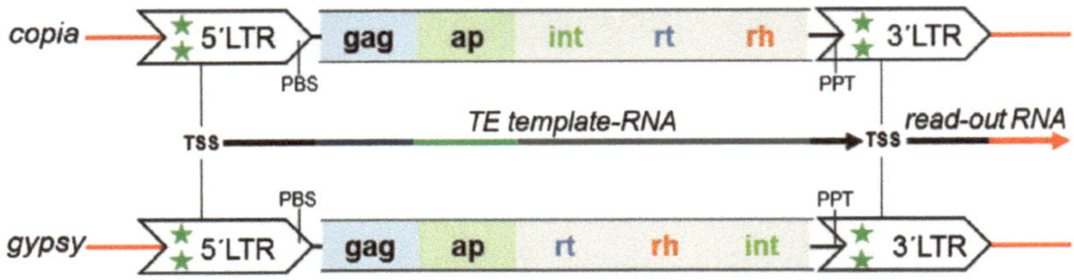

Figure 1 ***Schematic representation of the structure of copia and gypsy-like LTR-retrotransposons in plants.*** Upon activation, often mediated by regulatory motifs (green asterisks) located in the LTRs, transcription by the host Pol II starts at the transcriptional start sides (TSSs) that are located within in the 5′- and the 3′-LTR of the TE. Transcription from the 5′ LTR generates template transcripts that are either translated into the GAG capsid protein (gag), the polyprotein consisting of the aspartic proteinase (ap), the integrase (int), the reverse transcriptase (rt) and the RNaseH (rh) or used as a template for reverse transcription into TE-ecDNA. The primer binding site (PBS) and the polypurine tract (PPT) are indicated. Transcription originating in the 3′ LTR generates read-out RNA of flanking genomic regions. Depending on the sequential arrangement of int, rt and rh LTR retroTEs are subdivided into *copia* and *gypsy*-like elements. *Adapted from Grandbastien, M. A. (2015). LTR retrotransposons, handy hitchhikers of plant regulation and stress response.* Biochimica et Biophysica Acta, 1849, *403–416; Wicker, T., Sabot, F., Hua-Van, A., Bennetzen, J. L., Capy, P., Chalhoub, B., et al. (2007). A unified classification system for eukaryotic transposable elements.* Nature Reviews. Genetics, 8, *973–982. Scheme according to Thieme, M. (2017a).* Stress-induced mobilization of retrotransposons for plant breeding *(Doctoral dissertation). University of Basel.*

extrachromosomal complementary retroTE-DNA (ecDNA) that is potentially capable of entering into the nucleus and to integrate into genomic DNA (Schulman, 2013). As transcription of LTR-retroTEs starts and ends within the LTRs, the region upstream of the TSS in the 5′ LTR and downstream the terminator site in the 3′ LTR is missing in the initial transcript that serves as a template for reverse transcription (Fig. 1). Hence, these regions would get lost unless restored in a complex mechanism during cDNA synthesis. Based on the homology of the two LTRs, retroTEs have evolved a mechanism to restore missing sequence information at both extremities of the LTRs using the homologue sequence that adjoins the internal domain of the TE. Two specific domains are crucial during this process: primer binding site (PBS) and polypurine tract (PPT) (Fig. 1) (Schulman, 2013).

Depending on the sequential arrangement of the protein domains in the TE-body, LTR-TEs are further subdivided into Ty1-*copia* (Pseudoviridae) and Ty3-*gypsy* (Metaviridae)- retroTEs (Wicker et al., 2007) (Fig. 1).

While intact *copia* and *gypsy*– like retroTEs code for all domains needed for their transposition, several additional non-autonomous LTR-retroTEs are known in plants. These include LARDs (Large Retrotransposon

Derivatives) (Kalendar et al., 2004) TRIMs (Terminal-Repeat Retrotransposons In Miniatures) (Witte, Le, Bureau, & Kumar, 2001) and SMARTs (Small LTR Retrotransposons) (Gao, Chen, Chen, Meyers, & Jackson, 2012). All three families lack the coding regions needed to produce their own replication machinery. Instead, they carry an internal non-coding region that ranges in size from less than 0.3 kb (SMARTs and TRIMS) to up to 3.5 kb (LARDS). For their replication, these elements depend on proteins encoded by related autonomous elements. A descriptive case of this kind of functional trans-compensation of domains between related retrotransposons was shown for the *BARE-1 and 2 copia*-like LTR-retroTEs in barley. Due to a mutation in the open reading frame (ORF) of its structural GAG protein, *BARE-2* itself is considered to be non-autonomous. However, thanks to a high degree of homology, the *BARE-2*-retroTE can utilize *BARE-1*-GAG to fulfill its own lifecycle (Tanskanen, Sabot, Vicient, & Schulman, 2007). Although possibly biased by different class-dependent approaches of transposon annotations, retroTEs and more precisely LTR-retroTEs usually represent the largest class of TEs in currently known plant genomes (Fig. 2) (Vitte, Fustier, Alix, & Tenaillon, 2014).

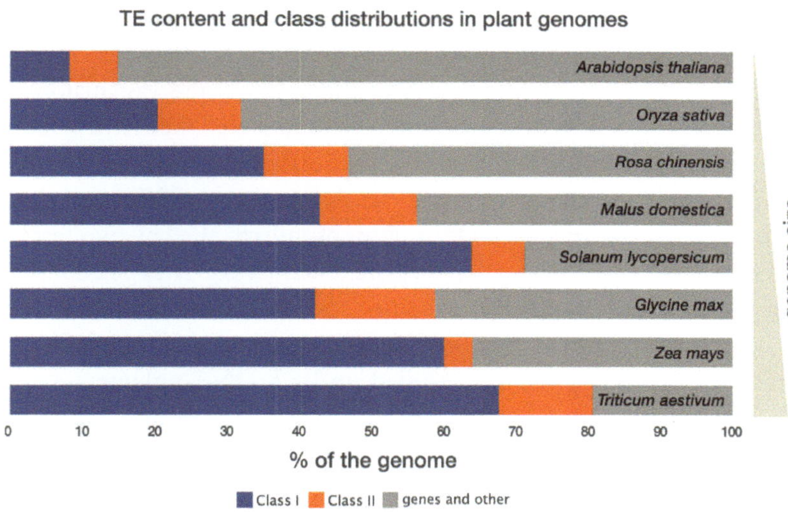

Figure 2 ***Percentage of plant genomes occupied by the two major classes of transposable elements.*** The TE content percentages for these plant species were obtained from: De La Chaux, Tsuchimatsu, Shimizu, & Wagner, 2012 (A.t.), Bruggmann et al., 2006 (O.s.), Hibrand Saint-Oyant et al., 2018 (R.c.), Daccord et al., 2017 (M.d.), Jouffroy, Saha, Mueller, Quesneville, & Maumus, 2016 (S.l.), Schmutz et al., 2010 (G.m.), Jiao et al., 2017 (Z.m.) and Clavijo et al., 2017 (T.a.). Plant species are ordered by genome size from small (top) to large (bottom).

Non-LTR retroTEs in plants are represented by autonomous LINEs (long interspersed nuclear elements) and non-autonomous SINEs (short interspersed nuclear elements (Schmidt, 1999). Transcription of LINEs is driven by an internal Pol II promoter located within in the 5′ UTR, making the process of self-amplification less complex. LINEs generally don't code for structural capsid proteins such as the GAG. Rather, they use an RNA-binding protein that assembles together with the RT and RH to form a ribonucleoprotein instead of a VLP. Furthermore, the mechanism of integration differs between LTR and non-LTR elements. While reverse transcription of LTR-retroTEs takes place in the cytoplasm and is hence spatially separated from integration, LINE-elements integrate by target-primed reverse transcription and without making use of an integrase (Schulman, 2013). In contrast to all previously described TEs, SINEs are derived from RNA-polymerase III transcripts such as 5S RNAs, 7SL RNA and tRNAs. They carry an internal Pol III promoter for transcription but depend on autonomous LINE-elements for their proliferation (Schulman, 2013; Sun, Fleurdepine, Bousquet-Antonelli, Caetano-Anolles, & Deragon, 2007; Wenke et al., 2011).

According to the current classification, DNA transposons generally comprise elements that do not rely on an RNA-intermediate for their transposition. With few exceptions such as *Helitrons* that transpose in a rolling circle-like mechanism (Kapitonov & Jurka, 2001) the vast majority of class II elements in plants move through "cut and paste", hence in a conservative mode of transposition. For this purpose, they solely encode for a single protein called the transposase. Analogous to the LTRs of class I elements, the internal region of DNA transposons is flanked by two terminal inverted repeat (TIR) sequences that are essential for transposition. The transposase catalyzes both the excision and the ligation of the DNA transposon into the target site. After integration, the sticky ends that are generated during the excision, are filled up by a DNA polymerase and a DNA ligase. This results in an element-specific target site duplication (TSD) at both termini of the TE. Both length of the TSD and the sequence of TIR and the sequence similarities of their transposases can be used to further subdivide class II elements. Accordingly, plant DNA-TEs are classified into six different subgroups: *Tc1-Mariner*, *hAT*, *Mutator*, *P*, *PIF-Harbinger* and *CACTA* (Wicker et al., 2007). Analogical to retroTEs, there are examples for non-autonomous DNA-transposons in plants. These include miniature inverted-repeat transposable elements (MITEs) that can be highly abundant in crop genomes (Vitte et al., 2014). In rice it was shown that non-autonomous MITEs depend on autonomous *Tc1-Mariner* elements for their mobility (Feschotte, Swamy, & Wessler, 2003).

Although DNA-TEs typically transpose in a conservative process and without using an RNA intermediate, they can occasionally increase in copy number if transposition happens during the S phase of cell replication (Chen, Greenblatt, & Dellaporta, 1987; Zhang, Zuo, Wang, & Peterson, 2014).

2.2 Impact of Transposable Elements on Genome Size and Architecture

As TEs represent substantial portions of plant genomes, they also play important roles as structural elements that shape genome architectures (Bennetzen & Wang, 2014). The structural analysis and comparison of various sequenced angiosperm genomes including *Arabidopsis*, soybean and rice, has revealed that there is a strong positive correlation between the TE content and the actual genome size (Tenaillon, Hollister, & Gaut, 2010). Besides drastic events such as whole genome duplications, TE-bursts are the main contributors to increases in genome size. With few exceptions, such as a reported burst of MITEs in rice (Naito et al., 2006), retroTE proliferation through copy and paste is the main driver of genome expansions in plants (Lee & Kim, 2014; Piegu et al., 2006; Zedek, Šmerda, Šmarda, & Bureš, 2010). Such expansions are often triggered by a massive amplification of a few TE-families or individual elements (Bennetzen & Wang, 2014). The potentially unlimited growth in genome size resulting from the copy and paste activity of active class I elements would ultimately lead to potentially adverse effects to the host (Bennett, 1972; Bennetzen, Ma, & Devos, 2005; Bennetzen & Kellogg, 1997; Diez et al., 2013; Schubert & Oud, 1997). Hence, unlimited accumulation of transposable elements in plant genomes is counteracted by illegitimate and homologous recombination events resulting in loss of DNA and genome shrinking (Devos, Brown, & Bennetzen, 2002; Hawkins, Proulx, Rapp, & Wendel, 2009; Vitte & Panaud, 2005). Sufficient sequence homology needed for homologous recombination can either exist between the two LTRs of one element but also between LTRs of two closely related retroTEs. This type of inter-element recombination can result in chromosome rearrangement and deletions of larger genomic fragments (Bennetzen & Wang, 2014). Isolated LTRs, also called soloLTRs are abundant remnants of such events in plant genomes (Devos et al., 2002; Ma, Devos, & Bennetzen, 2004; Shirasu, Schulman, Lahaye, & Schulze-Lefert, 2000). Besides their substantial role as regulators of genome size, plant TEs are pivotal for global genome architecture. For instance, there is evidence from rice and maize that TE insertions have been driving the evolution of the centromeric regions of chromosomes that are crucial during cell replication (Gao, Jiang, Wing, Jiang, & Jackson, 2015; Sharma, Wolfgruber, & Presting, 2013; Wolfgruber et al., 2009). On a smaller scale, active TEs can contribute to the evolution of new genes by mediating gene fragment transduction or exon shuffling (Elrouby & Bureau, 2010; Jiang, Bao, Zhang, Eddy, & Wessler, 2004). It has further been reported that mobile *Helitrons* and MULE-elements (so called Pack-MULEs) can capture and amplify functional genes in maize and rice (Jameson et al., 2008; Jiang et al., 2004).

3. REGULATION OF RETROTRANSPOSITION IN PLANTS

3.1 DNA Methylation

DNA methylation at cytosines in the form of 5-methylcytosine (^{m}C) is a key feature of epigenetic regulation and silencing of potentially mobile genetic elements in plants (Miura et al., 2001). The abundancy of ^{m}C has been shown to correlate with the presence of repetitive elements and transposons in the genome of *Arabidopsis* substantiating the role of DNA-methylation in regulating the activity of these elements (Cokus et al., 2008; Zhang et al., 2006). Depending on the sequence context (CG, CHG or CHH where H can be any base but G), it has been proposed that plants have evolved different families of specialized DNA-methyltransferases (DNMtases) to establish or maintain DNA-methylation (Henderson & Jacobsen, 2007). Global DNA-methylation levels in *Arabidopsis* have previously been determined to be approximately 24% (CG), 6.7.% (CHG) and 1.7% (CHH), respectively (Cokus et al., 2008).

3.1.1 Maintenance Methylation

Following DNA replication that results in hemimethylated DNA, the specific ^{m}C-pattern of the newly synthetized daughter strand has to be restored. In case of symmetric methylation in the CG-context, METHYLTRANSFERASE1 MET1 (Finnegan, Peacock, & Dennis, 1996; Kankel et al., 2003) which is a homologue of the mammalian maintenance DNA methyltransferase Dnmt1 (Finnegan & Dennis, 1993) adds methyl groups to the new unmethylated daughter strand. Symmetric CHG methylation in plants is catalyzed by the plant specific CHROMOMETHYLASE3 (CMT3) and in a reinforcing loop triggered by the activity of the histone 3 lysine 9 (H3K9) methyltransferase KRYPTONITE (KYP, also known as SUPPRESSOR OF VARIEGATION 3–9 HOMOLOGUE 4, SUVH4), SUVH5 and 6 (Ebbs & Bender, 2006; Jackson et al., 2002, 2004; Lindroth et al., 2001). Recently, MET2a, a less well characterized homologue of MET1 has also been reported to be important for CHG methylation exclusively at active TE families in *Arabidopsis* (Quadrana et al., 2016; Stroud, Greenberg, Feng, Bernatavichute, & Jacobsen, 2013). So far, two different mechanisms leading to specific methylation at asymmetric CHH sites are known in plants. While CMT2 recognizes H3K9me2, a heterochromatic mark co-localizing with long TEs (Gouil & Baulcombe, 2016), activity of the *de novo* methyltransferase DOMAINS REARRANGED METHYLTRANSFERASE 2 (DRM2) is guided by sequence specific small interfering (si)RNAs (Cao & Jacobsen, 2002).

3.1.2 RNA-Directed DNA Methylation (RdDM)

The potential of retroTEs to produce and integrate intact and unmethylated copies of themselves along the genome triggered the evolution of specialized silencing machineries in plants as well as in animals and fungi (Wheeler, 2013). Generally, these silencing mechanisms are based on the production of highly specific small non-coding RNA molecules that guide the silencing machinery to the corresponding locus in the genome. In plants, this process known as RNA-directed DNA methylation (RdDM) (Matzke, Kanno, & Matzke, 2015; Wassenegger, Heimes, Riedel, & Sänger, 1994) is based on the activity of two additional plant specific RNA-polymerases, NRPD (Pol IV) and NRPE (Pol V) that have evolved as specialized paralogs of NRPB (Pol II) (Herr, Jensen, Dalmay, & Baulcombe, 2005; Kanno et al., 2005; Matzke et al., 2015; Onodera et al., 2005; Pontier et al., 2005; Ream et al., 2009). In contrast to Pol II which is an essential enzyme producing messenger RNAs (mRNAs) in plants, mutants of Pol IV and V do not show any developmental defects in *Arabidopsis* under standard growth conditions (Pontier et al., 2005), whereas these mutants can drastically alter plant phenotype in some crops such as rice and maize – see the Chapter 11. The core of this complex silencing pathway, also called the "canonical RdDM", is now well investigated in *Arabidopsis* (Wendte & Pikaard, 2017).

Current models suggest that in a first step, RNA-DEPENDENT RNA POLYMERASE 2 (RDR2), converts non-coding RNAs produced by Pol IV into double stranded RNAs (dsRNAs) (Haag et al., 2012; Law, Vashisht, Wohlschlegel, & Jacobsen, 2011). These non-coding dsRNAs are subsequently processed by DCL3 (DICER-LIKE 3) into 24 nt siRNAs, stabilized by methylation of their 3′-OH groups (Yang, Ebright, Yu, & Chen, 2006) and loaded onto AGO4 (Zilberman, Cao, & Jacobsen, 2003) or AGO6 (Zheng, Zhu, Kapoor, & Zhu, 2007). Finally, presumably mediated by specific base pairing of AGO4-bound siRNAs with a nascent Pol V transcript (Wierzbicki, Ream, Haag, & Pikaard, 2009) or with genomic DNA (Lahmy et al., 2016), DRM2 is recruited resulting in a highly specific methylation of the Pol V transcribed loci in all three sequence contexts (Böhmdorfer et al., 2014).

Besides this well studied Pol IV-RDR2-dependent branch of the "canonical" RdDM pathway, there is emerging evidence for the importance of alternative routes of siRNA biogenesis resulting in Pol V-dependent TE-silencing (Cuerda-Gil & Slotkin, 2016) in plants. With few exceptions including the Pol IV-NERD RdDM pathway (Pontier et al., 2012), these so-called "non-canonical" RdDM pathways have in common that they are directly dependent on transcriptional activity of Pol II. Although reverse genetic approaches for studying Pol II-dependent silencing mechanisms are limited by a drastic loss of viability of hypomorphic Pol II-mutants (Zheng et al., 2009), there is increasing evidence describing silencing pathways acting downstream of Pol II (Fig. 3). Accordingly, it was shown that Pol II transcripts can be processed into primary 21−22 nt siRNAs in a RDR6-DCL2-DCL4-dependent manner (Nuthikattu et al., 2013). There is recent evidence that this pathway also termed RDR6-RdDM targets especially transcriptionally active, full length TEs in *Arabidopsis* (Panda et al., 2016). Importantly, primary siRNAs whose production can also be triggered by the DCL1-dependent formation of microRNAs (Creasey et al., 2014) can feed into an RNAi loop leading to a strong RDR6-dependend accumulation of secondary siRNAs and post transcriptional gene silencing (PTGS) (Cuerda-Gil & Slotkin, 2016). Besides the RDR6-DCL2-DCL4-dependent production of 21−22 nt siRNAs, Pol II transcripts can also be processed in a RDR6-DCL3-dependet manner resulting in 24 nt siRNAs, presumably reflecting partial redundancy and potential functional compensation between DCL2/4 and DCL3 in specifically silencing TEs present in high copy numbers in the genome (Gasciolli, Mallory, Bartel, & Vaucheret, 2005; Mari-Ordonez et al., 2013). Only recently, another Pol II-dependent "non-canonical" RdDM pathway that also relies on DCL3 but that functions independently of RDRs resulting in the production of 24 nt siRNAs has been identified in *Arabidopsis* (Panda et al., 2016). It is presumed that this kind of RDR-independent TE silencing is triggered by

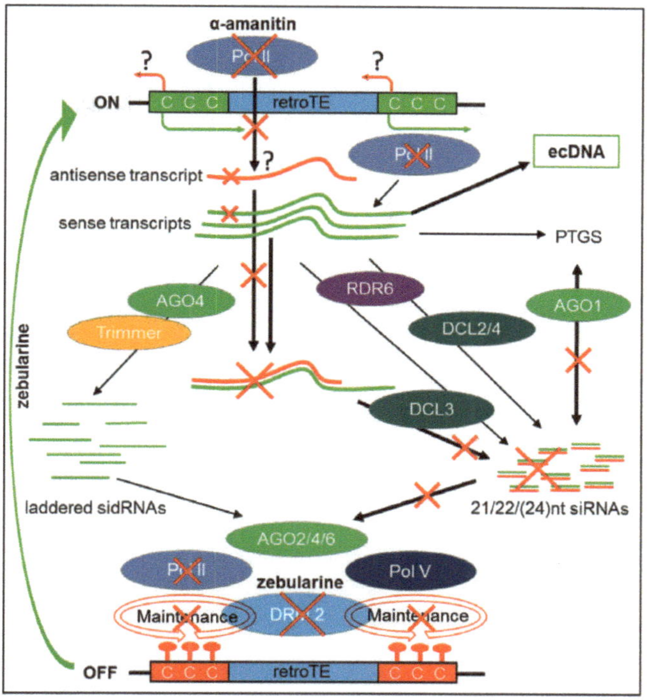

Figure 3 ***Updated model for the mechanism of the "non-canonical" expression-dependent RdDM leading to retroTE silencing in plants.*** The initial silencing of retroTE is presumably triggered by the Pol II-dependent production of antisense-transcripts. As previously suggested by (Panda et al., 2016) paired sense-antisense transcripts can be processed in a RDR6-independent manner resulting in the DCL3-dependent production of specific siRNAs. Inhibition of DNMtases (DRM2) with Z results in a loss of DNA-methylation of retroTE converting them into a transcriptional active (ON) state. As previously reported (Panda et al., 2016) this would lead to an increased (bold arrows) DCL3-dependent production of primary siRNAs. The additional general inhibition of Pol II with A reduces both sense and antisense-transcription, leading to less sense-antisense pairing of TE-transcripts, a reduced accumulation of primary siRNAs, a drop in PTGS and an increased production of retroTE-ecDNA. The specific targets of A and Z and resulting impacts on retroTE-silencing are marked with red crosses. Model according to (Thieme, 2017a).

the formation of imperfectly paired dsRNAs as they occur after transcription of inverted repeats. This scenario has been reported for the Mu killer locus that causes stable trans silencing of potentially active members of the MuDR-family in maize (Slotkin, Freeling, & Lisch, 2005). As transcription by Pol II directly precedes dicing by DCL3 it is self-evident that this pathway is also important in re-silencing TEs that are in a transcriptionally active state (Panda et al., 2016).

Another Pol II-dependent mechanism for the production of small RNAs triggering initial TE-silencing was found to be independent of DCL-enzymes (Ye et al., 2016). The authors of this study detected a new class of dicer independent small RNAs (sidRNAs) that ranged from 20–60 nt in size. Based on their observations they concluded that these sidRNAs are generated by stepwise trimming of the 3′ end of AGO4-bound Pol II or Pol IV transcripts originating from active loci including retroTEs.

Although the various routes of siRNA biogenesis described above can be disentangled based on their specific mechanistic properties, their actual biological function and importance is challenging to be determined. For instance, Panda and colleagues found evidence that a functional loss of the RDR6-RdDM pathway can be compensated by Pol IV-dependent silencing. More importantly, it seems that the transcriptional state of a TE predetermines which of the above described RdDM pathways becomes dominant in (re)-silencing certain TEs (Panda et al., 2016).

3.2 Chromatin State

The chromatin structure and its associated histone marks are key regulatory elements of gene regulation and TE-silencing in plants. Nucleosomes have approximatively 146 bp of DNA wrapped around a histone core complex and form the basic repeating units of eukaryotic chromatin. The core complex itself consists of eight histone proteins and more specifically of two dimers of H2A and H2B and a tetramer composed of two dimers of histones H3 and H4 (Luger, Mader, Richmond, Sargent, & Richmond, 1997).

Besides the reported distinct distribution of different histone variants (Stroud et al., 2012; Yelagandula et al., 2014), dynamic chemical modifications, such as acetylation, phosphorylation or methylation, of the N-terminal histone tails strictly correlate with the expressional state of chromosomal regions in *Arabidopsis* (Fuchs, Demidov, Houben, & Schubert, 2006). There is evidence for a very close connection between DNA-methylation pathways and certain types of histone modifications in plants. For instance, it was reported that CHG-methylation correlates with histone H3 lysine nine di-methylation (H3K9me2) (Bernatavichute, Zhang, Cokus, Pellegrini, & Jacobsen, 2008; Jackson et al., 2004) which is considered as an indicator for the presence of heterochromatin (Fuchs et al., 2006). A simultaneous knock-out of the histone methyltransferases SUVH4-6 resulted in the concomitant reduction of H3K9me2 and non-CG DNA methylation resulting in release of TE-transcription (Ebbs & Bender, 2006). Likewise, it was shown that TEs coincide with high levels of histone 3 lysine 27 monomethylation and that defects in the H3K27 monomethyltransferases ARABIDOPSIS TRITHORAX-RELATED PROTEIN 5 (ATXR5) and ATRX6 resulted in transcriptional up-regulation of TEs (Jacob et al., 2010).

Besides the addition of repressive marks, it was also reported that the active removal of activating marks, such as histone acetylation plays an important role for TE-silencing. For instance, knocking-out HISTONE DEACETYLASE 6 (HDA6) led to a loss of heterochromatin formation. It is assumed that an interaction between HDA6 and MET1 results in locus specific methylation in the CG-context thereby forming the basis for subsequent non-CG methylation resulting in silent chromatin (To et al., 2011). Consequently, it was found that a subset of TEs are transcriptionally reactivated in the *hda6* background (Hristova, Fal, Klemme, Windels, & Bucher, 2015; Liu et al., 2012; Probst et al., 2004).

A key-factor that mediates the crosstalk between the chromatin state and DNA-methylation in *Arabidopsis* is the chromatin-remodeling ATPase DECREASE IN DNA METHYLATION 1 (DDM1). By providing DNMtases access to heterochromatin (Zemach et al., 2013), DDM1 has been reported to play an important general role in maintaining CG and non-CG methylation (Jeddeloh, Stokes, & Richards, 1999; Vongs, Kakutani, Martienssen, & Richards, 1993) and TE-silencing (Lippman, May, Yordan, Singer, & Martienssen, 2003; Miura et al., 2001; Tsukahara et al., 2009). Mutating DDM1 results in a loss of H3K9me2, a global reduction of DNA methylation and a decrease of heterochromatin condensation at TEs in *Arabidopsis* (Gendrel, Lippman, Yordan, Colot, & Martienssen, 2002; Lippman et al., 2004). As a consequence, *ddm1* plants show a genome-wide upregulation of TEs (Lippman et al., 2004) and it was shown that inbreeding of *ddm1* mutant lines resulted in transposition of various class I and II TEs, including *gypsy* and *copia* elements (Tsukahara et al., 2009). Due to its central function in maintaining TE-silencing, several main discoveries elucidating expression-dependent TE-silencing pathways such as the RDR6-or the DCL3-dependent RdDM have been made in the *ddm1*-background where TEs are in a more "active state" (McCue et al., 2015; Nuthikattu et al., 2013; Panda et al., 2016; Teixeira et al. 2009).

4. TRANSPOSABLE ELEMENTS: LINKING THE GENOME AND THE EPIGENOME TO THE ENVIRONMENT

4.1 Stress-Dependent Activation of Transposable Elements in Plants

Already in the mid of the last century, when Barbara McClintock published her pioneering work on DNA-TEs in maize she described them as "controlling elements" (McClintock 1950, 1956) and linked their activation to the occurrence of a "challenge" or a stress (McClintock, 1984). She suggested that the occurrence of "genomic shocks" as for example of global mutations caused by irradiation could trigger the activation of transposition in plants (McClintock, 1984). Today her revolutionary hypotheses have been confirmed in a multitude of studies that clearly illustrate the role of both classes of TEs as potentially mobile regulatory elements and sensors of various stresses (Negi, Rai, & Suprasanna, 2016). Generally, retroTE-activity can be detected during different stages of their life cycle at the RNA and DNA levels (Schulman, 2013). As reviewed by Negi and colleagues, most of the currently known examples for the stress-dependent activation of class I elements are based on the detection of increased TE-transcript levels (Negi et al., 2016). However, transcript levels are only a first hint on the possible mobilization of TEs. In the case of retroTEs, extrachromosomal DNA resulting from the reverse transcription of such transcripts represent a much better proxy for *bona fide* mobilization of these TEs. Recently, an elegant method termed "mobilome sequencing" has been developed which allows the genome-wide detection of mobilized retroTEs (Lanciano et al., 2017).

4.2 Examples of Environmental and Developmental Stimuli

The list of conditions reported to induce TE-transcription comprises a broad panel of various biotic and abiotic stresses including tissue culture. In *Arabidopsis* there is evidence that elicitation with bacterial flagellin (Yu et al., 2013), heat (Pecinka et al., 2010; Tittel-Elmer et al., 2010), arsenic stress (Castrillo et al., 2013), salt stress, nutrition starvation as well as abscisic acid and cytokinin treatment (Duan et al., 2008; Zeller et al., 2009) can trigger the transcriptional activation of certain TEs. From what was found in other plant species including important crops, the list of activating environmental conditions leading to increased transcription of TEs can be expanded by treatment with the fungal elicitor cryptogein (*Nicotiana tabacum*) (Anca et al., 2014), wounding (*Quercus suber, Solanum chilense*)

(Rocheta, Carvalho, Viegas, & Morais-Cecílio, 2012; Tapia et al., 2005), aphids infestation (*Pinus sylvestris*) (Voronova, Belevich, Jansons, & Rungis, 2014), cold (*Citrus sinensis*) (Butelli et al., 2012), water deficiency (*Zea mays*) (Lu et al., 2011), UV-light (*Avena sativa and Cucumis melo)* (Kimura et al., 2001; Ramallo, Kalendar, Schulman, & Martinez-Izquierdo, 2008) and herbicide treatment with atrazine (*Oryza sativa*) (Zhang, Zhou, Song, Liu, & Yang, 2012). In rice, there is further evidence that inter-specific hybridization events can trigger retrotransposition of some class I elements (Wang et al., 2009).

Interestingly, there are also developmental aspects to the transcriptional activation and mobilization of TEs. A key tissue in this respective is the *Arabidopsis* endosperm: Here, genome-wide DNA demethylation (Hsieh et al., 2009) leads to the transcriptional activation of numerous TEs and concomitantly the production of heterochromatic small RNAs (Mosher et al., 2009). In line with these findings, using the highly sensitive mobilome sequencing approach, Lanciano *et al.* were able to demonstrate that *PopRice*, an LTR retroTE, was highly active in rice endosperm (Lanciano et al., 2017). Because TEs have also been found to potentially be active in other tissues, such as meristems (Yamashita & Tahara, 2006) it is now of great interest to systemically test the tissue-specific activity of certain TEs during normal plant development.

4.3 Mechanisms Underlying the Stress Response of LTR-Retrotransposons

As described above (Fig. 1) transcription of *copia* and *gypsy*-type class I elements initiates at the TSSs of both 5′ and 3′ LTRs. As reviewed by Galindo-Gonzalez and colleagues, there are some well-studied cases in plants elucidating the molecular mechanism behind the observed stress-dependency of retroTE-activation (Galindo-Gonzalez, Mhiri, Deyholos, & Grandbastien, 2017). It was shown for different stress-responsive class I elements that *cis*-regulatory motifs in the LTRs play a pivotal role in converting an environmental trigger into transcriptional upregulation and retroTE-mobilization (Casacuberta & Grandbastien, 1993; Takeda, Sugimoto, Otsuki, & Hirochika, 1998; Tapia et al., 2005). In this regard, a well-studied retroTE is the heat responsive *Ty1/copia*-like retroTE *ONSEN* (*AtCOPIA78*) in *Arabidopsis* (Ito et al. 2011; Pecinka et al., 2010; Tittel-Elmer et al., 2010). A closer examination of the promoter in the *ONSEN*-LTRs revealed that all eight copies in the Columbia accession of *Arabidopsis* contain a heat response element (HRE) with the consensus sequence nTTCnnGAAn (Cavrak et al., 2014). This study further provided evidence showing that binding of specific heat shock factors (HSFs) such as HSFA2 to these HREs mediates the heat-stress dependent activation of *ONSEN* transcription (Cavrak et al., 2014).

5. IMPACT ON GENE STRUCTURE AND EXPRESSION

5.1 Insertional Mutagenesis

Transposition of TEs can affect gene expression in numerous ways (Lisch, 2013). Loss of function mutations caused by insertions into regulatory regions, introns and exons of genes are commonly observed in plant genomes. Apart from obvious phenotypes caused by insertions of these elements mobilized in *Arabidopsis* mutants impaired in epigenetic regulation (Mirouze et al., 2009; Tsukahara et al., 2009) many examples are found among selected crop varieties. For instance, photoperiod-insensitivity in cultivars of *Glycine max* (Kanazawa, Liu, Kong, Arase, & Abe, 2009), parthenocarpy in varieties of *Malus domestica* (Yao, Dong, & Morris, 2001) and an observed dwarf phenotype in *O. sativa* (Chen et al., 2017) have been linked to mutations caused by novel TE insertions.

5.2 Impact on Splicing

Besides causing loss of function mutations, retroTE-insertions can also lead to changes in splicing. In case of three different alleles of the *waxy* gene involved in amylose biosynthesis in maize, it was shown that insertions of class I elements into introns can cause alternative splicing (Varagona, Purugganan, & Wessler, 1992). A recent study in oil palm further illustrated that also dynamic epigenetic modifications of inserted retroTEs can result in drastic phenotypic changes. Hence, it was reported that alternative splicing caused by hypomethylation of a LINE element inserted in the intron of the homeotic *DEFICIENS* gene underlies the mantled somaclonal variant of oil palm (Ong-Abdullah et al., 2015).

5.3 Expressional Changes in TE-Flanking Regions

Functional *copia* and *gypsy* elements are flanked by two identical LTRs in the same orientation. Hence, transcription initiating at the TSS of the 3′ LTR (Fig. 1) may also drive expression of flanking genomic regions (Chuong et al., 2017; Grandbastien, 2015). Depending on the *cis* regulatory elements of the retroTE, this so-called "read-through" transcription can therefore also be stress-dependent. In blood oranges, this kind of stress-dependent gene activation mediated by the 3′ LTR of a retroTE has been shown to account for the accumulation of anthocyanins following cold-stress. Mechanistically, the cold responsive promoter in the 3′ LTR drives the expression of a MYB transcriptional activator of anthocyanin production (Butelli et al., 2012). It was also reported that following a recombination event in some orange cultivars, the remaining soloLTR of the same retroTE is sufficient to cause the same cold-dependent phenotype.

Studies performed at the genome-wide level have demonstrated that TEs can have an important influence on their host's stress response. *mPing* is a DNA TE in rice that can be mobilized by *in vitro* culture (Naito et al., 2006). By combining whole genome sequencing to identify thousands of novel *mPing* insertions and transcriptomic analyses it was found that this TE can create novel stress-responsive gene regulatory pathways. A notable feature of this TE is that it is often found in promoter regions of genes. Because this TE acts as a cold-stress responsive enhancer for nearby genes, it renders them responsive to this stress, while these genes do not respond to stress in plants that do not carry such novel insertions (Naito et al., 2009). Striking evidence for the importance of various TE-super-families including retroTEs as mediators of abiotic stress responses were also found in maize. It was revealed that as much as 33% of genes that are only expressed in response to abiotic stresses are associated with upstream inserted TEs (Makarevitch et al., 2015). These observations raise the question as to the role of TEs in the evolution of stress response, not only in plants but also in animals (Chuong et al., 2017). Can novel stress-responsive gene regulatory pathways resulting from TE mobility contribute to the adaptation of the host?

Besides read-through transcription, the insertion of a retroTE close to a gene can also give rise to the recruitment and spreading of repressive epigenetic marks resulting in silencing of flanking regions (Sigman & Slotkin, 2016). A well-studied example for the alteration of the epigenetic landscape flanking a class I element (SINE-element) has been reported for the *FLOWERING WAGENINGEN (FWA)*-locus in *Arabidopsis* (Kinoshita et al., 2007). In this case, the closely inserted retroTE mediates establishment of DNA-methylation and hence forms the basis for imprinting of the *FWA*-locus (Fujimoto et al., 2008). TEs can thus bring endogenous genes under epigenetic control, indeed the vast majority of cases of stable heritable epigenetic variability has been directly linked to TEs (He et al., 2018; Ong-Abdullah et al., 2015).

5.4 Gene Movement and Duplications

It has long been known that TEs have the ability to capture genes or transcripts thereof. These captured genes can then be re-integrated elsewhere into the genome. Genes that have been duplicated via the activity of a retrotransposon can often be recognized by their missing introns. These shorter intron-less genes are the result of the reverse-transcription of spliced gene transcripts (Roy & Gilbert, 2005). Because of position effect variegation, the regulation and expression of a gene can be fundamentally changed if it is relocated into a different genomic region (Henikoff, 1990). There is increasing evidence that TEs can contribute to such major events in plants (Lisch, 2013). A striking example for a retroTE-mediated gene movement was found in *Solanum lycopersicum* where a 24.7 kB gene duplication event resulted in the evolution of the *SUN* locus that is underlying the oval shape of Roma tomatoes (van der Knaap, Sanyal, Jackson, & Tanksley, 2004; Xiao, Jiang, Schaffner, Stockinger, & van der Knaap, 2008).

6. TRANSPOSONS AS A TOOL FOR CROP IMPROVEMENT AND BREEDING

While class II DNA TEs also have a potential for applications in crop breeding, as we have discussed for rice, recent technological advances have specifically unlocked the monitoring and mobilization of class I retroTEs. In this context we would like to focus on retroTEs as a promising class of mobile elements that could be harnessed for breeding (Paszkowski, 2015). Indeed, ongoing retrotransposition events substantially contribute to phenotypic diversity of various crops (Vitte et al., 2014). With growing knowledge of the causes and the consequences of class I element mobility in crops, they are increasingly seen as a valuable resource to advance plant breeding (Mirouze & Vitte, 2014; Paszkowski, 2015). For instance, the evolution of major traits such as apical dominance in maize (Doebley, Stec, & Gustus, 1995) or the waxy phenotype in foxtail millet (Kawase, Fukunaga, & Kato, 2005) can be traced back to the selection of epi/genetic diversity caused by the mobility of transposable elements (Lisch, 2013). However, due to the sophisticated TE-silencing mechanism described above, there are relatively few cases where the frequency of retrotransposition was efficiently increased in a controlled manner in plants. Exceptions are found in cases where plants are exposed to major stresses such as de-differentiation in tissue culture (Hirochika, Sugimoto, Otsuki, Tsugawa, &

Kanda, 1996; Masuta et al., 2017) or when strong genomic stresses, such as inter-specific hybridization, occur (Wang et al., 2009).

Because TEs are important contributors to genome evolution, their mobility could be very useful in plant breeding. Indeed, for decades, active TEs have been used as a mutagen by their mobilization in tissue culture or via transgenesis. In comparison to chemical mutagenesis, mobilized TEs provide the advantage of labeling the insertion site, thus allowing the identification of genes of interest. However, de-differentiation of explants and the regeneration of these to full plants can be very challenging, depending on the plant species and variety. Because of these limitations, and because of the potentially adverse effects TEs can have on their host genome, they have for a long time only been considered as detrimental. Thus, TEs have not (at least not knowingly) been frequently used in plant and crop breeding programs as of yet.

Therefore, the ability to mobilize TEs in plants in a controlled manner would open up the possibility to study their importance in inducing genetic and epigenetic changes in response to external stimuli. One could then finally test in real-time whether TEs play an adaptive role in plants.

6.1 Detection of Retrotransposition Events

A prerequisite for harnessing TEs for plant breeding is their efficient detection during and after transposition (Vitte et al., 2014). Efficient classical approaches allow the quantification of TE copy numbers (quantitative real-time PCR and the now rarely used Southern blot) and PCR-based detection of TE insertion sites by "transposon display" (Ito et al., 2011). Advances in sequencing technologies and downstream data processing significantly accelerated the field of TE-research. Especially the availability of longer sequencing reads overcomes detection limitations of TE-associated structural variants caused by the repetitive nature of TEs (Debladis, Llauro, Carpentier, Mirouze, & Panaud, 2017). Recently developed methods to specifically sequence the active "mobilome" (mobile genetic elements in a cell) will help unlock the use of TEs in plant breeding (Lanciano et al., 2017; Thieme et al., 2017b), but also allow addressing fundamental questions in TE biology and epigenetics.

6.2 Approaches to Mobilize TEs for Plant Breeding

6.2.1 Genetic Engineering

Following the discovery that TEs could be mobilized in cell culture, some of these TEs have been cloned and transformed into other plant species. For instance, *Tnt1* which was initially isolated in *N. tabacum* (Grandbastien, Spielmann, & Caboche, 1989) was shown to be active in heterologous plant species such as *Arabidopsis* (Lucas, Feuerbach, Kunert, Grandbastien, & Caboche, 1995) *Medicago truncatula* (D'Erfurth et al., 2003) and finally *Cucumis sativus* (Zhang et al., 2018). Such transgenic TEs were found to be a useful mutagen allowing forward genetic mutant screens.

Another approach used to activate TEs in plants depends on the knock-out of genes involved in silencing TEs. In this regard, mutations in enzymes involved in the maintenance or *de novo* DNA methylation at TEs have been especially efficient. In *Arabidopsis*, it was shown that developmental defects observed in the *ddm1* mutants that were selfed over several generations could be in part attributed to the mobilization of the *CACTA* DNA transposon (Miura et al., 2001). Later, numerous other class 1 and class 2 TEs were found to be mobilized in *ddm1* (Tsukahara et al., 2009). For several years, the true biological function of the plant-specific RNA polymerases IV and V in *Arabidopsis* remained unclear. Detailed molecular studies clearly showed the role of these polymerases in RdDM targeting DNA methylation at TEs. As described above, it was then later found that in plants defective in Pol IV or Pol V, TEs could be mobilized under stress conditions (Ito et al., 2011) or by crossing these plants with *met1* (Mirouze et al., 2009). These reports unlocked the study of endogenous TEs in *Arabidopsis* silencing mutants.

In principle, the approach of using mutants defective in TE-silencing could also be promising to induce retrotransposition in crops. Indeed, there are examples for the analogical mobilization of retroTEs in crops. In rice for instance, the mobilization of the *copia*-like retroTE *Tos17* was enhanced by knocking out a H3K9 methyltransferase (Ding et al., 2007). However, the limited availability of mutant alleles (Paszkowski, 2015) that are in addition possibly accompanied by severe side-effects like observed in rice (Hu et al., 2014) or maize (Li et al., 2014), currently restricts the use of retroTEs in crop breeding. Yet, recent major advances in the field of genetic engineering and DNA-sequencing opens up completely new possibilities (Springer & Schmitz, 2017). Besides targeted mutagenesis of TE-silencers by means of zinc finger nucleases (ZFNs), transcription activator-like effector nucleases (TALENs) or CRISPR-Cas9 systems, especially specific "epigenome editing" using for example a nuclease deficient Cas9 protein (dCAS9) fused to a methyltransferase (Park, Keung, & Khalil, 2016) are catalyzing the progress in the field. However, in order to be applied, all these methods rely on transient or stable transformation. Depending on the species or variety, genetic engineering can be very elaborate or even impossible. Furthermore, the resulting plants may have to be legally considered as "genetically modified organisms" which can further complicate the downstream applications for breeding.

6.2.2 Transient Inhibition of Retrotransposon Silencing

An alternative approach to increase retroTE mobility in plants is to transiently inhibit enzymes involved in epigenetic silencing by applying different drugs (Pecinka & Liu, 2014). By targeting highly conserved silencing-pathways, the drug-mediated retroTE-activation does neither require prior knowledge of the genome sequence nor elaborate genetic engineering. Originally developed as agents in cancer therapy (Lyko & Brown, 2005), mainly inhibitors of DNMtases such as the cytidine-analogues 5-azacytidine (AZA) and zebularine (Z) are now also commonly used to induce epigenetic changes in plants (Griffin, Niederhuth, & Schmitz, 2016). During cell replication, both cytidine-analogues are incorporated into the DNA. In bacterial model systems, it was shown that the presence of these cytidine-analogues leads to a very stable (Z) (Champion et al., 2010) or even covalent (AZA) (Santi, Garrett, & Barr, 1983) binding of DNMtases to the DNA resulting in a depletion of active DNMtases and subsequent loss of DNA methylation. In plants, transient inhibition of DNMtases results in a reduction of methylation levels in all sequence contexts (Baubec, Pecinka, Rozhon, & Mittelsten Scheid, 2009; Griffin et al., 2016; Thieme et al., 2017b). In case of Z, which is considered to be more stable in aqueous solutions and less cytotoxic compared to AZA (Cheng et al., 2004) the observed reduced DNA-demethylation resulted in the transcriptional release of several TEs such as the non-LTR retroTE LINE1-4 in *Arabidopsis* (Baubec et al., 2009). Genome-wide DNA methylation analysis on plants treated with Z or AZA showed that both drugs induced very similar DNA methylation changes. Notably, both drugs had the strongest effect on the DNA methylation levels of *CACTA*-like DNA TEs (Griffin et al., 2016). While these drugs efficiently activated the transcription of TEs, we could not find reports on the actual mobilization of TEs by the application of epigenetic drugs on plants.

Because we wanted to be able to study TE bursts in wild type plants and use them for crop breeding, we tested if epigenetic drugs could be used to mobilize TEs (Thieme, 2017a; Thieme et al., 2017b). We used the well-studied heat-stress responsive *ONSEN* retroTE in *Arabidopsis* as a reporter for the drug treatments. Because numerous transposable elements are silenced not only by maintenance but also by RdDM-mediated DNA methylation (Matzke & Mosher, 2014; Yokthongwattana et al., 2010) we wanted to apply drugs that could affect both silencing pathways at different levels. The aforementioned Z is very efficient in inhibiting DNA methyltransferases, the last step in the TE silencing pathway. However, to our knowledge no drug had been reported to efficiently inhibit RdDM. Considering reports that RNA polymerase II (Pol II) transcripts may be the source of the initial highly specific signal targeting TEs (Cuerda-Gil & Slotkin, 2016), we wanted to test if this enzyme complex may be involved in limiting TE mobility. To test the involvement of Pol II in repressing TE mobility, we used two approaches: Plants partially defective in Pol II (*nrpb2-3*) (Zheng et al., 2009) and treatments with α-amanitin (A), a very potent inhibitor of Pol II (Lindell, Weinberg, Morris, Roeder, & Rutter, 1970). Testing drug combinations we found that combining Z with A could efficiently phenocopy a mutation on Pol IV/V in terms of *ONSEN* ecDNA-production. In descendants of plants treated with A, Z and heat-stress we were able to identify up to 75 novel *ONSEN* insertions per plant. Thus, the combined treatment with A&Z did not only lead to a very strong activation of *ONSEN* but also led to the massive amplification of this heat-responsive element in the wild type background (Thieme, 2017a; Thieme et al., 2017b). From the mechanistic point of view, we hypothesize that Pol II is involved in generating antisense transcripts starting within the LTRs that then lead to the production of siRNAs. These can then either target TE transcripts in PTGS or lead to specific DNA methylation at retroTE loci (Fig. 3 and Thieme et al., 2017b).

The treatment that resulted in many additional *ONSEN* copies led to a broad panel of phenotypes ranging from changes in seed color, chlorophyll content and flowering time. Generalizing our approach of the transient inhibition of TE-silencing, we showed that these drug treatments might also mobilize TEs in monocot rice. This suggests that this method could be used to increase phenotypic diversity in virtually any plant (Thieme et al., 2017b).

When performing these kinds of treatments, one should keep in mind that they may also induce DNA damage and heritable epigenetic changes that could also be responsible for the observed phenotypic changes. Stable epigenetic changes resulting from treatments with AZA have been reported in *Brassica rapa* (Amoah et al., 2012) and *Fragaria vesca* (Xu, Tanino, Horner, & Robinson, 2016). While in these plants, quantitative variations in traits such as seeds size and flowering time have been observed, genetic changes e.g. via the potential AZA-induced mobilization of TEs was not excluded.

6.3 Phenotypic Consequences of Induced Transposition in Plants

Within the realms of possibility, induced transposition has been used for decades in order to create phenotypic diversity and/or to identify gene functions (Gierl & Saedler, 1992; Naito et al., 2009). To obtain transposon-mediated diversity in a selected plant or crop, the target lines were either crossed with varieties that contain mobile TEs or these TEs were introduced by transformation. In most of these cases, class II DNA elements were used (e.g. Activator and Dissociation - Ac/Ds or Robertson's Mutator - Mu) (Settles, 2009). A great advantage of TE-mediated mutagenesis is that the mutated loci are then tagged with a known TE, thus simplifying the identification of the gene of interest. Together with T-DNA insertions this TE tagging approaches have been very useful in forward genetic mutant screens, allowing the identification of numerous genes. TE-mediated tagging is the preferred method compared to T-DNA insertions when plants are difficult to transform because a few efficient transformation events can lead to numerous independent TE insertions that can then be analyzed (Ramachandran & Sundaresan, 2001).

Our work on the drug-induced mobilization of a retroTE in *Arabidopsis* wild type plants has unlocked the so far little-used endogenous retroTEs for gene tagging. In *Arabidopsis,* we observed numerous phenotypic variations potentially resulting from the mobilization of *ONSEN*. To confirm that *ONSEN* indeed mutated genes, we chose one line with additional *ONSEN* insertions that displayed a very clear phenotype in the seeds. This line presented *transparent testa*, meaning that they were yellow as compared to dark brown in the wild type (Fig. 4). Using a candidate gene approach and genetic segregation patterns we were able to identify and confirm that an insertion of *ONSEN* in *TRANSPARENT TESTA 6* (TT6, *AT3G51240*) was responsible for the observed phenotype (Thieme et al., 2017b). This shows that it is now indeed possible to use endogenous retroTEs for forward genetic mutant screens in wild type plants.

Interestingly, we also noted several environment-dependent alterations of complex traits such as flowering time and biomass production that could be the result of various (epi)genetic changes caused by novel *ONSEN* insertions (Lisch, 2013). Even though we were able to clearly link the observed *transparent testa* phenotype to the homozygous insertion of *ONSEN* in the

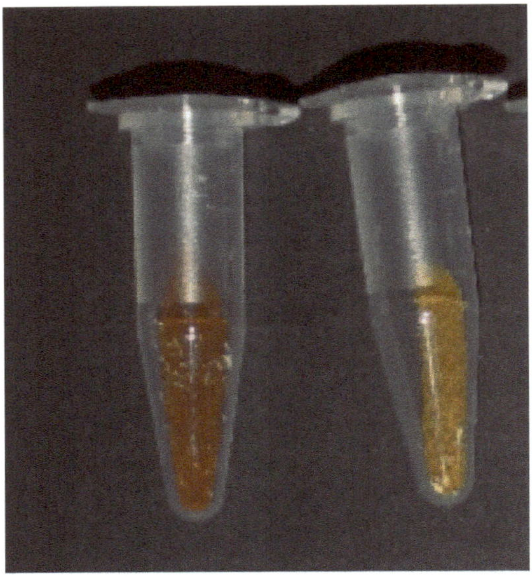

Figure 4 ***Phenotypic changes induced by the controlled mobilization of a retroTE.*** Photograph of seeds of wild-type Arabidopsis (left) and an *ONSEN*-tagged mutation in *TT6*. This mutation inhibits the biosynthesis of flavonoids in the seed coat.

TT6 locus it remains unclear whether transposition of other TEs or possible drug-induced epigenetic or genotoxic effects could underlie some of the observed phenotypes. For instance it was shown that a Z-treatment could induce DNA damage during strand synthesis in DNA replication (Liu et al., 2015) and lead to the induction of class II *CACTA*-elements in *Arabidopsis* (Baubec et al., 2009; Griffin et al., 2016).

In the light of trait stability, it is important to mention that the potential excision of mobilized DNA-TEs as well as inter and intra-element recombination in both classes of TEs could in principle lead to genomic and phenotypic alterations over generations (Bennetzen & Wang, 2014; Lisch, 2013; Vitte, Panaud, & Quesneville, 2007). In addition and as mentioned above, the massive amplifications of retroTE copy numbers followed by purifying selection can lead to phenotypic variation such as it has been observed in natural populations of the model plants *Arabidopsis* and *Brachypodium distachyon* (Quadrana et al., 2016; Stritt, Gordon, Wicker, Vogel, & Roulin, 2018). Although we did not observe any sudden massive gain or loss of *ONSEN*-copies over three generations of inbreeding of *ONSEN* high-copy *Arabidopsis* individuals (Thieme et al., 2017b) the question of trait stability after an induced burst of TEs in crop plants will be of utmost interest for breeders and has to be addressed in the near future.

6.4 Perspective: TE Mobilization in Crops and Applications in Breeding

As described above, experiments on TEs or epigenetic regulation in the model plant *Arabidopsis* are not necessarily valid in crops (Mirouze & Vitte, 2014). However, from a mechanistic point of view, the high degree of conservation of key enzymes involved in DNA methylation and especially of RNA-polymerase II in different plant species (Nawrath, Schell, & Koncz, 1990) and eukaryotes in general (Ream et al., 2009; Sweetser, Nonet, & Young, 1987), strongly suggested that A and Z could also interfere with TE-silencing other plant species. Indeed, we found that the double-treatment specifically activated the *Houba* retroTE in rice confirming this assumption (Thieme et al., 2017b). Notably, there are indications that *Houba* (Vitte et al., 2007), just like *ONSEN* (Quadrana et al., 2016), showed very recent natural transposition events in the rice and the *Arabidopsis* genomes, respectively. *Houba* is the most abundant retroTE of the *copia* family in rice and has been active in the last 500′000 years (Wicker & Keller, 2007). Based on these two cases, including distantly related species such as *Arabidopsis* and rice it can be speculated that similar expression dependent silencing mechanisms that directly depend on a silencing signal produced by Pol II generally exist in plants. Overall, these observations strongly suggest that it is now possible to mobilize endogenous retroTEs in virtually any crop plant. In addition, the ongoing elucidation of novel players in TE regulation that could be targeted for a controlled burst of TE-mobility as well as the methodical advances in tracking TEs will catalyze their use in plant breeding (Paszkowski, 2015).

6.4.1 Breeding Strategies

Because TEs can rewire gene regulatory pathways (Makarevitch et al., 2015), especially in response to stress, they could be a powerful tool for crop breeding. We have shown previously that the heat-stress activated *ONSEN* retroTE in *Arabidopsis* plants defective in the production of siRNAs can render flanking genes heat-stress responsive (Ito et al., 2011). Because of that interesting feature enabled by the stress responsive LTRs, the effect on the transcriptional regulation at the insertion site could also be related to the previous stress in wild type plants. To what extent different classes of TEs of crops depict a specific insertion bias will have to be determined. In this context one can envisage situations where such events may be beneficial to the plant resulting in a faster adaptation to changes in the environment (Ito et al., 2016; Lisch, 2013; Makarevitch et al., 2015). While adaptive powers have been attributed to TEs in other organisms (e.g. yeast (Adams & Oeller, 1986)), such effects still need to be demonstrated in plants.

Assuming that TEs can contribute to the adaptation to stresses, we propose to use the induced amplification of endogenous retroTEs for crop breeding. As an exemplary approach, the aforementioned method using inhibitors of retroTE-silencing (Thieme et al., 2017b) termed BUNGEE (breeding using jumping genes) is shown in Fig. 5. In order to breed improved stress tolerance in a plant, one first identifies retroTEs that respond to the wanted stress in a specific genetic background. If such a mobile

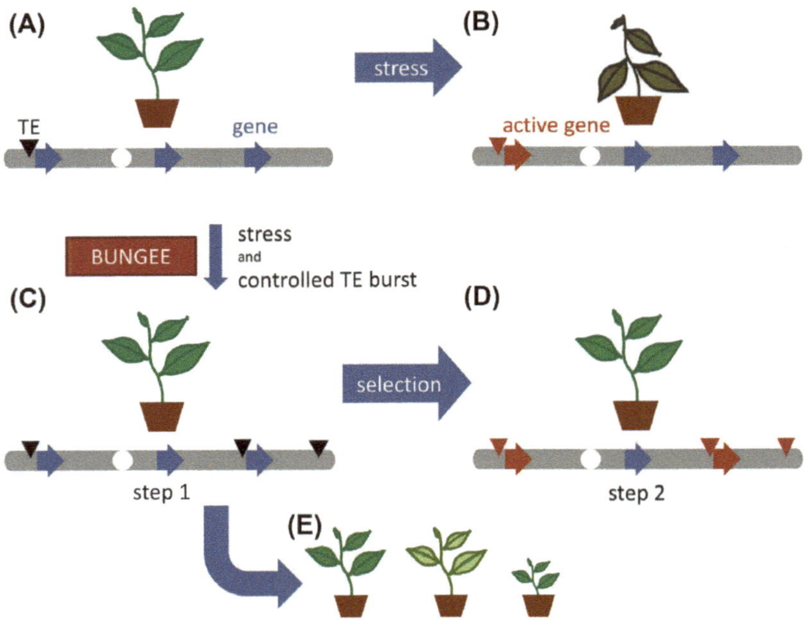

Figure 5 ***Example for using retroTEs in crop breeding. The BUNGEE crop breeding method.*** (A) Original crop plant and one of its chromosomes represented below (light grey) containing one retroTE insertion (black triangle). (B) The original crop does not withstand strong stresses because it only contains one gene responding to it (red arrow). (C) Novel crop variety generated by the BUNGEE and stress-mediated mobilization of retroTEs (additional black triangles on the chromosome) (D) Stress-tolerant individuals are detected through selection because they have acquired novel stress-responsive gene (additional red arrow on the chromosome) (E) BUNGEE also leads to phenotypic diversity and inducible phenotypes resulting from random mutagenesis.

element has successfully been identified, it is mobilized using a combination of drug and stress treatments. The progeny plants of this treatment can then be screened for phenotypic diversity and improved stress tolerance. This method has numerous advantages:

- Potential generation of gain-of function phenotypes
- In comparison to transformation and/or CRISPR-Cas9 editing, no prior knowledge about the resistance mechanism is required and no genetic engineering required.
- The retroTE insertions tag the genes of interest causing a specific phenotype, allowing their rapid identification.
- Trait stability due to the copy and paste mechanism of retroTEs.
- No reference genome is required.

One disadvantage of this method is that it will also generate mutations that can have a negative impact on yield (Fig. 5E). Such lines will be eliminated the selection step of the breeding process or can be (where possible) back-crossed to parental lines to segregate out unwanted TE insertions.

As retroTE-activity can differ between various varieties of the same crop as it was observed for *Houba* in rice (Yuzbasioglu, Yilmaz, Marakli, & Gozukirmizi, 2016), it would be very interesting to investigate and compare cultivar-specific mobilomes in response to stress. Using a similar approach as described for *Tnt1* in cucumber (Zhang et al., 2018), the rapid identification of stress responsive retroTEs using inhibitors of TE-silencing (Thieme, 2017a; Thieme et al., 2017b) could also be combined with targeted genetic engineering. For instance, cisgenesis could be used to transfer stress-responsive elements between different varieties (Hou, Atlihan, & Lu, 2014).

6.4.2 Outlook

As described above, our fundamental findings from the model plant *Arabidopsis* that are supported by the activation of the *Houba* retroTE in rice opened up new avenues to enhance plant breeding to face futures agronomic challenges. We anticipate that the current increase in understanding of TE-biology and epigenetics together with novel methods in genomics and bioinformatics will catalyze the active usage in TEs in breeding programs. Depending on the species and the breeding objective there will be different approaches to harness the huge epi/genetic resource of TEs in plant genomes (Gallusci et al., 2017).

REFERENCES

Abberton, M., Batley, J., Bentley, A., Bryant, J., Cai, H., Cockram, J., et al. (2016). Global agricultural intensification during climate change: A role for genomics. *Plant Biotechnology Journal, 14*, 1095–1098.

Adams, J., & Oeller, P. W. (1986). Structure of evolving populations of *Saccharomyces cerevisiae*: Adaptive changes are frequently associated with sequence alterations involving mobile elements belonging to the Ty family. *Proceedings of the National Academy of Sciences, 83*, 7124.

Amoah, S., Kurup, S., Rodriguez Lopez, C. M., Welham, S. J., Powers, S. J., Hopkins, C. J., et al. (2012). A Hypomethylated population of Brassica rapa for forward and reverse Epi-genetics. *BMC Plant Biology, 12*, 193.

Anca, I. A., Fromentin, J., Bui, Q. T., Mhiri, C., Grandbastien, M. A., & Simon-Plas, F. (2014). Different tobacco retrotransposons are specifically modulated by the elicitor cryptogein and reactive oxygen species. *Journal of Plant Physiology, 171*, 1533–1540.

Baubec, T., Pecinka, A., Rozhon, W., & Mittelsten Scheid, O. (2009). Effective, homogeneous and transient interference with cytosine methylation in plant genomic DNA by zebularine. *The Plant Journal: For Cell and Molecular Biology, 57*, 542–554.

Bennett, M. D. (1972). Nuclear DNA content and minimum generation time in Herbaceous plants. *Proceedings of the Royal Society of London. Series B. Biological Sciences, 181*, 109–135.

Bennetzen, J. L., & Kellogg, E. A. (1997). Do plants have a one-way ticket to genomic obesity? *The Plant Cell, 9*, 1509–1514.

Bennetzen, J. L., Ma, J., & Devos, K. M. (2005). Mechanisms of recent genome size variation in flowering plants. *Annals of Botany, 95*, 127–132.

Bennetzen, J. L., & Wang, H. (2014). The contributions of transposable elements to the structure, function, and evolution of plant genomes. *Annual Review of Plant Biology, 65*, 505–530.

Bernatavichute, Y. V., Zhang, X., Cokus, S., Pellegrini, M., & Jacobsen, S. E. (2008). Genome-wide association of histone H3 lysine nine methylation with CHG DNA methylation in *Arabidopsis thaliana*. *PLoS One, 3*, e3156.

Böhmdorfer, G., Rowley, M. J., Kuciński, J., Zhu, Y., Amies, I., & Wierzbicki, A. T. (2014). RNA-directed DNA methylation requires stepwise binding of silencing factors to long non-coding RNA. *The Plant Journal, 79*, 181–191.

Bruggmann, R., et al. (2006). Uneven chromosome contraction and expansion in the maize genome. *Genome Research, 16*, 1241–1251.

Butelli, E., Licciardello, C., Zhang, Y., Liu, J., Mackay, S., Bailey, P., et al. (2012). Retrotransposons control fruit-specific, cold-dependent accumulation of anthocyanins in blood oranges. *The Plant Cell Online, 24*, 1242–1255.

Cao, X., & Jacobsen, S. E. (2002). Locus-specific control of asymmetric and CpNpG methylation by the DRM and CMT3 methyltransferase genes. *Proceedings of the National Academy of Sciences of the United States of America, 99*(Suppl 4), 16491–16498.

Casacuberta, J. M., & Grandbastien, M. A. (1993). Characterisation of LTR sequences involved in the protoplast specific expression of the tobacco Tnt1 retrotransposon. *Nucleic Acids Research, 21*, 2087–2093.

Castrillo, G., Sanchez-Bermejo, E., de Lorenzo, L., Crevillen, P., Fraile-Escanciano, A., Tc, M., et al. (2013). WRKY6 transcription factor restricts arsenate uptake and transposon activation in Arabidopsis. *The Plant Cell Online, 25*, 2944–2957.

Cavrak, V. V., Lettner, N., Jamge, S., Kosarewicz, A., Bayer, L. M., & Mittelsten Scheid, O. (2014). How a retrotransposon exploits the plant's heat stress response for its activation. *PLoS Genetics, 10*, e1004115.

Champion, C., Guianvarc'h, D., Senamaud-Beaufort, C., Jurkowska, R. Z., Jeltsch, A., Ponger, L., et al. (2010). Mechanistic insights on the inhibition of c5 DNA methyltransferases by zebularine. *PLoS One, 5*, e12388.

Chen, J., Greenblatt, I. M., & Dellaporta, S. L. (1987). Transposition of Ac from the P locus of maize into unreplicated. *Chromosomal Sites Genetics, 117*, 109–116.

Cheng, J. C., Weisenberger, D. J., Gonzales, F. A., Liang, G., Xu, G.-L., Hu, Y.-G., et al. (2004). Continuous zebularine treatment effectively sustains demethylation in human bladder cancer cells. *Molecular and Cellular Biology, 24*, 1270–1278.

Chen, J., Zhao, H., Zheng, X., Liang, K., Guo, Y., & Sun, X. (2017). Recent amplification of Osr4 LTR-retrotransposon caused rice D1 gene mutation and dwarf phenotype. *Plant Diversity, 39*, 73–79.

Chuong, E. B., Elde, N. C., & Feschotte, C. (2017). Regulatory activities of transposable elements: From conflicts to benefits. *Nature Reviews. Genetics, 18*, 71–86.

Clavijo, B. J., et al. (2017). An improved assembly and annotation of the allohexaploid wheat genome identifies complete families of agronomic genes and provides genomic evidence for chromosomal translocations. *Genome Research, 27*, 885–896.

Cokus, S. J., Feng, S., Zhang, X., Chen, Z., Merriman, B., Haudenschild, C. D., et al. (2008). Shotgun bisulphite sequencing of the Arabidopsis genome reveals DNA methylation patterning. *Nature, 452*, 215–219.

Creasey, K. M., Zhai, J., Borges, F., Van Ex, F., Regulski, M., Meyers, B. C., et al. (2014). miRNAs trigger widespread epigenetically activated siRNAs from transposons in Arabidopsis. *Nature, 508*, 411–415.

Cuerda-Gil, D., & Slotkin, R. K. (2016). Non-canonical RNA-directed DNA methylation. *Nature Plants, 2*, 16163.

D'Erfurth, I., Cosson, V., Eschstruth, A., Lucas, H., Kondorosi, A., & Ratet, P. (2003). Efficient transposition of the Tnt1 tobacco retrotransposon in the model legume Medicago truncatula. *The Plant Journal, 34*, 95–106.

Daccord, N., et al. (2017). High-quality de novo assembly of the apple genome and methylome dynamics of early fruit development. *Nature Genetics, 49*, 1099–1106.

De La Chaux, N., Tsuchimatsu, T., Shimizu, K. K., & Wagner, A. (2012). The predominantly selfing plant *Arabidopsis thaliana* experienced a recent reduction in transposable element abundance compared to its outcrossing relative Arabidopsis lyrata. *Mobile DNA, 3*, 1–18.

Debladis, E., Llauro, C., Carpentier, M. C., Mirouze, M., & Panaud, O. (2017). Detection of active transposable elements in *Arabidopsis thaliana* using Oxford Nanopore Sequencing technology. *BMC Genomics, 18*, 537.

Devos, K. M., Brown, J. K. M., & Bennetzen, J. L. (2002). Genome size reduction through illegitimate recombination counteracts genome expansion in arabidopsis. *Genome Research, 12*, 1075–1079.

Diez, C. M., Gaut, B. S., Meca, E., Scheinvar, E., Montes-Hernandez, S., Eguiarte, L. E., et al. (2013). Genome size variation in wild and cultivated maize along altitudinal gradients. *The New Phytologist, 199*, 264–276.

Ding, Y., Wang, X., Su, L., Zhai, J., Cao, S., Zhang, D., et al. (2007). SDG714, a histone H3K9 methyltransferase, is involved in Tos17 DNA methylation and transposition in rice. *The Plant Cell Online, 19*, 9–22.

Doebley, J., Stec, A., & Gustus, C. (1995). Teosinte Branched1 and the origin of maize: Evidence for epistasis and the evolution of dominance. *Genetics, 141*, 333–346.

Duan, K., Ding, X., Zhang, Q., Zhu, H., Pan, A., & Huang, J. (2008). AtCopeg1, the unique gene originated from AtCopia95 retrotransposon family, is sensitive to external hormones and abiotic stresses. *Plant Cell Reports, 27*, 1065–1073.

Ebbs, M. L., & Bender, J. (2006). Locus-specific control of DNA methylation by the Arabidopsis SUVH5 histone methyltransferase. *The Plant Cell Online, 18*, 1166–1176.
Elrouby, N., & Bureau, T. E. (2010). Bs1, a new chimeric gene formed by retrotransposon-mediated exon shuffling in maize. *Plant Physiology, 153*, 1413–1424.
Feschotte, C., Swamy, L., & Wessler, S. R. (2003). Genome-wide analysis of mariner-like transposable elements in rice reveals complex relationships with stowaway miniature inverted repeat transposable elements (MITEs). *Genetics, 163*, 747–758.
Finnegan, E. J., & Dennis, E. S. (1993). Isolation and identification by sequence homology of a putative cytosine methyltransferase from *Arabidopsis thaliana*. *Nucleic Acids Research, 21*, 2383–2388.
Finnegan, E. J., Peacock, W. J., & Dennis, E. S. (1996). Reduced DNA methylation in *Arabidopsis thaliana* results in abnormal plant development. *Proceedings of the National Academy of Sciences of the United States of America, 93*.
Fischer, R. A., Byerlee, D., & Edmeades, G. O. (2014). *Crop yields and global food security: Will yield increase continue to feed the world?* Canberra: Australian Centre for International Agricultural Research.
Fuchs, J., Demidov, D., Houben, A., & Schubert, I. (2006). Chromosomal histone modification patterns–from conservation to diversity. *Trends in Plant Science, 11*, 199–208.
Fujimoto, R., Kinoshita, Y., Kawabe, A., Kinoshita, T., Takashima, K., Nordborg, M., et al. (2008). Evolution and control of imprinted FWA genes in the genus Arabidopsis. *PLoS Genetics, 4*, e1000048.
Galindo-Gonzalez, L., Mhiri, C., Deyholos, M. K., & Grandbastien, M. A. (2017). LTR-retrotransposons in plants: Engines of evolution. *Gene, 626*, 14–25.
Gallusci, P., Dai, Z., Genard, M., Gauffretau, A., Leblanc-Fournier, N., Richard-Molard, C., et al. (2017). Epigenetics for plant improvement: Current knowledge and modeling avenues. *Trends in Plant Science, 22*, 610–623.
Gao, D., Chen, J., Chen, M., Meyers, B. C., & Jackson, S. (2012). A highly conserved, small LTR retrotransposon that preferentially targets genes in grass genomes. *PLoS One, 7*, e32010.
Gao, D., Jiang, N., Wing, R. A., Jiang, J., & Jackson, S. A. (2015). Transposons play an important role in the evolution and diversification of centromeres among closely related species. *Frontiers in Plant Science, 6*, 216.
Gasciolli, V., Mallory, A. C., Bartel, D. P., & Vaucheret, H. (2005). Partially redundant functions of Arabidopsis DICER-like enzymes and a role for DCL4 in producing trans-acting siRNAs. *Current Biology: CB, 15*, 1494–1500.
Gendrel, A.-V., Lippman, Z., Yordan, C., Colot, V., & Martienssen, R. A. (2002). Dependence of heterochromatic histone H3 methylation patterns on the arabidopsis gene DDM1. *Science, 297*, 1871–1873.
Gierl, A., & Saedler, H. (1992). Plant-transposable elements and gene tagging. In R. A. Schilperoort, & L. Dure (Eds.), *10 Years plant molecular biology* (pp. 39–49). Dordrecht: Springer Netherlands.
Gouil, Q., & Baulcombe, D. C. (2016). DNA methylation signatures of the plant chromomethyltransferases. *PLoS Genetics, 12*, e1006526.
Grandbastien, M. A. (2015). LTR retrotransposons, handy hitchhikers of plant regulation and stress response. *Biochimica et Biophysica Acta, 1849*, 403–416.
Grandbastien, M.-A., Spielmann, A., & Caboche, M. (1989). Tnt1, a mobile retroviral-like transposable element of tobacco isolated by plant cell genetics. *Nature, 337*, 376–380.
Griffin, P. T., Niederhuth, C. E., & Schmitz, R. J. (2016). A comparative analysis of 5-azacytidine- and zebularine-induced DNA demethylation. *G3 (Bethesda), 6*, 2773–2780.
Haag, J. R., Ream, T. S., Marasco, M., Nicora, C. D., Norbeck, A. D., Pasa-Tolic, L., et al. (2012). In vitro transcription activities of Pol IV, Pol V, and RDR2 reveal coupling of Pol IV and RDR2 for dsRNA synthesis in plant RNA silencing. *Molecular Cell, 48*, 811–818.

Hawkins, J. S., Proulx, S. R., Rapp, R. A., & Wendel, J. F. (2009). Rapid DNA loss as a counterbalance to genome expansion through retrotransposon proliferation in plants. *Proceedings of the National Academy of Sciences of the United States of America, 106*, 17811–17816.

Henderson, I. R., & Jacobsen, S. E. (2007). Epigenetic inheritance in plants. *Nature, 447*, 418–424.

Henikoff, S. (1990). Position-effect variegation after 60 years. *Trends in Genetics, 6*, 422–426.

Herr, A. J., Jensen, M. B., Dalmay, T., & Baulcombe, D. C. (2005). RNA polymerase IV directs silencing of endogenous DNA. *Science, 308*, 118–120.

He, L., Wu, W., Zinta, G., Yang, L., Wang, D., Liu, R., et al. (2018). A naturally occurring epiallele associates with leaf senescence and local climate adaptation in Arabidopsis accessions. *Nature Communications, 9*, 460.

Hibrand Saint-Oyant, L., et al. (2018). A high-quality genome sequence of Rosa chinensis to elucidate ornamental traits. *Nature Plants, 4*, 473–484.

Hirochika, H., Sugimoto, K., Otsuki, Y., Tsugawa, H., & Kanda, M. (1996). Retrotransposons of rice involved in mutations induced by tissue culture. *Proceedings of the National Academy of Sciences of the United States of America, 93*, 7783–7788.

Hou, H., Atlihan, N., & Lu, Z.-X. (2014). New biotechnology enhances the application of cisgenesis in plant breeding. *Frontiers in Plant Science, 5*, 389.

Hristova, E., Fal, K., Klemme, L., Windels, D., & Bucher, E. (2015). HISTONE DEACETYLASE6 controls gene expression patterning and DNA methylation-independent euchromatic silencing. *Plant Physiology, 168*, 1298–1308.

Hsieh, T.-F., Ibarra, C. A., Silva, P., Zemach, A., Eshed-Williams, L., Fischer, R. L., et al. (2009). Genome-wide demethylation of arabidopsis endosperm. *Science, 324*, 1451.

Hu, L., Li, N., Xu, C., Zhong, S., Lin, X., Yang, J., et al. (2014). Mutation of a major CG methylase in rice causes genome-wide hypomethylation, dysregulated genome expression, and seedling lethality. *Proceedings of the National Academy of Sciences of the United States of America, 111*, 10642–10647.

Ito, H., Gaubert, H., Bucher, E., Mirouze, M., Vaillant, I., & Paszkowski, J. (2011). An siRNA pathway prevents transgenerational retrotransposition in plants subjected to stress. *Nature, 472*, 115–119.

Ito, H., Kim, J. M., Matsunaga, W., Saze, H., Matsui, A., Endo, T. A., et al. (2016). A stress-activated transposon in arabidopsis induces transgenerational abscisic acid insensitivity. *Scientific Reports, 6*, 23181.

Jackson, J. P., Johnson, L., Jasencakova, Z., Zhang, X., PerezBurgos, L., Singh, P. B., et al. (2004). Dimethylation of histone H3 lysine 9 is a critical mark for DNA methylation and gene silencing in *Arabidopsis thaliana*. *Chromosoma, 112*, 308–315.

Jackson, J. P., Lindroth, A. M., Cao, X., & Jacobsen, S. E. (2002). Control of CpNpG DNA methylation by the KRYPTONITE histone H3 methyltransferase. *Nature, 416*, 556–560.

Jacob, Y., Stroud, H., Leblanc, C., Feng, S., Zhuo, L., Caro, E., et al. (2010). Regulation of heterochromatic DNA replication by histone H3 lysine 27 methyltransferases. *Nature, 466*, 987–991.

Jameson, N., Georgelis, N., Fouladbash, E., Martens, S., Hannah, L. C., & Lal, S. (2008). Helitron mediated amplification of cytochrome P450 monooxygenase gene in maize. *Plant Molecular Biology, 67*, 295–304.

Jeddeloh, J. A., Stokes, T. L., & Richards, E. J. (1999). Maintenance of genomic methylation requires a SWI2/SNF2-like protein. *Nature Genetics, 22*, 94–97.

Jiang, N., Bao, Z., Zhang, X., Eddy, S. R., & Wessler, S. R. (2004). Pack-MULE transposable elements mediate gene evolution in plants. *Nature, 431*, 569–573.

Jiao, Y., et al. (2017). Improved maize reference genome with single-molecule technologies. *Nature, 546*, 524–527.

Jouffroy, O., Saha, S., Mueller, L., Quesneville, H., & Maumus, F. (2016). Comprehensive repeatome annotation reveals strong potential impact of repetitive elements on tomato ripening. *BMC Genomics, 17*, 624.

Kalendar, R., Vicient, C. M., Peleg, O., Anamthawat-Jonsson, K., Bolshoy, A., & Schulman, A. H. (2004). Large retrotransposon Derivatives: Abundant, conserved but nonautonomous retroelements of barley and related genomes. *Genetics, 166*, 1437–1450.

Kanazawa, A., Liu, B., Kong, F., Arase, S., & Abe, J. (2009). Adaptive evolution involving gene duplication and insertion of a novel ty1/copia-like retrotransposon in soybean. *Journal of Molecular Evolution, 69*, 164–175.

Kankel, M. W., Ramsey, D. E., Stokes, T. L., Flowers, S. K., Haag, J. R., Jeddeloh, J. A., et al. (2003). Arabidopsis MET1 cytosine methyltransferase mutants. *Genetics, 163*.

Kanno, T., Huettel, B., Mette, M. F., Aufsatz, W., Jaligot, E., Daxinger, L., et al. (2005). Atypical RNA polymerase subunits required for RNA-directed DNA methylation. *Nature Genetics, 37*, 761–765.

Kapitonov, V. V., & Jurka, J. (2001). Rolling-circle transposons in eukaryotes. *Proceedings of the National Academy of Sciences of the United States of America, 98*, 8714–8719.

Kawase, M., Fukunaga, K., & Kato, K. (2005). Diverse origins of waxy foxtail millet crops in East and Southeast Asia mediated by multiple transposable element insertions. *Molecular Genetics and Genomics, 274*, 131–140.

Kimura, Y., Tosa, Y., Shimada, S., Sogo, R., Kusaba, M., Sunaga, T., et al. (2001). OARE-1, a ty1-copia retrotransposon in oat activated by abiotic and biotic stresses. *Plant & Cell Physiology, 42*, 1345–1354.

Kinoshita, Y., Saze, H., Kinoshita, T., Miura, A., Soppe, W. J., Koornneef, M., et al. (2007). Control of FWA gene silencing in *Arabidopsis thaliana* by SINE-related direct repeats. *The Plant Journal : For Cell and Molecular Biology, 49*, 38–45.

Lahmy, S., et al. (2016). Evidence for ARGONAUTE4-DNA interactions in RNA-directed DNA methylation in plants. *Genes & Development, 30*, 2565–2570.

Lanciano, S., Carpentier, M. C., Llauro, C., Jobet, E., Robakowska-Hyzorek, D., Lasserre, E., et al. (2017). Sequencing the extrachromosomal circular mobilome reveals retrotransposon activity in plants. *PLoS Genetics, 13*, e1006630.

Law, J. A., Vashisht, A. A., Wohlschlegel, J. A., & Jacobsen, S. E. (2011). SHH1, a homeodomain protein required for DNA methylation, as well as RDR2, RDM4, and chromatin remodeling factors, associate with RNA polymerase IV. *PLoS Genetics, 7*, e1002195.

Lee, S. I., & Kim, N. S. (2014). Transposable elements and genome size variations in plants. *Genomics Inform, 12*, 87–97.

Li, Q., Eichten, S. R., Hermanson, P. J., Zaunbrecher, V. M., Song, J., Wendt, J., et al. (2014). Genetic perturbation of the maize methylome. *The Plant Cell Online, 26*, 4602–4616.

Lindell, T. J., Weinberg, F., Morris, P. W., Roeder, R. G., & Rutter, W. J. (1970). Specific inhibition of nuclear RNA polymerase II by α-amanitin. *Science, 170*, 447–449.

Lindroth, A. M., Cao, X., Jackson, J. P., Zilberman, D., McCallum, C. M., Henikoff, S., et al. (2001). Requirement of the CHROMOMETHYLASE3 for maintenance of CpXpG methylation. *Science, 292*, 2077–2080.

Lippman, Z., Gendrel, A.-V., Black, M., Vaughn, M. W., Dedhia, N., Richard McCombie, W., et al. (2004). Role of transposable elements in heterochromatin and epigenetic control. *Nature, 430*, 471–476.

Lippman, Z., May, B., Yordan, C., Singer, T., & Martienssen, R. (2003). Distinct mechanisms determine transposon inheritance and methylation via small interfering RNA and histone modification. *PLoS Biology, 1*, e67.

Lisch, D. (2013). How important are transposons for plant evolution? *Nature Reviews. Genetics, 14*, 49–61.
Liu, C. H., Finke, A., Diaz, M., Rozhon, W., Poppenberger, B., Baubec, T., et al. (2015). Repair of DNA damage induced by the cytidine analog zebularine requires ATR and ATM in arabidopsis. *The Plant Cell Online, 27*, 1788–1800.
Liu, X., Yu, C. W., Duan, J., Luo, M., Wang, K., Tian, G., et al. (2012). HDA6 directly interacts with DNA methyltransferase MET1 and maintains transposable element silencing in Arabidopsis. *Plant Physiology, 158*, 119–129.
Lucas, H., Feuerbach, F., Kunert, K., Grandbastien, M. A., & Caboche, M. (1995). RNA-mediated transposition of the tobacco retrotransposon Tnt1 in *Arabidopsis thaliana. The EMBO Journal, 14*, 2364–2373.
Lu, H.-F., Dong, H.-T., Sun, C.-B., Qing, D.-J., Li, N., Wu, Z.-K., et al. (2011). The panorama of physiological responses and gene expression of whole plant of maize inbred line YQ7-96 at the three-leaf stage under water deficit and re-watering. *TAG. Theoretical and Applied Genetics. Theoretische und angewandte Genetik, 123*, 943–958.
Luger, K., Mader, A. W., Richmond, R. K., Sargent, D. F., & Richmond, T. J. (1997). Crystal structure of the nucleosome core particle at 2.8[thinsp]A resolution. *Nature, 389*, 251–260.
Lyko, F., & Brown, R. (2005). DNA methyltransferase inhibitors and the development of epigenetic cancer therapies. *Journal of the National Cancer Institute, 97*, 1498–1506.
Ma, J., Devos, K. M., & Bennetzen, J. L. (2004). Analyses of LTR-retrotransposon structures reveal recent and rapid genomic DNA loss in rice. *Genome Research, 14*, 860–869.
Makarevitch, I., Waters, A. J., West, P. T., Stitzer, M., Hirsch, C. N., Ross-Ibarra, J., et al. (2015). Transposable elements contribute to activation of maize genes in response to abiotic stress. *PLoS Genetics, 11*, e1004915.
Mari-Ordonez, A., Marchais, A., Etcheverry, M., Martin, A., Colot, V., & Voinnet, O. (2013). Reconstructing de novo silencing of an active plant retrotransposon. *Nature Genetics, 45*, 1029–1039.
Masuta, Y., Nozawa, K., Takagi, H., Yaegashi, H., Tanaka, K., Ito, T., et al. (2017). Inducible transposition of a heat-activated retrotransposon in tissue culture. *Plant & Cell Physiology, 58*, 375–384.
Matzke, M. A., Kanno, T., & Matzke, A. J. (2015). RNA-directed DNA methylation: The evolution of a complex epigenetic pathway in flowering plants. *Annual Review of Plant Biology, 66*, 243–267.
Matzke, M. A., & Mosher, R. A. (2014). RNA-directed DNA methylation: An epigenetic pathway of increasing complexity. *Nature Reviews. Genetics, 15*, 394–408.
McClintock, B. (1950). The origin and behavior of mutable loci in maize. *Proceedings of the National Academy of Sciences of the United States of America, 36*, 344–355.
McClintock, B. (1956). Intranuclear systems controlling gene action and mutation. *Brookhaven Symposia in Biology, 8*, 58–74.
McClintock, B. (1984). The significance of responses of the genome to challenge. *Science, 226*, 792–801.
McCue, A. D., Panda, K., Nuthikattu, S., Choudury, S. G., Thomas, E. N., & Slotkin, R. K. (2015). ARGONAUTE 6 bridges transposable element mRNA-derived siRNAs to the establishment of DNA methylation. *The EMBO Journal, 34*, 20–35.
Mirouze, M., Reinders, J., Bucher, E., Nishimura, T., Schneeberger, K., Ossowski, S., et al. (2009). Selective epigenetic control of retrotransposition in Arabidopsis. *Nature, 461*, 427–430.
Mirouze, M., & Vitte, C. (2014). Transposable elements, a treasure trove to decipher epigenetic variation: Insights from arabidopsis and crop epigenomes. *Journal of Experimental Botany, 65*, 2801–2812.

Miura, A., Yonebayashi, S., Watanabe, K., Toyama, T., Shimada, H., & Kakutani, T. (2001). Mobilization of transposons by a mutation abolishing full DNA methylation in Arabidopsis. *Nature, 411*, 212–214.
Mosher, R. A., Melnyk, C. W., Kelly, K. A., Dunn, R. M., Studholme, D. J., & Baulcombe, D. C. (2009). Uniparental expression of PolIV-dependent siRNAs in developing endosperm of Arabidopsis. *Nature, 460*, 283–286.
Naito, K., Cho, E., Yang, G., Campbell, M. A., Yano, K., Okumoto, Y., et al. (2006). Dramatic amplification of a rice transposable element during recent domestication. *Proceedings of the National Academy of Sciences of the United States of America, 103*, 17620–17625.
Naito, K., Zhang, F., Tsukiyama, T., Saito, H., Hancock, C. N., Richardson, A. O., et al. (2009). Unexpected consequences of a sudden and massive transposon amplification on rice gene expression. *Nature, 461*, 1130–1134.
Nawrath, C., Schell, J., & Koncz, C. (1990). Homologous domains of the largest subunit of eucaryotic RNA polymerase II are conserved in plants. *Molecular and General Genetics MGG, 223*, 65–75.
Negi, P., Rai, A. N., & Suprasanna, P. (2016). Moving through the stressed genome: Emerging regulatory roles for transposons in plant stress response. *Frontiers in Plant Science, 7*, 1448.
Neumann, P., Požárková, D., & Macas, J. (2003). Highly abundant pea LTR retrotransposon Ogre is constitutively transcribed and partially spliced. *Plant Molecular Biology, 53*, 399–410.
Nuthikattu, S., McCue, A. D., Panda, K., Fultz, D., DeFraia, C., Thomas, E. N., et al. (2013). The initiation of epigenetic silencing of active transposable elements is triggered by RDR6 and 21-22 nucleotide small interfering RNAs. *Plant Physiology, 162*, 116–131.
Ohno, S. (1972). So much "junk" DNA in our genome. *Brookhaven Symposia in Biology, 23*, 366–370.
Ong-Abdullah, M., Ordway, J. M., Jiang, N., Ooi, S. E., Kok, S. Y., Sarpan, N., et al. (2015). Loss of Karma transposon methylation underlies the mantled somaclonal variant of oil palm. *Nature, 525*, 533–537.
Onodera, Y., Haag, J. R., Ream, T., Costa Nunes, P., Pontes, O., & Pikaard, C. S. (2005). Plant nuclear RNA polymerase IV mediates siRNA and DNA methylation-dependent heterochromatin formation. *Cell, 120*, 613–622.
Panda, K., Ji, L., Neumann, D. A., Daron, J., Schmitz, R. J., & Slotkin, R. K. (2016). Full-length autonomous transposable elements are preferentially targeted by expression-dependent forms of RNA-directed DNA methylation. *Genome Biology, 17*, 170.
Park, M., Keung, A. J., & Khalil, A. S. (2016). The epigenome: The next substrate for engineering. *Genome Biology, 17*, 183.
Paszkowski, J. (2015). Controlled activation of retrotransposition for plant breeding. *Current Opinion in Biotechnology, 32C*, 200–206.
Pecinka, A., Dinh, H. Q., Baubec, T., Rosa, M., Lettner, N., & Mittelsten Scheid, O. (2010). Epigenetic regulation of repetitive elements is attenuated by prolonged heat stress in Arabidopsis. *The Plant Cell Online, 22*, 3118–3129.
Pecinka, A., & Liu, C. H. (2014). Drugs for plant chromosome and chromatin research. *Cytogenetic and Genome Research, 143*, 51–59.
Piegu, B., Bire, S., Arensburger, P., & Bigot, Y. (2015). A survey of transposable element classification systems–a call for a fundamental update to meet the challenge of their diversity and complexity. *Molecular Phylogenetics and Evolution, 86*, 90–109.
Piegu, B., Guyot, R., Picault, N., Roulin, A., Sanyal, A., Kim, H., et al. (2006). Doubling genome size without polyploidization: Dynamics of retrotransposition-driven

genomic expansions in Oryza australiensis, a wild relative of rice. *Genome Research, 16*, 1262–1269.

Pontier, D., Picart, C., Roudier, F., Garcia, D., Lahmy, S., Azevedo, J., et al. (2012). NERD, a plant-specific GW protein, defines an additional RNAi-dependent chromatin-based pathway in Arabidopsis. *Molecular Cell, 48*, 121–132.

Pontier, D., Yahubyan, G., Vega, D., Bulski, A., Saez-Vasquez, J., Hakimi, M.-A., et al. (2005). Reinforcement of silencing at transposons and highly repeated sequences requires the concerted action of two distinct RNA polymerases IV in Arabidopsis. *Genes & Development, 19*, 2030–2040.

Probst, A. V., Fagard, M., Proux, F., Mourrain, P., Boutet, S., Earley, K., et al. (2004). Arabidopsis histone deacetylase HDA6 is required for maintenance of transcriptional gene silencing and determines nuclear organization of rDNA repeats. *The Plant Cell Online, 16*, 1021–1034.

Quadrana, L., Bortolini Silveira, A., Mayhew, G. F., LeBlanc, C., Martienssen, R. A., Jeddeloh, J. A., et al. (2016). The *Arabidopsis thaliana* mobilome and its impact at the species level. *Elife, 5*, e15716.

Ramachandran, S., & Sundaresan, V. (2001). Transposons as tools for functional genomics. *Plant Physiology and Biochemistry, 39*, 243–252.

Ramallo, E., Kalendar, R., Schulman, A. H., & Martinez-Izquierdo, J. A. (2008). Reme1, a Copia retrotransposon in melon, is transcriptionally induced by UV light. *Plant Molecular Biology, 66*, 137–150.

Ream, T. S., Haag, J. R., Wierzbicki, A. T., Nicora, C. D., Norbeck, A. D., Zhu, J. K., et al. (2009). Subunit compositions of the RNA-silencing enzymes Pol IV and Pol V reveal their origins as specialized forms of RNA polymerase II. *Molecular Cell, 33*, 192–203.

Rocheta, M., Carvalho, L., Viegas, W., & Morais-Cecílio, L. (2012). Corky, a gypsy-like retrotransposon is differentially transcribed in Quercus suber tissues. *BMC Research Notes, 5*, 432.

Roy, S. W., & Gilbert, W. (2005). Rates of intron loss and gain: Implications for early eukaryotic evolution. *Proceedings of the National Academy of Sciences of the United States of America, 102*, 5773–5778.

Santi, D. V., Garrett, C. E., & Barr, P. J. (1983). On the mechanism of inhibition of DNA-cytosine methyltransferases by cytosine analogs. *Cell, 33*, 9–10.

Schmidt, T. (1999). LINEs, SINEs and repetitive DNA: Non-LTR retrotransposons in plant genomes. *Plant Molecular Biology, 40*, 903–910.

Schmutz, J., et al. (2010). Genome sequence of the palaeopolyploid soybean. *Nature, 463*, 178–183.

Schnable, P. S., Ware, D., Fulton, R. S., Stein, J. C., Wei, F., Pasternak, S., et al. (2009). The B73 maize genome: Complexity, diversity, and dynamics. *Science, 326*, 1112–1115.

Schubert, I., & Oud, J. L. (1997). There is an upper limit of chromosome size for normal development of an organism. *Cell, 88*, 515–520.

Schulman, A. H. (2013). Retrotransposon replication in plants. *Current Opinion in Virology, 3*, 604–614.

Settles, A. M. (2009). Transposon tagging and reverse genetics. In A. L. Kriz, & B. A. Larkins (Eds.), *Molecular genetic approaches to maize improvement* (pp. 143–159). Berlin, Heidelberg: Springer.

Sharma, A., Wolfgruber, T. K., & Presting, G. G. (2013). Tandem repeats derived from centromeric retrotransposons. *BMC Genomics, 14*, 142.

Shirasu, K., Schulman, A. H., Lahaye, T., & Schulze-Lefert, P. (2000). A contiguous 66-kb barley DNA sequence provides evidence for reversible genome expansion. *Genome Research, 10*, 908–915.

Sigman, M. J., & Slotkin, R. K. (2016). The first rule of plant transposable element silencing: Location, location, location. *The Plant Cell Online, 28*, 304–313.

Slotkin, R. K., Freeling, M., & Lisch, D. (2005). Heritable transposon silencing initiated by a naturally occurring transposon inverted duplication. *Nature Genetics, 37*, 641–644.

Springer, N. M., & Schmitz, R. J. (2017). Exploiting induced and natural epigenetic variation for crop improvement. *Nature Reviews. Genetics, 18*, 563–575.

Stritt, C., Gordon, S. P., Wicker, T., Vogel, J. P., & Roulin, A. C. (2018). Recent activity in expanding populations and purifying selection have shaped transposable element landscapes across natural accessions of the mediterranean grass Brachypodium distachyon. *Genome Biology and Evolution, 10*, 304–318.

Stroud, H., Greenberg, M. V., Feng, S., Bernatavichute, Y. V., & Jacobsen, S. E. (2013). Comprehensive analysis of silencing mutants reveals complex regulation of the Arabidopsis methylome. *Cell, 152*, 352–364.

Stroud, H., Hale, C. J., Feng, S., Caro, E., Jacob, Y., Michaels, S. D., et al. (2012). DNA methyltransferases are required to induce heterochromatic re-replication in Arabidopsis. *PLoS Genetics, 8*, e1002808.

Sun, F. J., Fleurdepine, S., Bousquet-Antonelli, C., Caetano-Anolles, G., & Deragon, J. M. (2007). Common evolutionary trends for SINE RNA structures. *Trends in Genetics, 23*, 26–33.

Sweetser, D., Nonet, M., & Young, R. A. (1987). Prokaryotic and eukaryotic RNA polymerases have homologous core subunits. *Proceedings of the National Academy of Sciences of the United States of America, 84*, 1192–1196.

Takeda, S., Sugimoto, K., Otsuki, H., & Hirochika, H. (1998). Transcriptional activation of the tobacco retrotransposon Tto1 by wounding and methyl jasmonate. *Plant Molecular Biology, 36*, 365–376.

Tanskanen, J. A., Sabot, F., Vicient, C., & Schulman, A. H. (2007). Life without GAG: The BARE-2 retrotransposon as a parasite's parasite. *Gene, 390*, 166–174.

Tapia, G., Verdugo, I., Yanez, M., Ahumada, I., Theoduloz, C., Cordero, C., et al. (2005). Involvement of ethylene in stress-induced expression of the TLC1.1 retrotransposon from Lycopersicon chilense Dun. *Plant Physiology, 138*, 2075–2086.

Teixeira, F. K., et al. (2009). A role for RNAi in the selective correction of DNA methylation defects. *Science, 323*, 1600–1604.

Tenaillon, M. I., Hollister, J. D., & Gaut, B. S. (2010). A triptych of the evolution of plant transposable elements. *Trends in Plant Science, 15*, 471–478.

Thieme, M. (2017a). *Stress-induced mobilization of retrotransposons for plant breeding*. Doctoral dissertation. University of Basel.

Thieme, M., Lanciano, S., Balzergue, S., Daccord, N., Mirouze, M., & Bucher, E. (2017b). Inhibition of RNA polymerase II allows controlled mobilisation of retrotransposons for plant breeding. *Genome Biology, 18*, 134.

Tittel-Elmer, M., Bucher, E., Broger, L., Mathieu, O., Paszkowski, J., & Vaillant, I. (2010). Stress-induced activation of heterochromatic transcription. *PLoS Genetics, 6*, e1001175.

To, T. K., Kim, J. M., Matsui, A., Kurihara, Y., Morosawa, T., Ishida, J., et al. (2011). Arabidopsis HDA6 regulates locus-directed heterochromatin silencing in cooperation with MET1. *PLoS Genetics, 7*, e1002055.

Tsukahara, S., Kobayashi, A., Kawabe, A., Mathieu, O., Miura, A., & Kakutani, T. (2009). Bursts of retrotransposition reproduced in Arabidopsis. *Nature, 461*, 423–426.

van der Knaap, E., Sanyal, A., Jackson, S. A., & Tanksley, S. D. (2004). High-resolution fine mapping and fluorescence in situ hybridization analysis of sun, a locus controlling tomato fruit shape, reveals a region of the tomato genome prone to DNA rearrangements. *Genetics, 168*, 2127–2140.

Varagona, M. J., Purugganan, M., & Wessler, S. R. (1992). Alternative splicing induced by insertion of retrotransposons into the maize waxy gene. *The Plant Cell, 4*, 811–820.

Vitte, C., Fustier, M. A., Alix, K., & Tenaillon, M. I. (2014). The bright side of transposons in crop evolution. *Brief Functional Genomics, 13*, 276–295.

Vitte, C., & Panaud, O. (2005). LTR retrotransposons and flowering plant genome size: Emergence of the increase/decrease model. *Cytogenetic and Genome Research, 110*, 91–107.

Vitte, C., Panaud, O., & Quesneville, H. (2007). LTR retrotransposons in rice (*Oryza sativa*, L.): Recent burst amplifications followed by rapid DNA loss. *BMC Genomics, 8*, 218.

Vongs, A., Kakutani, T., Martienssen, R. A., & Richards, E. J. (1993). *Arabidopsis thaliana* DNA methylation mutants. *Science, 260*, 1926–1928.

Voronova, A., Belevich, V., Jansons, A., & Rungis, D. (2014). Stress-induced transcriptional activation of retrotransposon-like sequences in the Scots pine (*Pinus sylvestris* L.) genome. *Tree Genetics & Genomes, 10*, 937–951.

Wang, H., Chai, Y., Chu, X., Zhao, Y., Wu, Y., Zhao, J., et al. (2009). Molecular characterization of a rice mutator-phenotype derived from an incompatible cross-pollination reveals transgenerational mobilization of multiple transposable elements and extensive epigenetic instability. *BMC Plant Biol, 9*, 63.

Wassenegger, M., Heimes, S., Riedel, L., & Sänger, H. L. (1994). RNA-directed de novo methylation of genomic sequences in plants. *Cell, 76*, 567–576.

Wendel, J. F., Jackson, S. A., Meyers, B. C., & Wing, R. A. (2016). Evolution of plant genome architecture. *Genome Biology, 17*, 37.

Wendte, J. M., & Pikaard, C. S. (2017). The RNAs of RNA-directed DNA methylation. *Biochimica et Biophysica Acta, 1860*, 140–148.

Wenke, T., Dobel, T., Sorensen, T. R., Junghans, H., Weisshaar, B., & Schmidt, T. (2011). Targeted identification of short interspersed nuclear element families shows their widespread existence and extreme heterogeneity in plant genomes. *The Plant Cell Online, 23*, 3117–3128.

Wheeler, B. S. (2013). Small RNAs, big impact: Small RNA pathways in transposon control and their effect on the host stress response. *Chromosome Research: An International Journal on the Molecular, Supramolecular and Evolutionary Aspects of Chromosome Biology, 21*, 587–600.

Wicker, T., & Keller, B. (2007). Genome-wide comparative analysis of copia retrotransposons in Triticeae, rice, and Arabidopsis reveals conserved ancient evolutionary lineages and distinct dynamics of individual copia families. *Genome Research, 17*, 1072–1081.

Wicker, T., Sabot, F., Hua-Van, A., Bennetzen, J. L., Capy, P., Chalhoub, B., et al. (2007). A unified classification system for eukaryotic transposable elements. *Nature Reviews. Genetics, 8*, 973–982.

Wierzbicki, A. T., Ream, T. S., Haag, J. R., & Pikaard, C. S. (2009). RNA polymerase V transcription guides ARGONAUTE4 to chromatin. *Nature Genetics, 41*, 630–634.

Witte, C.-P., Le, Q. H., Bureau, T., & Kumar, A. (2001). Terminal-repeat retrotransposons in miniature (TRIM) are involved in restructuring plant genomes. *Proceedings of the National Academy of Sciences of the United States of America, 98*, 13778–13783.

Wolfgruber, T. K., Sharma, A., Schneider, K. L., Albert, P. S., Koo, D. H., Shi, J., et al. (2009). Maize centromere structure and evolution: Sequence analysis of centromeres 2 and 5 reveals dynamic loci shaped primarily by retrotransposons. *PLoS Genetics, 5*, e1000743.

Xiao, H., Jiang, N., Schaffner, E., Stockinger, E. J., & van der Knaap, E. (2008). A retrotransposon-mediated gene duplication underlies morphological variation of tomato fruit. *Science, 319*, 1527–1530.

Xu, J., Tanino, K. K., Horner, K. N., & Robinson, S. J. (2016). Quantitative trait variation is revealed in a novel hypomethylated population of woodland strawberry (*Fragaria vesca*). *BMC Plant Biol, 16*, 240.

Yamashita, H., & Tahara, M. (2006). A LINE-type retrotransposon active in meristem stem cells causes heritable transpositions in the sweet potato genome. *Plant Molecular Biology, 61*, 79–94.

Yang, Z., Ebright, Y. W., Yu, B., & Chen, X. (2006). HEN1 recognizes 21-24 nt small RNA duplexes and deposits a methyl group onto the 2' OH of the 3' terminal nucleotide. *Nucleic Acids Research, 34*, 667–675.

Yao, J., Dong, Y., & Morris, B. A. (2001). Parthenocarpic apple fruit production conferred by transposon insertion mutations in a MADS-box transcription factor. *Proceedings of the National Academy of Sciences of the United States of America, 98*, 1306–1311.

Ye, R., Chen, Z., Lian, B., Rowley, M. J., Xia, N., Chai, J., et al. (2016). A dicer-independent route for biogenesis of siRNAs that direct DNA methylation in arabidopsis. *Molecular Cell, 61*, 222–235.

Yelagandula, R., Stroud, H., Holec, S., Zhou, K., Feng, S., Zhong, X., et al. (2014). The histone variant H2A.W defines heterochromatin and promotes chromatin condensation in Arabidopsis. *Cell, 158*, 98–109.

Yokthongwattana, C., Bucher, E., Čaikovski, M., Vaillant, I., Nicolet, J., Mittelsten Scheid, O., et al. (2010). MOM1 and Pol-IV/V interactions regulate the intensity and specificity of transcriptional gene silencing. *The EMBO Journal, 29*, 340–351.

Yu, A., Lepere, G., Jay, F., Wang, J., Bapaume, L., Wang, Y., et al. (2013). Dynamics and biological relevance of DNA demethylation in Arabidopsis antibacterial defense. *Proceedings of the National Academy of Sciences of the United States of America, 110*, 2389–2394.

Yuzbasioglu, G., Yilmaz, S., Marakli, S., & Gozukirmizi, N. (2016). Analysis of Hopi/Osr27 and Houba/Tos5/Osr13 retrotransposons in rice. *Biotechnology & Biotechnological Equipment, 30*, 213–218.

Zedek, F., Šmerda, J., Šmarda, P., & Bureš, P. (2010). Correlated evolution of LTR retrotransposons and genome size in the genus eleocharis. *BMC Plant Biology, 10*, 265.

Zeller, G., Henz, S. R., Widmer, C. K., Sachsenberg, T., Ratsch, G., Weigel, D., et al. (2009). Stress-induced changes in the *Arabidopsis thaliana* transcriptome analyzed using whole-genome tiling arrays. *The Plant Journal : For Cell and Molecular Biology, 58*, 1068–1082.

Zemach, A., Kim, M. Y., Hsieh, P. H., Coleman-Derr, D., Eshed-Williams, L., Thao, K., et al. (2013). The Arabidopsis nucleosome remodeler DDM1 allows DNA methyltransferases to access H1-containing heterochromatin. *Cell, 153*, 193–205.

Zhang, Q., Du, H., Lv, D., Xiao, T., Pan, J., He, H., et al. (2018). Efficient transposition of the retrotransposon Tnt1 in cucumber (*Cucumis sativus* L.). *Horticultural Plant Journal, 4*, 111–116.

Zhang, X., Yazaki, J., Sundaresan, A., Cokus, S., Chan, S. W., Chen, H., et al. (2006). Genome-wide high-resolution mapping and functional analysis of DNA methylation in arabidopsis. *Cell, 126*, 1189–1201.

Zhang, J. J., Zhou, Z. S., Song, J. B., Liu, Z. P., & Yang, H. (2012). Molecular dissection of atrazine-responsive transcriptome and gene networks in rice by high-throughput sequencing. *Journal of Hazardous Materials, 219-220*, 57–68.

Zhang, J., Zuo, T., Wang, D., & Peterson, T. (2014). Transposition-mediated DNA re-replication in maize. *Elife, 3*, e03724.

Zhao, D., Ferguson, A. A., & Jiang, N. (2016). What makes up plant genomes: The vanishing line between transposable elements and genes. *Biochimica et Biophysica Acta, 1859*, 366–380.

Zheng, B., Wang, Z., Li, S., Yu, B., Liu, J. Y., & Chen, X. (2009). Intergenic transcription by RNA polymerase II coordinates Pol IV and Pol V in siRNA-directed transcriptional gene silencing in Arabidopsis. *Genes & Development, 23*, 2850–2860.

Zheng, X., Zhu, J., Kapoor, A., & Zhu, J.-K. (2007). Role of Arabidopsis AGO6 in siRNA accumulation, DNA methylation and transcriptional gene silencing. *The EMBO Journal, 26*, 1691–1701.

Zilberman, D., Cao, X., & Jacobsen, S. E. (2003). ARGONAUTE4 control of locus-specific siRNA accumulation and DNA and histone methylation. *Science, 299*, 716–719.

CHAPTER SEVEN

Roles of Epigenetic Mechanisms in Grafting and Possible Applications

Margot M.J. Berger, Philippe Gallusci and Emeline Teyssier[1]
UMR EGFV, Bordeaux Sciences Agro, INRA, Université de Bordeaux, Villenave d'Ornon, France
[1]Corresponding author: E-mail: emeline.teyssier@inra.fr

Contents

Advances in Botanical Research, Volume 88
ISSN 0065-2296
https://doi.org/10.1016/bs.abr.2018.10.003

Abstract

Grafting is a technic that allows combining the root system from one plant (rootstock) with the shoot of another plant (scion), with the aim to improve the plant agronomical characteristics. Several steps are required for a successful grafting interaction that involves the formation of a callus followed by the differentiation of vascular tissues that connect the rootstock to the scion. When successful, the rootstock-scion interaction results in a complex exchange of signals between the partners eventually leading to phenotypic variations. Several studies have now shown that in addition to nutritional and hormonal signals epigenetic regulations may also play an important role during the establishment of a successful graft interaction, and contribute to the numerous phenotypic consequences of rootstock-scion interactions. Here we summarize the most recent data indicating that both DNA methylation and siRNAs exchanges are essential components of the epigenetic dialogue between the graft partners, and could be the basis of strategies aiming at generating rootstock and scion phenotypic diversity in plants including those which are mainly clonally propagated.

1. INTRODUCTION

Grafting consists in joining together two complementary parts of plants, in order to create a chimerical organism functioning as a single plant. The upper portion of the grafted plant is called the scion, whereas the rootstock refers to the part that receives the scion and functions as the root system. Grafting has been used for millennia as testified by ancient texts (Mudge et al., 2009) and it is still widely exploited in horticulture, arboriculture, viticulture, and more recently its usage has been extended to vegetable production such as tomato or pepper (Kyriacou et al., 2017). Several reasons have contributed to the widespread usage of grafting. First grafting is used for clonal propagation of woody perennials and has allowed overcoming important threats in some plants such as phylloxera in grapevine (Kumar et al., 2017; Mudge et al., 2009). In addition to resistance to soil borne disease grafting also leads to enhancement of abiotic stress resistance and improvement of water and mineral nutrition (Albacete et al., 2015). Noteworthy grafting can drastically impact the development and physiology of the scion and depending on the graft combination may either enhance or limit scion vigor or impact fruit size, yield or quality (Koepke & Dhingra, 2013; Ollat et al., 2017; Warschefsky et al., 2016). The modifications of the scion could be correlated with rootstock specific properties as for example, the improvement of scion vigor associated to an enhanced ability to absorb and assimilate minerals such as nitrate (Albacete et al., 2015; Martínez-Ballesta et al., 2010).

However, in many cases the molecular bases involved in the rootstock-dependent scion modifications and *vice versa* are not clearly defined.

Studies aiming at unravelling the process by which scion and rootstock establish cellular and vascular continuity were initiated in 1980s using different species such as tomato (Jefree & Yeoman, 1983), *Prunus* spp. (Errea et al., 1994) or spruce (Miller & Barnett, 1993) but it is only recently that the use of *Arabidopsis thaliana* and the development of micrografting technics has allowed a more accurate understanding of the establishment of the graft junction (Fig. 1) (Matsuoka et al., 2016; Melnyk et al., 2015), as recently reviewed (Melnyk, 2016; Pina & Errea, 2005). So far, three main steps leading to the formation of a successful graft interaction have been described: (1) in a first step, tissue adhesion between rootstock and scion is initiated by the deposition of a homogeneous matrix containing pectins at the graft interface (Jefree & Yeoman, 1983; Miller & Barnett, 1993; Pina et al., 2012). (2) Cell divisions are then initiated and a callus is formed, connecting the scion with the rootstock (Pina et al., 2012). Cell expansion is also observed and participates in gap filing (Matsuoka et al., 2016). Later, the pectin layer becomes thinner and plasmodesmata form between rootstock and scion cells (Jefree & Yeoman, 1983; Kollmann et al., 1985; Kollmann

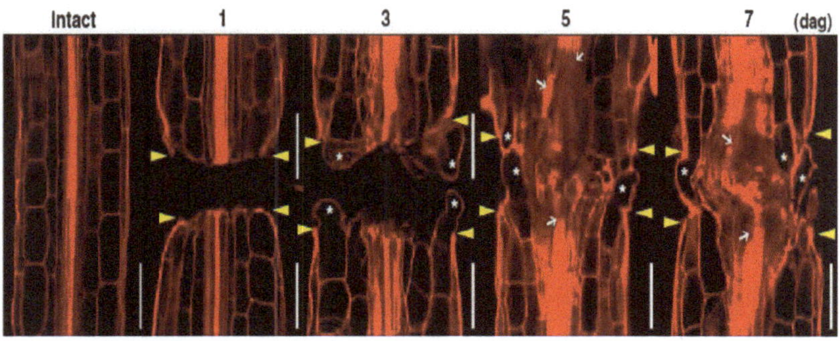

Figure 1 ***Arabidopsis micrografting: morphological changes during graft union development.*** Hypocotyl from 6 days old *Arabidopsis* plants were cut in the middle and reconnected by silicone tubing. Longitudinal sections were treated with modified pseudo-Schiff-propidium iodide (cell wall staining) and observed by confocal microscope. From left to right: intact hypocotyl before cutting; 1 day after grafting (dag); 3 dag; 5 dag and 7 dag. Yellow arrowheads indicate the positions of the cut surface, white asterisks, cell expansion in the cortex, and white arrows, a typical differentiated xylem vessel. Scale bars = 100 μm. *From Matsuoka, K., Sugawara, E., Aoki, R., Takuma, K., Terao-Morita, M., Satoh, S., et al. (2016). Differential cellular control by cotyledon-derived phytohormones involved in graft reunion of arabidopsis hypocotyls.* Plant & Cell Physiology, *57(12), 2620–2631.*

& Glockmann, 1991). (3) Finally a vascular cambium differentiates at the graft interface and initiates the production of secondary xylem and phloem, insuring the vascular reconnection between the scion and the rootstock.

Noteworthy the completion of grafting depends on the two partners. In the case of incompatible scion-rootstock combinations, a successful graft interaction cannot be achieved and grafting results in plant death that can occur either shortly or much later, months or years after grafting (for a review: Pina & Errea, 2005). In contrast, auto-graft, involving a scion and a rootstock from the same individual, and homo-graft, using a scion and a rootstock with the same genotype are generally successful, although they are not possible for some species, as for example Monocot species, which are refractory to grafting. Hetero-grafting, on the other hand, corresponds to the association of a scion and a rootstock with different genotypes, within the same variety or species or from different species. The success of hetero-grafting depends on the taxonomic distance between both partners: a close taxonomic relationship generally leads to a successful graft interaction, and grafting efficiency decreases with the taxonomic distance. But a few intergeneric combinations within the same family may also lead to successful grafts (Mudge et al., 2009). For example in *Solanaceae* grafting of eggplant and tomato leads to a compatible interaction (Wu et al., 2013). The molecular mechanisms explaining graft compatibility or incompatibility are not known, although graft incompatibility has been correlated with a decrease in symplasmic continuity between callus cells (Pina et al., 2012), an increase in phenol accumulation at the graft interface (Pina et al., 2012), and a decrease in the activity of reactive oxygen species (ROS) detoxifying enzymes (Irisarri et al., 2015). Moreover, since epigenetic mechanisms have been shown to play a role in the different cellular processes at play during graft healing, epigenetic plasticity is also likely critical for the establishment of grafting and may contribute to graft compatibility. Furthermore, evidence is growing that once the graft is healed, the scion and the rootstock are the sites of important epigenetic reprogramming in hetero-grafts that are guided by small RNA exchanges between partners (Dunoyer et al., 2010; Lewsey et al., 2016; Molnar et al., 2010).

The objective of this chapter is to review the current data pointing to a role of epigenetic mechanisms during grafting. We will describe the potential role of epigenetic mechanisms during the establishment of a successful graft union, and present evidence that epigenetic regulations contribute to the rootstock-mediated effects on the scion, and *vice versa*. Possible applications to breeding strategies will also be discussed.

2. ESTABLISHMENT OF THE GRAFT UNION

2.1 Establishment of the Graft Union: A Complex Process

Significant advances were recently made in describing the cellular events taking place during graft healing in the model plant *Arabidopsis* (Matsuoka et al., 2016; Melnyk et al., 2015). Furthermore an increasing number of transcriptomic analyses are published, describing the dynamic of the transcriptome at the graft interface in different species, suggesting different mechanisms involved in graft formation (Chen et al., 2017; Cookson et al., 2013; Melnyk et al., 2018; Qiu et al., 2016; Yin et al., 2012). However the cellular and molecular mechanisms required to produce a successful graft are not well defined at that time. Nevertheless all the aforementioned studies point to two important physiological aspects of graft healing: the induction of a stress response concomitantly with the establishment of the graft junction and the regulatory function of different plant hormones.

2.1.1 Graft Healing Is Associated With a Stress Response

Grafting generates multiple stresses for the plant that start with a mechanical injury generated by the initial cutting. This also creates an interruption of the vascular connections between the upper and basal parts of the plant that disturbs vertical hormonal and metabolites fluxes leading to an aberrant physiological situation. Recent data showed that starch accumulates at the graft interface, in the scion but not in the rootstock, and whereas glucose-responsive genes are induced in the scion, glucose-repressed genes are up-regulated in the rootstock (Melnyk et al., 2018), suggesting sugar accumulation and depletion in the grafted top and bottom respectively. This asymmetry although transitory, since it disappears upon vascular reconnection (Melnyk et al., 2018), may result in multiple stresses for the plant, including nutritional and hormonal stresses.

Transcriptomic analyses of grafted plants confirmed that a stress-response occurs shortly after grafting in different species including *Arabidopsis* (Melnyk et al., 2018; Yin et al., 2012), grape (Cookson et al., 2013) or litchi (Chen et al., 2017). Grafted and "cut but non-grafted" plants present very similar transcriptomic profiles characterized by the transient expression of stress related genes (Melnyk et al., 2018; Yin et al., 2012). This is consistent with the observation that mechanical wounding and grafting lead to common stress responses, such as callus formation and synthesis of stress-related hormones, jasmonic acid (JA) and ethylene. The production of JA was demonstrated in *Arabidopsis* (Matsuoka et al., 2018), whereas the synthesis of ethylene was suggested by the upregulation of ethylene biosynthesis genes in *Arabidopsis* and grape (Cookson et al., 2013; Yin et al., 2012). Wounding and grafting also trigger the production of reactive oxygen species (ROS). Based on the comparison of compatible and incompatible graft combinations, it was suggested that the redox status at the graft junction could control graft formation (Irisarri et al., 2015). Indeed graft compatibility was correlated with a decrease in the accumulation of ROS, and an increase in antioxidant activities or in the expression of genes coding for ROS detoxifying enzymes (Aloni et al., 2008; Irisarri et al., 2015).

Despite their high similarity, the transcriptomes of grafted and "cut but not grafted" *Arabidopsis* plants show some differences, indicating that a few genes are specifically induced during the graft interaction. These genes include stress-response genes (Melnyk et al., 2018; Yin et al., 2012), corresponding to immune- and chitin-responsive genes (Melnyk et al., 2018). More precisely, analysis of the tissues located immediately above and below the graft junction shows that the graft-related stress-response is transient, peaks 48 h after grafting, and occurs specifically in the rootstock (Melnyk et al., 2018). The graft-specific stress response seems more pronounced in hetero-grafts than in homo-grafts. This was suggested by the analysis of the transcriptomic dynamic at the graft interface in grapevine (Cookson et al., 2014). Contrary to the aforementioned studies that specifically focused on homo-grafting (Melnyk et al., 2018; Yin et al., 2012), Cookson et al. (2014) analyzed transcriptomes at the graft junction in homo- and hetero-grafts (Cookson et al., 2014). Unfortunately this study did not include the "cut but non-grafted" control, so that grafting-specific gene could not be identified. Nevertheless, it showed that many genes involved in stress response were induced at the graft interface in hetero-grafts when compared with autografted controls. More precisely genes related to biotic stress were strongly upregulated in hetero-grafted plants at three days after grafting, their expression, thereafter, gradually decreasing over time (Cookson et al., 2014).

Altogether these data indicate that the capacity of cells and tissues to cope with the grafting-induced stress and its deleterious effects is an important factor for grafting success.

2.1.2 Auxin, Cytokinins and Gibberellic Acid Control Grafting

Several studies have investigated the role of hormones in grafting, highlighting the prominent function of auxin, cytokinins and gibberellic acid in this process (for a review see Nanda & Melnyk, 2018).

Auxin is a well-known hormone with essential functions in plants that plays a central role during graft healing through its regulatory role in cell proliferation and vascular differentiation (Nieminen et al., 2015). As an illustration, using three *Solanaceae* species, Parkinson and Yeoman have shown that auxin application to the apical end of the internode is an absolute requirement for *in vitro* internode auto-grafting (Parkinson & Yeoman, 1982). Auxin accumulation at the graft junction both in the scion and rootstock was shown by direct quantification (Mo et al., 2018) and by the induction of auxin-responsive genes as well as of genes related to auxin transport and signaling shortly after grafting (Chen et al., 2017; Matsuoka et al., 2016; Melnyk et al., 2015, 2018; Mo et al., 2018; Yin et al., 2012; Zheng et al., 2009). Furthermore in *Arabidopsis* micrografts, cell proliferation and vascular reconnection was inhibited by the removal of cotyledons, which are major sites of auxin biosynthesis in juvenile seedlings or by plant treatments with triiodobenzoic acid (TIBA), an inhibitor of auxin polar transport (Matsuoka et al., 2016). Finally different *Arabidopsis* mutants affected in auxin perception or response (*aberrant lateral root formation 4* (*alf4*), *altered auxin response 1* (*axr1*), *indole-3-acetic acid factor 18* (*iaa18*) and the *transport inhibitor response 1, auxin signaling F-box 2 and 3* triple mutant (*tir1 afb2 afb3*)) showed delayed phloem reconnection (Melnyk et al., 2015).

Auxin often requires a precise interplay with cytokinins (for reviews: Nieminen et al., 2015; Schaller et al., 2015). First evidence that cytokinins also participate with auxin to the control of graft healing comes from the observation that they accumulate at the graft interface, as shown by direct quantification (Mo et al., 2018) or by the induction of cytokinin-responsive genes both in the scion and the rootstock (Melnyk et al., 2015). Furthermore the presence of cytokinins in the medium, although not an absolute requirement, increases the efficiency of vascular reconnection during *in vitro* internode grafting of *Solanaceae* species (Parkinson & Yeoman, 1982). Finally, cytokinins together with auxin are important regulators of cell proliferation (Schaller et al., 2014) and as such may participate to the control of cell division during graft-induced callogenesis, in a way similar to their role in callus formation during the wound healing process (Ikeuchi et al., 2017).

Using *Arabidopsis* micrografting, Matsuoka et al. have shown that cortex cell elongation was prevented in a gibberellin-deficient mutant or when the plant was treated with an inhibitor of gibberellin biosynthesis (Matsuoka et al., 2016), demonstrating a role for gibberellins in graft sealing.

Hence multiple hormones are present at the graft interface, potentially interacting with each other, and controlling graft healing.

2.2 Establishment of the Graft Union: Role of Epigenetic Mechanisms

As shown above, cells located at the graft junction, have to face very peculiar redox, metabolic and hormonal conditions. This specific context may trigger epigenetic regulations involved in the control of graft healing. Indeed there is evidence accumulating that epigenetic regulators are under the control of metabolic and redox cell status (Shen et al., 2016), and linked to hormonal signaling (for a review see Yamamuro et al., 2016). Stresses have also been shown to induce epigenetic changes that contribute to the adaptation of plants to potentially deleterious situations and epigenetic remodeling is believed to constitute part of the molecular bases of the memory of cells to stressful events, and might allow a better plant response to recurrent stresses (for reviews see: Annacondia et al., 2018; Bilichak & Kovalchuk, 2016; Avramova, 2015; Probst & Mittelsten Scheid, 2015; Chinnusamy & Zhu, 2009). In this context, it seems very likely that epigenetic processes will be involved in the complex cascade of molecular events that lead to a successful graft interaction. Moreover recent studies have shown that both DNA methylation and the control of histone post-translational modifications (PTMs) are involved in cellular processes that also occur during the interaction between the scion and the root stock and may therefore contribute to the graft healing.

2.2.1 DNA Methylation and Histone PTMs Are Components of Plant Responses to Wounding

As demonstrated for many abiotic stresses (Annacondia et al., 2018; Avramova, 2015; Bilichak & Kovalchuk, 2016; Chinnusamy & Zhu, 2009; Probst & Mittelsten Scheid, 2015), mechanical injuries generate DNA methylation changes, at a global level (Lewandowska-Gnatowska et al., 2014), and locally at specific loci (Lin et al., 2013). As an illustration, analysis of the global genomic content in 5-methylcytosine at different time points after maize leaves perforation demonstrated a transient decrease in global DNA methylation level (−20% to −30%) within 1 h after wounding (Lewandowska-Gnatowska et al., 2014). In sweet potato, leaf crushing leads to the transient down regulation of *LbMYB1* that encodes an activator of flavonoid biosynthesis, allowing for a metabolic rechanneling of phenylpropanoid precursors toward lignin biosynthesis. *LbMYB1* downregulation correlates with the induction of specific 22- and 24-nt small interfering RNAs (siRNAs), targeting *LbMYB1*, which were shown to be responsible for *LbMYB1* gene methylation (Lin et al., 2013), directly linking RNA-directed DNA methylation (RdDM) to the plant response to leaf injury.

Other examples of mechanical stresses are physical injuries provoked by herbivores. *Brassica rapa* leaves were shown to undergo DNA methylation changes when subjected to the caterpillar *Pieris brassicae* (Kellenberger et al., 2016). According to Methylation Sensitive Amplified Polymorphism (MSAP) analyses, these changes corresponded mostly to DNA demethylation events that were very similar to the modifications induced by the treatment of leaves with jasmonic acid (JA), a compound produced in response to herbivore damages that elicits similar plant defense responses (Bruinsma et al., 2009). As previously indicated (part 2.1.1), JA is produced at the graft interface. Noteworthy, JA but also ethylene, another stress related hormone are known to be associated with histone deacetylation and acetylation (reviewed in Yamamuro et al., 2016), suggesting that not only DNA methylation and demethylation events, but also histone PTMs may participate in graft healing. On one hand, the *HDA6* and *HDA19* genes encoding histone deacetylases (HDACs) are induced by JA and ethylene (Zhou et al., 2005) and downregulating these genes alters the expression of JA induced genes (Wu et al., 2008), and of JA and ethylene co-regulated-genes (Zhou et al., 2005) respectively. On the other hand, the transcriptional activation of ethylene responsive genes depends on the increase of H3K14Ac and H3K23Ac levels in their 5′ part (Zhang et al., 2017, 2016). Although such histone PTMs would require to be investigated in the context of grafting, the occurrence of JA and also possibly ethylene signals at the graft junction suggests that histone deacetylation/acetylation events take place locally at the graft interface, shortly after grafting and participate in the transcriptome reprogramming.

2.2.2 Cell Dedifferentiation Necessitates Epigenetic Reprogramming

One of the earliest events of grafting is the formation of a callus. Cytological studies suggested that cells responsible for callus formation originated mainly from the cambium, but also from the xylem ray parenchyma, the outer cortex and the pith (Hartmann et al., 2002). Callogenesis is a complex process necessitating profound modifications of adult somatic cells, which "reverse" to a less differentiated state and regain their competency for proliferation (for a review, see Ikeuchi et al., 2013). The mechanisms underlying callogenesis are being progressively deciphered using different models and an increasing number of data suggest that it also relies on epigenetic regulations. This has been the subject of a number of recent reviews

(Birnbaum & Roudier, 2017; de la Paz Sanchez et al., 2015; Jiang et al., 2015; Lee & Seo, 2018; Miguel & Marum, 2011), and we will mainly focus on aspects relevant to grafting.

2.2.2.1 In Vitro Callogenesis Requires Extensive Epigenetic Reprogramming

An important model for the analysis of callogenesis and cell dedifferentiation is *in vitro* culture of explants on auxin- and cytokinin-containing "Callus Inducing Medium" (CIM). A precise analysis of their organization and transcriptomes revealed that early steps of CIM-induced calli formation are likely sharing some of the regulatory pathways operating during lateral root meristem initiation (Atta et al., 2009; Sugimoto et al., 2010; Xu et al., 2012) and necessitate extensive epigenetic remodeling.

Evidence that *in vitro* callogenesis is associated with DNA methylation and histone PTM remodeling, comes from studies using *A. thaliana* mutants impaired in genes encoding various epigenetic writers. Many of the mutants with reduced callogenesis efficiency were affected in their ability to establish epigenetic marks associated with gene repression. They include the *DNA methyltransferase 1* (*met1*) mutants (Berdasco et al., 2008), which exhibit an almost complete elimination of *CG* methylation (Cokus et al., 2008; Lister et al., 2008; Zemach et al., 2013), the *kryptonite 2* (*kyp-2*) mutants (Grafi et al., 2007) characterized by a reduction in the levels of both methylated histone H3 on lysine 9 (H3K9me2), and CHG DNA methylation (Jackson et al., 2004; Johnson et al., 2002), the Polycomb Group *curly leaf—swinger* (*clf-swn*) double mutants and *embryonic flower 2* (*emf2*) mutants (He et al., 2012) affected for histone H3K27 trimethylation (Kim et al., 2012; Schubert et al., 2006; Yang et al., 2013) as well as the *histone deacetylase 9* (*hda9*) and *hd-tuins protein 1* (*hdt1*) (Lee et al., 2016) mutants showing an increase in histone H3 acetylation (Kim et al., 2013; Lawrence et al., 2004). However, the trithorax group (TrxG) mutants *arabidopsis trithorax-like protein 2* (*atxr2*) characterized by reduced H3K36 and H3K4 methylation (Saleh et al., 2008; Shafiq et al., 2014), two epigenetic marks associated with gene expression, were also less competent than WT for *in vitro* callogenesis (Lee et al., 2017). Accordingly the global level of the active marks, H3K4me3 and H3K36me3 increases during the leaf to callus transition (Lee et al., 2017). These results suggested that epigenetic remodeling was associated with callus induction in *Arabidopsis* as confirmed both by the genome wide analysis of DNA methylation and histone mark distributions (Berdasco et al., 2008; He et al., 2012; Lee et al., 2017). This was also demonstrated for DNA methylation in several crops including *Beta vulgaris* (Zakrzewski et al., 2017), *Brassica napus* (Gao et al., 2014), *Oryza sativa* (Stroud et al., 2013) and *Populus trichocarpa* (Vining et al., 2013).

Recent data revealed that sequential epigenetic modifications are required during *in vitro* callogenesis, in order to insure cell fate reprograming from mature differentiated cells to pluripotent root stem cells competent for regeneration. Indeed *in vitro* callogenesis from leaf explants proceeds through four steps, depending on the successive repression and induction of specific marker genes. These steps include (1) loss of mature leaf cell identity, associated with the repression of different genes such as *SAWTOOTH 1 (SAW1)* or *TEOSINTE BRANCHED CYCLOIDEA PCF TRANSCRIPTION FACTOR* genes (*TCPs)* (2) specification of pericycle competent cells, corresponding to the induction of *WUSCHEL RELATED HOMEOBOX 11 (WOX11)*, (3) promotion of cell divisions and formation of lateral root primordium-like tissues, linked to the induction of *LATERAL ORGAN BOUNDARIES DOMAIN* (*LBD)* genes (4) and finally acquisition of root stem cell identity, relying on the induction of *PLETHORA 1 (PLT1), PLT2*, and *WUSCHEL RELATED HOMEOBOX 5 (WOX 5)* (for a review see Lee & Seo, 2018). In this context, the addition of repressive marks, including DNA methylation mediated by DRM2 and MET1, and H3K27me3 catalyzed by the Polycomb Repressive Complexes 2 (PRC2), is first necessary to repress the expression of genes related to leaf identity. Berdasco et al. (2008) have identified three genes (*GLUTATHIONE S-TRANSFERASE TAU 10 [GSTU10],* MITOGEN-ACTIVATED PROTEIN KINASE 12 [*MAPK12*], *BETA-XYLOSIDASE 1* [*BXL1*]), which are downregulated during callogenesis and become concomitantly methylated in a DRM2/MET1 dependent manner (Berdasco et al., 2008), whereas the repression of genes playing a role in leaf like *SAW1* or *TCP 10* was associated with an increase in H3K27me3 (He et al., 2012). Noteworthy, a correlation was found between induction of the expression and decrease in H3K27me3 for auxin-related genes, or for genes specific to the root meristem like *WUSCHEL-RELATED HOMEOBOX 5* (*WOX5)* (He et al., 2012). Finally, the addition of epigenetic active marks is also necessary for the induction of callus induced-genes. For example the induction of the *LATERAL ORGAN BOUNDARIES DOMAIN* (LBD) genes depends on ATXR2 (Lee et al., 2017), which is responsible for the addition of an epigenetic mark associated with active transcription, H3K36me3, at their promoter (Lee et al., 2017). Interestingly ATRX2 is recruited to *LBD* promoters by the transcription factors AUXIN RESPONSE FACTOR 7 (ARF7) and ARF19 (Lee et al., 2017), which indicates that callogenesis is under auxin control.

Taken together these studies highlight the extensive epigenetic reprograming, required for *in vitro* callogenesis. Graft-dependent callogenesis may not involve the exact same cellular events as *in vitro* callogenesis, but it may also depend on large-scale epigenetic modifications.

2.2.2.2 Wound-Induced Callogenesis

Calli are also produced during tissue healing after a wound (Ikeuchi et al., 2013). Obviously callogenesis triggered by wounding or by CIM may share some mechanisms as revealed by the partial overlap of their transcriptome (Ikeuchi et al., 2017), and the identification of a few common regulators such as the transcription factor ETHYLENE RESPONSE FACTOR 115 (ERF115) (Heyman et al., 2016; Ikeuchi et al., 2017), but they also lead to different types of calli. Contrary to CIM-induced calli (Sugimoto et al., 2010), those induced after wounding are not organized like lateral root primordia, do not express root meristem markers, and their formation is not blocked by *solitary root* mutation which impairs lateral root initiation (Iwase et al., 2011).

Epigenetic remodeling during wound-induced cell dedifferentiation is first suggested by transcriptomic analysis of dedifferentiating cells after stem or hypocotyl incision and trunk debarking, the later corresponding to the elimination of bark and cambium from the trunk, while leaving differentiating xylem cells exposed. Ikeuchi et al. (2017) have performed the molecular characterization of *Arabidopsis* hypocotyls in a time course within 24 h after wounding by cutting (Ikeuchi et al., 2017). On the other hand, Zhang et al. (2011) have characterized transcriptomic changes occurring during cell dedifferentiation after *Populus* debarking (Zhang et al., 2011). In this case, secondary vascular tissue regeneration occurs within twelve days, starting with the dedifferentiation of xylem cells. Both models have revealed a very large number of Differentially Expressed Genes (DEGs) during cell dedifferentiation, including genes encoding epigenetic regulators involved in the control of DNA methylation and histone PTM levels. Of special interest genes coding for DNA methyltransferases (chromomethylases, CMTs, responsible for CHG maintenance and MET1-like DNA methyltransferases) were up-regulated both in dedifferentiated xylem (Zhang et al., 2011), and during wound-induced callus formation (Ikeuchi et al., 2017). Furthermore two genes encoding putative DNA GLYCOSYLASE-LYASE (DML) responsible for active DNA demethylation were down-regulated in dedifferentiated xylem (Zhang et al., 2011). Such modifications in gene expression profiles suggest important epigenetic remodeling during cell dedifferentiation after wounding, involving an increase in DNA methylation and modifications of histone PTMs.

A demonstration that epigenetic regulations are indeed important factors controlling wound-induced callogenesis comes from the characterization of the epigenetic status of genes coding for callogenesis central regulators *in Arabidopsis*, such as WOUND-INDUCED DEDIFFERENTIATION 3 (WIND3), PLT3, PLT5 and PLT7 or ERF115 (Ikeuchi et al., 2017). These genes that are induced during wound-related callogenesis are necessary for callus formation (Ikeuchi et al., 2017; Iwase et al., 2011). Interestingly their repressive status in unstressed plant is correlated by the presence of the repressive marks H3K27me3 in their promoter region as shown in intact ten-day-old seedlings (Ikeuchi et al., 2017). In line with this observation, callogenesis is initiated in an uncontrolled manner in PRC2 loss of function *swn clf* double mutants (Chanvivattana et al., 2004). Furthermore mitotic reactivation is observed even in highly differentiated cells such as root hair cells, concomitantly with the induction of *WIND3* in the PRC2 loss of function mutants, *clf swn* and *embryonic flower 2 vernalization 2 (emf vrn)* double mutants (Ikeuchi et al., 2015). This suggests that the presence of H3K27me3 marks at specific genes involved in the control of callus formation, such as *WIND3* is necessary to maintain cells in a differentiated state and to prevent callogenesis. In contrast, the removal of this mark would be required to allow the expression of these regulators and induce cell dedifferentiation and callogenesis. Of note, *WIND3* is associated with both H3K27me3 and H3K4me3 marks in intact plants (Ikeuchi et al., 2015). Hence *WIND3* transcriptional status is likely controlled by the balance between these two antagonistic marks, as also demonstrated for a number of other genes undergoing activation or repression at specific developmental stages, such as *AGAMOUS* (Saleh et al., 2007).

Noteworthy a few *JUMONJI* (*JMJ*) genes coding for histone demethylases, are induced during wound-triggered callus formation (Ikeuchi et al., 2017) and could be responsible for the active removal of H3K27me3 in response to wounding.

As a conclusion, although callogenesis induced by CIM and by a wound, do not involve the same regulatory pathways (Ikeuchi et al., 2017), both are correlated with epigenetic changes including the removal of H3K27me3 at a few specific loci, a process, which appears necessary to unlock cells from their differentiated state.

2.2.3 Epigenetic Reprogramming During Cell Differentiation

In addition to cell dedifferentiation and callogenesis, successful grafting also requires cell differentiation at the graft interface to generate epidermis, parenchyma, and vascular tissues. So far, studies describing cell differentiation during graft healing have only focused on the vascular reconnection process, as this is an absolute requirement for a successful grafting. Cellular events leading to vascular reconnection include the formation of a new cambium, and the subsequent production of secondary xylem and phloem. Accordingly, a recent transcriptomic analysis of tissues at the graft interface revealed a high transcriptional overlap with genes differentially expressed during vascular induction in leaf explants, with a sequential activation of genes associated with cambium, phloem, and xylem formation (Melnyk et al., 2018; Kondo et al., 2016, 2015). Whereas the genetic control of cambium initiation and functioning starts to be unraveled (for recent reviews see Barra-Jiménez & Ragni, 2017; Etchells et al., 2016; Nieminen et al., 2015), the possible involvement of epigenetic regulations in this process is still not described.

2.2.3.1 Evidences for Epigenetic Remodeling During Vascular Cell Differentiation

Although the genome-wide distribution of two histone marks, H3K4me3 and H3K27me3 has been characterized in *Eucalyptus* secondary xylem at an early developmental stage (Hussey et al., 2015), to our knowledge, there is no report about dynamic epigenetic changes during cambium initiation or, phloem and xylem formation. Yet, the recent description of specific epigenomes after plant dissection, suggests an important function for epigenetic regulations during cell differentiation, and in particular during vascular cell differentiation. As an illustration, the comparison of the methylomes of six different cell types extracted from *Arabidopsis* root meristem revealed that vascular cells are characterized by specific DNA methylation features (Kawakatsu et al., 2016). A hierarchical clustering based on differences in CG DNA methylation showed that cells originating from the stele and the endodermis form a distinct group compared with the other root cells (Kawakatsu et al., 2016). Furthermore the occurrence of specific methylation profiles in vascular tissues was demonstrated by the comparative MSAP analysis of ten different tissues and organs from poplar (Wang et al., 2016). Interestingly this analysis leads to the identification of seven genes which were differentially methylated in xylem tissues (developing and mature xylem) compared to non-xylem tissues (cambium, shoot apex, young and mature leaves, phloem, root, male and female inflorescences) and whose function was potentially related to xylogenesis (Wang et al., 2016). However, the relationship between the methylation status of these genes and their tissue-specific expression has not been demonstrated (Wang et al., 2016).

Finally, the transcriptomic analysis of vascular tissue differentiation in *Arabidopsis* roots also suggest that DNA methylation and other epigenetic mechanisms are likely important during vascular differentiation, by revealing an enrichment in epigenetic-related genes in developing xylem (Brady et al., 2007).

On another hand, evidence that epigenetic regulators are essential actors of cell differentiation is provided by the analysis of several *Arabidopsis* mutants affected in epigenetic control, using *in vitro* regeneration experiments.

2.2.3.2 In Vitro Differentiation From Calli Is More Efficient in Plants With Hypomethylated DNA

Most studies aiming at the analysis of epigenetic mechanisms during cell differentiation have used *in vitro* shoot induction from *Arabidopsis* callus, as a model. Calli formed *in vitro* on CIM, have the capacity to produce shoots when incubated on a cytokinin-rich medium, a process that presents important similarities with the reprogramming of a shoot apical meristem from a lateral root meristem.

Noteworthy, *in vitro* shoot regeneration is altered in mutants affected in epigenetic mechanisms (Li et al., 2011). A faster regeneration process has been associated with mutants characterized by reduced capacities in DNA methylation maintenance or in histone repressive mark establishment such as *met1* and *kyp-2* mutants, and with those showing an increase in active marks such as mutants of the histone demethylase gene *JUMONJI 14* (*JMJ14*) that present an increase in H3K4me3 (Deleris et al., 2010; Greenberg et al., 2013). On the contrary *histone acetyltransferase 1* (*hac1*) mutants (Earley et al., 2007) had a slower regeneration kinetics (Li et al., 2011). Acetylated histones being normally associated with gene transcription, this suggests that the addition of active marks is required for efficient shoot regeneration. In the case of *WUSCHEL*, a gene essential for the formation and maintenance of the shoot apical meristem (Laux et al., 1996), there is a decrease in both DNA methylation and H3K9me2, and an increase in the active marks H3K9ac and H3K4me3 at its regulatory regions, concomitantly with its induction during shoot induction (Li et al., 2011). As a conclusion, cell differentiation associated with shoot regeneration from CIM-induced calli necessitates the induction of specific genes involved in pluripotency acquisition, and shoot meristem initiation, that requires for their activation the removal of repressive marks and the subsequent addition of active marks. Although this remain to be shown, cell differentiation associated with vascular regeneration at the graft interface more probably also depends on epigenetic reprogramming.

More globally, although there is no description of epigenetic reprogramming during grafting, an increasing number of data support the idea of epigenetic regulations being involved in each of the three main processes taking place at the graft junction: stress response, cell dedifferentiation and vascular formation (Fig. 2). Epigenetic regulations involved in cell

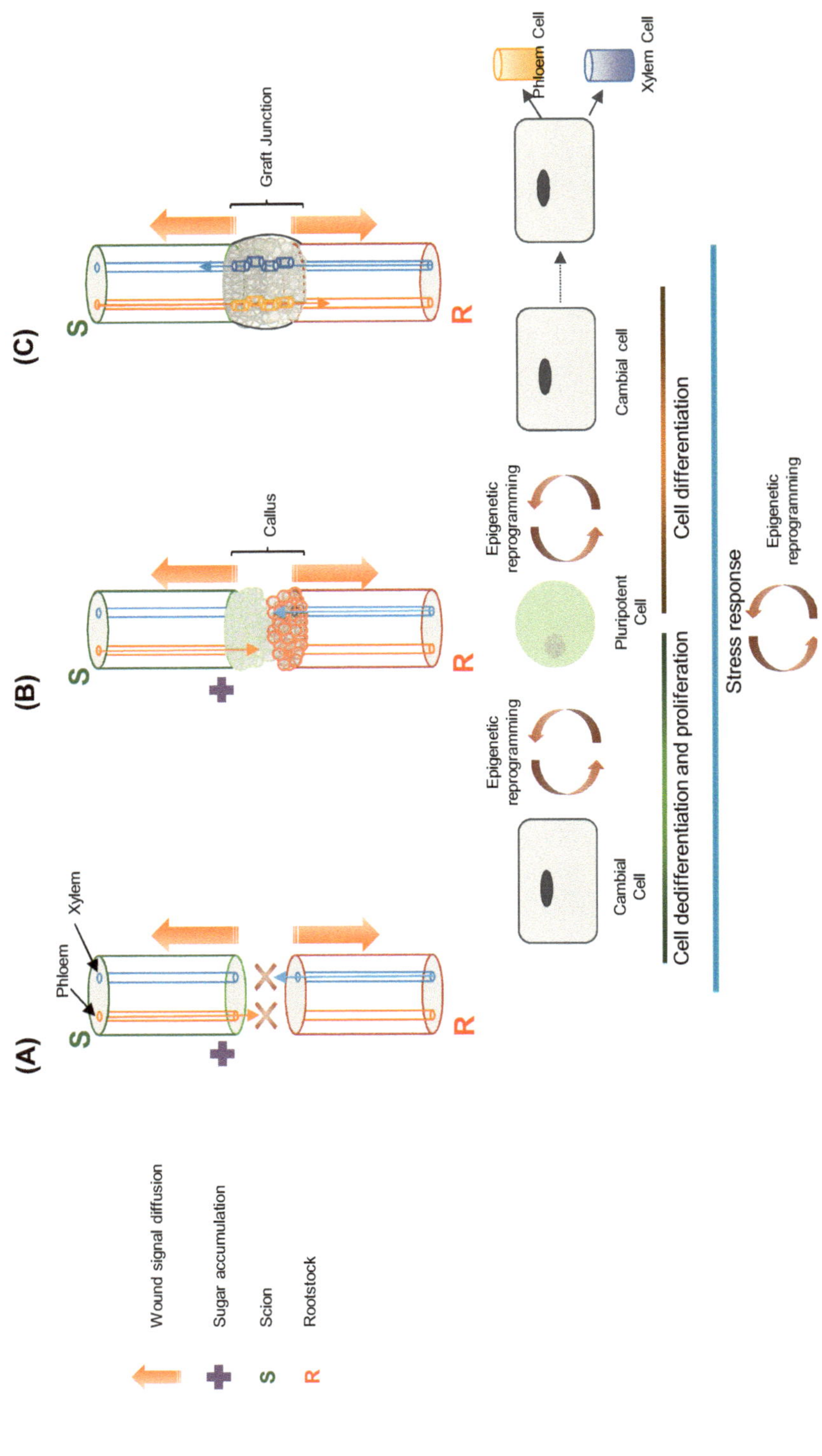
Wound signal diffusion
Sugar accumulation
S Scion
R Rootstock
(A)
Phloem
Xylem
S
R
(B)
Callus
S
R
(C)
Graft Junction
S
R
Cambial Cell
Epigenetic reprogramming
Pluripotent Cell
Epigenetic reprogramming
Cambial cell
Phloem Cell
Xylem Cell
Cell dedifferentiation and proliferation
Cell differentiation
Stress response
Epigenetic reprogramming

Figure 2 ***Early events during graft formation involve epigenetic remodeling.*** **(A) Initial situation.** Wounding and loss of plant integrity generate a stress response in the grafted plant, locally at the graft junction, and also potentially to the entire plant through long distance signaling. Hormonal and metabolic fluxes are interrupted, leading to an aberrant physiological situation at the graft interface as illustrated by sugar accumulation in the scion but not in the rootstock. Because epigenetic modifications are components of the stress response, grafting most probably induce stress-related epigenetic reprogramming. **(B) Callogenesis.** One of the earliest events during grafting is the formation of a callus between the scion and the rootstock: cells mainly originating from the cambium reverse to a less differentiated, pluripotent state and regain their competence for proliferation. An extensive number of data show that cell dedifferentiation requires transcriptional reprogramming depending on epigenetic remodeling. **(C) Vascular reconnection.** Vascular reconnection is considered as the basic requirement for a successful grafting. It relies on the formation of a new cambium and the production of secondary xylem and phloem from this cambium. Although the involvement of epigenetic regulations in cambium initiation and functioning has not been shown yet, different data point to a general role for epigenetic mechanisms during cell differentiation.

dedifferentiation and cell differentiation events are expected to be local and transient, impacting only a few specific cells at the graft junction. In contrast, stress-related epigenetic modifications may also affect distant part of the plant thanks to long-distance signaling (reviews: Gilroy et al., 2016; Savatin et al., 2014; Schilmiller & Howe, 2005). Whether grafting could trigger stable stress-related epigenetic modifications in the scion and/or the rootstock still await to be demonstrated. Additional systemic effects could be due to siRNAs exchanges between partners through the graft junction that can trigger epigenetic modifications in the grafted plant (part 4.2.2; Lewsey et al., 2016; Melnyk et al., 2011a; Dunoyer et al., 2010; Molnar et al., 2010) that could generate phenotypic changes on the scion and rootstock. Indeed grafting has been shown to impact the plant phenotype, as discussed in the following part.

3. ROOTSTOCK-MEDIATED EFFECTS ON THE SCION AND VICE-VERSA

3.1 Global Modification of Plant Development

Hetero-grafting is used in many species in order to improve scion development i.e. growth potential and fruit yield and quality (Koepke & Dhingra, 2013; Kyriacou et al., 2017; Ollat et al., 2017; Warschefsky et al., 2016), although the mechanisms underlying these effects are by far not yet understood. In some cases rootstock-conferred vigor has been linked to increased photosynthesis rate as for example in muskmelon hetero-grafts (Liu et al., 2011). Rootstock may also affect fruit quality, including fruit weight, metabolic content, firmness or color, as well as fruit productivity through the modification of different traits, such as flowering time, flower number, fruit set and biennial bearing (Koepke & Dhingra, 2013). Finally a few studies have described rootstock-scion interactions affecting scion disease resistance, as for example resistance of apple to fire blight (Jensen et al., 2012) although resistance to biotic stresses, in most cases, are related to intrinsic resistance of the rootstock, to soil borne diseases or pests (Koepke & Dhingra, 2013).

Until recently, most studies have focused on the impact of rootstock on scion, since the scion is the part of the plant producing marketable products. In contrast, the influence of scion on rootstock has been completely overlooked, although recent works reveal that scion can also influence root growth in grapevine and apple-tree. By comparing several graft combinations, significant modifications of the root system area and root length were observed depending on the scion cultivar (Harrison et al., 2016; Kocsis et al., 2016). Indeed, the scion is able to trigger invigorating effects on rootstock, indicating that the upper part of the plant should not be considered as a passive but together with the rootstock as an active partner that determines the growth characteristics of the grafted plant.

3.2 Inheritable Sporadic Phenotype Variations Affect Vegetative and Reproductive Plant Development After Hetero-Grafting

In addition to general effects observed in hetero-grafts, unexpected phenotypical modifications of phenotypes have been observed on scions after hetero-grafting only. Such modifications that occur with a low frequency and in a mosaic way mainly affect fruit shape and color or plant architecture. Most interestingly, some of these phenotypic variations were shown to be inheritable (Hirata, 1980; Taller et al., 1998; Yagishita, 1961). For example, in tomato, grafting a scion of the Jubilee cultivar (yellow skin and orange flesh) on a rootstock of the Tiny Tim cultivar (yellow skin and red flesh) led to changes in the scion fruit color: one out of sixteen plants produced fruits with red flesh instead of yellow (Hirata, 1980). The flowers formed on the scion with this modified phenotype were self-pollinated and the progeny was analyzed until the third generation. The new phenotype was clearly heritable but did not strictly follow Mendel's laws: notably a new plant type producing colorless-red fruits appeared in the first generation and some reversion events (increasing from 29% to 67% during the three first generations) were recorded. This suggests that phenotypic variations were due to epigenetic modifications rather than to genetic mutations, although mutations due to transposon mobility cannot be formally excluded.

Yagishita (1961) and Taller et al. (1998) made similar observations by grafting two distinct pepper cultivars, that presented several specific traits such as different fruit shape and fruit direction (erect or pendent fruits). Scions were obtained that presented new phenotypes. Successive grafts, using the same genotype as rootstock were performed with the progeny of the modified scions. In addition to modifications in fruit shape, scions showed variations in several traits including fruit direction, fruit clustering pungency, and color and shoot branching. These variations that appeared mostly in a mosaic state were either directly related to rootstock features, or did correspond to new traits, present neither in the stock nor in the scion. So far the causative molecular mechanisms underlying the differences between controls and variants have not been identified. However reversibility of the observed characters suggests that epigenetic mechanisms may be involved in the generation of such variants after hetero-grafting.

4. HETERO-GRAFTING-MEDIATED EFFECTS MAY BE DUE TO MULTIPLE FACTORS

Many reviews have described the different mechanisms possibly involved in the control by rootstocks of scion properties (Albacete et al., 2015; Gregory et al., 2013; Martínez-Ballesta et al., 2010; Ollat et al., 2017). Although the objective of this review is to focus on epigenetic mechanisms, it should be stressed out that many other mechanisms are also involved as briefly discussed below.

4.1 Physiological Parameters Play Important Roles in Rootstock-Scion Interactions

4.1.1 Nutrition and Hormonal Interactions

The improvement of crop yield and quality could first be linked to the capacity of the rootstock for nutrient uptake and assimilation (reviewed in Nawaz et al., 2016). However, rootstocks with a strong ability to absorb nutrients do not necessarily improve vine vigor as translocation of nutrients from the root to the scion is also essential. Hence, the interaction between rootstock and scion is as important as the scion and rootstock characteristics *per se*, although the underlying mechanisms remain poorly understood. Furthermore higher ion levels in the scion are not systematically correlated with an improved vigor as for example in kiwi (Thorp et al., 2007) indicating that other factors such as water nutrition, hormonal balance and long distance signaling might also be important, beside nutrient accumulation.

Considering water nutrition, several comparative analyses using woody species plants showed that dwarfing rootstock had reduced hydraulic conductivity as compare to more vigorous ones (Atkinson et al., 2003; Martínez-Alcántara et al., 2013). Mechanisms include a negative impact of limited hydraulic conductivity on photosynthesis by enhancing stomatal closure during high evaporative demand. As a consequence, plant growth is reduced by lowering net CO_2 assimilation rate, as was shown for grafted citrus trees (Martínez-Alcántara et al., 2013). In addition, a principal component analysis (PCA) using scions grafted on rootstocks from a Recombinant Imbred Line (RIL) population revealed an association between water use efficiency, photosynthesis, plant biomass, but also hormonal traits (Cantero-Navarro et al., 2016). Some of the graft combinations shown in this study resulted both in improved fruit yield and water use efficiency by up to 40% compared to self-grafted plants, and, interestingly, the analysis of scion xylem sap in this study revealed that hormonal factors were implicated in this response. This is in accordance with previous work in which both ionic and phytohormones concentrations found in xylem sap seem to be directly influenced by the rootstock used, and to impact photosynthesis, leaf growth, and leaf biomass (Albacete et al., 2009).

Grafting also disturbs the hormonal balance of both partners inducing phenotypical and physiological consequences (reviewed in Albacete et al., 2015; Aloni et al., 2010). An illustration is provided by the comparison of auxin and cytokinin transport rates by rootstock xylem of grafted and un-grafted *prunus*. Interestingly the most invigorating rootstock in this study showed the highest rates of cytokinin transport in xylem sap, suggesting that the plant stimulating property might be linked to rootstock capacity to synthesize and supply enough cytokinin to the shoot (Sorce et al., 2002).

As seen above, physiological changes induced by grafting are extensive and susceptible to impact plant growth potential, but the precise underlying mechanisms remain poorly understood. In addition to the physiological factors mentioned above, systemic movements of macromolecules such as proteins, mRNAS and small RNAs take place in the grafted plants and may also induce changes in the scion and the rootstock.

4.1.2 Exchange of Proteins, Messenger RNAs and Small RNAs

4.1.2.1 Messenger RNAs and Proteins

Characterization of messenger RNA (mRNA) exchanges across graft junction by high throughput RNA sequencing analyses made on rootstock and scion tissues and phloem sap, assessed long distance bi-directional movement of transcripts in grafted plants (Kanehira et al., 2010; Kehr & Kragler, 2018; Notaguchi et al., 2014; Thieme et al., 2015; Toscano-Morales et al., 2016; Yang et al., 2015; Zhang et al., 2013). However, mechanisms underlying the regulation of mRNA translocation from cells to sieve element, transport from sieve elements to destination cells, and if and how mRNA are selected to be delivered from one cell to another remain unknown (for reviews: Morris, 2018; Notaguchi & Okamoto, 2015). Genotypes, graft combinations, and growth environments seem to impact on the direction of mRNA movement through the graft junction, and determine the number and species of mobile mRNAs being exchanged. Indeed, a recent study based on detection and quantification of mobile RNAs in grafted grapevine revealed that field-grown mature grafts had much fewer genes transmitting mRNAs than the *in vitro* young grafts. Moreover, mRNA movements differed between the two types of graft, with more rootstock mRNAs moving into the scion in the *in vitro* grafts, and the converse in the field grafts. Furthermore, the mobile mRNA populations were different in field grafts compared to *in vitro* grafts (Yang et al., 2015).

Protein and peptide long distance and bi-directional movement across graft junctions have also been shown using GFP-tagged proteins in both hetero- and homo-grafting (Paultre et al., 2016; Toscano-Morales et al., 2016, 2014). But, as for mRNAs, mechanisms controlling protein transports through graft junctions are unknown.

Noteworthy, both mRNA and protein translocation through graft have been reported as inducing developmental changes in hetero-grafts including modified leaves architecture, or development of adventitious roots (Haywood et al., 2005; Toscano-Morales et al., 2016).

4.1.2.2 Small RNAs

Finally there are mounting evidences for long-range mobility of small RNAs (sRNAs), which are responsible for targeted gene silencing and may thereby influence plant phenotypes (as reviewed in Pyott & Molnar, 2015; Parent et al., 2012; Molnar et al., 2011). Different classes of sRNAs have been identified in plants, which are associated with Post Transcriptional Gene Silencing (PTGS) and/or Transcriptional Gene Silencing (TGS). Micro RNAs (miRNAs) are 21-nt long sRNAs generated mainly by the RNAse III nuclease, Dicer-like 1 (DCL1) from single stranded RNAs harboring partial double strand structures responsible for PTGS. After loading into ARGONAUTE 1 (AGO1), they form the so called RNA silencing induced complex (RISC), which is responsible for sequence specific degradation of homologous mRNAs and/or translation inhibition (reviews: Rogers & Chen, 2013, this book: Chapter 2). Specific miRNAs have been detected in the phloem sap (reviews: Kehr & Kragler, 2018; Mermigka et al., 2016; Marín-González & Suárez-López, 2012) and different studies suggest that their movement may play a role in developmental regulation such as tuberization in potato (Bhogale et al., 2014; Martin et al., 2009) and environmental responses like adaptation to phosphate deprivation (Pant et al., 2008).

Although miRNAs are generally considered as acting mostly in a cell autonomous manner (Parent et al., 2012), more investigations would be required to establish the contribution of their systemic movement to grafting-induced changes in scion or rootstock phenotypes. Small interfering RNAs (siRNAs) of different sizes (21 to 24-nt) and origins have also been characterized. Briefly, two main types of siRNAs can be distinguished depending on their mode of action. 21/22-nt long siRNAs are produced through DCL2 or DCL4 activity from transgenes or specific genes called *TAS* genes and they most often induce degradation of their target mRNAs through PTGS, although they can also trigger TGS at some specific loci, as shown for active TE (Bond & Baulcombe, 2015; McCue et al., 2015; Nuthikattu et al., 2013; Panda et al., 2016; Pontier et al., 2012; Zhao et al., 2016). In contrast 24-nt siRNAs are generated from heterochromatic regions or transgene through successive actions of RNA polymerase IV, RNA-dependent RNA polymerase RDRD2, and DCL3. They trigger methylation of homologous regions through a process known as RNA-directed DNA methylation (RdDM), leading to TGS (Matzke & Mosher, 2014, see Chapter 2 this book).

Different results suggest that siRNAs are transported in the phloem and could be responsible for systemic gene silencing spreading in the plant. First siRNAs have been detected in the phloem sap (Buhtz et al., 2008; Yoo et al., 2004). Second the characterization of siRNA populations by sRNA-SEQ, in *Arabidopsis* micrografting experiments using mutants impaired in siRNAs biogenesis, revealed that all type of siRNA 21-, 22-, 23- and 24-nt siRNAs are able to move from the scion to the rootstock, according to a source-sink flux (Melnyk et al., 2011a; Dunoyer et al., 2010; Molnar et al., 2010; part 4.2.2.1). As described in paragraphs 4.2.2.1 and 4.2.2.2, siRNA exchanges between the scion and the rootstock can induce epigenetic changes in the grafted plant, potentially leading to modifications of the phenotype.

Altogether the exchanges of proteins, miRNAs and mRNAs across the graft junction can impact the scion and rootstock physiology and development, and participate in rootstock-scion interactions. In addition, siRNAs are mobile through the graft junction and can mediate epigenetic modifications and chromatin remodeling, leading to transcriptional reprogramming. Such epigenetic modifications are stable and perpetuated through cell divisions, and eventually meiosis, thereby allowing stable and heritable phenotype modifications. They could therefore contribute to the changes observed after hetero-grafting (part 3.2).

4.2 Evidence for Epigenetic Reprogramming as an Explanation for Hetero-Grafting-Mediated Effects

4.2.1 Indirect Evidence: Scions Show Transcriptional Reprogramming When Grafted on a Rootstock From a Different Genotype

4.2.1.1 Transcriptional Reprogramming Occurs in Hetero-Grafts

To our knowledge, there is no report comparing the transcriptomes of non-grafted and homo-grafted plants. As a consequence, the effect of grafting by itself on the scion transcriptome cannot be ascertained. In contrast, several studies have compared scion transcriptomes after homo-grafting and hetero-grafting using different rootstocks. Such analyses revealed that hetero-grafting induces transcriptional reprogramming in the scions when compared to homo-grafting. Moreover in hetero-grafts, the scion transcriptome may depend on the rootstock genotype, especially when the rootstocks have contrasted effects on scion development, as shown in sweet cherry (Prassinos et al., 2009), apple tree (Jensen et al., 2012, 2010), watermelon (Liu et al., 2016), 'Shatangju' mandarin (Liu et al., 2017), or grape (Corso et al., 2016). Such rootstock-dependent transcriptome reprogramming could explain differences in plant phenotypes. For example Jensen et al., (2010) identified transcripts whose expression in apple tree shoot tips was correlated with tree size. Of interest, the *SDH* gene, encoding a sorbitol dehydrogenase, was systematically more expressed in larger trees (Jensen et al., 2010). Because the sorbitol dehydrogenase converts sorbitol, the primary form of transported carbon in apple tree (sorbitol) to fructose, increased *SDH* expression in shoot apices could contribute to improve carbon sink capacity, and therefore plant vigor (Jensen et al., 2010). Interestingly the difference in *SDH* expression in shoot tips was detected both in mature trees, and in young trees, 6 weeks after grafting (Jensen et al., 2010), suggesting that the mechanisms responsible for the regulation of *SDH* expression operate shortly after grafting and is maintained over time. However, underlying mechanisms were not studied, neither was the possible role of epigenetic regulations.

4.2.1.2 Transcriptional Changes in Hetero-Grafts May Accumulate as Secondary Consequences of Rootstock-Mediated Effects on Scion Physiology and Development

Transcriptional reprogramming in hetero-grafts could operate in the scions a short time after grafting, but also on a long term, during plant growth, as secondary consequences of rootstock-dependent changes in the physiological status of the scion, or consequently to interactions with the environment. As an illustration, several studies suggest that the divergence between scion transcriptomes may increase with time, after grafting on different rootstocks. Notably the comparison of two different grapevine hetero-grafts combining *Vitis vinifera* cv. Cabernet Sauvignon as a scion with two different rootstocks (Riparia Gloire de Montpellier or 1103 Paulsen) four months after grafting did not reveal any transcriptomic difference in shoot apices, although the two compared grafts were associated with different scion vigor (Cookson & Ollat, 2013). Unfortunately this report did not include a later sampling point and there is no evidence that differences would appear after a longer time. Jensen et al. (2010) also mentioned that in apple trees relatively few genes expressed in the scion were affected by the rootstock six weeks after grafting whereas more differences were observed when apple trees grafted ten years before were used (Jensen et al., 2010).

As a consequence, in order to point to transcriptional changes which are direct consequences of the grafting process by itself (homo- or hetero-grafting), and not secondary consequences of rootstock-scion interactions, analyses performed a short time after grafting should be considered at first.

4.2.1.3 Transcriptional Changes in Hetero-Grafts Suggest a Role for the Epigenetic Machinery

Noteworthy, large transcriptomic differences were observed in grape shoot tips by comparing hetero-graft and homo-graft shortly after grafting (Cookson & Ollat, 2013), and among the 2040 DEGs, 19 were identified as genes encoding proteins related to "RNA regulation of transcription/chromatin remodeling factors", so that this functional category was overrepresented in the genes upregulated in the shoot apex of hetero-compared to auto-grafts. Among them, three genes coding for putative CMTs, and three others for putative histone methyltransferases (HMTs) were specifically upregulated in the hetero-grafts (Cookson & Ollat, 2013). Similarly, genes encoding enzymes involved in DNA methylation were shown to be deregulated in tomato grafted on eggplant as compared to un-grafted tomato. However the expression profiles of these genes in tomato homo-grafts control were not shown (Wu et al., 2013). These results indicate that transcriptomic reprogramming takes place in the shoot apex of hetero-grafted plants shortly after grafting, and impacts the expression of genes involved in epigenetic remodeling of plants.

Interestingly Cookson et al. (2013) have noticed an overlap between DEGs identified in the aforementioned study and those associated with hybrid vigor suggesting that the interaction between two different genomes may trigger comparable transcriptional responses (Cookson & Ollat, 2013; Shen et al., 2012). This is quite surprising because the two situations seem totally different. In hybrids, the two genomes are simultaneously present in the same cells, whereas in grafted plant, they are located in different cells, and interactions between the two partners only occur at the graft junction. However, exchange of genetic material at the graft interface cannot be totally excluded as showed by the presence of cells with novel combination of genetic material at the graft junction (Stegemann & Bock, 2009), but might be of limited impact. The transcriptional analogy between hetero-grafts and hybrids would suggest that similar molecular mechanisms are operating in both types of organisms that could act either locally in single cells or in a systemic way. Such mechanisms may involve epigenetic regulatory pathways and signaling as suggested by an increasing number of recent studies showing that hybrids undergo trans-chromosomal methylation and demethylation events, where the methylation state of one parental allele is altered to resemble the other allele originating from the other parent (Greaves et al., 2015, 2016; He et al., 2013; Shivaprasad et al., 2011; Zhu et al., 2017).

As described in the following parts, two series of experimental data have demonstrated that grafting indeed triggers epigenetic remodeling and more precisely changes in DNA methylation.

4.2.2 Locus Specific Methylation Changes Depending on Mobile siRNAs in Grafted Plants

The mobility of siRNAs and their impact on gene silencing has been the subject of a number of recent reviews (Melnyk et al., 2011b; Mermigka et al., 2016; Molnar et al., 2011; Tamiru et al., 2018). The following part focuses on results concerning specifically transcriptional gene silencing (TGS) diffusion across the graft junction.

4.2.2.1 TGS Induced by a Transgene Can Spread Through the Graft Junction

TGS refers to locus-specific RNA-directed DNA methylation (RdDM), depending mainly on Domain rearranged DNA methyltransferases (DRMs) guided by 24-nt siRNAs (reviews: Matzke & Mosher, 2014; Zhang & Zhu, 2011; Law & Jacobsen, 2010) although 21/22-nt siRNAs may also be involved (Bond & Baulcombe, 2015; McCue et al., 2015; Nuthikattu et al., 2013; Panda et al., 2016; Pontier et al., 2012; Zhao et al., 2016). TGS was initially characterized using two-component transgene systems composed of (1) a silencing-inducing construct producing hairpin RNAs with inverted repeat sequences (here below referred to as 'silencing transgene') and (2) a transgene (here below referred to as 'target transgene') with promoter/enhancer regulatory sequences homologous to the 'silencing transgene' (Daxinger et al., 2009; Kanno et al., 2008; Mette et al., 2000; Sijen et al., 2001). Alternatively 'silencing transgenes' targeting endogenous sequences of the plant under study were employed (Cigan et al., 2005; Heilersig et al., 2006; Kasai et al., 2016; Melquist & Bender, 2003; Sijen et al., 2001). In these systems, silencing of the target gene relies on the siRNA of 24-nt produced from the 'silencing transgene' (Daxinger et al., 2009; Heilersig et al., 2006) and is accompanied by the methylation of the enhancer/promoter of the 'target transgene' or endogenous gene (Cigan et al., 2005; Daxinger et al., 2009; Heilersig et al., 2006; Melquist & Bender, 2003; Mette et al., 2000; Sijen et al., 2001).

Such two-component transgene systems were also used to demonstrate that TGS can spread through the graft junction in a systemic manner in micrografts (Bai et al., 2011; Kasai et al., 2016; Melnyk et al., 2011a), by combining a TGS-inducing scion containing the 'silencing transgene' with a rootstock containing the target transgene but not the 'silencing transgene', and vice-versa (Bai et al., 2011; Kasai et al., 2016; Melnyk et al., 2011a). Target transgene silencing was observed a few weeks after grafting, with a gradual loss of expression (Bai et al., 2011; Melnyk et al., 2011a). Furthermore, as expected for TGS, it was accompanied by an increase in the methylation of the target promoter (Bai et al., 2011; Kasai et al., 2016; Melnyk et al., 2011a). Clearly silencing of the 'target transgene' was more efficient when the 'silencing transgene' was expressed in the scion and the 'target transgene' in the rootstock, compared with the reciprocal situation. As an illustration, in *Arabidopsis* Melnyk et al. (2011a and 2011b) could not observe any GFP 'target gene' silenced in the scion, when a 35S promoter-driven 'silencing transgene' was expressed in the rootstock (over 32 assays). In contrast, in *Nicotiana*, Bai et al. (2011) could demonstrate silencing of the GFP 'target gene' in the scion, but under specific conditions: the 'silencing transgene' was under the control of a companion cell-specific promoter in order to maximize siRNA concentrations in the phloem. Furthermore the efficiency of the silencing was dependent on the sink potential of the scion: the elimination of all developed leaves three weeks after grafting was necessary to detect a clear TGS in the stem and in newly developed leaves, around the veins (Bai et al., 2011). Hence the efficiency of siRNAs movement between the scion and the rootstock appears as a key parameter controlling TGS spreading in grafted plants. In accordance with the siRNAs being transported in the phloem (part 4.1.2.2), this movement is a source to sink movement, therefore depending on the metabolic relationships between organs.

Consistent with the major role of 24-nt siRNAs in TGS, this size class of siRNAs appears to be essential for 'T transgene' transcriptional silencing in the rootstock (Melnyk et al., 2011a). Indeed graft-dependent silencing is abolished in the rootstock when *dicer-like 3* (*dcl3*) scions unable to produce 24-nt siRNAs (Henderson et al., 2006; Xie et al., 2004) are used. Nevertheless both 21- and 24-nt siRNAs are mobile in the grafted plants, although their amount in the recipient tissues is not proportional to their amount in the producing tissues. Both Melnyk et al. and Kasai et al. showed that whereas 21-nt siRNAs were predominantly synthesized from the silencing transgene in the scion, 21-nt and 24-nt siRNAs were detected in roughly the same amount in the silenced rootstock (Kasai et al., 2016; Melnyk et al., 2011a). Therefore a role for 21-nt siRNAs in graft-dependent targeted TGS cannot be formally excluded. Of note TGS does not require a high accumulation of siRNAs in the recipient tissues, as the quantity of silencer-specific mobile RNAs detected in the rootstock corresponded to only 1% of the quantity estimated in the silencer expressing shoot (Melnyk et al., 2011a).

Interestingly Kasai et al. (2016) have shown, using tobacco-potato hetero-grafting, that grafting—dependent TGS is stably maintained during plant development and asexual reproduction in the absence of inducing siRNAs (Kasai et al. 2016). In this experiment the endogenous *GRANULE-BOUND STARCH SYNTHASE I* (*GBSSI*) gene was silenced in potato rootstocks by tobacco scions expressing the corresponding 'silencing transgene'. Both silenced roots and silenced micro-tubers were recovered from the rootstocks, and *GBSSI* silencing was further assessed in shoot regenerated from silenced roots and in shoots produced from silenced micro-tubers even though they contain no detectable inducing siRNAs as controlled by sRNA deep sequencing. Furthermore silencing was also maintained in the micro-tubers produced from these silenced shoots (Kasai et al., 2016).

4.2.2.2 Endogenous Repeated Sequences Present in the Scion Induce DNA Methylation Events in the Rootstock

Endogenous sequences are also able to produce mobile siRNAs, as shown for two specific loci corresponding to near perfect inverted repeats, and associated with siRNAs of different sizes *in Arabidopsis* (Dunoyer et al., 2010; Molnar et al., 2010). The mobility and functionality of the siRNAs produced from these near perfect repeats were investigated by micrografting two contrasted genotypes: a genotype harboring the intact near perfect inverted repeats was used as a scion, whereas another genotype with only one repeat or an incomplete repeat, unable to produce siRNAs, was used as rootstock. Both studies showed that endogenous inverted repeats predominantly generate mobile siRNAs forms of 24-nt (Dunoyer et al., 2010; Molnar et al., 2010). Dunoyer et al. (2010), furthermore, demonstrated that the siRNAs mobility was associated with the methylation of the homologous sequence in the rootstock albeit not at the same level as for the WT scion (Dunoyer et al., 2010).

On a larger scale, the comparative analysis of rootstock siRNA populations and DNA methylation for different graft combinations using *dicer-like 2, 3, 4* (*dcl 2,3,4*) triple mutants impaired for the production of 22- to 24-nt siRNA (Henderson et al., 2006), allowed the identification of thousands of loci with mobile-siRNA-associated DNA methylation in *Arabidopsis* (Lewsey et al., 2016; Molnar et al., 2010). More precisely, these loci correspond to sequences which are methylated in the rootstocks of all graft combination tested: C24/C24, Col0/Col0, C24/Col0, or C24/*dcl 2,3,4* grafts, except in grafts associating two mutants impaired in siRNA production as in the *dcl 2,3,4*/*dcl 2,3,4* graft combination (C24 and Col0 are two different *Arabidopsis* accessions; *dcl 2,3,4 correspond to dicer-like 2, 3, 4* triple mutant in a Col0 background and graft combinations are indicated as scion/rootstock). The mobile-siRNA-associated DNA methylation events identified in this experiment do not correspond to transgressive methylation events, as they are similarly found in both homo- (C24/C24, Col0/Col0) and hetero-grafts (C24/Col0 and converse). They rather corresponds to a constitutive mechanism responsible for the reinforcement of methylation at loci which are already methylated by siRNAs produced in the rootstock. The methylated loci identified in this study were significantly associated with transposon elements (TEs) corresponding to specific superfamilies of TEs including those typically found in gene-rich regions (Lewsey et al., 2016). The mobile siRNAs—dependent methylation of these TEs was interpreted as a mechanism aiming at reinforcing their silencing. Interestingly the loci with mobile-siRNA-associated DNA methylation were also targeted by 21-nt/22-nt siRNAs, suggesting that 21-nt/22-nt siRNAs may also play a role in graft dependent RdDM (McCue et al., 2015), although they are far less abundant than mobile 24-nt siRNAs (Lewsey et al., 2016).

Alternatively the loci targeted by 24-nt and 21-/22-nt siRNAs may be regulated by both TGS and PTGS mechanisms.

4.2.3 Hetero-Grafting Is Accompanied by Changes in DNA Methylation Profiles

Mobile-siRNA could induce *de novo* transgressive methylation events in hetero-grafts (Lewsey et al., 2016). This was observed in *Arabidopsis* hetero-grafts formed by a C24 scion fused to a Col0 rootstock when siRNAs produced in the scion, move to the rootstock and find homologous sequences which are normally un-methylated in the rootstock, as demonstrated at three independent loci by Lewsey et al. (2016). These transgressive methylation events are reminiscent of trans-chromosomal methylation events observed in hybrids (Greaves et al., 2012, 2015, 2016; Shivaprasad et al., 2011) and it would be necessary to evaluate the frequency of such transgressive chromosomal events in hetero-grafts in *Arabidopsis* and in other species.

Epigenetic changes after grafting were also reported in different crop species, *Solanaceae* (Wu et al., 2013), *Cucurbitaceae* (Avramidou et al., 2015) and in *Hevea trees* (Uthup et al., 2018), although the underlying mechanisms and the role of siRNA mobility were not deciphered. In these studies, changes in DNA methylation profiles were demonstrated in scions of hetero-grafts. The analysis of graft-dependent changes in *Solanaceae* suggests that interspecific hetero-grafting (tomato on eggplant and *vice versa*) but not homo-grafting triggers epigenetic reprogramming in the scions (Wu et al., 2013). The comparison of Methylation Sensitive Amplified Polymorphism (MSAP) profiles from expanded leaves harvested on tomato and eggplant scions revealed variations between hetero-grafted and non-grafted plants whereas none or very infrequent alterations were detected in homo-grafts compared to non-grafted controls. This result reinforces the hypothesis that connecting two individuals with different genomes induces a specific molecular response in the scion. Furthermore it demonstrates that epigenetic mechanisms, more specifically DNA methylation and demethylation, are linked to this process. Interestingly among 15 selected MSAP markers, which were tested for inheritability after self-pollination, 11 showed complete or high level of inheritance (Wu et al., 2013), revealing that part of the graft-induced changes in DNA methylation are stable through meiosis. They present features of epimutations that may explain part of the phenotypical modifications associated with hetero-grafting.

Furthermore such graft-dependent epimutations could potentially be used for crop improvement, when associated with agronomic traits of interest.

Interestingly small genome differences between the graft partners seem sufficient to trigger DNA methylation changes in the scion. This was shown in rubber tree by grafting five scions issued from the same rubber tree on five genetically divergent rootstocks from the same species (Uthup et al., 2018). Differences in scion DNA methylation were first revealed by differences in plant MSAP profiles and further confirmed by BS-sequencing three of the differentially methylated fragments. Unfortunately the extent of the genetic differences existing between the different rootstocks and the scions was not characterized in this study, and the five plants used as rootstocks, as well as the unique plant used to produce the five scions were only described as coming from open pollinated seeds collected from the field (Uthup et al., 2018).

Altogether these studies show that the transcriptomic reprogramming observed in hetero-grafts is correlated with an epigenetic reprogramming that involves changes in DNA methylation patterns, although the causal link between DNA methylation and transcriptional reprogramming has not been demonstrated yet (Fig. 3). Notably in *Arabidopsis* mobile siRNAs have been shown to target TE superfamilies, including TE enriched in euchromatin regions (Lewsey et al., 2016). If a similar mechanism exists in crop species, we may expect that graft-dependent changes in methylation in crop species may affect together with TE, nearby genes and therefore have a large impact on gene expression, because most crop species are characterized by TE rich genomes, where genes are often present in the vicinity of TEs.

5. CONCLUSION AND POSSIBLE BREEDING APPLICATIONS

Epigenetic mechanisms induce stable changes in gene expression without affecting the underlying genomic DNA sequence. They could therefore explain part of graft-induced effects, through stable modifications of the plant transcriptomes, in response to rootstock-scion interactions. Such epigenetic-based modifications would be relevant for breeding applications because many epigenetic modifications are stable and furthermore they are sometimes transmitted across generations allowing heritable phenotypic variation in the absence of DNA sequence change.

At this time, although studies aiming at unravelling the relevance of epigenetic mechanisms in grafting have remained scarce, they provide strong

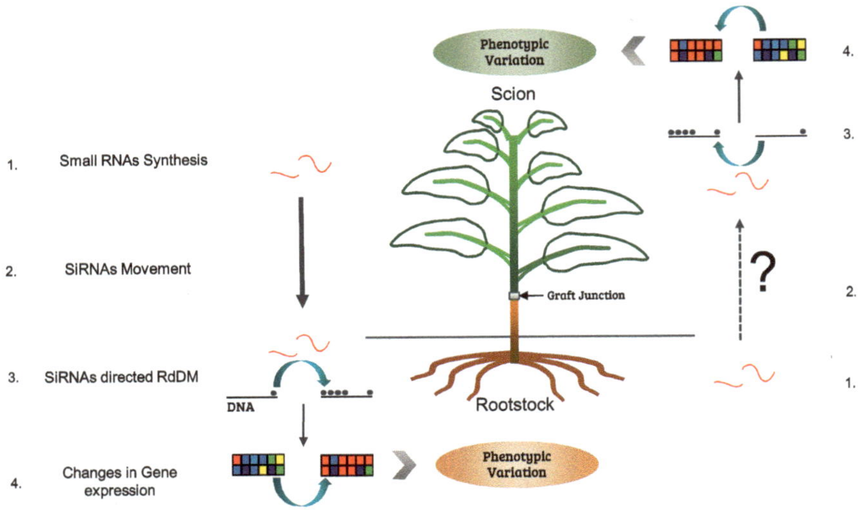

Figure 3 ***Epigenetic reprogramming as an explanation for hetero-grafting-mediated effects.*** The exchanges of small interfering RNAs (siRNAs) of all types (21–24 nt) between the scion and the rootstock have been demonstrated, characterized by a much higher efficiency for the source to sink direction, corresponding usually to the scion to rootstock direction. Exchanged siRNAs work as guides and trigger *de novo* methylation at homologous loci in the recipient tissue through RNA-directed DNA methylation (RdDM), requiring at least 24 nt long siRNAs. Modifications in cytosine methylation status can lead to changes in gene expression when affecting potentially transcriptionally active regions. Moreover, in hetero-grafts, siRNAs could find new, normally un-methylated targets, leading to so called "transgressive methylation events" potentially leading to new phenotypic features. Whereas scion-dependent *de novo* methylation events have been shown to occur in the rootstock, reciprocal events (rootstock-dependent *de novo* methylation in the scion) still need to be demonstrated.

evidence that hetero-grafting can lead to significant DNA methylation changes, through the RdDM pathway, following movements of 24-nt siRNA across the graft junction (Fig. 3). Whether histone post-translational modifications are also triggered after hetero-grafting is presently unknown. The generation of epigenetic modifications after homo-grafting is also not documented, but it cannot be excluded that the physiological stress induced by the grafting procedure also triggers stable epigenetic-changes.

In addition, shortly after grafting, massive transcriptomic reprogramming occurs locally at the graft junction, that may be associated with epigenome modifications. Indeed, the early cellular and molecular events associated with graft healing, including cell dedifferentiation and differentiation to form specific tissues such as vascular tissues (Pina & Errea, 2005) are crucial for the success of the graft. Because these cellular events rely on the reprogramming of the epigenomes of the cells located at the graft interface, it is very likely that the epigenetic machineries of the scion and the rootstock are important factors contributing to a successful grafting process.

The epigenetic modifications induced by grafting could impact the plant phenotype, especially if they affect meristematic cells. To date there is no demonstration for a functional link between a grafted-dependent epigenetic modification and any physiological/biological modification. Such a demonstration is challenging because grafting induce many different physiological changes, in addition to epigenetic changes, and it is difficult to associate a precise cause for any specific modification of traits. The situation is even more intricate as the plant physiological status has been shown to impact epigenetic mechanisms: the plant hormonal status interacts with epigenetic regulatory processes (Yamamuro et al., 2016) as does the plant metabolic status (Shen et al., 2016). These two parameters which are impacted by grafting, may influence the epigenetic homeostasis of plants, in addition to possible direct epigenetic interactions between the two partners, mainly mediated by small RNA exchanges.

Hetero-grafting appears as a method of choice for the production of epigenetic modifications whether these modifications are produced through siRNAs exchanges or as a secondary consequence of physiological changes. Many experimental data suggest that siRNAs represent key actors for epigenetic reprogramming through RdDM in hetero-grafts. The efficiency of siRNAs movement from the rootstock to the scion in crop species still need to be measured, but methylome reprogramming in the scion by siRNAs emanating from the rootstock should definitely be considered as it might shed new light on rootstock dependent effects on scion. For example it could provide explanation to the heritable developmental modifications observed in scions: *de novo* DNA methylation (RdDM) at selected loci could occur in scion when grafted to a plant with a different genome, because small RNAs produced by the rootstock could move to the scion and work as guide for RdDM in the scion cells (Fig. 3). Whether such DNA methylation induced gene silencing generated by grafting can trigger stable epimutation and consequently stable phenotypic modifications such as those observed in the tomato *colorless non-ripening (cnr)* epimutant (Manning et al., 2006) and many other plants (Cubas et al., 1999; Ong-Abdullah et al., 2015; Zhang et al., 2012) still needs to be demonstrated. Furthermore, the demonstration of graft dependent *de novo* methylation events also suggest that scion impacts rootstock epigenetic characteristics, hence potentially rootstock development and physiology, an aspect which is usually not considered.

The stability of graft-dependent epigenetic modifications still need to be assessed but preliminary observations suggest that part of them could be stably inherited and therefore could be used as a source for phenotypic diversity. Because plants used for grafting are often clonally propagated, the phenotypic diversity produced through epigenetic-based mechanisms may be conserved through selection and asexual amplification of the modified scions. Such epigenetic variations present the advantage to generate phenotypic diversity without affecting the genetic identity of the plants. Hence epigenetic editing approaches may be developed in the future, based on the grafting between a siRNA producing scion and a "to be modified rootstock," or *vice-versa*. The feasibility of this approach has already been demonstrated in potato (Kasai et al., 2016).

REFERENCES

Albacete, A., Martínez-Andújar, C., Martínez-Pérez, A., Thompson, A. J., Dodd, I. C., & Pérez-Alfocea, F. (2015). Unravelling rootstock×scion interactions to improve food security. *Journal of Experimental Botany, 66*(8), 2211–2226.

Albacete, A., Martinez-Andujar, C., Ghanem, M. E., Acosta, M., Sanchez-Bravo, J., Asins, M. J., et al. (2009). Rootstock-mediated changes in xylem ionic and hormonal status are correlated with delayed leaf senescence, and increased leaf area and crop productivity in salinized tomato. *Plant, Cell & Environment, 32*(7), 928–938.

Aloni, B., Cohen, R., Karni, L., Aktas, H., & Edelstein, M. (2010). Hormonal signaling in rootstock-scion interactions. *Scientia Horticulturae, 127*(2), 119–126.

Aloni, B., Karni, L., Deventurero, G., Levin, Z., Cohen, R., Katzir, N., et al. (2008). Physiological and biochemical changes at the rootstock-scion interface in graft combinations between Cucurbita rootstocks and a melon scion. *The Journal of Horticultural Science and Biotechnology, 83*(6), 777–783.

Annacondia, M. L., Magerøy, M. H., & Martinez, G. (2018). Stress response regulation by epigenetic mechanisms: Changing of the guards. *Physiologia Plantarum, 162*(2), 239–250.

Atkinson, C. J., Else, M. A., Taylor, L., & Dover, C. J. (2003). Root and stem hydraulic conductivity as determinants of growth potential in grafted trees of apple (Malus pumila Mill.). *Journal of Experimental Botany, 54*(385), 1221–1229.

Atta, R., Laurens, L., Boucheron-Dubuisson, E., Guivarc'h, A., Carnero, E., Giraudat-Pautot, V., et al. (2009). Pluripotency of *Arabidopsis* xylem pericycle underlies shoot regeneration from root and hypocotyl explants grown in vitro. *The Plant Journal, 57*(4), 626–644.

Avramidou, E., Kapazoglou, A., Aravanopoulos, F. A., Xanthopoulou, A., Ganopoulos, I., Tsaballa, A., et al. (2015). Global DNA methylation changes in Cucurbitaceae inter-species grafting. *Crop Breeding and Applied Biotechnology, 15*(2), 112–116.

Avramova, Z. (2015). Transcriptional "memory" of a stress: Transient chromatin and memory (epigenetic) marks at stress-response genes. *The Plant Journal, 83*(1), 149–159.

Bai, S., Kasai, A., Yamada, K., Li, T., & Harada, T. (2011). A mobile signal transported over a long distance induces systemic transcriptional gene silencing in a grafted partner. *Journal of Experimental Botany, 62*(13), 4561–4570.

Barra-Jiménez, A., & Ragni, L. (2017). Secondary development in the stem: When *Arabidopsis* and trees are closer than it seems. *Current Opinion in Plant Biology, 35*, 145–151.

Berdasco, M., Alcázar, R., García-Ortiz, M. V., Ballestar, E., Fernández, A. F., Roldán-Arjona, T., et al. (2008). Promoter DNA hypermethylation and gene repression in undifferentiated arabidopsis cells. *PLoS One, 3*(10).

Bhogale, S., Mahajan, A. S., Natarajan, B., Rajabhoj, M., Thulasiram, H. V., & Banerjee, A. K. (2014). MicroRNA156: A potential graft-transmissible MicroRNA that modulates plant architecture and tuberization in Solanum tuberosum ssp. andigena. *Plant Physiology, 164*(2), 1011–1027.

Bilichak, A., & Kovalchuk, I. (2016). Transgenerational response to stress in plants and its application for breeding. *Journal of Experimental Botany, 67*(7), 2081–2092.

Birnbaum, K. D., & Roudier, F. (2017). Epigenetic memory and cell fate reprogramming in plants. *Regeneration, 4*(1), 15–20.

Bond, D. M., & Baulcombe, D. C. (2015). Epigenetic transitions leading to heritable, RNA-mediated de novo silencing in *Arabidopsis thaliana. Proceedings of the National Academy of Sciences, 112*(3), 917–922.

Brady, S. M., Orlando, D. A., Lee, J.-Y., Wang, J. Y., Koch, J., Dinneny, J. R., et al. (2007). A high-resolution root spatiotemporal map reveals dominant expression patterns. *Science, 318*, 802–806.

Bruinsma, M., Pang, B., Mumm, R., Van Loon, J. J. A., & Dicke, M. (2009). Comparing induction at an early and late step in signal transduction mediating indirect defence in *Brassica oleracea. Journal of Experimental Botany, 60*(9), 2589–2599.

Buhtz, A., Springer, F., Chappell, L., Baulcombe, D. C., & Kehr, J. (2008). Identification and characterization of small RNAs from the phloem of Brassica napus. *The Plant Journal, 53*(5), 739–749.

Cantero-Navarro, E., Romero-Aranda, R., Fernández-Muñoz, R., Martínez-Andújar, C., Pérez-Alfocea, F., & Albacete, A. (2016). Improving agronomic water use efficiency in tomato by rootstock-mediated hormonal regulation of leaf biomass. *Plant Science, 251*, 90–100.

Chanvivattana, Y., Bishopp, A., Schubert, D., Stock, C., Moon, Y.-H., Sung, Z. R., et al. (2004). Interaction of Polycomb-group proteins controlling flowering in *Arabidopsis. Development, 131*(21), 5263–5276.

Chen, Z., Zhao, J., Hu, F., Qin, Y., Wang, X., & Hu, G. (2017). Transcriptome changes between compatible and incompatible graft combination of Litchi chinensis by digital gene expression profile. *Scientific Reports,* 7(1), 1–12.

Chinnusamy, V., & Zhu, J. K. (2009). Epigenetic regulation of stress responses in plants. *Current Opinion in Plant Biology, 12*(2), 133–139.

Cigan, M. A., Unger-Wallace, E., & Haug-Collet, K. (2005). Transcriptional gene silencing as a tool for uncovering gene function in maize. *The Plant Journal, 43*(6), 929–940.

Cokus, S. J., Feng, S., Zhang, X., Chen, Z., Merriman, B., Haudenschild, C. D., et al. (2008). Shotgun bisulphite sequencing of the Arabidopsis genome reveals DNA methylation patterning. *Nature, 452*(7184), 215–219.

Cookson, S. J., Clemente Moreno, M. J., Hevin, C., Nyamba Mendome, L. Z., Delrot, S., Magnin, N., et al. (2014). Heterografting with nonself rootstocks induces genes involved in stress responses at the graft interface when compared with autografted controls. *Journal of Experimental Botany, 65*(9), 2473–2481.

Cookson, S. J., Clemente Moreno, M. J., Hevin, C., Nyamba Mendome, L. Z., Delrot, S., Trossat-Magnin, C., et al. (2013). Graft union formation in grapevine induces transcriptional changes related to cell wall modification, wounding, hormone signalling, and secondary metabolism. *Journal of Experimental Botany, 64*(10), 2997–3008.

Cookson, S. J., & Ollat, N. (2013). Grafting with rootstocks induces extensive transcriptional re-programming in the shoot apical meristem of grapevine. *BMC Plant Biology, 13*(1).

Corso, M., Vannozzi, A., Ziliotto, F., Zouine, M., Maza, E., Nicolato, T., et al. (2016). Grapevine rootstocks differentially affect the rate of ripening and modulate auxin-related genes in cabernet Sauvignon berries. *Frontiers in Plant Science, 7*, 1–14.

Cubas, P., Vincent, C., & Coen, E. (1999). An epigenetic mutation responsible for natural variation in floral symmetry. *Nature, 401*(6749), 157–161.

Daxinger, L., Kanno, T., Bucher, E., Van Der Winden, J., Naumann, U., Matzke, A. J. M., et al. (2009). A stepwise pathway for biogenesis of 24-nt secondary siRNAs and spreading of DNA methylation. *The EMBO Journal, 28*(1), 48–57.

de la Paz Sanchez, M., Aceves-García, P., Petrone, E., Steckenborn, S., Vega-León, R., Álvarez-Buylla, E. R., et al. (2015). The impact of Polycomb group (PcG) and Trithorax group (TrxG) epigenetic factors in plant plasticity. *New Phytologist, 208*(3), 684–694.

Deleris, A., Greenberg, M. V. C., Ausin, I., Law, R. W. Y., Moissiard, G., Schubert, D., et al. (2010). Involvement of a Jumonji-C domain-containing histone demethylase in DRM2-mediated maintenance of DNA methylation. *EMBO Reports, 11*(12), 950–955.

Dunoyer, P., Brosnan, C. A., Schott, G., Wang, Y., Jay, F., Alioua, A., et al. (2010). An endogenous, systemic RNAi pathway in plants. *The EMBO Journal, 29*(10), 1699–1712.

Earley, K. W., Shook, M. S., Brower-Toland, B., Hicks, L., & Pikaard, C. S. (2007). *In vitro* specificities of *Arabidopsis* co-activator histone acetyltransferases: Implications for histone hyperacetylation in gene activation. *The Plant Journal, 52*(4), 615–626.

Errea, P., Felipe, A., & Herrero, M. (1994). Graft establishment between compatible and incompatible *Prunus* spp. *Journal of Experimental Botany, 45*(3), 393–401.

Etchells, J. P., Smit, M. E., Gaudinier, A., Williams, C. J., & Brady, S. M. (2016). A brief history of the TDIF-PXY signalling module: Balancing meristem identity and differentiation during vascular development. *New Phytologist, 209*(2), 474–484.

Gao, Y., Ran, L., Kong, Y., Jiang, J., Sokolov, V., & Wang, Y. (2014). Assessment of DNA methylation changes in tissue culture of *Brassica napus*. *Russian Journal of Genetics, 50*(11), 1186–1191.

Gilroy, S., Białasek, M., Suzuki, N., Górecka, M., Devireddy, A. R., Karpiński, S., et al. (2016). ROS, calcium, and electric signals: Key mediators of rapid systemic signaling in plants. *Plant Physiology, 171*(3), 1606–1615.

Grafi, G., Ben-Meir, H., Avivi, Y., Moshe, M., Dahan, Y., & Zemach, A. (2007). Histone methylation controls telomerase-independent telomere lengthening in cells undergoing dedifferentiation. *Developmental Biology, 306*(2), 838–846.

Greaves, I. K., Eichten, S. R., Groszmann, M., Wang, A., Ying, H., Peacock, W. J., et al. (2016). Twenty-four–nucleotide siRNAs produce heritable trans-chromosomal methylation in F1 *Arabidopsis* hybrids. *Proceedings of the National Academy of Sciences, 113*(44), E6895–E6902.

Greaves, I. K., Gonzalez-Bayon, R., Wang, L., Zhu, A., Liu, P.-C., Groszmann, M., et al. (2015). Epigenetic changes in hybrids. *Plant Physiology, 168*(4), 1197–1205.

Greaves, I. K., Groszmann, M., Ying, H., Taylor, J. M., Peacock, W. J., & Dennis, E. S. (2012). Trans chromosomal methylation in *Arabidopsis* hybrids. *Proceedings of the National Academy of Sciences, 109*(9), 3570–3575.

Greenberg, M. V. C., Deleris, A., Hale, C. J., Liu, A., Feng, S., & Jacobsen, S. E. (2013). Interplay between active chromatin marks and RNA-directed DNA methylation in *Arabidopsis thaliana*. *PLoS Genetics, 9*(11).

Gregory, P. J., Atkinson, C. J., Bengough, A. G., Else, M. A., Fernández-Fernández, F., Harrison, R. J., et al. (2013). Contributions of roots and rootstocks to sustainable, intensified crop production. *Journal of Experimental Botany, 64*(5), 1209–1222.

Harrison, N., Barber-Perez, N., Pennington, B., Cascant-Lopez, E., & Gregory, P. J. (2016). Root system architecture in reciprocal grafts of apple rootstock-scion combinations. *Acta Horticulturae, 1130*, 409–414.

Hartmann, H. T., Kester, D. E., Davies, F. T., & Geneve, R. L. (2002). Principles of grafting and budding. In *Plant propagation* (7th ed., pp. 411–460). Upper Saddle River, N.J.: Prentice Hall.

Haywood, V., Yu, T. S., Huang, N. C., & Lucas, W. J. (2005). Phloem long-distance trafficking of GIBBERELLIC ACID-INSENSITIVE RNA regulates leaf development. *The Plant Journal, 42*(1), 49–68.

He, C., Chen, X., Huang, H., & Xu, L. (2012). Reprogramming of H3K27me3 is critical for acquisition of pluripotency from cultured arabidopsis tissues. *PLoS Genetics, 8*(8), 1–13.

He, G., He, H., & Deng, X. W. (2013). Epigenetic variations in plant hybrids and their potential roles in heterosis. *Journal of Genetics and Genomics, 40*(5), 205–210.

Heilersig, B. H. J. B., Loonen, A. E. H. M., Janssen, E. M., Wolters, A. M. A., & Visser, R. G. F. (2006). Efficiency of transcriptional gene silencing of GBSSI in potato depends on the promoter region that is used in an inverted repeat. *Molecular Genetics and Genomics, 275*(5), 437–449.

Henderson, I. R., Zhang, X., Lu, C., Johnson, L., Meyers, B. C., Green, P. J., et al. (2006). Dissecting *Arabidopsis thaliana* DICER function in small RNA processing, gene silencing and DNA methylation patterning. *Nature Genetics, 38*(6), 721–725.

Heyman, J., Cools, T., Canher, B., Shavialenka, S., Traas, J., Vercauteren, I., et al. (2016). The heterodimeric transcription factor complex ERF115-PAT1 grants regeneration competence. *Nature Plants, 2*, 1–7.

Hirata, Y. (1980). Graft-induced in tomato changes in skin and flesh color (*Lycopersicon esculentum* Mill.). *Journal of the Japanese Society for Horticultural Science, 49*(2), 211–216.

Hussey, S. G., Mizrachi, E., Groover, A., Berger, D. K., & Myburg, A. A. (2015). Genome-wide mapping of histone H3 lysine 4 trimethylation in *Eucalyptus grandis* developing xylem. *BMC Plant Biology, 15*(1), 1–14.

Ikeuchi, M., Iwase, A., Rymen, B., Harashima, H., Shibata, M., Ohnuma, M., et al. (2015). PRC2 represses dedifferentiation of mature somatic cells in Arabidopsis. *Nature Plants, 1*(July), 1–7.

Ikeuchi, M., Iwase, A., Rymen, B., Lambolez, A., Kojima, M., Takebayashi, Y., et al. (2017). Wounding triggers callus formation via dynamic hormonal and transcriptional changes. *Plant Physiology, 175*. pp.01035.2017.

Ikeuchi, M., Sugimoto, K., & Iwase, A. (2013). Plant callus: Mechanisms of induction and repression. *The Plant Cell Online, 25*(9), 3159–3173.

Irisarri, P., Binczycki, P., Errea, P., Martens, H. J., & Pina, A. (2015). Oxidative stress associated with rootstock-scion interactions in pear/quince combinations during early stages of graft development. *Journal of Plant Physiology, 176*, 25–35.

Iwase, A., Mitsuda, N., Koyama, T., Hiratsu, K., Kojima, M., Arai, T., et al. (2011). The AP2/ERF transcription factor WIND1 controls cell dedifferentiation in arabidopsis. *Current Biology, 21*(6), 508–514.

Jackson, J. P., Johnson, L., Jasencakova, Z., Zhang, X., PerezBurgos, L., Singh, P. B., et al. (2004). Dimethylation of histone H3 lysine 9 is a critical mark for DNA methylation and gene silencing in *Arabidopsis thaliana*. *Chromosoma, 112*(6), 308–315.

Jefree, C. E., & Yeoman, M. M. (1983). Development of intercellular connections between opposing cells in a graft union. *New Phytologist, 93*(4), 491–509.

Jensen, P. J., Halbrendt, N., Fazio, G., Makalowska, I., Altman, N., Praul, C., et al. (2012). Rootstock-regulated gene expression patterns associated with fire blight resistance in apple. *BMC Genomics, 13*(1), 9.

Jensen, P. J., Makalowska, I., Altman, N., Fazio, G., Praul, C., Maximova, S. N., et al. (2010). Rootstock-regulated gene expression patterns in apple tree scions. *Tree Genetics and Genomes, 6*(1), 57–72.

Jiang, F., Feng, Z., Liu, H., & Zhu, J. (2015). Involvement of plant stem cells or stem cell-like cells in dedifferentiation. *Frontiers in Plant Science, 6*, 1–6.

Johnson, L., Cao, X., & Jacobsen, S. (2002). Interplay between two epigenetic marks. DNA methylation and histone H3 lysine 9 methylation. *Current Biology: CB, 12*(16), 1360–1367.

Kanehira, A., Yamada, K., Iwaya, T., Tsuwamoto, R., Kasai, A., Nakazono, M., et al. (2010). Apple phloem cells contain some mRNAs transported over long distances. *Tree Genetics and Genomes, 6*(5), 635–642.

Kanno, T., Bucher, E., Daxinger, L., Huettel, B., Böhmdorfer, G., Gregor, W., et al. (2008). A structural-maintenance-of-chromosomes hinge domain-containing protein is required for RNA-directed DNA methylation. *Nature Genetics, 40*(5), 670–675.

Kasai, A., Bai, S., Hojo, H., & Harada, T. (2016). Epigenome editing of potato by grafting using transgenic tobacco as siRNA donor. *PLoS One, 11*(8), 1–12.

Kawakatsu, T., Stuart, T., Valdes, M., Breakfield, N., Schmitz, R. J., Nery, J. R., et al. (2016). Unique cell-type-specific patterns of DNA methylation in the root meristem. *Nature Plants, 2*(May), 16058.

Kehr, J., & Kragler, F. (2018). Long distance RNA movement. *New Phytologist, 218*(1), 29–40.

Kellenberger, R. T., Schlüter, P. M., & Schiestl, F. P. (2016). Herbivore-Induced DNA demethylation changes floral signalling and attractiveness to pollinators in *Brassica rapa. PLoS One, 11*(11), 1–17.

Kim, W., Latrasse, D., Servet, C., & Zhou, D. X. (2013). Arabidopsis histone deacetylase HDA9 regulates flowering time through repression of AGL19. *Biochemical and Biophysical Research Communications, 432*(2), 394–398.

Kim, S. Y., Lee, J., Eshed-Williams, L., Zilberman, D., & Sung, Z. R. (2012). EMF1 and PRC2 cooperate to repress key regulators of arabidopsis development. *PLoS Genetics, 8*(3).

Kocsis, L., Tarczal, E., & Molnár Kocsisné, G. (2016). Grape rootstock-scion interaction on root system development. *Acta Horticulturae, 1136*, 27–32.

Koepke, T., & Dhingra, A. (2013). Rootstock scion somatogenetic interactions in perennial composite plants. *Plant Cell Reports, 32*(9), 1321–1337.

Kollmann, R., & Glockmann, C. (1991). Studies on graft unions - III. On the mechanism of secondary formation of plasmodesmata at the graft interface. *Protoplasma, 165*(1–3), 71–85.

Kollmann, R., Yang, S., & Glockmann, C. (1985). Studies on graft unions - II. Continuous and half plasmodesmata in different regions of the graft interface. *Protoplasma, 126*(1–2), 19–29.

Kondo, Y., Fujita, T., Sugiyama, M., & Fukuda, H. (2015). A novel system for xylem cell differentiation in *Arabidopsis thaliana. Molecular Plant, 8*(4), 612–621.

Kondo, Y., Nurani, A. M., Saito, C., Ichihashi, Y., Saito, M., Yamazaki, K., et al. (2016). Vascular cell induction culture system using arabidopsis leaves (VISUAL) reveals the sequential differentiation of sieve element-like cells. *The Plant Cell Online, 28*(6), 1250–1262.

Kumar, P., Rouphael, Y., Cardarelli, M., & Colla, G. (2017). Vegetable grafting as a tool to improve drought resistance and water use efficiency. *Frontiers in Plant Science, 8*.

Kyriacou, M. C., Rouphael, Y., Colla, G., Zrenner, R., & Schwarz, D. (2017). Vegetable grafting: The implications of a growing agronomic imperative for vegetable fruit quality and nutritive value. *Frontiers in Plant Science, 8*, 1–23.

Laux, T., Mayer, K. F., Berger, J., & Jürgens, G. (1996). The WUSCHEL gene is required for shoot and floral meristem integrity in Arabidopsis. *Development, 122*, 87–96.

Law, J. A., & Jacobsen, S. E. (2010). Establishing, maintaining and modifying DNA methylation patterns in plants and animals. *Nature Reviews. Genetics, 11*(3), 204–220.

Lawrence, R. J., Earley, K., Pontes, O., Silva, M., Chen, Z. J., Neves, N., et al. (2004). A concerted DNA methylation/histone methylation switch regulates rRNA gene dosage control and nucleolar dominance. *Molecular Cell, 13*(4), 599–609.

Lee, K., Park, O. S., Jung, S. J., & Seo, P. J. (2016). Histone deacetylation-mediated cellular dedifferentiation in Arabidopsis. *Journal of Plant Physiology, 191*, 95–100.
Lee, K., Park, O. S., & Seo, P. J. (2017). Arabidopsis ATXR2 deposits H3K36me3 at the promoters of LBD genes to facilitate cellular dedifferentiation. *Science Signaling, 10*(507), 1–11.
Lee, K., & Seo, P. J. (2018). Dynamic epigenetic changes during plant regeneration. *Trends in Plant Science, 23*(3), 235–247.
Lewandowska-Gnatowska, E., Polkowska-Kowalczyk, L., Szczegielniak, J., Barciszewska, M., Barciszewski, J., & Muszyńska, G. (2014). Is DNA methylation modulated by wounding-induced oxidative burst in maize? *Plant Physiology and Biochemistry, 82*, 202–208.
Lewsey, M. G., Hardcastle, T. J., Melnyk, C. W., Molnar, A., Valli, A., Urich, M. A., et al. (2016). Mobile small RNAs regulate genome-wide DNA methylation. *Proceedings of the National Academy of Sciences, 113*(6), E801–E810.
Li, W., Liu, H., Cheng, Z. J., Su, Y. H., Han, H. N., Zhang, Y., et al. (2011). DNA methylation and histone modifications regulate de novo shoot regeneration in arabidopsis by modulating wuschel expression and auxin signaling. *PLoS Genetics,* 7(8).
Lin, J. S., Lin, C. C., Li, Y. C., Wu, M. T., Tsai, M. H., Hsing, Y. I. C., et al. (2013). Interaction of small RNA-8105 and the intron of IbMYB1 RNA regulates IbMYB1 family genes through secondary siRNAs and DNA methylation after wounding. *The Plant Journal, 75*(5), 781–794.
Lister, R., O'Malley, R. C., Tonti-Filippini, J., Gregory, B. D., Berry, C. C., Millar, A. H., et al. (2008). Highly integrated single-base resolution maps of the epigenome in arabidopsis. *Cell, 133*(3), 523–536.
Liu, X. Y., Li, J., Liu, M. M., Yao, Q., & Chen, J. Z. (2017). Transcriptome profiling to understand the effect of citrus rootstocks on the growth of "Shatangju" Mandarin. *PLoS One, 12*(1), 1–22.
Liu, Y. F., Qi, H. Y., Bai, C. M., Qi, M. F., Xu, C. Q., Hao, J. H., et al. (2011). Grafting helps improve photosynthesis and carbohydrate metabolism in leaves of muskmelon. *International Journal of Biological Sciences,* 7(8), 1161–1170.
Liu, N., Yang, J., Fu, X., Zhang, L., Tang, K., Guy, K. M., et al. (2016). Genome-wide identification and comparative analysis of grafting-responsive mRNA in watermelon grafted onto bottle gourd and squash rootstocks by high-throughput sequencing. *Molecular Genetics and Genomics, 291*(2), 621–633.
Manning, K., Tör, M., Poole, M., Hong, Y., Thompson, A. J., King, G. J., et al. (2006). A naturally occurring epigenetic mutation in a gene encoding an SBP-box transcription factor inhibits tomato fruit ripening. *Nature Genetics, 38*(8), 948–952.
Marín-González, E., & Suárez-López, P. (2012). "And yet it moves": Cell-to-cell and long-distance signaling by plant microRNAs. *Plant Science, 196*, 18–30.
Martin, A., Adam, H., Diaz-Mendoza, M., Zurczak, M., Gonzalez-Schain, N. D., & Suarez-Lopez, P. (2009). Graft-transmissible induction of potato tuberization by the microRNA miR172. *Development, 136*(17), 2873–2881.
Martínez-Alcántara, B., Rodriguez-Gamir, J., Martínez-Cuenca, M. R., Iglesias, D. J., Primo-Millo, E., & Forner-Giner, M. A. (2013). Relationship between hydraulic conductance and citrus dwarfing by the Flying Dragon rootstock (*Poncirus trifoliata* L. Raft var. monstruosa). *Trees - Structure and Function, 27*(3), 629–638.
Martínez-Ballesta, M. C., Alcaraz-López, C., Muries, B., Mota-Cadenas, C., & Carvajal, M. (2010). Physiological aspects of rootstock-scion interactions. *Scientia Horticulturae, 127*(2), 112–118.
Matsuoka, K., Sugawara, E., Aoki, R., Takuma, K., Terao-Morita, M., Satoh, S., et al. (2016). Differential cellular control by cotyledon-derived phytohormones involved in graft reunion of arabidopsis hypocotyls. *Plant & Cell Physiology, 57*(12), 2620–2631.

Matsuoka, K., Yanagi, R., Yumoto, E., Yokota, T., Yamane, H., Satoh, S., et al. (2018). RAP2.6L and jasmonic acid—responsive genes are expressed upon Arabidopsis hypocotyl grafting but are not needed for cell proliferation related to healing. *Plant Molecular Biology, 96*(6), 531—542.

Matzke, M. A., & Mosher, R. A. (2014). RNA-directed DNA methylation: An epigenetic pathway of increasing complexity. *Nature Reviews. Genetics, 15*(6), 394—408.

McCue, A. D., Panda, K., Nuthikattu, S., Choudury, S. G., Thomas, E. N., & Slotkin, R. K. (2015). ARGONAUTE 6 bridges transposable element mRNA-derived siRNAs to the establishment of DNA methylation. *The EMBO Journal, 34*(1), 20—35.

Melnyk, C. W. (2016). Plant grafting: Insights into tissue regeneration. *Regeneration, 4*(1), 3—14.

Melnyk, C. W., Gabel, A., Hardcastle, T. J., Robinson, S., Miyashima, S., Grosse, I., et al. (2018). Transcriptome dynamics at *Arabidopsis* graft junctions reveal an intertissue recognition mechanism that activates vascular regeneration. *Proceedings of the National Academy of Sciences, 201718263*.

Melnyk, C. W., Molnar, A., Bassett, A., & Baulcombe, D. C. (2011a). Mobile 24 nt small RNAs direct transcriptional gene silencing in the root meristems of *Arabidopsis thaliana*. *Current Biology, 21*(19), 1678—1683.

Melnyk, C. W., Molnar, A., & Baulcombe, D. C. (2011b). Intercellular and systemic movement of RNA silencing signals. *The EMBO Journal, 30*(17), 3553—3563.

Melnyk, C. W., Schuster, C., Leyser, O., & Meyerowitz, E. M. (2015). A developmental framework for graft formation and vascular reconnection in arabidopsis thaliana. *Current Biology, 25*(10), 1306—1318.

Melquist, S., & Bender, J. (2003). Transcription from an upstream promoter controls methylation signaling from an inverted repeat of endogenous genes in Arabidopsis. *Genes & Development, 17*(16), 2036—2047.

Mermigka, G., Verret, F., & Kalantidis, K. (2016). RNA silencing movement in plants. *Journal of Integrative Plant Biology, 58*(4), 328—342.

Mette, M. F., Aufsatz, W., van der Winden, J., Matzke, M. A., & Matzke, A. J. (2000). Transcriptional silencing and promoter methylation triggered by double- stranded RNA. *The EMBO Journal, 19*(19), 5194—5201.

Miguel, C., & Marum, L. (2011). An epigenetic view of plant cells cultured in vitro: Somaclonal variation and beyond. *Journal of Experimental Botany, 62*(11), 3713—3725.

Miller, H., & Barnett, J. R. (1993). The structure and composition of bead-like projections on Sitka Spruce callus cells formed during grafting and in culture. *Annals of Botany, 72*, 441—448.

Mo, Z., Feng, G., Su, W., Liu, Z., & Peng, F. (2018). Transcriptomic analysis provides insights into grafting union development in pecan (*Carya illinoinensis*). *Genes, 9*(2).

Molnar, A., Melnyk, C. W., Bassett, A., Hardcastle, T. J., Dunn, R., Baulcombe, D. C., et al. (2010). Small silencing RNAs in plants are mobile and direct epigenetic modification in recipient cells. *Science, 328*(5980), 872—875.

Molnar, A., Melnyk, C., & Baulcombe, D. C. (2011). Silencing signals in plants: A long journey for small RNAs. *Genome Biology, 12*(1), 1—8.

Morris, R. J. (2018). On the selectivity, specificity and signalling potential of the long-distance movement of messenger RNA. *Current Opinion in Plant Biology, 43*, 1—7.

Mudge, K., Janick, J., Scofield, S., & Goldschmidt, E. E. (2009). A history of grafting. *Horticultural Reviews, 35*, 437—493.

Nanda, A. K., & Melnyk, C. W. (2018). The role of plant hormones during grafting. *Journal of Plant Research, 131*(1), 49—58.

Nawaz, M. A., Imtiaz, M., Kong, Q., Cheng, F., Ahmed, W., Huang, Y., et al. (2016). Grafting: A technique to modify ion accumulation in horticultural crops. *Frontiers in Plant Science,* 7(October), 1—15.

Nieminen, K., Blomster, T., Helariutta, Y., & Mähönen, A. P. (2015). Vascular cambium development. *The Arabidopsis Book, 13*, e0177.

Notaguchi, M., Higashiyama, T., & Suzuki, T. (2014). Identification of mRNAs that move over long distances using an RNA-seq analysis of Arabidopsis/Nicotiana benthamiana heterografts. *Plant & Cell Physiology, 56*(2), 311–321.

Notaguchi, M., & Okamoto, S. (2015). Dynamics of long-distance signaling via plant vascular tissues. *Frontiers in Plant Science, 6*(March), 1–10.

Nuthikattu, S., McCue, A. D., Panda, K., Fultz, D., DeFraia, C., Thomas, E. N., et al. (2013). The initiation of epigenetic silencing of active transposable elements is triggered by RDR6 and 21-22 nucleotide small interfering RNAs. *Plant Physiology, 162*(1), 116–131.

Ollat, N., Cookson, S. J., Lauvergeat, V., Marguerit, E., Barrieu, F., Gambetta, G., et al. (2017). Grapevine roots: the dark side. In *Acta Horticulturae* (pp. 213–226). Leuven, Belgium: International Society for Horticultural Science (ISHS).

Ong-Abdullah, M., Ordway, J. M., Jiang, N., Ooi, S. E., Kok, S. Y., Sarpan, N., et al. (2015). Loss of Karma transposon methylation underlies the mantled somaclonal variant of oil palm. *Nature, 525*(7570), 533–537.

Panda, K., Ji, L., Neumann, D. A., Daron, J., Schmitz, R. J., & Slotkin, R. K. (2016). Full-length autonomous transposable elements are preferentially targeted by expression-dependent forms of RNA-directed DNA methylation. *Genome Biology, 17*(1), 1–19.

Pant, B. D., Buhtz, A., Kehr, J., & Scheible, W. R. (2008). MicroRNA399 is a long-distance signal for the regulation of plant phosphate homeostasis. *The Plant Journal, 53*(5), 731–738.

Parent, J.-S., Martínez de Alba, A. E., & Vaucheret, H. (2012). The origin and effect of small RNA signaling in plants. *Frontiers in Plant Science, 3*(August), 1–9.

Parkinson, M., & Yeoman, M. M. (1982). Graft formation in cultured, explanted internodes. *New Phytologist, 91*(4), 711–719.

Paultre, D. S. G., Gustin, M.-P., Molnar, A., & Oparka, K. J. (2016). Lost in transit: Long-distance trafficking and phloem unloading of protein signals in arabidopsis homografts. *The Plant Cell Online, 28*(9), 2016–2025.

Pina, A., & Errea, P. (2005). A review of new advances in mechanism of graft compatibility-incompatibility. *Scientia Horticulturae, 106*(1), 1–11.

Pina, A., Errea, P., & Martens, H. J. (2012). Graft union formation and cell-to-cell communication via plasmodesmata in compatible and incompatible stem unions of Prunus spp. *Scientia Horticulturae, 143*, 144–150.

Pontier, D., Picart, C., Roudier, F., Garcia, D., Lahmy, S., Azevedo, J., et al. (2012). NERD, a plant-specific GW protein, defines an additional RNAi-dependent chromatin-based pathway in arabidopsis. *Molecular Cell, 48*(1), 121–132.

Prassinos, C., Ko, J. H., Lang, G., Iezzoni, A. F., & Han, K. H. (2009). Rootstock-induced dwarfing in cherries is caused by differential cessation of terminal meristem growth and is triggered by rootstock-specific gene regulation. *Tree Physiology, 29*(7), 927–936.

Probst, A. V., & Mittelsten Scheid, O. (2015). Stress-induced structural changes in plant chromatin. *Current Opinion in Plant Biology, 27*, 8–16.

Pyott, D. E., & Molnar, A. (2015). Going mobile: Non-cell-autonomous small RNAs shape the genetic landscape of plants. *Plant Biotechnology Journal, 13*(3), 306–318.

Qiu, L., Jiang, B., Fang, J., Shen, Y., Fang, Z., Saravana Kumar, R. M., et al. (2016). Analysis of transcriptome in hickory (Carya cathayensis), and uncover the dynamics in the hormonal signaling pathway during graft process. *BMC Genomics, 17*(1), 1–13.

Rogers, K., & Chen, X. (2013). Biogenesis, turnover, and mode of action of plant MicroRNAs. *The Plant Cell Online, 25*(7), 2383–2399.

Saleh, A., Al-Abdallat, A., Ndamukong, I., Alvarez-Venegas, R., & Avramova, Z. (2007). The *Arabidopsis* homologs of trithorax (ATX1) and enhancer of zeste (CLF) establish

"bivalent chromatin marks" at the silent *AGAMOUS* locus. *Nucleic Acids Research, 35*(18), 6290–6296.
Saleh, A., Alvarez-Venegas, R., Yilmaz, M., Le, O., Hou, G., Sadder, M., et al. (2008). The highly similar arabidopsis homologs of trithorax ATX1 and ATX2 encode proteins with divergent biochemical functions. *The Plant Cell Online, 20*(3), 568–579.
Savatin, D. V., Gramegna, G., Modesti, V., & Cervone, F. (2014). Wounding in the plant tissue: the defense of a dangerous passage. *Frontiers in Plant Science, 5*, 3–11.
Schaller, G. E., Bishopp, A., & Kieber, J. J. (2015). The Yin-Yang of hormones: Cytokinin and auxin interactions in plant development. *The Plant Cell Online, 27*(1), 44–63.
Schaller, G. E., Street, I. H., & Kieber, J. J. (2014). Cytokinin and the cell cycle. *Current Opinion in Plant Biology, 21*, 7–15.
Schilmiller, A. L., & Howe, G. A. (2005). Systemic signaling in the wound response. *Current Opinion in Plant Biology, 8*(4), 369–377.
Schubert, D., Primavesi, L., Bishopp, A., Roberts, G., Doonan, J., Jenuwein, T., et al. (2006). Silencing by plant Polycomb-group genes requires dispersed trimethylation of histone H3 at lysine 27. *The EMBO Journal, 25*(19), 4638–4649.
Shafiq, S., Berr, A., & Shen, W. H. (2014). Combinatorial functions of diverse histone methylations in *Arabidopsis thaliana* flowering time regulation. *New Phytologist, 201*(1), 312–322.
Shen, H., He, H., Li, J., Chen, W., Wang, X., Guo, L., et al. (2012). Genome-wide analysis of DNA methylation and gene expression changes in two arabidopsis ecotypes and their reciprocal hybrids. *The Plant Cell Online, 24*(3), 875–892.
Shen, Y., Issakidis-Bourguet, E., & Zhou, D. X. (2016). Perspectives on the interactions between metabolism, redox, and epigenetics in plants. *Journal of Experimental Botany, 67*(18), 5291–5300.
Shivaprasad, P. V., Dunn, R. M., ACM, S. B., Bassett, A., & Baulcombe, D. C. (2011). Extraordinary transgressive phenotypes of hybrid tomato are influenced by epigenetics and small silencing RNAs. *The EMBO Journal, 12*, 1–10.
Sijen, T., Vijn, I., Rebocho, A., van Blokland, R., Roelofs, D., Mol, J. N., et al. (2001). Transcriptional and posttranscriptional gene silencing are mechanistically related. *Current Biology: CB, 11*, 436–440.
Sorce, C., Massai, R., Picciarelli, P., & Lorenzi, R. (2002). Hormonal relationships in xylem sap of grafted and ungrafted Prunus rootstocks. *Scientia Horticulturae, 93*(3–4), 333–342.
Stegemann, S., & Bock, R. (2009). Exchange of genetic material between cells in plant tissue grafts. *Science, 324*(5927), 649–651.
Stroud, H., Ding, B., Simon, S. A., Feng, S., Bellizzi, M., Pellegrini, M., et al. (2013). Plants regenerated from tissue culture contain stable epigenome changes in rice. *ELife, 2013*(2), 1–14.
Stroud, H., Greenberg, M. V. C., Feng, S., Bernatavichute, Y. V., & Jacobsen, S. E. (2013). Comprehensive analysis of silencing mutants reveals complex regulation of the Arabidopsis methylome. *Cell, 152*(1–2), 352–364.
Sugimoto, K., Jiao, Y., & Meyerowitz, E. M. (2010). Arabidopsis regeneration from multiple tissues occurs via a root development pathway. *Developmental Cell, 18*(3), 463–471.
Taller, J., Hirata, Y., Yagishita, N., Kita, M., & Ogata, S. (1998). Graft-induced genetic changes and the inheritance of several characteristics in pepper. *TAG. Theoretical and Applied Genetics. Theoretische und angewandte Genetik*, (97), 705–713.
Tamiru, M., Hardcastle, T. J., & Lewsey, M. G. (2018). Regulation of genome-wide DNA methylation by mobile small RNAs. *New Phytologist, 217*(2), 540–546.
Thieme, C. J., Rojas-Triana, M., Stecyk, E., Schudoma, C., Zhang, W., Yang, L., et al. (2015). Endogenous Arabidopsis messenger RNAs transported to distant tissues. *Nature Plants, 1*, 1–8.

Thorp, T. G., Boyd, L. M., Barnett, A. M., Lowe, R. G., Hofstee, B. J., Blattmann, P. J., et al. (2007). Effect of inter-specific rootstocks on inorganic nutrient concentrations and fruit quality of "Hortl6A" kiwifruit (*Actinidia chinensis* Planch, var. chinensis). *The Journal of Horticultural Science and Biotechnology, 82*(6), 829–838.

Toscano-Morales, R., Xoconostle-Cázares, B., Martinez-Navarro, A. C., & Ruiz-Medrano, R. (2014). Long distance movement of an arabidopsis translationally controlled tumor protein (AtTCTP2) mRNA and protein in tobacco. *Frontiers in Plant Science, 5*, 1–10.

Toscano-Morales, R., Xoconostle-Cázares, B., Martinez-Navarro, A. C., & Ruiz-Medrano, R. (2016). AtTCTP2 mRNA and protein movement correlates with formation of adventitious roots in tobacco. *Plant Signaling & Behavior, 11*(3), 1–5.

Uthup, T. K., Karumamkandathil, R., Ravindran, M., & Saha, T. (2018). Heterografting induced DNA methylation polymorphisms in *Hevea brasiliensis*. *Planta, 248*(3).

Vining, K., Pomraning, K. R., Wilhelm, L. J., Ma, C., Pellegrini, M., Di, Y., et al. (2013). Methylome reorganization during in vitro dedifferentiation and regeneration of *Populus trichocarpa*. *BMC Plant Biology, 13*(1), 1.

Wang, Q., Ci, D., Li, T., Li, P., Song, Y., Chen, J., et al. (2016). The role of DNA methylation in xylogenesis in different tissues of poplar. *Frontiers in Plant Science, 7*, 1–12.

Warschefsky, E. J., Klein, L. L., Frank, M. H., Chitwood, D. H., Londo, J. P., von Wettberg, E. J. B., et al. (2016). Rootstocks: Diversity, domestication, and impacts on shoot phenotypes. *Trends in Plant Science, 21*(5), 418–437.

Wu, R., Wang, X., Lin, Y., Ma, Y., Liu, G., Yu, X., et al. (2013). Inter-species grafting caused extensive and heritable Alterations of DNA methylation in Solanaceae plants. *PLoS One, 8*(4).

Wu, K., Zhang, L., Zhou, C., Yu, C. W., & Chaikam, V. (2008). HDA6 is required for jasmonate response, senescence and flowering in Arabidopsis. *Journal of Experimental Botany, 59*(2), 225–234.

Xie, Z., Johansen, L. K., Gustafson, A. M., Kasschau, K. D., Lellis, A. D., Zilberman, D., et al. (2004). Genetic and functional diversification of small RNA pathways in plants. *PLoS Biology, 2*(5), 642–652.

Xu, K., Liu, J., Fan, M., Xin, W., Hu, Y., & Xu, C. (2012). A genome-wide transcriptome profiling reveals the early molecular events during callus initiation in Arabidopsis multiple organs. *Genomics, 100*(2), 116–124.

Yagishita, N. (1961). Studies on graft hybrids of *Capsicun annuum* L. II. Variation in fruit shape caused by grafting for three generations and the effects in the progeny. *The Botanical Magazine, Tokyo, 74*, 480–489.

Yamamuro, C., Zhu, J. K., & Yang, Z. (2016). Epigenetic modifications and plant hormone action. *Molecular Plant, 9*(1), 57–70.

Yang, C., Bratzel, F., Hohmann, N., Koch, M., Turck, F., & Calonje, M. (2013). VAL-and AtBMI1-Mediated H2Aub initiate the switch from embryonic to postgerminative growth in arabidopsis. *Current Biology, 23*(14), 1324–1329.

Yang, Y., Mao, L., Jittayasothorn, Y., Kang, Y., Jiao, C., Fei, Z., et al. (2015). Messenger RNA exchange between scions and rootstocks in grafted grapevines. *BMC Plant Biology, 15*(1), 251.

Yin, H., Yan, B., Sun, J., Jia, P., Zhang, Z., Yan, X., et al. (2012). Graft-union development: A delicate process that involves cell-cell communication between scion and stock for local auxin accumulation. *Journal of Experimental Botany, 63*(11), 4219–4232.

Yoo, B.-C., Kragler, F., Varkonyi-Gasic, E., Haywood, V., Archer-Evans, S., Lee, Y. M., et al. (2004). A systemic small RNA signaling system in plants. *The Plant Cell Online, 16*(1979–2000).

Zakrzewski, F., Schmidt, M., Van Lijsebettens, M., & Schmidt, T. (2017). DNA methylation of retrotransposons, DNA transposons and genes in sugar beet (*Beta vulgaris* L.). *The Plant Journal, 90*(6), 1156–1175.

Zemach, A., Kim, M. Y., Hsieh, P. H., Coleman-Derr, D., Eshed-Williams, L., Thao, K., et al. (2013). The arabidopsis nucleosome remodeler DDM1 allows DNA methyltransferases to access H1-containing heterochromatin. *Cell, 153*(1), 193–205.

Zhang, L., Cheng, Z., Qin, R., Qiu, Y., Wang, J.-L., Cui, X., et al. (2012). Identification and characterization of an epi-allele of FIE1 reveals a regulatory linkage between two epigenetic marks in rice. *The Plant Cell Online, 24*(11), 4407–4421.

Zhang, W. N., Duan, X. W., Ma, C., Harada, T., & Li, T. Z. (2013). Transport of mRNA molecules coding NAC domain protein in grafted pear and transgenic tobacco. *Biologia Plantarum, 57*(2), 224–230.

Zhang, J., Gao, G., Chen, J. J., Taylor, G., Cui, K. M., & He, X. Q. (2011). Molecular features of secondary vascular tissue regeneration after bark girdling in Populus. *New Phytologist, 192*(4), 869–884.

Zhang, F., Qi, B., Wang, L., Zhao, B., Rode, S., Riggan, N. D., et al. (2016). EIN2-dependent regulation of acetylation of histone H3K14 and non-canonical histone H3K23 in ethylene signalling. *Nature Communications, 7*, 1–14.

Zhang, F., Wang, L., Qi, B., Zhao, B., Ko, E. E., Riggan, N. D., et al. (2017). EIN2 mediates direct regulation of histone acetylation in the ethylene response. *Proceedings of the National Academy of Sciences, 114*(38), 201707937.

Zhang, H., & Zhu, J. K. (2011). RNA-directed DNA methylation. *Current Opinion in Plant Biology, 14*(2), 142–147.

Zhao, J. H., Fang, Y. Y., Duan, C. G., Fang, R. X., Ding, S. W., & Guo, H. S. (2016). Genome-wide identification of endogenous RNA-directed DNA methylation loci associated with abundant 21-nucleotide siRNAs in *Arabidopsis*. *Scientific Reports, 6*, 1–13.

Zheng, B. S., Chu, H. L., Jin, S. H., Huang, Y. J., Wang, Z. J., Chen, M., et al. (2009). CDNA-AFLP analysis of gene expression in hickory (*Carya cathayensis*) during graft process. *Tree Physiology, 30*(2), 297–303.

Zhou, C., Zhang, L., Duan, J., Miki, B., & Wu, K. (2005). HISTONE DEACETYLASE19 is involved in jasmonic acid and ethylene signalling of pathogen response in Arabidopsis. *The Plant Cell Online, 17*, 1196–1204.

Zhu, A., Greaves, I. K., Dennis, E. S., & Peacock, W. J. (2017). Genome-wide analyses of four major histone modifications in Arabidopsis hybrids at the germinating seed stage. *BMC Genomics, 18*(1), 1–9.

CHAPTER EIGHT

Hybrid Vigor: Importance of Epigenetic Processes and Consequences for Breeding

Naomi Miyaji and Ryo Fujimoto[1]

Graduate School of Agricultural Science, Kobe University, Rokkodai, Nada-ku, Kobe, Japan

[1]Corresponding author: E-mail: leo@people.kobe-u.ac.jp

Contents

Abstract

Heterosis or hybrid vigor is a phenomenon in which a hybrid progeny exhibits superior performance compared to their parental inbred lines. F_1 hybrid seed production system is used for many crops and vegetables because of the high yield due to hybrid vigor. It has been over 100 years since the discovery of the hybrid vigor phenomenon; however, the molecular mechanism of hybrid vigor is still a mystery. Understanding the mechanism of hybrid vigor will allow developing more efficient breeding systems and contribute to food security. At this time, hybrid vigor research has mainly used genetic approaches providing a number of findings in many plant species. High throughput sequencing will make genetic analysis more efficient, and large-scale quantitative trait analysis or genome wide association analysis has been performed in some plant

Advances in Botanical Research, Volume 88
ISSN 0065-2296
https://doi.org/10.1016/bs.abr.2018.10.001

species. However recent works have revealed that epigenetic regulations also contribute to hybrid vigor. In this chapter, we will discuss the most recent findings suggesting that epigenetic regulation may play important roles in hybrid vigor.

1. INTRODUCTION

Hybrid vigor or heterosis is a phenomenon where F_1 hybrids have superior performance relative to their parental inbred lines. Already in 1876, Charles Darwin recorded that the progeny of cross-pollinated maize were 25% taller than the progeny of self-pollinated maize and this was the first description of the hybrid vigor in plants (Duvick, 2001). In 1908, George Harrison Shull and Edward Murray East rediscovered the phenomenon of hybrid vigor in maize independently (Duvick, 2001). Shull introduced the term "heterosis" in 1914 to replace the more cumbersome word "heterozygosis", which did not express the superior performance of the hybrids (Shull, 1948).

As the superior performance in F_1 hybrids appears in important agronomical traits such as biomass, yield, and abiotic and biotic stress tolerance, this phenomenon was applied to commercial production of F_1 hybrid seeds not only in maize but also in many other crops and vegetables. Nowadays, breeders develop two elite inbred parental lines, and the F_1 hybrid seeds are harvested for commercial use (Fig. 1). The F_1 hybrids show hybrid vigor, giving extra yield, and uniformity in growth and flowering, which allows machine harvesting, and reproducibility (a guarantee of quality each year) because of the production by crosses of uniform inbred lines. Self-incompatibility, cytoplasmic male sterility, or photoperiod/thermo-sensitive genic male sterility makes it possible to develop a systematic/effective F_1 hybrid seed production system (Fig. 1). Hybrid corn was the main crop in the 1950s in the United States, and yield advanced from 1 ton per hectare in 1930 to 4 tons per hectare in 1960 and approximately 12 tons per hectare in 2017 (Crow, 1998; Duvick, 2001). Hybrid rice was accepted in China, especially in southern China, because of significantly increased yield and suitable climatic conditions. As a consequence about half of the total area of rice in China was planted with F_1 hybrids by the early 2000s (Cheng, Zhuang, Fan, Du, & Cao, 2007; Feng, Li, Qin, Liao, & Siddique, 2017). Hybrids are also used in crops such as sorghum, sunflower, canola, and in a number of vegetables. Attempts have been made to produce hybrid seeds in wheat by developing male sterile lines (Kempe, Rubtsova, & Gils, 2014).

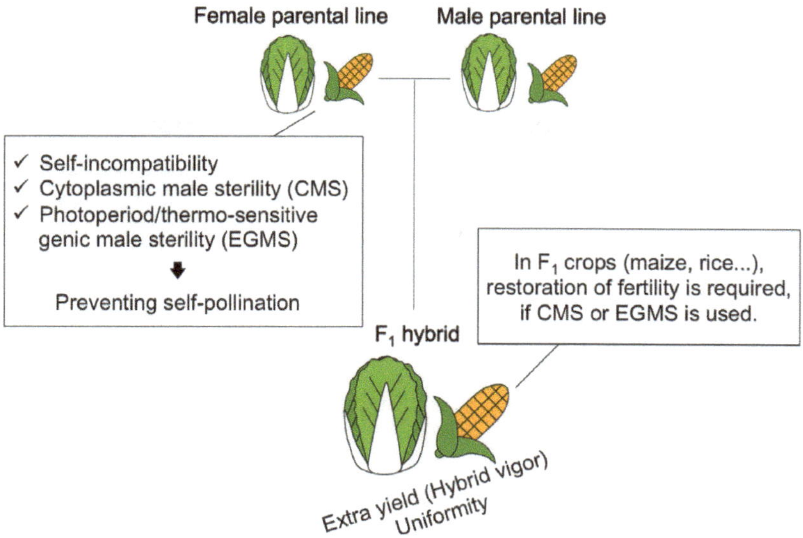

Figure 1 The strategy for production of F_1 hybrid seeds. Breeders develop two elite parental lines for making F_1 hybrid seeds. For producing F_1 hybrid seeds effectively, self-incompatibility (preventing self-pollination) is used, or else a cytoplasmic male sterile (CMS) or environment genic male sterile (EGMS) line is used as a maternal line. Photoperiod or thermo-sensitive genic male sterility is known as EGMS. To produce seeds of F_1 hybrids such as crops, fertility needs to be restored.

Attempts to understand the genetic basis of hybrid vigor began in the 1990s and several hypotheses including dominance, overdominance, and epistasis were put forward to explain the mechanism of hybrid vigor. The dominance hypothesis is that hybrid vigor results in the suppression/complementation of deleterious recessive alleles from one parent by dominant alleles from the other parent (Fig. 2) (Bruce, 1910; Crow, 1998; Davenport, 1908; Jones, 1917). The overdominance hypothesis attributes heterozygosity at individual loci as leading to hybrid vigor (Fig. 2) (Crow, 1998; East, 1936; Shull, 1908), and the epistasis hypothesis suggests interactions between two or more non-allelic genes derived from the parental lines generate hybrid vigor (Fig. 2) (Powers, 1944; Richey, 1942; Williams, 1959). Genetic approaches using quantitative trait loci (QTL) analysis or genome-wide association studies (GWASs) have been performed in maize, rice, sorghum, tomato, rapeseed, and cotton in attempts to understand the genetic basis of hybrid vigor (Giraud et al., 2017; Huang et al., 2016; Li et al., 2001, 2017; Lippman & Zamir, 2007; Lu, Romero-Severson, & Bernardo, 2003; Luo et al., 2001; Stuber, Lincoln, Wolff, Helentjaris, &

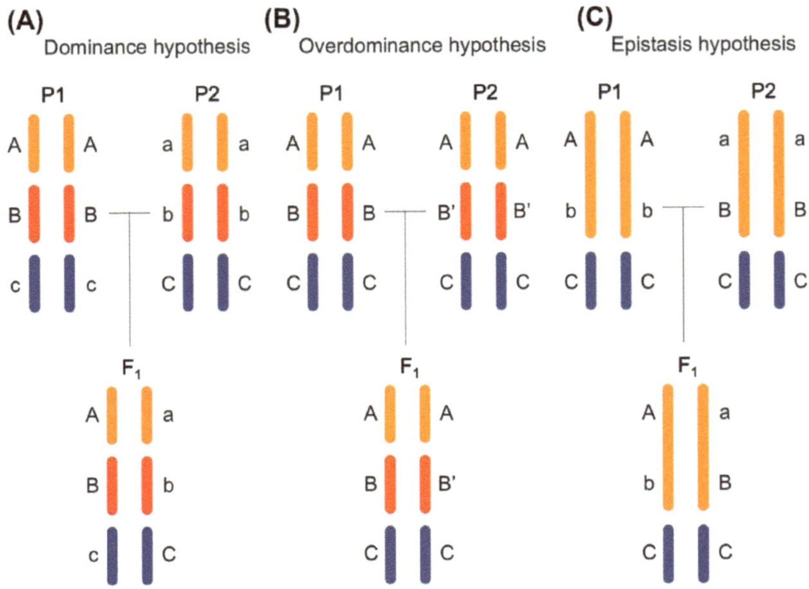

Figure 2 Three hypotheses for explaining the genetic basis of hybrid vigor. Parent 1 (P1) and P2 have homozygous allelic genes. (A) In the dominance hypothesis, hybrid vigor results in the suppression/complementation of recessive alleles (a, b, or c) of one parent by the dominant alleles (A, B, or C) from the other parent. (B) In the overdominance hypothesis, heterozygosity at individual loci (BB′) leads to hybrid vigor. (C) In the epistasis hypothesis, the interactions between two or more non-allelic genes (A and B) are important for hybrid vigor.

Lander, 1992; Yang et al., 2017; Zhen et al., 2017). However, the hypotheses supported by each study were different, and there is so far no consensus view.

Hybrid vigor contributes to increased yield in many crops and vegetables, thus understanding the underlying biological mechanisms of hybrid vigor could lead to more precise, less labor intensive, and more advanced methods than current breeding methods. Recent study has revealed the possibility of an epigenetic contribution to hybrid vigor. This chapter describes recent research on epigenetic processes in hybrid vigor.

2. HYBRID VIGOR PHENOTYPES

The intensity of hybrid vigor is described by means of two indices, the mid-parent heterosis (MPHs) and the best-parent heterosis (BPHs). MPH is the performance of a hybrid relative to the mean value of its parental lines, whereas BPH is the performance of hybrids relative to the parent having the best value for the trait (Springer & Stupar, 2007). BPH is important to select useful lines for F_1 hybrid breeding in crops and vegetables.

2.1 Hybrid Vigor in Reproductive Traits

In maize, hybrids obtained by crossing B73 and Mo17 showed greater yield (mg/ha), seed weight (g/ear), and ear length than the parental lines. Plant height and 11-days seedling height also showed hybrid vigor, but at a lower level than for yield, demonstrating a variation in the intensity of hybrid vigor depending on the trait analyzed (Springer & Stupar, 2007). Variations of BPH between traits were also observed using multi-parental hybrids (i.e., by crossing 293 diverse maize inbred lines as male plants to B73), and weak correlations of BPH were found for different traits such as between plant height and days to flowering or leaf width (Flint-Garcia, Buckler, Tiffin, Ersoz, & Springer, 2009; Stupar et al., 2008). Whether there is any positive correlation between genetic distance and hybrid vigor has been debated (Birchler, Yao, Chudalayandi, Vaiman, & Veitia, 2010). Initially, in maize, it was accepted that crosses using more genetically divergent parental lines lead to greater hybrid vigor, suggesting a positive correlation between genetic distance and hybrid vigor. Yet, a negative correlation was observed between genetic distance and hybrid vigor when the genetic distance was extremely high, indicating that the genetic distance between parental lines should exceed certain limits (Moll, Lonnquist, Vélez Fortuno, & Johnson, 1965). In addition, even when the genetic distance is not too high, positive correlation is not always observed between genetic distance and hybrid vigor not only in maize but also in other plant species (Barth, Busimi, Friedrich Utz, & Melchinger, 2003; Flint-Garcia et al., 2009; Girke, Schierholt, & Becker, 2012; Kawamura et al., 2016; Yang et al., 2017).

In rice, hybrids display a 10%–20% increase in yield as compared with their parental inbred lines. Hybrid vigor is observed in roots, seedlings, or yield (including yield-related traits such as spikelet number per panicle and effective panicle number) (Li et al., 2016; Zhai et al., 2013). Grain number also showed hybrid vigor in many hybrids, and heading date tended to be earlier in high-yielding hybrids (Huang et al., 2015; Zhen et al., 2017). Large-scale genomic and phenotypic data analysis in hybrid rice showed that overall heterozygosity did not contribute to hybrid vigor (Huang et al., 2015). In contrast the accumulation of heterozygosity at some specific loci is important for hybrid vigor, and a correlation was observed between the number of superior alleles and yield increase. The hybrid vigor phenotypes share some common yield-related traits such as flowering time or plant architecture, and hybrid rice cultivars tended to be heterozygous at functional and non-functional genes involved in flowering time such as *Heading date 3a* (*Hd3a*) or *Grain number, plant height, and heading date 8* (*Ghd8*) or involved in the blanching of panicles such as *SQUAMOSA PROMOTER BINDING PROTEIN-LIKE 14* (*OsSPL14*) (Huang et al., 2016; Li et al., 2016). By QTL analysis using 10,074 F_2 lines derived from 17 hybrids, positive dominance effects tended to be overrepresented in the grain-yield-related traits, with the contribution of male alleles being larger than female alleles in most of the grain-yield QTLs (Huang et al., 2016).

Unlike maize and rice, hybrid wheat is rarely cultivated. However, a high-yielding potential in hybrid wheat has been shown and a cost-efficient hybrid seed production system is under development (Jiang, Schmidt, Zhao, & Reif, 2017; Kempe et al., 2014; Zhao et al., 2015). In soybean, commercial hybrid varieties have been released since 2002, and hybrid vigor is observed in yield-related traits (Zhang et al., 2017). In canola, increased silique number and grain yield relative to parental lines were observed in hybrids. In this plant, increased silique number is independent from flowering time (Shen et al., 2017). In tomato, hybrid vigor was only detected in yield-related traits such as seed number per plant, fruit number, total yield, and biomass (Semel et al., 2006). A single heterozygous gene, *SINGLE FLOWER TRUSS* (*SFTs*), which is orthologous to the floral activator *FLOWERING LOCUS T* (*FT*) in *Arabidopsis thaliana*, contributes to the increase in fruit yield (increase in canonical inflorescence number or total flowers per plant), and heterozygosity of this gene results in a shift in the balance of vegetative to reproductive growth, which is due to dosage-dependent suppression of growth termination imposed by *SELF PRUNING* (*SP*), an ortholog of floral repressor *TERMINAL FLOWER 1* (*TFL1*) in *A. thaliana*. This is the first example of a single overdominant gene in hybrid vigor in plants (Krieger, Lippman, & Zamir, 2010).

2.2 Hybrid Vigor in Vegetative Growth

Hybrid vigor was also observed in the model plant *A. thaliana* as demonstrated using 169 F_1 hybrids derived from crosses between 63 accessions and three lines (Columbia-0 (Col), C24, and Nd), that displayed hybrid vigor in shoot biomass; 29 and 23 out of 52 crosses showed significantly high MPH and BPH, respectively (Meyer, Törjék, Becher, & Altmann, 2004). In another study using five F_1 hybrids of *A. thaliana* (C24 × Aa, C24 × Ws, Ws × Col, Col × C24 and Col × Wei), biomass yield and rosette diameter showed high MPH and BPH, while flowering time and seed yield showed low MPH and BPH (Barth et al., 2003). Some *A. thaliana* F_1 hybrids analyzed in these two studies and in others also showed hybrid vigor for shoot biomass (Barth et al., 2003; Fujimoto, Taylor, Shirasawa, Peacock, & Dennis, 2012a; Groszmann et al., 2014; Kawanabe et al., 2016; Meyer et al., 2004, 2012). Noteworthy, *A. thaliana* F_1 hybrids obtained after various crosses including C24 x Col and C24 x Landsberg *erecta* (L*er*), developed cotyledon with an increased size from only a few days after sowing (DAS) (Fujimoto et al., 2012a; Groszmann et al., 2014; Meyer et al., 2012), and a different pattern of vegetative growth was observed between the hybrids and from the parental lines (Groszmann et al., 2014; Kawanabe et al., 2016). Greater shoot biomass in F_1 hybrids was due to an increased leaf area due to increased cell size and number, with relative levels and ratios between F_1 hybrids and parental lines that varied between the hybrids (Groszmann et al., 2014).

As the F_1 hybrids between C24 and Col, which are often used for studies of hybrid vigor, showed extreme late-flowering phenotype (Fig. 3), this may explain the increase shoot biomass because of longer period of vegetative growth (Chen, 2010). Flowering time is controlled by the combination of the alleles at the *FRIGIDA* (*FRI*) and *FLOWERING LOCUS C* (*FLC*) loci (Koornneef, Blankestijn-de Vries, Hanhart, Soppe, & Peeters, 1994); *FLC* encodes a repressor of flowering, with its activity regulated by *FRI* (Sheldon et al., 1999). In the hybrids, C24 contributes to an active *FRI* allele and Col to a strong *FLC* allele (Michaels & Amasino, 1999), resulting in high *FLC* activity and consequent late flowering (Fig. 3). To avoid any growth effects due to late flowering, flowering time was modified using nonfunctional alleles of *FLC* in the hybrid. These new hybrids presented flowering times that were similar to one of the parents ($\sim$24 days after sowing (DAS)) and rosette diameters still showed hybrid vigor (up to 21 DAS) (Fig. 3), indicating that flowering time does not influence hybrid vigor at the development stages that precede the transition to flowering (Fujimoto et al., 2012a).

A commercial F_1 hybrid cultivar of Chinese cabbage (*Brassica rapa* var. *pekinensis*), 'W39', showed vegetative hybrid vigor at early developmental stages and a 25% yield increase as compared to the best parent (Fig. 4) (Saeki et al., 2016). Increased cotyledon size in 'W39' was obvious from 2 to 4 DAS, and the first two leaves of the 'W39' at 14 DAS were larger and wider than those of the largest parent (Saeki et al., 2016). Early plant size and yield were examined in 32 F_1 hybrids of Chinese cabbage. Some F_1 hybrids

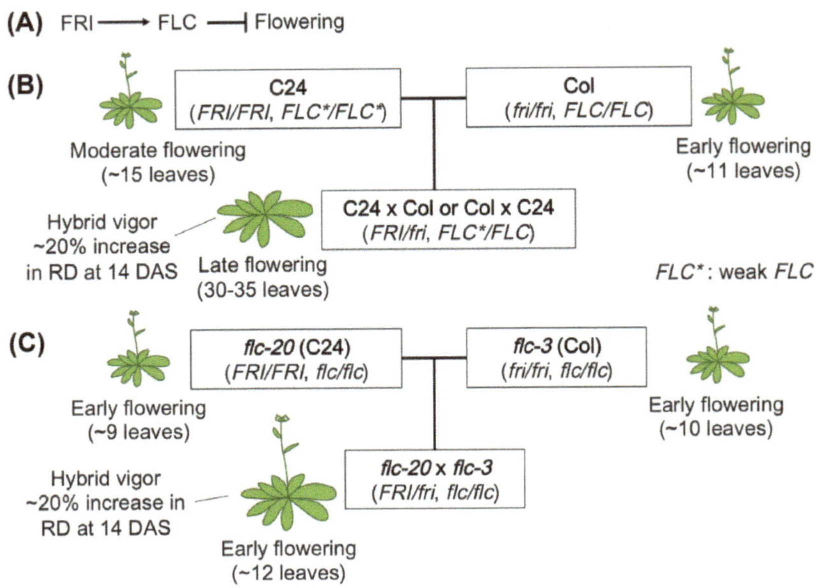

Figure 3 Hybrid vigor is independent of flowering time. (A) Flowering pathway. FRIGIDA (FRI) promotes *FLOWING LOCUS C* (*FLC*) expression, and FLC suppresses flowering. (B) *FRI* and *FLC* genotypes of wild type C24, Columbia-0 (Col), and their F_1 hybrids (C24 x Col or Col x C24 hybrid). *FLC** represents a weak *FLC* allele. F_1 hybrids have functional *FRI* and *FLC*, so they show later flowering than their parental lines. F_1 hybrids show a 20% increase in rosette diameter (RD) at 14 days after sowing (DAS). (C) The F_1 hybrids between *flc-20* in C24 and *flc-3* in Col showed early flowering and F_1 hybrids showed 20% increase in RD at 14 DAS, similar to the wild type C24 x Col hybrid.

showed hybrid vigor for cotyledon size, leaf size, or yield. There was a moderate correlation in MPH between leaf size at 21DAS and yield but not in BPH (Kawamura et al., 2016).

Increased leaf area provides a greater capacity for photosynthesis, and the importance of the link between chlorophyll biosynthesis/photosynthesis and heterotic biomass has been reported. Treatment of seedlings at 3 DAS with norflurazon, an inhibitor of phytoene desaturase and hence of chlorophyll synthesis and photosynthesis, resulted in an absence of the heterotic response of hybrids at 21 DAS in *A. thaliana* (Fujimoto et al., 2012a). When Chinese cabbage seeds were grown on Murashige and Skoog (MS) medium for one week, and subsequently transferred to MS medium with 1 μM norflurazon and grown for a further two weeks, the absence of chlorophyll did not block hybrid vigor that was clearly observed in the first and

Figure 4 The phenotype of hybrid vigor in Chinese cabbage at early and late developmental stages. Cotyledon at 4 days after sowing (DAS) (A), seedlings at 14 DAS (B), and harvested heads at ~ 2.5 months (C) are shown. S27 and R29 are female and male parental lines, respectively, and 'W39' is a commercial F_1 hybrid between S27 and R29.

second leaves of the 'W39' hybrid. In contrast, seeds grown on MS medium with 1 μM norflurazon for one week and transferred to MS medium without norflurazon for three weeks did not show any hybrid vigor in the third and fourth leaves of the 'W39' hybrids, which were green (Saeki et al., 2016). These results suggest that chlorophyll biosynthesis or photosynthesis is important during the first week of cotyledon growth and even after for increased leaf size in 'W39' hybrids. This increased leaf area in F_1 hybrids could lead to an increased photosynthetic capacity per individual plant, and consequently to a higher yield (Saeki et al., 2016).

2.3 Hybrid Vigor in Stress Tolerance

Most research in hybrid vigor has been focused on growth vigor or increased yield because these traits are directly linked to the breeding of F_1 hybrid cultivar. There is little research on increased stress tolerance in hybrids. As this question becomes more important due to current environmental changes, stress tolerance in F_1 hybrid cultivars will be a more important issue to consider in the future. Of course, there is already some evidence of hybrid vigor in stress tolerance at this time. In *A. thaliana*, two F_1 hybrids (Col x L*er*, Col x Sei-0) showed hybrid vigor in *Pseudomonas syringae* pv. *tomato* DC3000 resistance, while the other 18 F_1 hybrids analyzed in this study did not show any (Yang L. et al., 2015). Hybrid vigor in freezing tolerance in reciprocal F_1 hybrids between C24 and Col accessions was observed in both acclimated and non-acclimated plants. The content of soluble sugars such as fructose, glucose, raffinose, and sucrose in leaves was more increased in the F_1 hybrids than in the parental lines in the acclimated state (Rohde, Hincha, & Heyer, 2004). Further combinations of *A. thaliana* accessions were tested, and hybrid vigor in freezing tolerance was observed in some F_1 hybrids (Korn, Peterek, Mock, Heyer, & Hincha, 2008). Introduction of the hybrid vigor in stress tolerance will be a new breeding approach for producing better F_1 hybrid cultivars.

3. TRANSCRIPTOMIC ANALYSIS IN HYBRID VIGOR

The underlying hypothesis for a transcriptomic approach is that genes whose expression changes in F_1 hybrids may be involved in hybrid vigor. Many transcriptomic analyses initially using microarray technology and more recently, RNA-sequencing (RNA-seq) have been performed to compare parental lines with their F_1 hybrids in many plant species (Lippman & Zamir, 2007). Annotated transcriptomes were compared between F_1 hybrids and each parent or between F_1 hybrids and average level of parental gene expression (termed mid-parent value; MPV). Often, to differentiate between unexpected and expected changes in gene expression in the F_1 hybrids, gene expression levels are classified in additive or non-additive. Non-additive gene expression level corresponds to situation when gene expression level of F_1 hybrids is either higher or lower than MPV, whereas additive gene expression level is where gene expression levels in F_1 hybrids is equal to the MPV. Gene expression is further classified into high-parent

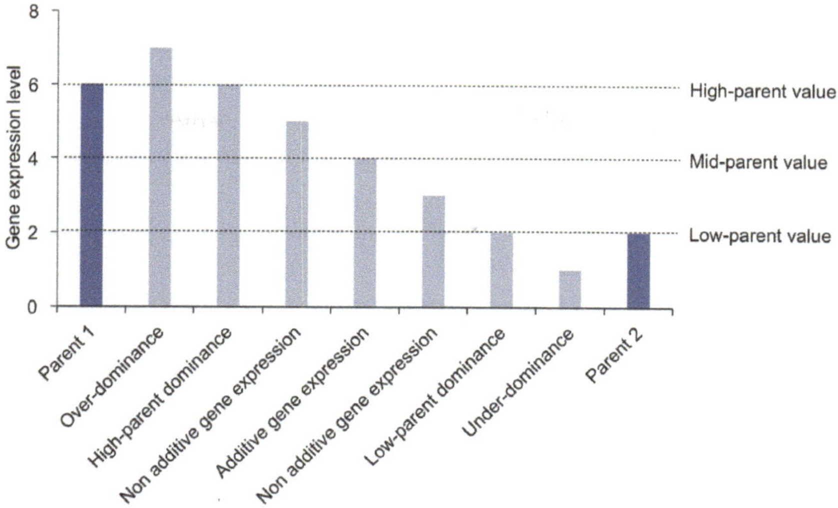

Figure 5 Gene expression patterns in an F_1 hybrid. In non-additive gene expression, the gene expression level of F_1 hybrid is either higher or lower than the mid-parent value (MPV). The gene expression level in an F_1 hybrid is equal to the MPV in additive gene expression. In high- or low-parent dominance, the gene expression level in an F_1 hybrid is equal to one parent but is significantly different from the other parent. In over- or under-dominance, the gene expression level in an F_1 hybrid is outside the expression level of both parental lines.

dominance, low-parent dominance, over-dominance, and under-dominance depending on the relative expression of hybrids and parental lines: high- or low-parent dominance correspond to situation where expression levels in hybrids are equal to one parent but are significantly different from the second parent, whereas over- and under-dominance is when gene expression levels in hybrids are outside the parental range (Fig. 5).

Studies in *A. thaliana* using crosses between C24 x Col or between C24 x L*er* showed that genes involved in chlorophyll biosynthesis and photosynthesis were highly represented among the genes with increased transcription in F_1 hybrids at 3 DAS (Fujimoto et al., 2012a; Zhu et al., 2016). At 10 DAS in the aerial tissue, genes categorized into 'Response to biotic stimulus' tended to be up-regulated and genes categorized into 'Response to stress' tended to be either up- or down-regulated in C24 x Col hybrids. Only 12 genes overlapped between the 3 and 10 DAS sets of non-additively expressed genes in C24 x Col hybrids. At 3 DAS, up-regulated genes showed an over-dominance mode, while non-additively expressed genes showed high-parent or low-parent dominance at 10 DAS in C24 x Col hybrids (Fujimoto et al., 2012a). Transcriptomes of Col, C24, and the reciprocal hybrids at 4, 6, and 10 DAS were also compared (Meyer et al., 2012). Seventy-five percent of differentially expressed genes (DEGs) showed differences in the transcript levels at only one time point, and 10% of genes were significantly different at all time points (Meyer et al., 2012). In C24 x L*er* hybrids, transcriptome analyses in early seedlings at 0, 3, 5, and 7 DAS identified 454 to 1316 non-additively expressed genes, but most non-additively expressed genes did not overlap between time points (Zhu et al., 2016). These results indicate that the non-additive gene expression profile is drastically changed during plant developmental even when stages that differ only by a few days are analyzed (Fujimoto et al., 2012a; Meyer et al., 2012; Zhu et al., 2016).

In other plant species such as maize, rice, and canola, transcriptome analyses have been performed in different tissues or developmental stages (Hu et al., 2016; Jahnke et al., 2010; Li et al., 2016; Paschold et al., 2012; Shen et al., 2017; Stupar et al., 2008; Wei et al., 2009; Zhai et al., 2013). In maize seedlings, transcriptome analysis using microarrays on the parental inbred lines, B73 and Mo17, and their heterotic F_1 hybrids showed all possible modes of gene expression in F_1 hybrids, with most genes having additive gene expression profiles (Swanson-Wagner et al., 2006). This is in agreement with results from other plant species, showing that the majority of genes showed additive gene expression levels, with only a small proportion of genes showing non-additive levels (Springer & Stupar, 2007), and the number of DEGs between parental lines was higher than the number of those identified between parental lines and F_1 hybrids. Thus, some authors have suggested that the mid-parent expression pattern in F_1 hybrids might play a role in hybrid vigor (Springer & Stupar, 2007). There is no clear conclusion as to which is more important, additive or non-additive expression for generating hybrid vigor, and both classes of genes probably contribute. In addition, DEGs between F_1 hybrids and their parental lines or non-additively expressed genes showed tissue/stage specificity, which makes it difficult to identify any key genes or novel expression pattern associated with hybrid vigor phenotype by transcriptome analysis (Hochholdinger & Hoecker, 2007; Springer & Stupar, 2007). However, transcriptome analysis can inform about global trends of gene expression and RNA-seq enables us to distinguish transcripts between the parental alleles in F_1 hybrids at the whole genome level (Saeki et al., 2016).

4. EPIGENETIC PROCESSES IN HYBRID VIGOR

Epigenetics can be associated with changes in the expression of genes that do not involve changes in DNA sequence and epigenetic control is known to play an essential role in normal plant development (Fujimoto et al., 2012b; Quadrana & Colot, 2016). In addition, an increasing number of reports are revealing the association between epigenetic regulation and agricultural traits. Vernalization in Brassicaceae, fruit ripening in tomato, sex determination in melon, salt-tolerance in wheat, and drought tolerance in rice are some of the traits known to be epigenetically regulated or influenced (Groszmann et al., 2011b; Karan, DeLeon, Biradar, & Subudhi, 2012; Liu et al., 2015; Martin et al., 2009; Shea et al., 2018; Wang et al., 2014). Populations of epigenetic recombinant inbred lines (epi-RILs) constructed with parents, which differed only in epigenetic marks, such as hybrids between *methyltransferase 1* (*met1*; involved in maintenance of CG methylation) and wild type (WT) Col, or between *decrease in DNA methylation 1* (*ddm1*; chromatin remodeling factor and is involved in the maintenance of DNA methylation) and WT Col showed variation of plant phenotypes in *A. thaliana* (Johannes et al., 2009; Reinders et al., 2009; Teixeira et al., 2009). Stable inheritance of complex traits such as flowering time, primary root length, plant height, and biomass has been observed in these epiRIL populations, providing evidence that epigenetic variation can contribute to such traits including those important in agriculture (Cortijo et al., 2014; Johannes et al., 2009; Reinders et al., 2009). Thus, epigenetics in agriculture is becoming increasingly important, and recent progress has pointed to the possibility that epigenetic regulation might contribute to hybrid vigor.

4.1 Whole Genome Comparison of Epigenetic States Between F_1 Hybrid and Its Parental Lines

Like transcriptome analysis, small RNA expression and populations, DNA methylation, or histone modifications have been compared between heterotic F_1 hybrids and their parental lines at the whole genome levels by small RNA-sequencing, whole genome bisulfite sequencing (WGBS), or chromatin immunoprecipitation sequencing (ChIP-seq), respectively.

Small RNAs are involved in a broad range of processes; micro RNAs (miRNAs), which are predominantly 21 nucleotides (nt) in length, are involved in post transcriptional gene silencing (PTGS, Cui, You, & Chen, 2017; Guo, Liu, Smith, Liang, & Wang, 2016), whereas small interfering RNA (siRNA) of 24-nt are involved in transcriptional gene silencing

(TGS) including heterochromatin maintenance and silencing of transposable elements (TEs) (Wendte & Pikaard, 2017). 24-nt siRNAs are necessary to guide *de novo* DNA methylation in a process called RNA-directed DNA methylation (RdDM). Two plant specific RNA polymerases, RNA Polymerase IV (Pol IV) and Pol V, along with RNA-DEPENDENT RNA POLYMERASE 2 (RDR2), DICER-LIKE 3 (DCL3), and ARGONAUTE 4(AGO4) all function in this RdDM pathway in *A. thaliana* (Fig. 6) (Wendte & Pikaard, 2017). Pol IV and RDR2 are involved in the biogenesis of 24-nt siRNAs and are physically associated in both *A. thaliana* and maize (Haag et al., 2014, 2012; Law, Vashisht, Wohlschlegel, & Jacobsen, 2011).

Several reports have examined global small RNA profiles using heterotic F_1 hybrids and their parental lines in maize, rice, canola and *A. thaliana*, and differences of 24-nt siRNA levels between heterotic hybrids and their parental lines were observed (Barber et al., 2012; Groszmann et al., 2011a; He et al., 2010; Li, Varala, Moose, & Hudson, 2012; Shen et al., 2012, 2017). Change of 24-nt siRNAs may result in changes of DNA

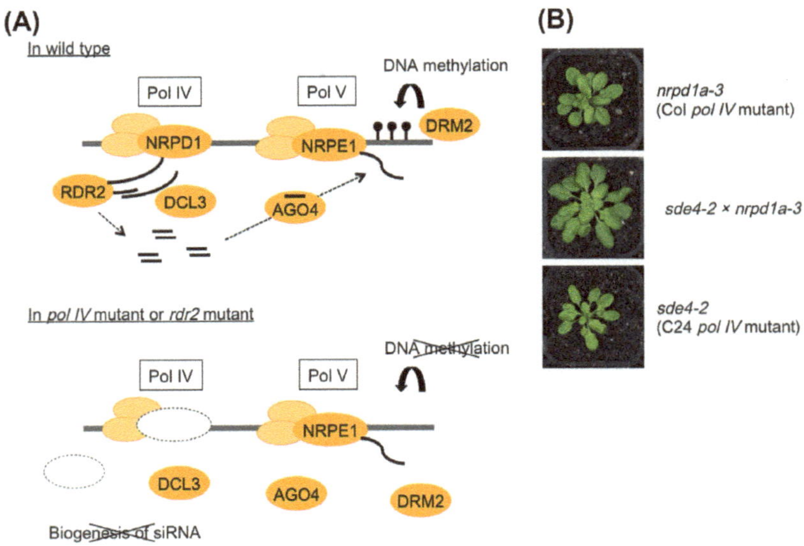

Figure 6 Pol IV function in RNA-directed DNA methylation (RdDM) and hybrid vigor. (A) Schematic diagram of RdDM. In *pol IV* (*nrpd1a* or *sde4*) or *rdr2* mutants, 90% of small interfering RNAs (siRNAs) and DNA methylation mediated through these genes is lost. (B) Plant phenotype of F_1 hybrid without Pol IV function at 28 days after sowing. F_1 hybrids lacking Pol IV function showed hybrid vigor, indicating that Pol IV function is independent from hybrid vigor. ARGONAUTE 4 (AGO4), DICER-LIKE 3 (DCL3), DOMAINS REARRANGED METHYLTRANSFERASE 2 (DRM2), NUCLEAR RNA POLYMERASE D1A (NRPD1), NUCLEAR RNA POLYMERASE D1B (NRPE1), RNA Polymerase IV (Pol IV), RNA-DEPENDENT RNA POLYMERASE 2 (RDR2). SILENCING DEFECTIVE 4 (SDE4).

methylation levels and consequently in the expression level of TEs and of some protein-coding genes. As siRNAs have the ability to modify epigenetic marks in *trans*, siRNA derived from one parental allele may modify the epigenetic state of the other parental allele in hybrids (Greaves et al., 2015; Groszmann, Greaves, Fujimoto, Peacock, & Dennis, 2013).

DNA methylation states have been compared between heterotic F_1 hybrids and parental lines in rice, *Brassica napus*, and *A. thaliana* (Chodavarapu et al., 2012; Greaves et al., 2012; He et al., 2010; Shen et al., 2012, 2017; Zhang et al., 2016a). In *B. napus*, F_1 hybrids had a slightly higher level of DNA methylation in all three sequence contexts, CG, CHG, and CHH sites (H is A, T, or C), and increased 24-nt siRNA clusters were observed in F_1 hybrids, suggesting that increased 24-nt siRNAs are directly correlated with increased DNA methylation levels (Shen et al., 2017). In rice, DNA methylation levels in the F_1 hybrids were intermediate to parental methylation levels, and differential methylation was strongly associated with regions where siRNA production was different between the two parents (Chodavarapu et al., 2012). In the case of *A. thaliana* using 15-day-old seedlings of C24 and L*er*, and their reciprocal hybrids, increased DNA methylation levels were observed in all three sequence contexts in both reciprocal hybrids relative to parental lines. F_1 hybrids had increased DNA methylation relative to parental lines at both protein-coding genes and TEs. Increased DNA methylation in F_1 hybrids occurred predominantly in regions covering siRNA clusters, especially the regions that were differentially methylated between the parents (Greaves et al., 2012; Shen et al., 2012). Change of DNA methylation states in the F_1 hybrids resulted from Trans-Chromosomal Methylation (TCM) (an increase in methylation at a locus with a previously low methylation allele gaining methylation to resemble the more heavily methylated allele) or Trans-Chromosomal deMethylation (TCdM) (loss of methylation at a genomic segment) (Fig. 7). Both TCM and TCdM events in hybrids are largely dependent on 24-nt siRNAs (Greaves et al., 2012).

To clarify the relationship between siRNAs and hybrid vigor or between TCM/TCdM and hybrid vigor, mutants impaired in the biogenesis of 24-nt siRNAs or in RdDM components were analyzed in maize and *A. thaliana* (Barber et al., 2012; Kawanabe et al., 2016; Zhang et al., 2016a). Hybrid vigor was fully maintained in hybrids homozygous for *mediator of paramutation 1* (*mop1*), the ortholog of *RDR2* in maize, or homozygous for *rdr2*, for *nuclear RNA polymerase IV, subunit 1* (*nrpd1*; largest subunits of Pol IV) and

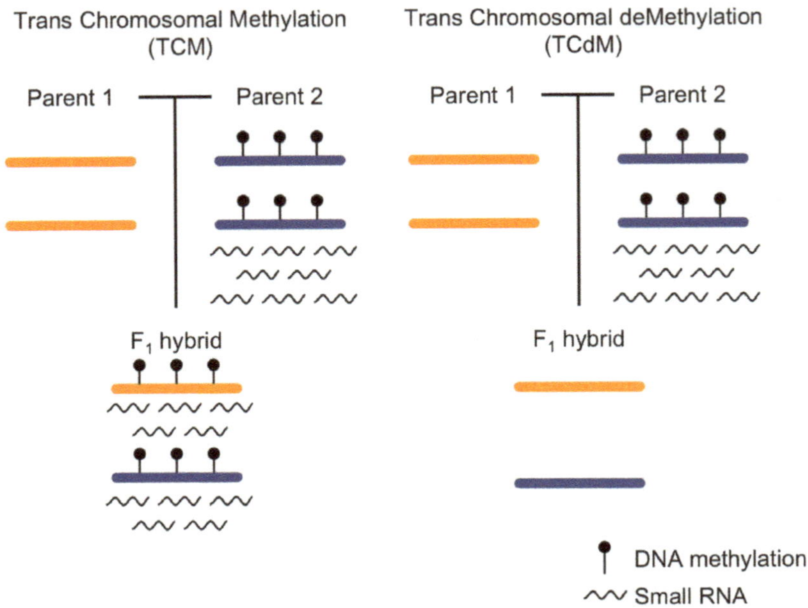

Figure 7 Interaction of DNA methylation between parental alleles in the F_1 hybrids. Trans-chromosomal methylation (TCM) means that a locus with a low methylated allele from one parent gains methylation to resemble the more heavily methylated allele from the other parent by the RdDM pathway. Trans-chromosomal demethylation (TCdM) means that a locus with heavily methylated allele from one parent loses methylation and becomes like the less methylated allele from the other parent.

for genes involved in the RdDM machinery or in Pol V function in *A. thaliana* (Fig. 6) (Barber et al., 2012; Kawanabe et al., 2016; Zhang et al., 2016b, 2016a). In hybrids with mutations in both *nrpd1* (*pol iv* mutant) and *nrpe1* (*pol v* mutant), TCM and TCdM were abolished, but hybrid vigor was maintained (Zhang et al., 2016a). These results indicate that the RdDM pathway is not necessary for the establishment of hybrid vigor in maize or *A. thaliana*, and TCM and TCdM are not necessary for hybrid vigor.

Genome wide histone modification was also compared between heterotic F_1 hybrids and parental lines in rice and *A. thaliana*. The distribution and abundance of histone modifications such as tri-methylation of lysine 4 on histone H3 (H3K4me3) or tri-methylation of lysine 27 on histone H3 (H3K27me3) were similar between parental lines, and most of genes showed an additive pattern in F_1 hybrids in both species (He et al., 2010; Moghaddam et al., 2011; Zhu, Greaves, Dennis, & Peacock, 2017), suggesting that histone modification is not critical for hybrid vigor either.

4.2 Difference of Epigenetic States Between Parental Lines May Cause Growth Vigor

However, it has been suggested that heterozygosity of epigenetic states (epiHybrid) might be important for hybrid vigor as enhanced growth was observed in several of the hybrids between WT and specific epiRIL lines. Thirty-four selected epiRILs derived from hybrids between *met1* and WT were crossed with WT Col. Among the 34 epiHybrids, 30 did not exceed the best parent in rosette size, whilst the remaining 4 showed increased rosette size until bolting. The epiHybrid obtained after crossing one of the tested epiRIL, epi31 with WT (epi31 x WT) showed consistent vigorous growth, whereas those obtained from the reciprocal cross (WT x epi31) did not, suggesting that parent-of-origin effect can influence hybrid growth (Dapp et al., 2015).

Similarly, in the case of epiRILs derived from hybrids obtained by crossing *A. thaliana ddm1* mutant and WT plants, the epiHybrids, WT Col (female) and 19 epiRILs (male), were generated. Three of the 19 epiHybrids showed greater than 15% increased leaf area compared with the best parent. Phenotypical variations were observed between the 19 epi-Hybrids, and these hybrids were generated using WT Col as the female line, suggesting that these variations could be due to paternal effects (Lauss et al., 2018). In epiHybrids derived from *ddm1* and WT crosses, there was no significant correlation between paternal DNA methylation states and hybrid performance in leaf area among 19 crosses, suggesting that hybrid performance could be due to specific methylation differences between epiRILs (Lauss et al., 2018). These two reports using epiHybrids support the involvement of epigenetic regulation and interactions in hybrid vigor. Hybrid vigor generated from crosses between epigenetically divergent lines may be a general phenomenon (Dapp et al., 2015; Lauss et al., 2018).

The other example is that enhanced vigor was observed in F_4 generations derived from hybrids generated from crosses between WT and the *msh1* mutant in *A. thaliana* (Virdi et al., 2015). *MutS HOMOLOG1* (*MSH1*) encodes a protein dually targeted to mitochondria and plastids and is involved in organelle genome stability (Abdelnoor et al., 2003; Xu et al., 2011). Disruption of MSH1 causes changes of DNA methylation in *A. thaliana* (Virdi et al., 2015), and causes altered phenotype not only in *A. thaliana* but also in other species such as sorghum, tomato, and soybean (Xu et al., 2012). *MSH1*-RNA interference (RNAi) suppression in sorghum or tomato showed altered phenotypes, which were inherited even in transgene-null offspring. Crossing between WT and transgene-null plants with developmental alterations, results in enhanced heritable growth vigor in the progeny lines in sorghum and tomato (de la Rosa Santamaria et al., 2014; Yang X. et al., 2015). These results suggest that epigenetic reprogramming can result in enhanced growth patterns and could therefore be used for breeding applications in crop plants (Virdi et al., 2015; Yang X. et al., 2015).

4.3 Epigenetic Regulation of Clock Genes Might Be Involved in Hybrid Vigor

Altered expression of the circadian clock genes in the F_1 results in an increased expression level of downstream genes involved in energy and metabolism including photosynthesis and starch pathways, and was suggested to lead to the increase of biomass observed in hybrid vigor (Ni et al., 2009). In *A. thaliana*, *CIRCADIAN CLOCK ASSOCIATED 1* (*CCA1*) and *LATE ELONGATED HYPOCOTYL* (*LHY*), which encodes MYB-domain transcription factors, accumulate at dawn and negatively regulate *TIMING OF CAB EXPRESSION 1* (*TOC1*) expression by binding to the evening element on the *TOC1* promoter. In the evening, *TOC1* is highly expressed and represses *CCA1* and *LHY* expression. This circular feedback regulation affects central oscillation that maintains the rhythms, amplitude and/or phase of the circadian clock. The progeny of allotetraploids generated by inter-specific hybridization between *A. thaliana* and *Arabidopsis arenosa* showed growth vigor and non-additive expression of over 1400 genes. Some non-additively expressed genes contain at least one CCA1 binding site or evening element in their promoter region, a DNA regulatory element that has a role in circadian-regulated gene expression. Noteworthy, *CCA1* and *LHY* expression levels in allotetraploids were lower than the MPV at Zeitgeber time (ZT; hours since dawn) 6−12 and higher than the MPV at ZT15; amplitude but not the phase of the circadian clock is altered in allotetraploids. Expression levels in *CCA1* and *LHY* were associated with the acetylation of lysine 9 on histone H3 (H3K9Ac) and H3K4me2 levels (Ni et al., 2009). Similar results were observed in an intra-specific hybrid between C24 and Col accessions of *A. thaliana*. As correct circadian regulation enhances fitness and metabolism, the authors suggest that altered expression of circadian oscillator genes is associated with hybrid vigor in C24 x Col hybrids as well as in allotetraploids (Ni et al., 2009).

In hybrids between C24 and L*er*, expression levels of *CCA1* were different between reciprocal hybrids; *CCA1* expression peaks measured in C24 x L*er* were lower than those measured in L*er* x C24 hybrids at ZT6 (e.g. ZT6 is 6 h from dawn). The DNA methylation level at CHH sites in the promoter region of *CCA1* in L*er* x C24 hybrids was higher than that in C24 x L*er* hybrids, suggesting there was negative correlation between DNA methylation level and *CCA1* expression. In addition, L*er* x *ago4* (L*er*) hybrids showed lower *CCA1* expression peaks, higher CHH methylation levels in the promoter region of *CCA1*, and greater dry weight than those measured in *ago4* (L*er*) x L*er* hybrids. These results suggest that parent-of-origin effect on CHH methylation via RdDM affects *CCA1* expression amplitude and leads to growth difference between reciprocal hybrids (Ng et al., 2014). It is not clear how this model is related to the hybrid vigor shown in both reciprocal hybrids (C24 x L*er* or L*er* x C24 hybrids) or in C24 x Col hybrids.

Unlike in *A. thaliana*, in maize, there is a temporal shift of the binding of ZmCCA1 to targets from ZT9 to ZT3 (early morning); the proportion of ZmCCA1 proteins bound to target sites was higher at ZT3 than at ZT9 or ZT15 in hybrids whereas no difference between ZT3 and ZT9 was observed in parental lines. ZmCCA1 target genes were up-regulated in the morning at ZT3 in hybrids as compared with parental lines. From these results, a phase-shift model for hybrid vigor was predicted; the altered temporal binding of ZmCCA1 to photosynthetic and metabolic genes causes non-additive gene expression levels and increased carbon fixation capacity, leading to hybrid vigor (Ko et al., 2016).

It is certain that circadian rhythm is important for the growth of plants (Ni et al., 2009), and these reports suggest that the circadian rhythm is a key factor of hybrid vigor. However, factors causing growth vigor are different in *A. thaliana* and maize. Further research will be necessary to examine whether the different models, alteration of the amplitude in *A. thaliana* and phase-shift of the circadian rhythm in maize, are due to a difference in the mechanism of hybrid vigor between these two species.

4.4 DDM1 Plays an Important Role in Hybrid Vigor

Recently two groups showed that DDM1 is a major regulator of hybrid vigor in *A. thaliana* (Kawanabe et al., 2016; Zhang et al., 2016b). The F_1 hybrids having homozygous mutations in *ddm1* (termed *ddm1* C24 x Col hybrids) had reduced vegetative hybrid vigor in C24 x Col hybrids (Fig. 8) (Kawanabe et al., 2016; Zhang et al., 2016b). A reduction of vegetative hybrid vigor by defects in DDM1 function was also observed in the C24 x Cvi and the Col x Cvi hybrids, the level of reduction of hybrid vigor being dependent on the combination of parental lines (Kawanabe et al., 2016).

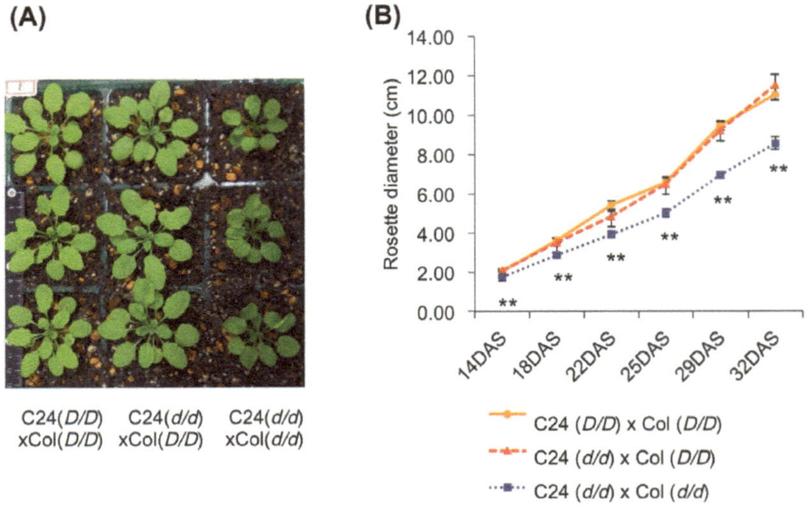

Figure 8 Function of DDM1 in hybrid vigor. (A) The phenotypes of C24 (*D/D*) x Columbia-0 (Col) (*D/D*), C24 (*d/d*) x Col (*D/D*), and C24 (*d/d*) x Col (*d/d*) at 25 days after sowing (DAS). (B) Rosette diameter of C24 (*D/D*) x Col (*D/D*), C24 (*d/d*) x Col (*D/D*), and C24 (*d/d*) x Col (*d/d*) hybrids at 14, 18, 22, 25, 29 and 32 DAS. *D*, wild type *DDM1*; *d*, *ddm1* mutant. Asterisk indicates $p < 0.05$ (Student's t test).

Non-additively expressed genes involved in SA related pathway were identified in WT and *ddm1* C24 x Col hybrids (Zhang et al., 2016b). SA concentrations in Col, C24, and C24 x Col hybrids were lower than those in *ddm1* (Col), *ddm1* (C24), and *ddm1* C24 x Col hybrids, respectively (Zhang et al., 2016b). The order from the higher SA concentrations in these six lines was *ddm1* (C24), C24, *ddm1* C24 x Col hybrids, *ddm1* (Col), C24 x Col hybrids, and Col (Fig. 9) (Zhang et al., 2016b). The authors suggest that Salicylic acid (SA) concentration in C24 x Col hybrids was in appropriate range for growth vigor performance, but SA concentration in Col and *ddm1* C24 x Col hybrids were lower or higher than the appropriate range, respectively (Fig. 9); a model of hermetic modulation of hybrid vigor by SA was suggested (Zhang et al., 2016b). However, the difference of SA concentration between WT hybrids and Col or *ddm1* C24 x Col hybrids was not great (Fig. 9) (Zhang et al., 2016b). Further study will be required to confirm this hypothesis.

In the hybrid between a heterozygous *ddm1-9* mutant in C24 and a homozygous *ddm1-1* mutant in Col, plants homozygous for the *ddm1*

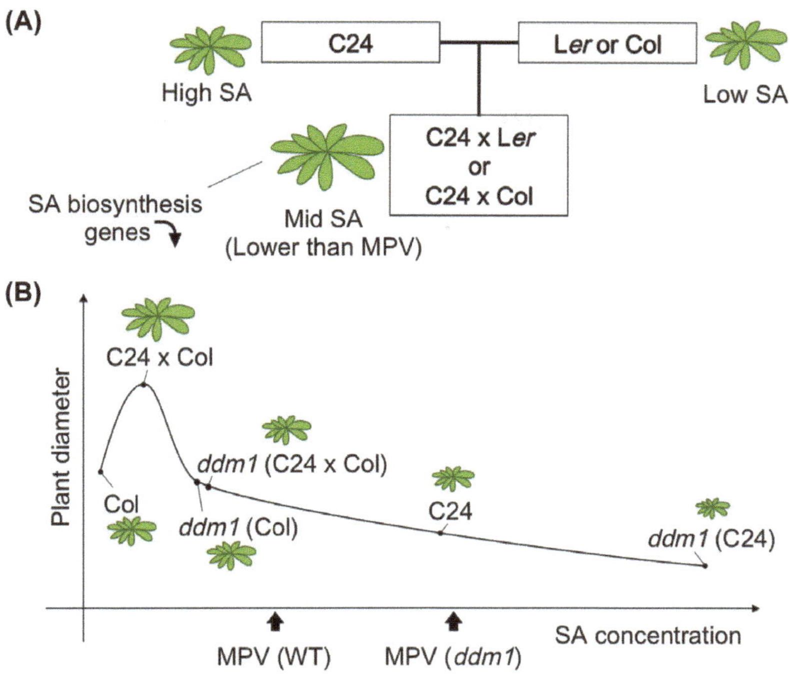

Figure 9 Salicylic acid (SA) concentration and plant diameter of wild type and *ddm1* mutant. (A) SA concentration of C24, Landsberg *erecta* (L*er*), Columbia-0 (Col), and their F_1 hybrid C24 x L*er* and C24 x Col. SA concentration lower than mid-parent value (MPV) is observed in F_1 hybrids. Genes involved in SA biosynthesis are down-regulated in C24 x L*er* or C24 x Col hybrids compared with MPV. (B) SA concentration and plant diameter of Col, C24, C24 x Col, and their *ddm1* mutant, *ddm1* (Col), *ddm1* (C24), and *ddm1* (C24 x Col). C24 x Col hybrids are considered to have appropriate SA concentration for hybrid vigor compared with *ddm1* hybrids, *ddm1* (C24 x Col). *WT*, wild type.

mutation were smaller than heterozygotes. However, some plants homozygous for *ddm1* showed hybrid vigor to levels similar to the plants having a heterozygous *ddm1* mutation. Some plants heterozygous for the *ddm1* mutation showed hybrid vigor reduced to the level of the homozygous *ddm1* hybrid plants (Fig. 10) (Kawanabe et al., 2016). In both cases there was an identical genetic background, and it may be similar to the growth vigor seen in epiHybrids (Dapp et al., 2015; Lauss et al., 2018). These effects may result from the previous methylation state of the genome in the *ddm1* parent, and genes or chromosomal segments important for hybrid vigor coming from the *ddm1* parent having an altered level of DNA methylation.

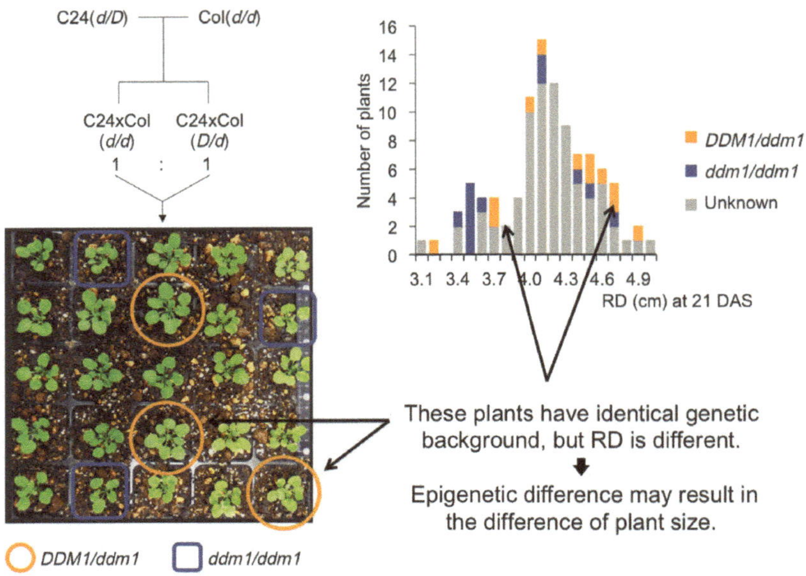

Figure 10 *DDM1* genotype does not correlate with their plant size. F_1 hybrids between C24 (*d/D*) and Columbia-0 (Col) (*d/d*) segregate their genotype, *d/d* and *D/d* in 1 to 1 ratio. The phenotype of the F_1 hybrid is shown by the photo at 21 days after sowing (DAS). F_1 hybrids (*D/d*) have identical genetic background, but rosette diameter (RD) is variable within this population and some F_1 hybrids (*D/d*) are smaller than F_1 hybrids (*d/d*). *D*, wild type DDM1; *d*, *ddm1* mutant.

5. PERSPECTIVE

The F_1 hybrid cultivars that benefit from hybrid vigor have contributed to increasing crop yields during the last century. However, we still cannot predict the intensity of hybrid vigor before the F_1 hybrids have been produced, which renders the breeding process of F_1 hybrid cultivars still laborious, time-consuming, and costly. Recent research in *A. thaliana* suggests that epigenetic regulation may be involved in hybrid vigor; the vigorous growth in F_1 hybrids between epiRILs (Dapp et al., 2015; Lauss et al., 2018) and WT and the reduction of hybrid vigor levels by disruption of DDM1 function suggest that alterations in DNA methylation affect the level of hybrid vigor (Fujimoto et al. 2018; Kawanabe et al., 2016; Zhang et al. 2016b). These studies may represent a breakthrough in elucidating the molecular mechanism underlying hybrid vigor.

In order to apply the knowledge gained from *A. thaliana* to crops, the creation of mutants in genes involved in epigenetic regulation such as

DDM1 or *MET1* in both parental lines is one possible approach. However, in many cases, paralogous genes are found in crops, which make it difficult to generate real loss of function mutants. In addition, as some of these genes might be essential for plant growth, mutants that have a complete loss of gene function might not be viable. However, methods do exist that allow introducing mutation of one parent to the other parent by backcrossing as successfully shown for the *mop1* mutant in maize (Barber et al., 2012). Further research progress will be needed to clarify the contribution of epigenetics to hybrid vigor in crops.

ACKNOWLEDGEMENTS

We thank Dr. Elizabeth S. Dennis (CSIRO and University of Technology, Sydney, Australia) and Dr. Kenji Osabe (Okinawa Institute of Science and Technology Graduate University, Japan) for helpful comments and manuscript editing.

REFERENCES

Abdelnoor, R. V., Yule, R., Elo, A., Christensen, A. C., Meyer-Gauen, G., & Mackenzie, S. A. (2003). Substoichiometric shifting in the plant mitochondrial genome is influenced by a gene homologous to MutS. *Proceedings of the National Academy of Sciences of the United States of America, 100*, 5968–5973.

Barber, W. T., Zhang, W., Win, H., Varala, K. K., Dorweiler, J. E., Hudson, M. E., et al. (2012). Repeat associated small RNAs vary among parents and following hybridization in maize. *Proceedings of the National Academy of Sciences of the United States of America, 109*, 10444–10449.

Barth, S., Busimi, A. K., Friedrich Utz, H., & Melchinger, A. E. (2003). Heterosis for biomass yield and related traits in five hybrids of *Arabidopsis thaliana* L. Heynh. *Heredity, 91*, 36–42.

Birchler, J. A., Yao, H., Chudalayandi, S., Vaiman, D., & Veitia, R. A. (2010). Heterosis. *The Plant Cell, 22*, 2105–2112.

Bruce, A. B. (1910). The mendelian theory of heredity and the augmentation of vigor. *Science, 32*, 627–628.

Chen, Z. J. (2010). Molecular mechanisms of polyploidy and hybrid vigor. *Trends in Plant Science, 15*, 57–71.

Cheng, S. H., Zhuang, J. Y., Fan, Y. Y., Du, J. H., & Cao, L. Y. (2007). Progress in research and development on hybrid rice: A super-domesticate in China. *Annals of Botany, 100*, 959–966.

Chodavarapu, R. K., Feng, S., Ding, B., Simon, S. A., Lopez, D., Jia, Y., et al. (2012). Transcriptome and methylome interactions in rice hybrids. *Proceedings of the National Academy of Sciences of the United States of America, 109*, 12040–12045.

Cortijo, S., Wardenaar, R., Colomé-Tatché, M., Gilly, A., Etcheverry, M., Labadie, K., et al. (2014). Mapping the epigenetic basis of complex traits. *Science, 343*, 1145–1148.

Crow, J. F. (1998). 90 years ago: The beginning of hybrid maize. *Genetics, 148*, 923–928.

Cui, J., You, C., & Chen, X. (2017). The evolution of microRNAs in plants. *Current Opinion in Plant Biology, 35*, 61–67.

Dapp, M., Reinders, J., Bédiée, A., Balsera, C., Bucher, E., Theiler, G., et al. (2015). Heterosis and inbreeding depression of epigenetic *Arabidopsis* hybrids. *Nature Plants, 1*, 15092.

Davenport, C. B. (1908). Degeneration, albinism and inbreeding. *Science, 28*, 454–455.
de la Rosa Santamaria, R., Shao, M. R., Wang, G., Nino-Liu, D. O., Kundariya, H., Wamboldt, Y., et al. (2014). *MSH1*-induced non-genetic variation provides a source of phenotypic diversity in *Sorghum bicolor*. *PLoS One, 9*, e108407.
Duvick, D. N. (2001). Biotechnology in the 1930s: The development of hybrid maize. *Nature Reviews. Genetics, 2*, 69–74.
East, E. M. (1936). Heterosis. *Genetics, 21*, 375–397.
Feng, F., Li, Y., Qin, X., Liao, Y., & Siddique, K. H. M. (2017). Changes in rice grain quality of Indica and Japonica type varieties released in China from 2000 to 2014. *Frontiers in Plant Science, 8*, 1863.
Flint-Garcia, S. A., Buckler, E. S., Tiffin, P., Ersoz, E., & Springer, N. M. (2009). Heterosis is prevalent for multiple traits in diverse maize germplasm. *PLoS One, 4*, e7433.
Fujimoto, R., Sasaki, T., Ishikawa, R., Osabe, K., Kawanabe, T., & Dennis, E. S. (2012b). Molecular mechanisms of epigenetic variation in plants. *International Journal of Molecular Sciences, 13*, 9900–9922.
Fujimoto, R., Taylor, J. M., Shirasawa, S., Peacock, W. J., & Dennis, E. S. (2012a). Heterosis of *Arabidopsis* hybrids between C24 and Col is associated with increased photosynthesis capacity. *Proceedings of the National Academy of Sciences of the United States of America, 109*, 7109–7114.
Fujimoto, R., Uezono, K., Ishikura, S., Osabe, K., Peacock, W. J., & Dennis, E. S. (2018). Recent research on the mechanism of heterosis is important for crop and vegetable breeding systems. *Breeding Science, 68*, 145–158.
Giraud, H., Bauland, C., Falque, M., Madur, D., Combes, V., Jamin, P., et al. (2017). Reciprocal genetics: Identifying QTLs for general and specific combining abilities in hybrids between multiparental populations from two maize (*Zea mays* L.) heterotic groups. *Genetics, 207*, 1167–1180.
Girke, A., Schierholt, A., & Becker, H. C. (2012). Extending the rapeseed gene pool with resynthesized *Brassica napus* II: Heterosis. *TAG. Theoretical and Applied Genetics. Theoretische und angewandte Genetik, 124*, 1017–1026.
Greaves, I. K., Gonzalez-Bayon, R., Wang, L., Zhu, A., Liu, P. C., Groszmann, M., et al. (2015). Epigenetic changes in hybrids. *Plant Physiology, 168*, 1197–1205.
Greaves, I. K., Groszmann, M., Ying, H., Taylor, J. M., Peacock, W. J., & Dennis, E. S. (2012). Trans chromosomal methylation in *Arabidopsis* hybrids. *Proceedings of the National Academy of Sciences of the United States of America, 109*, 3570–3575.
Groszmann, M., Gonzalez-Bayon, R., Greaves, I. K., Wang, L., Huen, A. K., Peacock, W. J., et al. (2014). Intraspecific Arabidopsis hybrids show different patterns of heterosis despite the close relatedness of the parental genomes. *Plant Physiology, 166*, 265–280.
Groszmann, M., Greaves, I. K., Albert, N., Fujimoto, R., Helliwell, C. A., Dennis, E. S., et al. (2011b). Epigenetics in plants-vernalisation and hybrid vigour. *Biochimica et Biophysica Acta (BBA) - Gene Regulatory Mechanisms, 1809*, 427–437.
Groszmann, M., Greaves, I. K., Albertyn, Z. I., Scofield, G. N., Peacock, W. J., & Dennis, E. S. (2011a). Changes in 24-nt siRNA levels in Arabidopsis hybrids suggest an epigenetic contribution to hybrid vigor. *Proceedings of the National Academy of Sciences of the United States of America, 108*, 2617–2622.
Groszmann, M., Greaves, I. K., Fujimoto, R., Peacock, W. J., & Dennis, E. S. (2013). The role of epigenetics in hybrid vigour. *Trends in Genetics, 29*, 684–690.
Guo, Q., Liu, Q., Smith, N. A., Liang, G., & Wang, M. B. (2016). RNA silencing in plants: Mechanisms, technologies and applications in horticultural crops. *Current Genomics, 17*, 476–489.

Haag, J. R., Brower-Toland, B., Krieger, E. K., Sidorenko, L., Nicora, C. D., Norbeck, A. D., et al. (2014). Functional diversification of maize RNA Polymerase IV and V subtypes via alternative catalytic subunits. *Cell Reports, 9*, 378–390.

Haag, J. R., Ream, T. S., Marasco, M., Nicora, C. D., Norbeck, A. D., Pasa-Tolic, L., et al. (2012). In vitro transcription activities of Pol IV, Pol V, and RDR2 reveal coupling of Pol IV and RDR2 for dsRNA synthesis in plant RNA silencing. *Molecular Cell, 48*, 811–818.

He, G., Zhu, X., Elling, A. A., Chen, L., Wang, X., Guo, L., et al. (2010). Global epigenetic and transcriptional trends among two rice subspecies and their reciprocal hybrids. *The Plant Cell, 22*, 17–33.

Hochholdinger, F., & Hoecker, N. (2007). Towards the molecular basis of heterosis. *Trends in Plant Science, 12*, 427–432.

Huang, X., Yang, S., Gong, J., Zhao, Y., Feng, Q., Gong, H., et al. (2015). Genomic analysis of hybrid rice varieties reveals numerous superior alleles that contribute to heterosis. *Nature Communications, 6*, 6258.

Huang, X., Yang, S., Gong, J., Zhao, Q., Feng, Q., Zhan, Q., et al. (2016). Genomic architecture of heterosis for yield traits in rice. *Nature, 537*, 629–633.

Hu, X., Wang, H., Diao, X., Liu, Z., Li, K., Wu, Y., et al. (2016). Transcriptome profiling and comparison of maize ear heterosis during the spikelet and floret differentiation stages. *BMC Genomics, 17*, 959.

Jahnke, S., Sarholz, B., Thiemann, A., Kühr, V., Gutiérrez-Marcos, J. F., Geiger, H. H., et al. (2010). Heterosis in early seed development: A comparative study of F_1 embryo and endosperm tissues 6 days after fertilization. *TAG. Theoretical and Applied Genetics. Theoretische und angewandte Genetik, 120*, 389–400.

Jiang, Y., Schmidt, R. H., Zhao, Y., & Reif, J. C. (2017). A quantitative genetic framework highlights the role of epistatic effects for grain-yield heterosis in bread wheat. *Nature Genetics, 49*, 1741–1746.

Johannes, F., Porcher, E., Teixeira, F. K., Saliba-Colombani, V., Simon, M., Agier, N., et al. (2009). Assessing the impact of transgenerational epigenetic variation on complex traits. *PLoS Genetics, 5*, e1000530.

Jones, D. F. (1917). Dominance of linked factors as a means of accounting for heterosis. *Genetics, 2*, 466–479.

Karan, R., DeLeon, T., Biradar, H., & Subudhi, P. K. (2012). Salt stress induced variation in DNA methylation pattern and its influence on gene expression in contrasting rice genotypes. *PLoS One, 7*, e40203.

Kawamura, K., Kawanabe, T., Shimizu, M., Nagano, A. J., Saeki, N., Okazaki, K., et al. (2016). Genetic distance of inbred lines of Chinese cabbage and its relationship to heterosis. *Plant Gene, 5*, 1–7.

Kawanabe, T., Ishikura, S., Miyaji, N., Sasaki, T., Wu, L. M., Itabashi, E., et al. (2016). Role of DNA methylation in hybrid vigor in *Arabidopsis thaliana. Proceedings of the National Academy of Sciences of the United States of America, 113*, E6704–E6711.

Kempe, K., Rubtsova, M., & Gils, M. (2014). Split-gene system for hybrid wheat seed production. *Proceedings of the National Academy of Sciences of the United States of America, 111*, 9097–9102.

Koornneef, M., Blankestijn-de Vries, H., Hanhart, C., Soppe, W., & Peeters, T. (1994). The phenotype of some late-flowering mutants is enhanced by a locus on chromosome 5 that is not effective in the Landsberg *erecta* wild-type. *The Plant Journal, 6*, 911–919.

Korn, M., Peterek, S., Mock, H. P., Heyer, A. G., & Hincha, D. K. (2008). Heterosis in the freezing tolerance, and sugar and flavonoid contents of crosses between *Arabidopsis thaliana* accessions of widely varying freezing tolerance. *Plant, Cell and Environment, 31*, 813–827.

Ko, D. K., Rohozinski, D., Song, Q., Taylor, S. H., Juenger, T. E., Harmon, F. G., et al. (2016). Temporal shift of circadian-mediated gene expression and carbon fixation contributes to biomass heterosis in maize hybrids. *PLoS Genetics, 12*, e1006197.

Krieger, U., Lippman, Z. B., & Zamir, D. (2010). The flowering gene *SINGLE FLOWER TRUSS* drives heterosis for yield in tomato. *Nature Genetics, 42*, 459–463.

Lauss, K., Wardenaar, R., Oka, R., van Hulten, M. H. A., Guryev, V., Keurentjes, J. J. B., et al. (2018). Parental DNA methylation states are associated with heterosis in epigenetic hybrids. *Plant Physiology, 176*, 1627–1645.

Law, J. A., Vashisht, A. A., Wohlschlegel, J. A., & Jacobsen, S. E. (2011). SHH1, a homeodomain protein required for DNA methylation, as well as RDR2, RDM4, and chromatin remodeling factors, associate with RNA polymerase IV. *PLoS Genetics, 7*, e1002195.

Li, D., Huang, Z., Song, S., Xin, Y., Mao, D., Lv, Q., et al. (2016). Integrated analysis of phenome, genome, and transcriptome of hybrid rice uncovered multiple heterosis-related loci for yield increase. *Proceedings of the National Academy of Sciences of the United States of America, 113*, E6026–E6035.

Li, Z. K., Luo, L. J., Mei, H. W., Wang, D. L., Shu, Q. Y., Tabien, R., et al. (2001). Overdominant epistatic loci are the primary genetic basis of inbreeding depression and heterosis in rice. I. Biomass and grain yield. *Genetics, 158*, 1737–1753.

Lippman, Z. B., & Zamir, D. (2007). Heterosis: Revisiting the magic. *Trends in Genetics, 23*, 60–66.

Liu, R., How-Kit, A., Stammitti, L., Teyssier, E., Rolin, D., Mortain-Bertrand, A., et al. (2015). A DEMETER-like DNA demethylase governs tomato fruit ripening. *Proceedings of the National Academy of Sciences of the United States of America, 112*, 10804–10809.

Li, Y., Varala, K., Moose, S. P., & Hudson, M. E. (2012). The inheritance pattern of 24 nt siRNA clusters in Arabidopsis hybrids is influenced by proximity to transposable elements. *PLoS One, 7*, e47043.

Li, H., Yang, Q., Fan, N., Zhang, M., Zhai, H., Ni, Z., et al. (2017). Quantitative trait locus analysis of heterosis for plant height and ear height in an elite maize hybrid zhengdan 958 by design III. *BMC Genetics, 18*, 36.

Luo, L. J., Li, Z. K., Mei, H. W., Shu, Q. Y., Tabien, R., Zhong, D. B., et al. (2001). Overdominant epistatic loci are the primary genetic basis of inbreeding depression and heterosis in rice. II. Grain yield components. *Genetics, 158*, 1755–1771.

Lu, H., Romero-Severson, J., & Bernardo, R. (2003). Genetic basis of heterosis explored by simple sequence repeat markers in a random-mated maize population. *TAG. Theoretical and Applied Genetics. Theoretische und angewandte Genetik, 107*, 494–502.

Martin, A., Troadec, C., Boualem, A., Rajab, M., Fernandez, R., Morin, H., et al. (2009). A transposon-induced epigenetic change leads to sex determination in melon. *Nature, 461*, 1135–1138.

Meyer, R. C., Törjék, O., Becher, M., & Altmann, T. (2004). Heterosis of biomass production in Arabidopsis. Establishment during early development. *Plant Physiology, 134*, 1813–1823.

Meyer, R. C., Witucka-Wall, H., Becher, M., Blacha, A., Boudichevskaia, A., Dörmann, P., et al. (2012). Heterosis manifestation during early Arabidopsis seedling development is characterized by intermediate gene expression and enhanced metabolic activity in the hybrids. *The Plant Journal, 71*, 669–683.

Michaels, S. D., & Amasino, R. M. (1999). *FLOWERING LOCUS C* encodes a novel MADS domain protein that acts as a repressor of flowering. *The Plant Cell, 11*, 949–956.

Moghaddam, A. M. B., Roudier, F., Seifert, M., Bérard, C., Magniette, M. L., Ashtiyani, R. K., et al. (2011). Additive inheritance of histone modifications in *Arabidopsis thaliana* intra-specific hybrids. *The Plant Journal, 67*, 691–700.

Moll, R. H., Lonnquist, J. H., Vélez Fortuno, J., & Johnson, E. C. (1965). The relationship of heterosis and genetic divergence in maize. *Genetics, 52*, 139–144.
Ng, D. W. K., Miller, M., Yu, H. H., Huang, T. Y., Kim, E. D., Lu, J., et al. (2014). A role for CHH methylation in the parent-of-origin effect on altered circadian rhythms and biomass heterosis in *Arabidopsis* intraspecific hybrids. *The Plant Cell, 26*, 2430–2440.
Ni, Z., Kim, E. D., Ha, M., Lackey, E., Liu, J., Zhang, Y., et al. (2009). Altered circadian rhythms regulate growth vigour in hybrids and allopolyploids. *Nature, 457*, 327–331.
Paschold, A., Jia, Y., Marcon, C., Lund, S., Larson, N. B., Yeh, C. T., et al. (2012). Complementation contributes to transcriptome complexity in maize (*Zea mays* L.) hybrids relative to their inbred parents. *Genome Research, 22*, 2445–2454.
Powers, L. (1944). An expansion of Jones's theory for the explanation of heterosis. *The American Naturalist, 78*, 275–280.
Quadrana, L., & Colot, V. (2016). Plant transgenerational epigenetics. *Annual Review of Genetics, 50*, 467–491.
Reinders, J., Wulff, B. B. H., Mirouze, M., Marí-Ordóñez, A., Dapp, M., Rozhon, W., et al. (2009). Compromised stability of DNA methylation and transposon immobilization in mosaic *Arabidopsis* epigenomes. *Genes & Development, 23*, 939–950.
Richey, F. D. (1942). Mock-dominance and hybrid vigor. *Science, 96*, 280–281.
Rohde, P., Hincha, D. K., & Heyer, A. G. (2004). Heterosis in the freezing tolerance of crosses between two *Arabidopsis thaliana* accessions (Columbia-0 and C24) that show differences in non-acclimated and acclimated freezing tolerance. *The Plant Journal, 38*, 790–799.
Saeki, N., Kawanabe, T., Ying, H., Shimizu, M., Kojima, M., Abe, H., et al. (2016). Molecular and cellular characteristics of hybrid vigour in a commercial hybrid of Chinese cabbage. *BMC Plant Biology, 16*, 45.
Semel, Y., Nissenbaum, J., Menda, N., Zinder, M., Krieger, U., Issman, N., et al. (2006). Overdominant quantitative trait loci for yield and fitness in tomato. *Proceedings of the National Academy of Sciences of the United States of America, 103*, 12981–12986.
Shea, D. J., Itabashi, E., Takada, S., Fukai, E., Kakizaki, T., Fujimoto, R., et al. (2018). The role of *FLOWERING LOCUS C* in vernalization of *Brassica*: The importance of vernalization research in the face of climate change. *Crop & Pasture Science, 69*, 30–39.
Sheldon, C. C., Burn, J. E., Perez, P. P., Metzger, J., Edwards, J. A., Peacock, W. J., et al. (1999). The *FLF* MADS box gene: A repressor of flowering in Arabidopsis regulated by vernalization and methylation. *The Plant Cell, 11*, 445–458.
Shen, H., He, H., Li, J., Chen, W., Wang, X., Guo, L., et al. (2012). Genome-wide analysis of DNA methylation and gene expression changes in two *Arabidopsis* ecotypes and their reciprocal hybrids. *The Plant Cell, 24*, 875–892.
Shen, Y., Sun, S., Hua, S., Shen, E., Ye, C. Y., Cai, D., et al. (2017). Analysis of transcriptional and epigenetic changes in hybrid vigor of allopolyploid *Brassica napus* uncovers key roles for small RNAs. *The Plant Journal, 91*, 874–893.
Shull, G. H. (1908). The composition of a field of maize. *Journal of Heredity, 4*, 296–301.
Shull, G. H. (1948). What is "heterosis"? *Genetics, 33*, 439–446.
Springer, N. M., & Stupar, R. M. (2007). Allelic variation and heterosis in maize: How do two halves make more than a whole? *Genome Research, 17*, 264–275.
Stuber, C. W., Lincoln, S. E., Wolff, D. W., Helentjaris, T., & Lander, E. S. (1992). Identification of genetic factors contributing to heterosis in a hybrid from two elite maize inbred lines using molecular markers. *Genetics, 132*, 823–839.
Stupar, R. M., Gardiner, J. M., Oldre, A. G., Haun, W. J., Chandler, V. L., & Springer, N. M. (2008). Gene expression analyses in maize inbreds and hybrids with varying levels of heterosis. *BMC Plant Biology, 8*, 33.

Swanson-Wagner, R. A., Jia, Y., DeCook, R., Borsuk, L. A., Nettleton, D., & Schnable, P. S. (2006). All possible modes of gene action are observed in a global comparison of gene expression in a maize F_1 hybrid and its inbred parents. *Proceedings of the National Academy of Sciences of the United States of America, 103*, 6805–6810.

Teixeira, F. K., Heredia, F., Sarazin, A., Roudier, F., Boccara, M., Ciaudo, C., et al. (2009). A role for RNAi in the selective correction of DNA methylation defects. *Science, 323*, 1600–1604.

Virdi, K. S., Laurie, J. D., Xu, Y. Z., Yu, J., Shao, M. R., Sanchez, R., et al. (2015). *Arabidopsis* MSH1 mutation alters the epigenome and produces heritable changes in plant growth. *Nature Communications, 6*, 6386.

Wang, M., Qin, L., Xie, C., Li, W., Yuan, J., Kong, L., et al. (2014). Induced and constitutive DNA methylation in a salinity-tolerant wheat introgression line. *Plant and Cell Physiology, 55*, 1354–1365.

Wei, G., Tao, Y., Liu, G., Chen, C., Luo, R., Xia, H., et al. (2009). A transcriptomic analysis of superhybrid rice *LYP9* and its parents. *Proceedings of the National Academy of Sciences of the United States of America, 106*, 7695–7701.

Wendte, J. M., & Pikaard, C. S. (2017). The RNAs of RNA-directed DNA methylation. *Biochimica et Biophysica Acta (BBA) – Gene Regulatory Mechanisms, 1860*, 140–148.

Williams, W. (1959). Heterosis and the genetics of complex characters. *Nature, 184*, 527–530.

Xu, Y. Z., Arrieta-Montiel, M. P., Virdi, K. S., de Paula, W. B. M., Widhalm, J. R., Basset, G. J., et al. (2011). MutS HOMOLOG1 is a nucleoid protein that alters mitochondrial and plastid properties and plant response to high light. *The Plant Cell, 23*, 3428–3441.

Xu, Y. Z., de la Rosa Santamaria, R., Virdi, K. S., Arrieta-Montiel, M. P., Razvi, F., Li, S., et al. (2012). The chloroplast triggers developmental reprogramming when MUTS HOMOLOG1 is suppressed in plants. *Plant Physiology, 159*, 710–720.

Yang, X., Kundariya, H., Xu, Y. Z., Sandhu, A., Yu, J., Hutton, S. F., et al. (2015). MutS HOMOLOG1-derived epigenetic breeding potential in tomato. *Plant Physiology, 168*, 222–232.

Yang, L., Li, B., Zheng, X. Y., Li, J., Yang, M., Dong, X., et al. (2015). Salicylic acid biosynthesis is enhanced and contributes to increased biotrophic pathogen resistance in *Arabidopsis* hybrids. *Nature Communications, 6*, 7309.

Yang, M., Wang, X., Ren, D., Huang, H., Xu, M., He, G., et al. (2017). Genomic architecture of biomass heterosis in *Arabidopsis*. *Proceedings of the National Academy of Sciences of the United States of America, 114*, 8101–8106.

Zhai, R., Feng, Y., Wang, H., Zhan, X., Shen, X., Wu, W., et al. (2013). Transcriptome analysis of rice root heterosis by RNA-Seq. *BMC Genomics, 14*, 19.

Zhang, C., Lin, C., Fu, F., Zhong, X., Peng, B., Yan, H., et al. (2017). Comparative transcriptome analysis of flower heterosis in two soybean F1 hybrids by RNA-seq. *PLoS One, 12*, e0181061.

Zhang, Q., Li, Y., Xu, T., Srivastava, A. K., Wang, D., Zeng, L., et al. (2016a). The chromatin remodeler DDM1 promotes hybrid vigor by regulating salicylic acid metabolism. *Cell Discovery, 2*, 16027.

Zhang, Q., Wang, D., Lang, Z., He, L., Yang, L., Zeng, L., et al. (2016b). Methylation interactions in *Arabidopsis* hybrids require RNA-directed DNA methylation and are influenced by genetic variation. *Proceedings of the National Academy of Sciences of the United States of America, 113*, E4248–E4256.

Zhao, Y., Li, Z., Liu, G., Jiang, Y., Maurer, H. P., Würschum, T., et al. (2015). Genome-based establishment of a high-yielding heterotic pattern for hybrid wheat breeding. *Proceedings of the National Academy of Sciences of the United States of America, 112*, 15624–15629.

Zhen, G., Qin, P., Liu, K. Y., Nie, D. Y., Yang, Y. Z., Deng, X. W., et al. (2017). Genome-wide dissection of heterosis for yield traits in two-line hybrid rice populations. *Scientific Reports, 7*, 7635.

Zhu, A., Greaves, I. K., Dennis, E. S., & Peacock, W. J. (2017). Genome-wide analyses of four major histone modifications in Arabidopsis hybrids at the germinating seed stage. *BMC Genomics, 18*, 137.

Zhu, A., Greaves, I. K., Liu, P. C., Wu, L., Dennis, E. S., & Peacock, W. J. (2016). Early changes of gene activity in developing seedlings of Arabidopsis hybrids relative to parents may contribute to hybrid vigour. *The Plant Journal, 88*, 597–607.

CHAPTER NINE

The Control of Bud Break and Flowering Time in Plants: Contribution of Epigenetic Mechanisms and Consequences in Agriculture and Breeding

Amanda Malvessi Cattani[*,a], Tiago Sartor[*,a], Vítor da Silveira Falavigna[§], Diogo Denardi Porto[¶], Carolina Pereira Silveira[||], Paulo Ricardo Dias de Oliveira[||] and Luís Fernando Revers[*,||,1]

[*]Graduate Program in Cell and Molecular Biology, Centro de Biotecnologia, Universidade Federal do Rio Grande do Sul, Porto Alegre, Brazil
[§]AGAP, Univ. Montpellier, CIRAD, INRA, Montpellier SupAgro, Montpellier, France
[¶]Centro de Pesquisa Agropecuária do Trópico Semiárido, Empresa Brasileira de Pesquisa Agropecuária, Petrolina, Brazil
[||]Centro Nacional de Pesquisa de Uva e Vinho, Empresa Brasileira de Pesquisa Agropecuária, Bento Gonçalves, Brazil
[1]Corresponding author: E-mail: luis.revers@embrapa.br

Contents

[a] These authors contributed equally to this work.

Advances in Botanical Research, Volume 88
ISSN 0065-2296
https://doi.org/10.1016/bs.abr.2018.10.002

Abstract

In perennial plants, the release of bud dormancy, with subsequent flowering, resembles the vernalization process of *Arabidopsis thaliana* and cereals. Especially for perennial crops from temperate regions, dormancy is an important adaptive trait for both survival and growth. Exposure to sufficient chilling during winter dormancy ensures the normal phenological traits in subsequent growing periods. Here, we compile research data on mechanisms controlling the overlapping developmental processes that define dormancy induction, maintenance and release, bud burst and flowering. Recent findings highlight the relevance of genome-wide epigenetic modifications related to dormancy events, and more specifically the epigenetic regulation of *DORMANCY-ASSOCIATED MADS-box*, *FLOWERING LOCUS C* and *FLOWERING LOCUS T* genes, key integrators of vernalization effectors on flowering. The roles of plant growth regulators in controlling bud break and flowering are discussed in relation to epigenetic mechanisms. A growing body of knowledge demonstrates that epigenetic regulation plays a key role in these processes in perennial horticultural and forestry plants. We discuss the most relevant molecular and genomics research that contribute to better understanding of the dormancy process and pave the way to precise manipulation of dormancy-related horticultural traits, such as flowering time. Finally, some of the challenges for further research in bud dormancy and consequences in agriculture are discussed within the context of global climate change.

1. INTRODUCTION

Most of perennial tree species in temperate areas are cultivated in regions with well-differentiated seasons. In addition, perennials differ from annual and biennial lifestyles in their ability to turn on/off growth in response to environmental and seasonal changes. As an adaptation to long periods of unfavorable environmental conditions, including cold winters and low light availability, many perennial plants have developed a mechanism of protection called dormancy. The first definition of dormancy was proposed by Lang, Early, Martin, and Darnell (1987) as "the absence of visible growth in any plant structure containing a meristem". Later, Rohde and Bhalerao (2007) proposed another definition aiming to accurate the expression "visible growth" and the theory that dormancy is not only absence of growth, but also include the inability to resume it. Dormancy was therefore redefined as "the inability to initiate growth from meristems (and other organs and cells with the capacity to resume growth) under favorable conditions" (Rohde & Bhalerao, 2007). This definition mainly describes the meristem states including axillary dormant buds with correlative inhibition or apical dominance, but does not include physical dormancy or ecodormancy (Rohde & Bhalerao, 2007), one of the three physiological stages of bud dormancy defined by Lang et al. (1987). Indeed, considering ecodormancy, growth inhibition is determined by adverse external stimuli and once environmental conditions become favorable, the meristematic activity restarts. The two other dormancy stages defined by Lang et al. (1987) are paradormancy and endodormancy. In paradormancy, distal organs signalize growth suppression (apical dominance), whereas in endodormacy, plant internal signals are determinant for growth arrest. Endodormant buds are not capable of resuming growth even when exposed to growth-promoting conditions. Visible changes can be observed in plants during dormancy establishment, maintenance and release. These include the cessation of apical growth, bud development, acquisition of cold/desiccation tolerance, and leaf senescence, as observed in deciduous species (Cooke, Eriksson, & Junttila, 2012). Initially, leaf primordia are modified to form hard scales during bud set (Horvath, 2010). When plants start resuming growth, bud burst can be observed as a swelling (green tip), followed by emission of young leaves and could be related to B and C stages of Fleckinger phenological scale (EPPO, 1984).

Within this context, buds are defined as "the primary shoot-producing meristematic organs for dicotyledonous plants" (Horvath, 2010) and are largely responsible for ensuring a new reproductive cycle in perennial species. Complex and orchestrated mechanisms that involve environmental stimulus, molecular pathways, hormonal signaling, and epigenetic modifications are responsible for both bud set and bud break. As a rule of thumb, photoperiod and temperature are the major signals that stimulate bud set, although their relative importance depends on the species (Cooke et al., 2012). In species of *Betula* (birch), *Populus* (aspen) and *Salix* (willow tree), photoperiod is the major stimulus (Howe, Gardner, Hackett, & Furnier, 1996; Junttila, 1976), whereas for *Malus* (apple), *Pyrus* (pear) and *Sorbus* (mountain ash) species, low temperatures are more determinant (Heide, 2011;

Heide & Prestrud, 2005). Dormancy is a cyclic and dynamic process and in some cases, the same stimuli that induce bud set are essential for dormancy release and bud break. In *Malus* species, for example, exposure to low temperatures during the autumn induces bud set, and continuous chilling exposure over the winter triggers the release from dormancy, indicating that the meristem responds to temperature changes during its endogenous development. The molecular mechanisms that control dormancy establishment and release are still far from a complete elucidation, although different models have been proposed (Campoy, Ruiz, & Egea, 2011; Horvath, 2009; Rinne et al., 2011).

Epigenetic modifications play a crucial role during dormancy regulation, from bud set to bud break, and are generally divided into active or repressive modifications. Active modifications include trimethylation of histone H3 lysine 4 (H3K4me3), di- or trimethylation of histone H3 lysine 36 (H3K36me2/me3), and acetylation of histone H3 lysine 9 (H3K9ac). Repressive modifications include histone deacetylation, trimethylation of histone H3 lysine 27 (H3K27me3), methylation of histone H3 lysine 9, and depending on its position relative to a gene, DNA methylation (see chapter 2 of this book). Nucleosome remodeling may also act to activate or repress gene expression (Guo et al., 2015; He, 2012).

In this chapter, we will discuss the already known regulatory mechanisms that regulate dormancy, bud break and flowering, focusing on the epigenetics events that could directly or indirectly influence these pathways. Potential biotechnological applications for breeding purposes are also presented.

2. CONTROL OF DORMANCY AND FLOWERING TIME

2.1 Chilling Requirement

Although the molecular mechanisms responsible for cold perception and rest completion in dormant buds are far from being fully understood, empirical knowledge about the effects of cold exposure on bud phenology is being employed successfully since a long time in dormancy research as well as for fruit production. Early conceptual models that are still in use today postulate that individual buds somehow record the amount of time of exposure to temperatures typically in the range from 0 to 7.2°C (Chandler, Kimball, Philp, Tufts, & Weldon, 1937; Lamb, 1948). These first models already provided a good correlation with bud phenology and proved to be useful for researchers and growers. Temperature records from meteorological stations, often in hourly intervals, could be used to estimate dormancy status in tree orchards, and the concept of chilling hours (hours of exposure to temperatures that contribute to dormancy progression) became an important tool for bud dormancy research in trees.

Further analysis of the interaction between climate and bud phenology showed that under slightly warmer temperatures, bud dormancy still progresses, although not in the same pace as with lower temperatures. Towards warmer environments (nearly 15°C), bud dormancy remain unchanged, and when temperature is close to 20°C, the effect of cold is negated: the record of cold exposure is progressively erased and further exposure to chilling is necessary to compensate the period of time under warm temperatures (Erez, Couvillon, & Hendershott, 1979; Young, 1992). These discoveries were implemented in new bud dormancy completion models in which, instead of chilling hours, the exposure to cold temperatures is counted as chilling units, whereas the duration under warmer temperature is either considered of no effect or subtracted from the chilling unit count (Richardson, Seeley, & Walker, 1974; Shaltout, Unrath, & Akademiya, 1983). The exact ranges of temperatures are specific to each model.

All the aforementioned models based on ambient temperature data in temperate regions provide a good prediction of bud phenology. However, when applied to sites with warm winters, the predictive power of these models is not accurate enough. With that in mind, Erez, Fishman, Linsley-Noakes, and Allan (1990) developed a more elaborate conceptual model based on a two-step process for the memory of cold exposure. First, chilling (higher efficiencies occurring between 6°C and 8°C) induces the accumulation of a precursor, which is thermally labile. Therefore, this first reaction is reversible after sufficient time of exposure to higher temperatures. Second, when the precursor accumulates to a threshold value, a second reaction takes place that is not reversible. It produces, in conceptual terms, a stable factor that signals a fixed amount of cold exposure. This process is repeated several times during winter, until the number of dormancy-releasing factors (chilling units) reaches a critical level. The performance of this so called Dynamic Model has been shown to be superior to previous models in warm winter regions (Dennis, 2003; Erez et al., 1990).

The total amount of chilling hours (or chilling units) required for bud endodormancy release is referred to as chilling requirement (CR), and is genetically controlled (Hauagge & Cummins, 1991; Labuschagné, Louw, Schmidt, & Sadie, 2002; Ruiz, Campoy, & Egea, 2007). CR is variable across cultivars of commercial tree species, which is highly valuable for breeding, allowing crops to achieve high yields in diverse climates.

2.2 QTLs for Bud Dormancy

The heritable nature of CR has been extensively used to map loci significantly contributing to this trait (Allard et al., 2016; Celton et al., 2011; Frewen et al., 2000; Graham et al., 2009; Urrestarazu et al., 2017; van Dyk, Soeker, Labuschagne, & Rees, 2010). Bud dormancy, at first glance, appears as a set of discrete phenotypes (*e.g.* endodormant or ecodormant buds) although in genetic research, it is considered as a quantitative trait. The phenotypic data used as input to quantitative trait loci (QTL) mapping is typically the time period, measured in days, from a reference date until half of all buds of each individual burst during the growing season. The reference date used in most cases is the time of bud flush of the early blooming parental genotype, and is directly related to the CR of each sibling.

One of the best characterized loci controlling bud dormancy in trees, however, was not mapped as a quantitative trait, but rather as a categorical trait with two discrete phenotypes: setting or not setting buds before winter. The peach genotype named "evergrowing" is unable to set cold-hardened buds, and the analysis of a population derived from a cross between *evergrowing* and a high CR peach cultivar allowed the identification of the *EVG* locus located in peach chromosome 1 as responsible for the trait (Wang, Georgi, Reighard, Scorza, & Abbott, 2002). Sequencing and annotation of the *EVG* locus revealed the presence of six MADS-box genes, whose sequences are similar to the *A. thaliana SHORT VEGETATIVE PHASE* (*SVP*) gene, which acts as a transcriptional repressor in distinct flowering and hormonal-signaling pathways (see Section 4, Gregis et al., 2013). These genes were named *DORMANCY-ASSOCIATED MADS-box* (*DAM*) genes 1 to 6 (Bielenberg et al., 2008) and their expression patterns were later shown to correlate with several phases of the dormancy-cycle, strongly suggesting them as important players for dormancy regulation in peach (Li, Reighard, Abbott, & Bielenberg, 2009). *DAM* genes were further identified in other perennials such as leafy spurge (Horvath, Sung, Kim, Chao, & Anderson, 2010), apricot (Sasaki et al., 2011; Yamane, Kashiwa, Ooka, Tao, & Yonemori, 2008), pear (Ubi et al., 2010) and apple (Falavigna et al., 2014; Mimida et al., 2015; Porto et al., 2016).

QTL mapping of CR in peach also pointed to the *EVG* locus as contributing to bud dormancy variation (Fan et al., 2010). Fine mapping of the

EVG locus, combined to DNA sequencing of extreme phenotypes from segregating populations, is in agreement with previous data showing that two peach *DAM* genes, *PpeDAM5* and *PpeDAM6*, are playing major roles during bud dormancy (Zhebentyayeva et al., 2014). Furthermore, sequence data showed a tight association between large insertions in *PpeDAM5* and *PpeDAM6* first introns and low CR (Zhebentyayeva et al., 2014).

Besides peach, QTLs for bud dormancy in regions syntenic to the *EVG* locus were found in other *Prunus* species (Castède et al., 2015; Olukolu et al., 2009; Sánchez-Pérez, Dicenta, & Martínez-Gómez, 2012). However, no *DAM* homologs could be found in major QTLs for CR in apple (Celton et al., 2011; Urrestarazu et al., 2017; van Dyk et al., 2010), pear (Gabay et al., 2017) or poplar (Rohde et al., 2011), although minor QTLs were identified coinciding with the position of *DAM* genes in apple and pear (Allard et al., 2016; Gabay et al., 2017). Possibly, *DAM* genes are more conserved in these species, or the role played by the *DAM* homologs in apple, pear and poplar is not the same as in *Prunus* species (Romeu et al., 2014).

2.3 Epigenetic Modifications of *FLC* x Dynamic Model of CR

Similarly to the variation in CR of tree crop cultivars, the model plant *A. thaliana* show natural variation in flowering time (Bloomer & Dean, 2017; Coustham et al., 2012). Interestingly, up to 70% of this variation is due to polymorphisms of only two genes, *FRIGIDA* (*FRI*) and *FLOWERING LOCUS C* (*FLC*) (Bloomer & Dean, 2017). The role of FLC in the regulation of flowering time in *A. thaliana* is very well demonstrated in the literature. FLC is a floral repressor and is highly expressed in most tissues during the vegetative phase (Choi, Hyun, & Kang, 2009). During winter, if the plant is exposed to sufficient cold, *FLC* expression is permanently silenced, unlocking downstream pathways of floral induction.

In an elegant approach, Angel, Song, Dean, and Howard (2011) proposed that the dynamics of *FLC* silencing by epigenetic modifications is the main mechanism underlying the memory of cold exposure. During winter, the cold-induced plant-homeodomain (PHD) protein VERNALIZATION INSENSITIVE 3 (VIN3) accumulates in a region spanning the first exon and a portion of the first intron of the *FLC* gene, named nucleation region. VIN3 recruits POLYCOMB REPRESSIVE COMPLEX 2 (PRC2) proteins, which modify the chromatin in the nucleation region by adding the H3K27me3 mark (trimethylation of Lysine 27 of Histone H3), a repressive mark stable over cell cycles (Feng, Jacobsen, & Reik, 2010). After returning to warmer temperatures, the remainder of the *FLC* chromatin is silenced by the same mechanism under a positive feedback process (Angel et al., 2011). The model predicts that *FLC* gene expression in each individual cell is bistable, that is, can either be activated or silenced. The quantitative signal transduction is achieved when in a population of cells, each cell respond autonomously to cold exposure. Later on, cold-induced *FLC* antisense transcripts (*COOLAIR*) were demonstrated to participate in the silencing process during the first weeks of chilling, but the overall model has been shown to be accurate (Fig. 1; Bloomer & Dean, 2017).

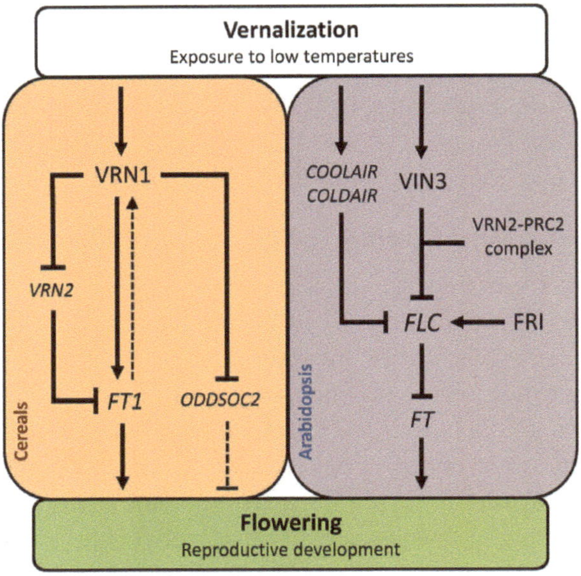

Figure 1 ***Vernalization pathways in Arabidopsis and cereals.*** Vernalization is the process that renders a plant competent to flower after a period of exposure to chilling temperatures. In cereals, such as wheat and barley, cold exposure triggers the accumulation of VRN1 protein, which directly binds to the flower-inducing *FT1* promoter and induces its expression. FT1 might promote the expression of *VRN1* in a feedback loop regulation. In parallel, VRN1 also binds to the promoter and represses two flowering repressors, *VRN2* (a repressor of *FT1* in cereals) and *ODDSOC2* (Andrés & Coupland, 2012; Deng et al., 2015). In *Arabidopsis*, the flower inhibitor *FLC* gene is initially repressed via cold-induced transcription of the *FLC* antisense transcript *COOLAIR*. After 3 weeks of continuous chilling, a *FLC* sense transcript from intron 1 (*COLDAIR*) is also induced. VIN3, a plant homeodomain (PHD) protein, accumulates during continuous cold exposure and associates with the VRN2-containing PRC2 complex to form a PHD-PRC2 complex. This complex promotes trimethylation of histone H3 lysine 27 (H3K27me3) at the *FLC* chromatin to epigenetically repress the *FLC* gene, allowing *FT* to be expressed and induce flowering (Ietswaart, Wu, & Dean, 2012; Li & Cui, 2016; Song et al., 2012a).

A. thaliana accessions differing in flowering time were screened for variation in *FLC* sequence (Li et al., 2014). Interestingly, most variation occurred in noncoding regions of the *FLC* alleles and the predicted FLC protein was identical over all tested genes. Moreover, polymorphism in the nucleation region showed the strongest association with *FLC* silencing phenotypes (Li et al., 2014). One of the early flowering accessions, *Lov-1*, revealed a single nucleotide polymorphism (SNP) that disrupts one of two conserved binding sites for B3 transcriptional regulators (Coustham et al., 2012). This site was later found to be targeted by the transcriptional repressor VIVIPAROUS1/ABI3-LIKE (VAL1), which is necessary for the nucleation of PHD-PRC2 proteins (Qüesta, Song, Geraldo, An, & Dean, 2016).

Epigenetic regulation is regarded as a strong candidate molecular mechanism for bud dormancy progression (Ríos, Leida, Conejero, & Badenes, 2014). Conceptually, the similarities between the Dynamic model and the *FLC*-VIN3-PRC2 regulon are worth noting. The intermediate thermolabile precursor and the stable factor preconized by the model show high resemblance to the VIN3 concentration and histone modifications in the *FLC* locus, respectively. *VIN3* expression is cold-induced, but its protein concentration rapidly declines in the warm (Angel et al., 2011). After the trimethylation of histones at the *FLC* locus, the cold signal is stably and irreversibly transduced in each cell. Individual cells autonomously and stochastically responding to chilling resembles the Dynamic Model of CR, where the independant accumulation of still unknown dormancy-releasing factors is necessary in order to lead to endodormancy release.

From a conceptual point of view, the characterization of epigenetic modifications seems a very promising approach to better understand bud dormancy regulation and flowering at the molecular level. The literature on this subject, especially in recent years, has grown significantly (see next section). *DAM* genes are the main candidates for the role of *FLC* in dormant buds, mainly from studies on *Prunus* species. Although recent findings were able to identify *FLC*-like genes in fruit tree species (Kumar et al., 2017, 2016; Niu et al., 2016; Porto et al., 2016; Takeuchi et al., 2018) to this date, well established homologs of *FLC* remain to be found. On the other hand, homologs of the flowering integrator gene *FLOWERING LOCUS T* (*FT*) were already identified in several tree genomes (Kotoda et al., 2010). An overview of the *FT* role in flowering-time regulation is present on Section 4 of this chapter. Interestingly in poplar, two *FT* paralogs, *FT1* and *FT2*, control dormancy transitions (Böhlenius et al., 2006; Hsu et al., 2011). The first is induced by chilling and signals dormancy release, while the second one promotes vegetative growth and inhibition of bud set during summer (Hsu et al., 2011). The identification of upstream regulators of each *FT* ortholog can provide valuable information over the control of bud dormancy.

3. EPIGENETIC AS A KEY MECHANISM FOR DORMANCY REGULATION

3.1 Chromatin Modifications during Dormancy and Bud Break

Although bud dormancy is a well-characterized process at the physiological level, the molecular mechanisms controlling this process are only recently starting to be unveiled. Indeed, as demonstrated in the previous section, it is exactly the similarities to other better established phenomena in model plant species such as seed dormancy and flowering that guided the studies over bud dormancy control. As expected, recent studies have provided evidences that bud dormancy is under complex regulatory pathways including, hormonal control, the overlap of cold and photoperiod pathways, miRNAs regulation, and several epigenetic mechanisms. At present, a huge focus is being given to the contribution of epigenetics during plant developmental processes (reviewed in Banerjee, Wani, & Roychoudhury, 2017; Gallusci et al., 2017; Richards et al., 2017), with a special attention to bud dormancy (Ríos et al., 2014). The first clue suggesting that dormancy is epigenetically regulated comes from the study of the major role of cold in this process. Both induction and overcoming of the trait are directly linked to cold exposure, in a way similar to the "vernalization memory" described in *A. thaliana* that involves epigenetic mechanisms (Banerjee et al., 2017). In this context, histone modifications, changes in DNA methylation patterns, and the regulatory role of small non-coding RNAs (sncRNAs) have already been observed during dormancy in several perennial species.

For example, early studies in chestnut using suppression subtractive hybridization were able to identify genes involved in phosphorylation of histone H3 (H3 kinase *CsaAUR3*) during bud burst, and in H2B monoubiquitination (histone mono-ubiquitinase *CsaHUB2*) and acetylation of histone H3 serine (histone acetyltransferase *CsaGCN5L*) in dormant apical buds (Santamaría, Rodríguez, Cañal, & Toorop, 2011). The same group already reported that DNA methylation levels were higher in dormant apical chestnut buds, while H4 histone acetylation levels were higher in non-dormant ones. Interestingly, the same methylation pattern was not observed in axillary buds, confirming that the DNA methylation levels observed in apical buds is a seasonal trend (Santamaría et al., 2009). In a study using full-genome microarray analysis, the regulation of genes encoding components of the DNA methylation and chromatin remodeling machineries were investigated at distinct stages of the active–dormancy cycle in hybrid aspen (Karlberg et al., 2010). It was demonstrated that short-day and low temperature treatments induced the expression of several histone deacetylases (*HDA14, HDA08, HDA9, SIN3*), a histone lysine methyltransferase (*SUVR3*) and histone ubiquitination (*HUB2*) genes. Simultaneously, a *DEMETER*-like gene, which encodes a DNA glycosylase, was downregulated by short-day treatment, possibly leading to a further increase in DNA methylation and chromatin compaction. On the other hand, long-term low temperature exposure induced a *DML* (*DEMETER*-related) gene, probably leading to demethylation and activation of genes that contribute to dormancy release.

The first evidence demonstrating that changes in cytosine methylation patterns may play a regulatory role in the control of dormancy release was obtained from MSAP (Methylation Sensitive Amplified Polymorphism) and RNA-seq analyses in apple. This study also showed that chilling availability appears to modulate the expression of genes involved in active DNA methylation and demethylation, with a progressive decrease in DNA methylation levels being observed during dormancy release. Using the same approach, high expression levels of the histone acetyltransferase *HCA1* and the histone deacetylases *HDA14* and *HDA19* where identified towards the initiation of active growth, while downregulation of other histone deacetylases like *HDA06* and *HDA08* was observed in the same time points (Kumar et al., 2017). Another study showed that dormancy release triggered the induction of two types of histones, the canonical H2A and an H2A.Z histone variant (Falavigna et al., 2014). In *A. thaliana*, H2A.Z-containing nucleosomes coordinate one of the thermosensory responses (Kumar & Wigge, 2010), being usually more expressed during the S phase of the cell cycle (March-Díaz & Reyes, 2009). The expression of these genes during dormancy release suggests that the dynamic balance in their opposing activities is necessary to regulate the transcriptional levels of target genes.

In poplar, epigenetic changes seem to take place during dormancy in order to control differential gene expression. The balance of epigenetic marks related to low transcriptional activity during winter (i.e. DNA methylation and histone H4 hypoacetylation) with chromatin modifications compatible with increased gene expression during growth resumption (i.e. DNA hypomethylation and histone H4 acetylation) were observed (Conde, González-Melendi, & Allona, 2013). Further studies demonstrated that growth reactivation is preceded by DNA demethylation mediated by *DML*, thus promoting the necessary changes to apical bud physiology in order to start the bud break process (Conde et al., 2017b, 2017a). The DNA methylation in poplar is a widespread process in non-condensed chromatin and is associated to tissue-specific gene expression. Moreover, in the shoot apical meristem (SAM), this process responds to variations of water availability and affects the expression of genes involved in hormonal pathways (Lafon-Placette et al., 2013, 2018). Taken together, these findings indicate that the dynamics of genomic DNA methylation levels could be involved in the regulation of dormancy–growth cycle.

Recently, the ABA-mediated regulation of dormancy in response to photoperiod has been proposed for growth cessation (Tylewicz et al., 2018). In this model, ABA accumulation in the buds triggers plasmodesmata closure, ensuring bud growth arrest until sufficient chilling has been accumulated. The role of ABA in the induction of several epigenetic marks such as histone modifications, DNA methylation and short interfering RNA pathways was already documented for several development processes and stress responses in plants (Chinnusamy, Gong, & Zhu, 2008). In strawberry, dormancy establishment induced global DNA methylation in young leaves, and these changes were synchronized with endogenous ABA levels (Zhang et al., 2012c). However, how ABA is able to modulate epigenetic changes during the bud dormancy process is yet to be discovered.

3.2 Epigenetic Regulations of *DAM* and *EBB* Genes

Some of the most emblematic genes associated to bud dormancy regulation, the *DAM* genes, are also under strong epigenetic control. Both H3K4me3 and H3K27me3, two epigenetic marks related to activation and repression of transcription, respectively, are recurrently found over *DAM* genes in the different species that have been analyzed. The leafy spurge *EesDAM1* and peach *PpeDAM6* genes showed both an increase of the H3K27me3 and a decrease of the H3K4me3 levels in promoter regions when comparing eco-to endodormancy (Horvath et al., 2010; Leida, Conesa, Llácer, Badenes, & Ríos, 2012). In peach, the same epigenetic changes were also identified at the transcriptional start site (TSS) and in the largest intron of *PpeDAM6* (Leida et al., 2012). Interestingly, these two *DAM* genes displayed a seasonal expression profile during dormancy, with a peak of transcription during endodormancy that correlates to the levels of the aforementioned epigenetic marks. Additionally, the abundance of acetylation of histone H3 (H3ac), another epigenetic mark associated with gene expression, in the promoter of *PpeDAM6* decreased during dormancy release (Leida et al., 2012). A sequential chain of events affecting *PpeDAM6* chromatin status was identified: loss of H3K4me3 and H3ac modifications during dormancy release would contribute to gene repression, while increase of H3K27me3 at the same time point would enable long-term gene inactivation (Leida et al., 2012). These two pioneering studies have provided the first evidence that epigenetic regulation of *DAM* genes may be operating during dormancy in a manner similar to the regulation of the *Arabidopsis FLC* gene during vernalization.

Further studies in peach were aimed to identify putative regulations of the *DAM* locus, which is responsible for the *EVG* phenotype (Bielenberg et al., 2008). Genes involved in chromatin remodeling were identified in QTLs for nine traits related to bud dormancy, flowering and fruit harvest, emphasizing a prominent role of chromatin modification pathways in this process (Romeu et al., 2014). Thereafter, de la Fuente, Conesa, Lloret, Badenes, and Ríos (2015) identified a significant enrichment of H3K27me3 during dormancy release concomitantly with reduced expression at different regions of the *DAM* locus, and this may be one of the controlling mechanisms responsible for the differential transcriptional profiles observed for *PpeDAM* genes during bud dormancy (Li et al., 2009). In pear, the *DAM* gene *PpyMADS13-1* displayed a seasonal expression pattern that resembles the one identified for *EesDAM1* and *PpeDAM6* (Saito et al., 2013). Besides a reduction of H3K4me3 prior to endodormancy release, the authors identified a tendency for *PpyMADS13-1* promoter to lose the histone variant H2A.Z during dormancy progression (Saito et al., 2015). Within this context, the decrease of *PpyMADS13-1* expression during dormancy progression, along with the removal of this histone variant, add a new layer of complexity to the epigenetic regulation on dormancy-related genes. Finally, recent studies demonstrated that DNA methylation and small interference RNAs (siRNAs) may also participate to the regulation of the sweet cherry *DAM* genes *PavMADS1* and *PavMADS2*. Near dormancy release, *PavMADS1* promoter presented an increase in DNA methylation as well as in the abundance of matching siRNAs (Rothkegel et al., 2017). These findings suggest that both DNA methylation and siRNAs are one of the mechanisms responsible for the seasonal expression profile of these genes during the dormancy process. Taken together, these results indicate that a complex regulatory network involving epigenetic marks is acting over the *DAM* genes during dormancy.

Another important dormancy regulator gene is called *EARLY BUD-BREAK 1* (*EBB1*), which controls the timing of bud break in poplar by playing an integrative role in the reactivation of the SAM after dormancy (Yordanov, Ma, Strauss, & Busov, 2014). To date, the molecular regulation of bud break and its epigenetic control are poorly understood. Interestingly, transcripts of *DML* gene were localized in poplar apices, similar to that reported for *EBB1* (Conde et al., 2017b; Yordanov et al., 2014). However, the direct target genes of *DML*-dependent DNA demethylation and *EBB1* are not related, indicating that they may act by separate pathways to control bud break. In pear, the orthologous *EBB1* gene *PpyEBB* was highly induced prior to bud break in flower buds, and the analysis of its chromatin status revealed high levels of active histone modifications (H3K4me3) during ecodormancy in relation to endodormancy (Anh Tuan et al., 2016). However, no changes were detected in the levels of H3K27me3 in the same time points. Interestingly, two of the most important dormancy regulators discovered so far, *DAM* and *EBB* genes, are under strong epigenetic control. Taken together, these evidences show that epigenetic mechanisms play an important role in the regulation of bud dormancy, and a better understanding of these processes may help developing a valuable resource to genetically manipulate this trait.

4. *FT*: A MASTER SWITCH CONTROLLING FLOWERING IN ANGIOSPERMS

4.1 The Florigen *FT* Integrates Signals from Different Floral Pathways

The decision to flower is one of the most important steps for plant reproductive success and survival and represents a major life cycle transition in which meristem cells must commit with reproductive development. The flowering pathways have been extensively studied in the model plant *A. thaliana* (Amasino, 2005, 2010; Andrés et al., 2014; Lee et al., 2006; Mateos et al., 2015; Michaels & Amasino, 1999; Moon, Lee, Kim, & Lee, 2005). Five major floral pathways were characterized so far and they finely tune the time of flowering via endogenous (gibberellins, autonomous, and aging), as well as environmental factors (photoperiod and vernalization)

(Blümel, Dally, & Jung, 2015; Khan, Ai, & Zhang, 2014; Lucas-Reina et al., 2016; Song, Ito, & Imaizumi, 2013).

The vernalization and the autonomous pathways control flowering via repression of the *FLC* gene. The photoperiodic pathway involves the perception of the day length by the circadian clock that induces the expression of *CONSTANS* (*CO*). In *Arabidopsis*, flowering is anticipated during long days and delayed during short days, making it a facultative long day plant. The gibberellins (GA) pathway is a hormonal control of flowering and studies have reported that GA levels increase in the leaves and meristems of *Arabidopsis* plants, inducing the expression of the *FT* gene (Andrés et al., 2014; Song et al., 2013). The aging pathway is regulated by the balance of two microRNAs, *miR156* and *miR172*, that regulate transition from juvenile to adult and reproductive phase. All of the flower signaling pathways in *Arabidopsis* are integrated by the *FT* gene, which ultimately regulates the expression of the floral meristem identity genes *LEAFY* (*LFY*) and *APETALA 1* (*AP1*). These major floral pathways are summarized in Fig. 2. Other recently discovered floral pathways include regulation of flowering via ambient temperature, stress, nutrients and sugar balance (for detailed reviews on the flowering pathways see Andrés & Coupland, 2012; Blümel et al., 2015; Cho, Yoon, & An, 2017; Khan et al., 2014; Lucas-Reina et al., 2016; Song et al., 2013).

Different plant species may present different signaling mechanisms for the same endogenous and environmental factors that trigger vegetative-to-reproductive meristem transition, meaning that there is a functional diversification in floral pathways among angiosperms (Blümel et al., 2015; Lucas-Reina et al., 2016). For example, wheat, barley, and *Arabidopsis* all have a cold (vernalization) requirement to induce flowering. However, while in *Arabidopsis* vernalization causes the repression of the flowering-inhibitor *FLC* gene, in wheat and barley vernalization induces the expression of the *VERNALIZATION 1* (*VRN1*) gene that directly binds to *FT* and promotes inflorescence formation (Fig. 1; Andrés & Coupland, 2012; Deng et al., 2015).

The hormone GA seems to affect flower formation differently in *Arabidopsis* compared to several perennial fruit trees including apple, mango, cherry, peach, apricot, almond, and lemon (Khan et al., 2014; Turnbull, Anderson, & Winston, 1996; Upreti et al., 2013; Zhang et al., 2016). While GA is a potent inducer of flowering in *Arabidopsis*, it has been reported that exogenous application of GA usually leads to a reduction of flower formation in fruit trees (Khan et al., 2014; Turnbull et al., 1996; Zhang et al., 2016).

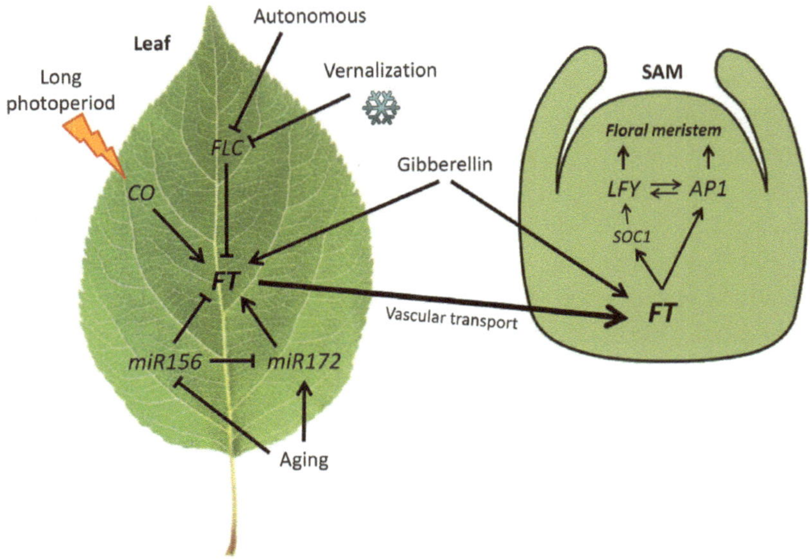

Figure 2 ***Major floral pathways in Arabidopsis.*** Different signaling pathways converge to the *FT* gene, whose protein and mRNA are translocated to the shoot apical meristem (SAM) via vascular transport (Andrés & Coupland, 2012; Jackson & Hong, 2012). The FT protein in the SAM induces the expression of *SOC1*, an activator of *LFY*. At the same time, FT protein also induces the expression of *AP1*, another floral meristem identity gene. The signal initiated by *FT* is amplified via feedforward regulation: *FT* induces *LFY* and *AP1*, and *LFY* and *AP1* continue to induce each other. This mechanism ensures that after *FT* activation, the meristem becomes committed with reproductive development and the process can no longer be reversed.

Moreover, application of paclobutrazol, an inhibitor of GA biosynthesis, was able to even increase the flowering rate of apple and promote early-flowering in mango (Upreti et al., 2013; Zhang et al., 2016). Zhang et al. (2016) have proposed a model in which GA acts by reducing cytokinin (CK) levels in apple shoots, and that a high CK/GA ratio is necessary to promote flower formation. Notwithstanding, the pathways by which GA affects flowering differently in *Arabidopsis* and in most perennials, are not fully understood. This highlights the difficulty of describing a universal model for floral signaling pathways in angiosperms and has led many review papers to focus on the flowering pathways of specific groups of plants (Deng et al., 2015; Guo et al., 2015; Hanke, Flachowsky, Peil, & Hättasch, 2007; Ramírez & Davenport, 2010; Sun et al., 2014; Weller & Ortega, 2015).

Despite all of the efforts aiming at understanding how different species perceive environmental cues and endogenous signals to promote flower formation, only scarce information concerning the molecular signaling of reproductive meristem formation is available for perennials with a winter dormancy period. This may be due to the fact that differentiation of vegetative meristems into flower organs usually occurs concomitantly with the formation of buds for the next growing cycle (known as flower buds) (Foster, Johnston, & Seleznyova, 2003; Hanke et al., 2007) and also because plants tend to resume flowering soon or immediately after bud break, which makes dormancy the main limiting factor for the cultivation of temperate plants in warmer climates.

Noteworthy, regardless of the pathway that a given plant may utilize to promote floral development, the integration of the flowering pathways seems to converge to the *FT* gene and this appears to be highly conserved among higher plants, between dicots and monocots, and between annual and perennial species (Blümel et al., 2015; Khan et al., 2014; Lucas-Reina et al., 2016; Putterill & Varkonyi-Gasic, 2016). For instance, in both rice and *Arabidopsis*, the *FT* gene is expressed in leaves and the protein and mRNA are translocated to the apical meristem, acting as a strong promoter of flowering (Amasino & Michaels, 2010; Andrés & Coupland, 2012; Jackson & Hong, 2012; Sun et al., 2014). Similarly, studies involving woody perennial dicots and the annual herbaceous plant *Arabidopsis* revealed the conserved role of *FT* genes on flowering (Kotoda et al., 2010; Tränkner et al., 2010). More specifically, the constitutive expression of the apple *MdoFT1* gene was able to induce flowering of *in vitro* shoot cultures of apple and *Populus* (Kotoda et al., 2010; Tränkner et al., 2010). Ectopic expression of the apple *MdoFT1* or *MdoFT2* genes in the annual plant *Arabidopsis* was also able to induce an early-flowering phenotype under long days in wild-type background plants (Kotoda et al., 2010; Tränkner et al., 2010). Besides, several other reports involving the homologous and heterologous expression of *FT* and *FT*-like genes from different plants show that *FT* is able to promote flowering regardless of the species (Blümel et al., 2015; Putterill & Varkonyi-Gasic, 2016; Wickland & Hanzawa, 2015). Thus, the *FT* signaling network appears to constitute a highly conserved convergence hub for the flowering pathways in higher plants (Fig. 2).

4.2 Epigenetic Control of *FT* Expression

The expression of a given gene is affected by different factors, including the genetic information that is contained in the DNA sequence (*e.g. cis*-elements in a promoter), and by epigenetic modifications in the chromatin that can, in some cases, be transmitted to progeny (*e.g.* histone modifications and DNA methylation; see chapters 1 and 2). Like many other genes, *FT* chromatin is also subjected to epigenetic modifications that participate to the control of gene expression, thereby determining flowering time, as illustrated below.

As part of the ambient temperature flowering pathway, *Arabidopsis* plants growing under long days in temperatures of 27°C show accelerated flowering compared to plants growing at 23°C. This has been attributed to PHYTOCHROME INTERACTING FACTOR 4 (PIF4) binding to the *FT* promoter under high temperatures (27°C). At 23°C, binding of PIF4 to the *FT* promoter is hampered by the occupancy of histone H2A variant H2A.Z at the *FT* locus. Exposure of plants to high temperatures of 27°C triggers the eviction of H2A.Z nucleosomes occupying the *FT* locus. The rate of histone H2A.Z occupancy in the chromatin is one of the factors that mediate temperature responses in *Arabidopsis* and its presence on the *FT* locus decreases accordingly under high temperatures, allowing the binding of PIF4 to the *FT* promoter and flowering (Fig. 3; Lucas-Reina et al., 2016; Song et al., 2013).

The protein AT-HOOK MOTIF NUCLEAR LOCALIZED 22 (AHL22) was described as a repressor of flowering. Overexpression of *AHL22* causes a late-flowering phenotype by reducing the levels of *FT* mRNA. AHL22 binds to AT-rich DNA sequences within the *FT* locus where it recruits histone deacetylases (HDACs), HDA1/HDA19, HDA6, and HDA9. These HDACs remove acetyl groups from the N-terminal tail of histones H3, thus promoting the silencing of the *FT* gene. Evidence also suggests that AHL22 mediates dimethylation of histone H3K9 by recruiting specific histone methyltransferases, as suggested by the increase of H3K9me2 mark abundance in lines overexpressing *AHL22* (Fig. 3; Yun, Kim, Jung, Seo, & Park, 2012).

In *Arabidopsis*, it has been shown that activation of the *FT* gene by CO in the photoperiodic pathway is dependent on the direct interaction between MORF RELATED GENE (MRG) group proteins, MRG1 and MRG2, with the CO protein. The proteins MRG1 and MRG2 recognize H3K4me3 and H3K36me3 histone marks on the *FT* chromatin and allow the binding of CO to the *FT* promoter (Fig. 3). The *mrg1 mrg2* double mutant has reduced *FT* mRNA levels and a late-flowering phenotype associated with impaired binding of CO to the *FT* promoter under long photoperiods (Bu et al., 2014).

Presence of H3K4me3 close by the TSS is generally associated with active transcription (Guo et al., 2015; He, 2012). Absence of two redundant H3K4me3 demethylases, *Jumonji 4* (*AtJmj4*) and *EARLY FLOWERING 6*

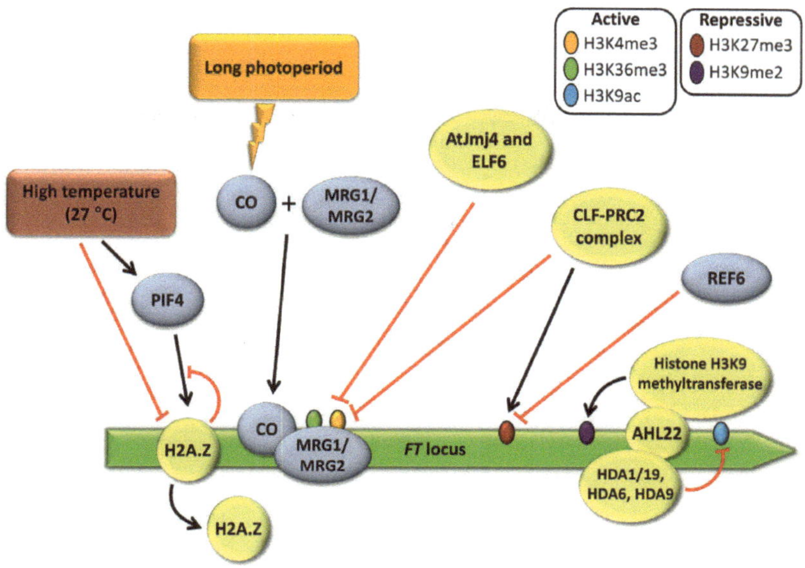

Figure 3 ***Epigenetic mechanisms controlling*** **FT** ***expression in Arabidopsis.*** The binding site of PIF4 to the *FT* promoter is occupied by the histone variant H2A.Z. Exposure to high temperatures causes the eviction of H2A.Z from the *FT* locus, allowing the binding of PIF4. MRG1 and MRG2 recognize methylated H3K4 and H3K36 histone marks, allowing the binding of CO to the *FT* promoter. The binding of CO can be counteracted by the activity of H3K4me3 demethylases (AtJmj4 and ELF6). The CLF-PRC2 complex silences *FT* via trimethylation of histone H3K27 and via demethylation of H3K4me3 marks. REF6 removes H3K27me3 marks, acting against the CLF-PRC2 complex to allow *FT* transcription. AHL22 represses *FT* by recruiting histone deacetylases (HDA1/19, HDA6, and HDA9) and histone H3K9 methyltransferases. Active and repressive histone modifications are indicated in the upper right, respectively. Proteins acting to activate *FT* transcription are depicted in blue, whereas proteins whose function is to silence *FT* expression are pictured in yellow.

(*ELF6*), causes an early-flowering phenotype in *Arabidopsis* under both short and long days (Fig. 3; Jeong et al., 2009). In another study, CURLY LEAF (CLF), an Enhancer of zeste protein constitutive of PRC2, was shown to be a strong repressor of *FT* by mediating the trimethylation of H3K27 and by partially suppressing H3K4me3 marks into the *FT* chromatin (Jiang, Wang, Wang, & He, 2008). In contrast, RELATIVE OF EARLY FLOWERING 6 (REF6) was shown to be an important histone H3K27 demethylase promoting *FT* expression, acting dynamically against the CLF-PRC2 complex (Fig. 3; He, 2012).

Interestingly, it has been demonstrated that deposition of active and repressive histone modifications occur independently of each other and are mutually antagonistic (Jiang et al., 2008; Shafiq, Berr, & Shen, 2014). In other words, activation or repression of *FT* expression in *Arabidopsis* relies on the relative abundance of active or repressive epigenetic marks. The implications of this finding is that, in order to fully understand how plants control the time of flowering, it may be necessary to look further into the mechanisms that regulate the balance of active and repressive epigenetic marks on the *FT* chromatin.

In rice, chromatin methylation also appears to play an important role in the expression pattern of two *FT* homologs, *Heading date 3a* (*Hd3a*) and *RICE FLOWERING LOCUS T 1* (*RFT1*). Knockdown plants with reduced synthesis of S-Adenosyl-l-methionine, a universal methyl group donor, show low levels of histone H3K4me3 and DNA methylation at the *Hd3a* and *RFT1* loci, which was associated with reduced expression of these genes and late-flowering phenotype (Sun et al., 2014). Two genes that are responsible for histone H3K36 di- and tri-methylation, *SET DOMAIN GENE 724* (*SDG724*) and *SDG725*, were also shown to be important for flowering time regulation in rice. A loss-of-function mutant for *SDG724* and a knockdown plant for *SDG725* were associated with a reduction of H3K36 methylation and transcription of *RFT1* and *Hd3a*, and of other flower-related genes (Sun et al., 2014). The presence of H3K9ac around the TSS of *RFT1* was also correlated with active gene transcription and flowering in rice (Sun et al., 2014).

As mentioned above, little information regarding the flowering pathways of perennial plants is available, and even less or no information at all on the epigenetic regulation of *FT* genes of perennials might exist. Nevertheless, a study with sexual dimorphism in poplar has shown that DNA methylation is an important regulatory factor for flower development in *Populus tomentosa* (Song et al., 2012b). In another work using the early-flowering trifoliate orange *Poncirus trifoliate*, the authors treated plants with 5-azacytidine, a DNA methyltransferase inhibitor, and observed several abnormalities related to vegetative development, without noticeable changes in the flowering time of the already precocious trifoliate orange (Zhang, Mei, Liu, Khan, & Hu, 2014). Notwithstanding, it was observed an increase in *CiFT* transcript accumulation with increasing concentrations of 5-azacytidine. The expression level of other flowering genes *CiLFY*, *CiAP1*, *TERMINAL FLOWER 1* (*CiTFL1*), and *CiFLC* was highest at 250 μM of 5-azacytidine and then decreased at higher concentrations (Zhang et al., 2014). Because a decrease in the overall DNA methylation pattern led to an increase in the expression of five major floral genes, the authors concluded that DNA methylation plays an important role in the regulation of the early-flowering trait of trifoliate orange (Zhang et al., 2014).

The knowledge regarding how plants flower is of utmost importance for breeding programs, especially for perennial fruit trees, in which a long juvenile phase represents a significant hindrance. Activation of *FT* and *FT*-like genes through homologous or heterologous expression in juvenile perennials is sufficient to accelerate entrance into the reproductive phase. *FT* is regarded as the master switch that must be turned on by angiosperm plants to initiate flower formation and complete their reproductive cycle. Epigenetics is a rapidly growing field and we address here that a direct link between epigenetics and the *FT* gene is essential to flowering. As a future perspective, we believe that understanding the genetic mechanisms underlying the epigenetic control of *FT* expression (and how this is linked to repression of *FT* expression in juvenile perennials) could help breeding programs seek important characteristics to shorten the juvenile phase in perennial fruit trees.

5. CHROMATIN REMODELING AND HORMONAL STIMULUS: AN INHERENT PART OF THE FLOWERING PROCESS

The regulation of flowering time depends on many factors including environmental, genetic, and epigenetic mechanisms as detailed in the previous sections. Hormonal regulation plays a central role in this process, integrating internal and external stimulus. In this section, we will explore the importance of well-known phytohormones in the flowering process, connecting them with epigenetic modifications that occur in the chromatin during floral transition (Campos-Rivero et al., 2017; Yamamuro, Zhu, & Yang, 2016).

5.1 Gibberellins

In the model plant *A. thaliana*, the timing of flowering mediated by GA signaling has been extensively studied, leading to the elucidation of some of the underlying molecular mechanisms. The induction of flowering starts with the perception of bioactive GAs and is mediated, among other factors, by nuclear proteins called DELLA. Briefly, DELLA are negative regulators of GA signaling that sequestrate transcriptional activators, inhibiting flowering (Achard & Genschik, 2009; Harberd, 2003). In leaves, the transcriptional activators of *FT*, like CO and PIF4 are sequestered by

DELLA proteins, preventing them from binding to DNA and activating *FT* transcription (Conti, 2017; de Lucas et al., 2008; Tiwari et al., 2010; Wang et al., 2016; Xu et al., 2016). In the SAM, DELLAs bind to SQUAMOSA PROMOTER BINDING PROTEIN-LIKE (SPL) transcriptional regulators and prevent their function as transactivators on target genes (Conti, 2017; Hyun et al., 2016; Yu et al., 2012). When GA is present, the GID1 receptor undergoes conformational changes, increasing its affinity for DELLA proteins (Griffiths et al., 2006; Willige et al., 2007). This interaction (DELLA-GID1) favors the binding of the E3 Ubiquitin ligase SLEEPY1 (SLY1) that marks DELLA for degradation by proteasome 26 (Dill, Thomas, Hu, Camille, & Steber, 2004). Without DELLA proteins, the activators CO and PIF4 are free to promote *FT* expression. In the SAM, SPLs are able to activate *LFY, FRUITFUL* (*FUL*), *SUPPRESSOR OF OVEREXPRESSION OF CONSTANS 1* (*SOC1*) and *AP1* by directly binding to their promoter regions. Expression of these genes contribute to the amplification of the FT-FD signal and causes the activation of the floral meristem identity genes (Abe et al., 2005; Conti, 2017; Melzer et al., 2008; Moon et al., 2003). In contrast to *Arabidopsis*, application of exogenous GAs were shown to inhibit flower formation in some perennial crops like apple (Marcelle & Sironval, 1963; Zhang et al., 2016), peach (Southwick & Glozer, 2000) and citrus (Goldberg-Moeller et al., 2013; Guardiola, Monerri, & Agusti, 1982; Tan & Swain, 2006), as previously described in this chapter. Some works reporting the molecular link between chromatin modifications and GA responses have been already published and will be now more thoroughly discussed.

PKL genes code for an CHD3-family ATP-dependent chromatin remodeling factor, that promote the trimethylation of H3K27, contributing for tissue-specific gene repression in plants (Ogas, Kaufmann, Henderson, & Somerville, 1999; Zhang et al., 2008, 2012b). Interestingly, analyses of *pkl* mutant plants showed a defective ability to respond to GA, demonstrating the important role of methylation in the induction of GA transcriptional network (Rider et al., 2003; Zhang et al., 2008). A recent study showed that PKL is required to promote flowering in Arabidopsis, probably due to its close interaction with DELLA proteins (Park et al., 2017). Another interesting point is that histone modifications are able to regulate genes involved in GA biosynthesis. Some examples as *GA REQUIRING1* (*GA1*), *GA2, GA3, ENT-KAURENOIC ACID HYDROXYLASE1* (*KAO1*) and *KAO2* have already been reported to undergo H3K27ac modification (Charron, He, Elling, & Deng, 2009).

Dioxygenase genes (*GA2ox7* and *GA3ox2*) are the major targets for light regulation in the GA metabolism (Kamiya & García-Martínez, 1999) and showed to be downregulated by H3K27me3. The higher transcript accumulation of *GA2ox7* and *GA3ox2* correlates with the reduction of H3K27me3 abundance over the promoter regions and first exons of dioxygenase genes, removing the repressive effect of this modification (Charron et al., 2009). On the other hand, genes that are not affected by the GA signaling have no differences in H3K27me3 levels (Charron et al., 2009). These results suggest an import role of epigenetics modifications in GA biosynthesis, signaling and response, influencing flowering time.

5.2 Jasmonic Acid

Jasmonic acid (JA) is a fatty acid-derived molecule that orchestrates different responses to abiotic and biotic stress in plants (Browse, 2009; Stintzi & Browse, 2000) and plays important functions in the regulation of floral development in *Arabidopsis* (Park et al., 2002; Stintzi & Browse, 2000), rice (Cai et al., 2014; Xiao et al., 2014), maize (Acosta et al., 2009; Yan et al., 2012), tomato (Li et al., 2004) and tobacco (Stitz, Hartl, Baldwin, & Gaquerel, 2014). The currently accepted model proposes that, in the absence of JA, repressors of JASMONATE-ZIM domain (JAZ) family interact with the transcriptional factors MYB21 and MYB24, preventing them to activate responsive JA genes. In contrast, the JA signal makes JAZ proteins a potential target for the F-box protein CORONATINE INSENSITIVE PROTEIN 1 (COI1) to ubiquitination and subsequent degradation (Chini et al., 2007). As soon as JAZ is degraded, these factors are released and can activate genes involved in anther development and filament elongation (Song et al., 2011).

Studies developed in *Arabidopsis* have now revealed the importance of HDA6, a RPD3-type histone deacetylase, in JA response, senescence and flowering (Wu, Zhang, Zhou, Yu, & Chaikam, 2008). The interaction of HDA6 with COI1, has been demonstrated (Devoto et al., 2002), suggesting an important connection between JA response and histone modifications. The analysis of *axe1-5*, a splice site mutant of *AtHDA6* (Murfett, Wang, Hagen, & Guilfoyle, 2001) and HDA6-RNAi plants have showed that the JA responsive genes, *PDF1.2*, *VSP2*, *JIN1*, and *ERF1* were downregulated in these plants, reinforcing the idea that HDA6 is necessary to the JA-responsive pathway (Wu et al., 2008). Furthermore, *axe1-5* and HDA6-RNAi plants displayed increased leaf longevity and late flowering compared to wild type plants. These plants also showed an hyperacetylation of histones at the *FLC* gene and an increase in its expression, suggesting that HDA6 is required for *FLC* deacetylation and repression (Wu et al., 2008). Little is known about epigenetics modification and JA response in perennial plants, revealing a huge gap that remains to be investigated.

5.3 Brassinosteroids

Brassinosteroids (BR) are a class of polyhydroxylated steroidal phytohormones that are involved in many aspects of plant biology, including cell elongation, cell division, root growth, photo-morphogenesis, stomatal and vascular differentiation and seed germination (Gudesblat & Russinova, 2011; Wei & Li, 2016). Phenotype analyses of BRs defective mutants revealed a delayed flowering time, indicating a positive role of BRs in the establishment of flowers (Domagalska et al., 2007; Li, Li, Chen, & An, 2010). A higher level of *FLC* transcripts and increased levels of histone H3 acetylation normally associated with active chromatin at the *FLC* locus were observed in these mutants, suggesting that BRs could contribute to a silenced epigenetic state at the *FLC* promoter, helping to downregulated it and promote flowering (Domagalska et al., 2007).

The transcriptional factor BRI1-EMS suppressor 1 (BES1), one of the major regulators of BR signaling that coordinates several genes in response to BR (Yin et al., 2005), directly interacts with REF6 and its homolog, EARLY FLOWERING 6 (EFL6), both jumonji N/C (JmjN/C) domain-containing proteins essential to catalytic demethylation of histones (Agger, Christensen, Cloos, & Helin, 2008). REF6 and EFL6 (recruited by BES1) have been shown to be involved in BR signaling by affecting histone methylation in the promoters of BR-responsive genes (Noh et al., 2004; Yu et al., 2008). BES1 can also recruit a histone lysine methyltransferase called SET DOMAIN GROUP 8 (SDG8), which is implicated in H3K36 di- and trimethylation (Wang et al., 2014; Xu et al., 2008).

Analysis of the *rice plasticity 1* (*rpl1*) mutant that is impaired in a gene involved in plastic response of rice plants to environmental changes, have also demonstrated an interesting connection between BRs and histone modification (Zhang et al., 2012a). The *rpl1* mutant showed an increased in DNA methylation level and a decrease in overall histone acetylation. Responses to various plant phytohormones including BRs were also negatively affected. Noteworthy, the putative rice BR receptor *OsaBRI1*, a key hormone signaling gene, was extremely downregulated in the *rlp1* mutant concomitantly to changes in histone marks in the *OsaBRI1* gene locus (Zhang et al., 2012a), reinforcing the importance of epigenetic regulations in BR perception and signaling.

5.4 Ethylene

Ethylene (ET) is a volatile organic chemical compound involved in fruit ripening, senescence of leaves, response to stress and a floral repressor in *Arabidopsis* (Achard et al., 2006). The transcription factor ETHYLENE INSENSITIVE 3 (EIN3) and EIN3-like (EIL) mediate ET transcriptional responses (Guo & Ecker, 2003). EIN3 accumulation delays flowering by activating the *ETHYLENE RESPONSE 1* (*ERF1*) related genes. The negative role of ET in flowering through the EIN3-ERF1 is attributed to reduced bioactive GA levels, causing enhanced accumulation of DELLA proteins (Achard et al., 2007). In the work reported by Zhou, Zhang, Duan, Miki, and Wu (2005), the importance of HDA19, a RPD3-type histone deacetylase, in integrating hormonal stimulus and epigenetic regulations was analyzed in *Arabidopsis*. By overexpressing *HDA19*, an increased expression of *ERF1* was detected, showing that histone deacetylation positively regulates *ERF1*, contributing to an increased delay in flowering (Zhou et al., 2005). These observations contributed to elucidate the roles of histone modification and ET response. However much more efforts need to be done to completely understand this hormonal-epigenetics regulation.

5.5 Salicylic Acid

Salicylic acid (SA) is an important phytohormone that belongs to the group of phenolic compounds and plays distinct roles in many aspects of plant growth and development, as well as in disease resistance (Vicente & Plasencia, 2011). The importance of SA in the flowering process is already described (Cleland & Ajami, 1974; Lee & Skoog, 1965) but the knowledge about its role in the floral regulatory network is still very limited. Arabidopsis SA-deficient plants show a late-flowering phenotype and an increased expression of *FLC*. Low levels of *FT* transcripts were also observed in these mutants compared to wild type, under short day or long day conditions (Martínez, Pons, Prats, & León, 2004).

On the other hand, the *CO* and *SOC1* genes showed a different profile under the same daylength conditions in SA-deficient plants (Martínez et al., 2004). *CO* and *SOC1* expression decreased in long day conditions, while in short day-grown plants, amounts of *CO* transcripts increased and *SOC1* remained stable (Martínez et al., 2004). Given that, *CO* and *SOC1* are strictly regulated by light changes (Suárez-López et al., 2001), suggesting that SA regulates flowering by interacting with the photoperiod dependent pathway (Martínez et al., 2004; Vicente & Plasencia, 2011).

Sumoylation is a post-translational modification that conjugates small ubiquitin modifier peptides (SUMO) to protein substrates (Mahajan, Delphin, Guan, Gerace, & Melchior, 1997; Matunis, Coutavas, & Blobel, 1996). In mammals and yeasts sumoylation has been included in the list of histone modifications associated to transcriptional repression (Garcia-Dominguez & Reyes, 2009; Nathan et al., 2006; Shiio & Eisenman, 2003). In plants, it is related to biotic and abiotic stress responses, flowering and development (Chosed, Mukherjee, Lois, & Orth, 2006; Jin et al., 2008; Novatchkova, Budhiraja, Coupland, Eisenhaber, & Bachmair, 2004). Analysis of plant mutants in the *SUMO E3 ligase* (*SIZ1*) gene, which play crucial role during SUMO steps, revealed an early flowering time phenotype under short day (Jin et al., 2008). It has also showed elevated SA levels, suggesting an important connection between SUMO modification and SA response (Jin et al., 2008).The authors concluded that under low levels of SA an increased *SIZ1* expression is observed, contributing to sumoylation of *FLOWERING LOCUS D* (*FLD*) and its consequent low expression. The *FLD* gene encodes a plant homolog to an important protein in mammals histone deacetylase complexes (He, Michaels, & Amasino, 2003). Mutations in *FLD* and its consequent low expression promote hyperacetylation of histones in *FLC* chromatin, leading to its transcriptional activation and delayed flowering time (He et al., 2003). Nevertheless, when high levels of SA are perceived, *SIZ1* is repressed, which increases FLD levels and results in lower *FLC* transcript amounts, allowing floral transition (Campos-Rivero et al., 2017; Jin et al., 2008).

5.6 Cytokinins

CK are plant hormones involved, in cell proliferation, differentiation and many other related processes (Werner et al., 2008; Werner & Schmülling, 2009). Their role in the progress of dormancy to flowering is still unclear, although studies showed that cell division is higher in SAM during floral transition (Jacqmard, Gadisseur, & Bernier, 2003). The literature about epigenetic regulation and the role of CK in flowering induction is still sparse, but some insights have been described (Godge, Kumar, & Kumar, 2008; Li et al., 2008; Meijón, Jesús Cañal, Valledor, Rodríguez, & Feito, 2011; Tanaka et al., 1997). The study of Li et al. (2008) analyzed *A. thaliana* plants mutated for the *S-adenosyl-L-homocysteine hydrolase* (*SAHH*) gene, which codes for a key enzyme in the stabilization of methylation potential in cells (Palmer & Abeles, 1979; Tanaka et al., 1997). They also investigated RNAi lines with reduced *AtSAHH2* gene expression and observed a global reduction of DNA methylation levels and high levels of endogenous CK, even when compared to CK overproducers (Catterou et al., 2002; Chang, Jones, Banowetz, & Clark, 2003).

Similarly, increased levels of CK were observed in *sahh1-1* tobacco mutants (Tanaka et al., 1997). In tobacco plants, it was also observed that CK upregulate genes encoding Cytosine DNA methyltransferases (Li et al., 2008). The overexpression of *SAHH1* and *Adenosine kinase 1* (*ADK1*) in response to CK was similarly detected (Li et al., 2008; Pereira et al., 2007). Taken together, these results indicate a 'feedback-regulatory loop between cytokinin and DNA methylation: reduced DNA methylation potential in plant cells seems to lead to an increased level of CK, which in turn stimulates the genes encoding enzymes of the DNA methylation machinery' (Li et al., 2008). Moreover, the delayed flowering time found in Arabidopsis mutants knockdown for *SAHH* could be related to the accumulation of CK in response to reduction of global DNA methylation (Campos-Rivero et al., 2017; Li et al., 2008; Rocha et al., 2005). In the same way, the treatment with 9-(S)-(2,3-dihydroxypropyl)-adenine (DHPA), an inhibitor of SAHH in tobacco plants resulted in global DNA hypomethylation, flower morphology alterations, reduced fertility and upregulation of floral organ identity genes (Fulneček et al., 2011). The analysis of a gene coding for a CK binding protein, *PETUNIA CYTOKININ BINDING PROTEIN (PETCBP)* from *Petunia hybrida* cv. Mitchell, revealed a highly similar sequence (85%—90%) to the *SAHH* gene from several plant species (Godge et al., 2008). Authors suggest that PETCBP could be a positive regulator of CK response, mainly by modulating CK signal transduction. They also report that antisense expression of this gene resulted in plants with delayed flowering time (Godge et al., 2008).

Another work reported by Meijón et al. (2011) investigated if the treatment with GA inhibitors applied to improve flower production in ornamental azaleas plants could also influence DNA methylation and other phytohormones (as CK) during floral transition. They found that before flower set, low DNA methylation levels were seen, but when floral organs formation was achieved, genomic DNA became hypermethylated. They also observed low levels of GAs (as expected) and high endogenous CK amounts during floral transition. The authors suggested that CK could induce DNA demethylation, thereby contributing to the induction of gene expression during flowering transition (Meijón et al., 2011).

5.7 Abscisic Acid

ABA is a phytohormone associated mostly with drought stress and important to coordinate an adaptive response during water deprivation (Shinozaki & Yamaguchi-Shinozaki, 2007). Moreover, ABA plays important roles in plant development, even without stress stimuli, like induction of seed and bud dormancy (Barrero et al., 2005). During the transition of flowering, both positive and negative effects of ABA have been reported (Riboni, Test, Galbiati, Tonelli, & Conti, 2016, 2013; Shu et al., 2016; Wang et al., 2013; and reviewed by; Shu, Luo, Meng, & Yang, 2018). Briefly, the delay of *A. thaliana* floral transition is strictly related to *ABSCISIC ACID-INSENSITIVE* (*ABI*) genes. Studies analyzing the *Arabidopsis abi3*, *abi4* and *abi5* mutants revealed an early flowering phenotype; consistently to that, transgenic plants overexpressing those genes (*ABI3*, *ABI4* and *ABI5*) showed delayed floral transition (Shu et al., 2016; Wang et al., 2013). *ABI4* and *ABI5* directly induces *FLC* transcription, while FLC further represses expression of *FT* in Arabidopsis, regulating the flowering process in a negative way (Shu et al., 2018; Wang et al., 2013).

Recently, Shu et al. (2016) proposed that ABI4 negatively controls GA biogenesis, suggesting that it may repress flower formation through the GA flowering pathway. Positive regulation of flowering mediated by ABA response occurs especially under environmental stress conditions, such as drought (Riboni, Galbiati, Tonelli, & Conti, 2013, 2016). It is already described that drought stress causes early flowering in various plant species and high level of ABA biosynthesis (Budak, Hussain, Khan, Ozturk, & Ullah, 2015; Munemasa et al., 2015; Riboni, Test, Galbiati, Tonelli, & Conti, 2014). Molecular mechanisms required for ABA response in order to promote flowering and overpass the drought stress is still under evaluation, but some key factors were already described. Riboni et al. (2016) proposed that in Arabidopsis, ABA induces the drought escape response by promoting *FT* expression, for which *CO* and flower-promoting gene *GIGANTEA* (*GI*) genes are required in a photoperiod dependent manner. In citrus species, water deficit promotes flowering, and the higher level of ABA was also detected during the floral inductive stage (Li et al., 2017). Furthermore, in litchi species, drought stress combined with low temperature remarkably promoted flowering; however, the detailed relationship between ABA biosynthesis and/or signaling and flowering in litchi needs further investigation (Shen, Xiao, Qiu, Chen, & Chen, 2016).

In the epigenomic context, the nuclear interaction of SWI/SNF (chromatin remodeling complex) with *ABA HYPERSENSITIVE* (*HAB1*) gene, involved in negative regulation of ABA signaling and flower induction in Arabidopsis, was reported as necessary for ABA-responsive gene regulation (Rodriguez, Leube, & Grill, 1998; Saez, Rodrigues, Santiago, Rubio, & Rodriguez, 2008).

Another work showed that Arabidopsis mutants impaired in the *ABA OVERLY-SENSITIVE* (*ABO4*) gene that encodes an important regulator of ABA plant responses (Liu et al., 2010), showed an early flowering phenotype. *abo4-1* plants showed lower expression of *FLC* and high amounts of *FT* transcripts. An altered histone modification pattern in these two loci were also observed (Yin et al., 2009). Thirunavukkarasu et al., (2014) reported that stress-responsive genes involved in the ABA signaling promote floral transition and are regulated by epigenetic events under water stress condition in maize. Altogether, these findings suggest an important crosstalk between epigenetics, ABA responses, drought stress and flowering.

5.8 Auxins

Auxins are phytohormones involved in many aspects of plant biology such as vascular tissue formation, adventitious root initiation, tropistic responses, apical dominance and fruit development (Goldfarb, Lanz-Garcia, Lian, & Whetten, 2003; Luo, Zhou, & Zhang, 2018; Sundberg & Østergaard, 2009). The first link between auxin and floral transition was reported when the auxin transport mutant *pin1* (*PIN-FORMED1*) was characterized (Gälweiler et al., 1998). The *PIN1* gene encodes for a member of the putative auxin efflux regulator proteins, important in polar auxin transport (Gälweiler et al., 1998) and have been related to plant organ formation by regulating auxin distribution (Benková et al., 2003; Furutani et al., 2004). The most intriguing characteristic of the *pin1* mutant was the formation of inflorescence without flowers. Two other mutants showed very similar phenotypes: *pinoid* (Bennett, Alvarez, Bossinger, & Smyth, 1995) and *monopteros* (*mp*) (Przemeck, Mattsson, Hardtke, Sung, & Berleth, 1996). PINOID is a serine/threonine protein kinase and have been reported to be involved in signaling and polar auxin transport (Benjamins, Quint, Weijers, Hooykaas, & Offringa, 2001). MP/ARF5 is a member of a family of transcription factors called auxin response factors (ARFs), and are essential to auxin signaling and response (Hardtke &

Berleth, 1998; Quint & Gray, 2006). Interestingly, these mutants could progress from vegetative to reproductive phase while the inflorescence is formed, however no flower is observed (Cheng & Zhao, 2007). These findings demonstrate the important role of auxin in flower formation, even if the complete pathways connecting auxins and formation of floral primordial remain unclear (Cheng & Zhao, 2007). The link between epigenetic, auxin response and floral initiation was reported in a recent study which identified an auxin hormone-regulated chromatin state switch in order to promote floral primordial development in *Arabidopsis* (Wu et al., 2015). Based on these findings a model for cell auxin sensing and recruitment of chromatin remodeling factors was proposed. The presence of low auxin levels triggers a repressor system for auxin responsive genes that is orchestrated by monopteros (MP) and auxin sensitive Aux/IAA proteins (ASP). In this way, ASP binds to the MP factor associated with its target loci. ASP directly recruits the TOPLESS (TPL) repressor and the histone deacetylase HDA19, preventing gene expression (Guilfoyle & Hagen, 2012; Szemenyei, Hannon, & Long, 2008; Wu et al., 2015; Yamaguchi et al., 2013). The system also prevent recruitment of SWI/SNF ATPase subgroup BRAHMA (BRM) and SPLAYED (SYD) chromatin remodeling complexes, creating an inactive chromatin state (Wu et al., 2015). However, when high levels of auxin are present, ASPs are degraded and prevent the action of TPL and HD19. The recruiting of BRM or SYD complexes is possible and an open chromatin state is achieved, enabling the activation of the general transcriptional machinery (Wu et al. 2015). In loss of function studies, the authors also showed that the SWI/SNF ATPase activity is essential for flower primordium initiation, revealing a 'simple and elegant mechanism for small-signaling-molecule-regulated chromatin state switch' that respond to auxin signal and is essential for flower development (Wu et al., 2015).

6. BUD BREAK AND FLOWERING: CONSEQUENCES IN AGRICULTURE AND BREEDING

Throughout the evolution, temperate fruit trees have developed bud dormancy as an adaption to seasonality (Campoy et al., 2011; Ionescu, Møller, & Sánchez-Pérez, 2017). Bud break and subsequent flowering happen after chilling and heating requirements have been fulfilled during winter and spring, respectively. Within the context of global climate change, there are a number of evidences that increasing temperatures results in shifting the timing of phenological events, mainly in springtime, when plants resume growth (Hänninen & Tanino, 2011; Legave, Guédon, Malagi, El Yaacoubi, & Bonhomme, 2015; El Yaacoubi, Malagi, Oukabli, Hafidi, & Legave, 2014). Early bud break and flowering have many implications such as an increased risk of frost damage (Cannell & Smith, 1986; Vitasse, Lenz, & Karner, 2014) affecting the photosynthetic capacity of the trees (Ensminger, Schmidt, & Lloyd, 2007), bud burst delay combined with low burst and poor fruit set (Abbott, Zhebentyayeva, Barakat, & Liu, 2015; Celton et al., 2011; Dirlewanger et al., 2012; Erez, 2000), fertility abnormalities due to mis-synchronization of flowering of self-incompatible cultivars (Dirlewanger et al., 2012), changes in flower size and timing of anthesis (Scaven & Rafferty, 2013) and modifications of fruit harvesting periods and fruit marketing logistics (Dirlewanger et al., 2012). The production of floral scent, nectar and pollen can also be affected by temperature. Altered floral scent emission at higher temperatures affect the detectability of flowers by pollinating insects, such as moths, that rely on long-distance cues to locate floral resources (Kevan & Baker, 1983; Yuan, Himanen, Holopainen, Chen, & Stewart, 2009). Changes in nectar production and composition have immediate effects on pollinators activity and fitness (Burkle & Irwin, 2009), especially for those insects, such as some lepidopterans and wasps, that rely on nectar for amino acids as well as for sugars (Kevan & Baker, 1983). Similarly, decreased pollen production affect the reproductive success of many bees, which may need to collect pollen from a large number of plants to successfully rear their offspring (Müller et al., 2006). Along with the effects of warming on floral traits, elevated temperatures can alter other plant characteristics such as vegetative growth. It has been observed that plants exposed to winter warming of 1.5°C via open top chambers were several centimeters taller than plants in control chambers (Liu, Mu, Niklas, Li, & Sun, 2012). In contrast to the negative effects of the warming temperatures over the floral traits, increased vegetative growth may represent a benefit for agriculture in certain circumstances, such as wood and fruit crop production in sub-optimal environments. For example, the Brazilian apple cultivar 'Castel Gala', which is originated from a spontaneous bud sport mutation from a 'Gala Standard' tree, features a low CR when compared to the original cultivar, exhibits a precocious growth cycle that starts 25 days earlier and produces fruits that are anatomically and nutritionally equivalent (Denardi & Seccon, 2005). Due to its earlier bud burst, 'Castel Gala' also presents taller and more vigorous plants, reaching full production capacity earlier and can be planted in warmer climate sites than its parental type. Examples like this and the abovementioned information help pave the way to better analyze the challenging context of breeding strategies to deal with bud phenology traits in order to generate adapted cultivars to the new climate scenarios for agriculture.

Bud dormancy can still be considered a biological black box, in which the outputs to certain inputs can be predicted with some degree of certainty, but the internal workings remain obscure. Epigenetic regulation seems to play roles in some central steps of the bud dormancy cycles, including gene expression adjustments that remain stable over cell cycles and the quantification of cold exposure. Although the participation of epigenetic modifications in bud dormancy is suggested by many research data in the literature, the genes responsible for transducing epigenetic marks into phenological behaviour remain to be fully characterized. *DAM* and *EBB* genes, at this time, are the most promising candidates; however, further studies on repressors of *FT* transcription in temperate woody plants may uncover new pieces of this puzzle.

Exploiting epigenetic variations (DNA methylation and histone PTMs) for breeding applications depends on the plant propagation strategies (sexual versus clonal, McKey, Elias, Pujol, & Duputié, 2010). Because DNA methylation patterns can be transmitted after mitosis and meiosis, DNA methylation marks could be useful in all crops, irrespective to their propagation mode. By contrast, histone modifications are reset during meiosis (Ingouff et al., 2010; Xiao & Wagner, 2015), therefore, they would be of little benefit for breeding purposes in sexually propagated crops. However, they could be applicable for clonally propagated crops, such as in perennial fruit crops, because new epigenetic patterns could be maintained in meristems for grafting propagation (Bräutigam et al., 2013; Douhovnikoff & Dodd, 2014). As examples of heritable epigenetic states, epimutations, irrespective of their origin (induced or natural epialleles), are generally stable in plant populations and can be selected as classical phenotypes in breeding schemes. An extensive review about the use of epigenetics for breeding purposes is described by Gallusci et al. (2017).

A successful example of the use of epigenetic research and its application for breeding purposes has been demonstrated in the African oil palm *Elaeis guineensis,* where a flower and fruit abnormality known as 'mantled' can develop in oil palm cultivars derived from tissue culture and the resulting mantled palms can become unproductive (Ong-Abdullah et al., 2015). By performing a genome-wide, unbiased, DNA methylation analysis to look for loci epigenetically associated with the mantled phenotype, it was discovered that hypomethylation of a single *Karma* family retrotransposon embedded in the intron of the homeotic gene *DEFICIENS* is common to all mantled clones and associated with aberrant splicing and termination of the gene transcript. Loss of methylation — dubbed the *Bad Karma* epiallele — predicts a loss of oil palm yield and this property enable screening for higher-performing clones at the plantlet stage. This finding is likely to provide a way to detect unproductive palms much earlier than was previously possible, enabling their timely replacement in plantations, which besides the obvious economic importance, also benefits the environment. The results found for the 'mantled' phenotype in *E. guineensis* have also other key implications. For instance, a well-planned and performed genome-wide methylation mapping can pinpoint precise spots in the genome of a non-model organism that are responsible for a trait of interest.

Unveiling the molecular mechanisms responsible for the regulation of dormancy completion and flowering time in woody perennials is an important front for both basic and applied research. The compatibility between CR and the local climate is one of the main factors determining the success of a temperate fruit production enterprise. Conventional tree breeding, especially for dormancy traits, is costly and highly time-consuming because of protracted generation times. The advances in genomic research in woody perennials has now opened the opportunity to assist the breeders with tools generated through extensive sequencing, genome wide association studies and high density genotyping arrays that lead to the discovery of highly predictive molecular markers for traits of horticultural interest (Laurens et al., 2018). Therefore, markers linked to genes proved to participate in dormancy and flowering can be an invaluable asset for assisted breeding strategies. In combination with a biotechnological approach, master controllers of flowering time and bud dormancy could be used as tools to shorten the juvenile phase of target plant materials in order to accelerate breeding programs. This type of approach is already being suggested with the employment of transgenes that have the ability to induce flowering, such as *FT* (Yamagishi, Kishigami, & Yoshikawa, 2014), *TFL* (Freiman et al., 2012) and *BpMADS4* (Flachowsky et al., 2011; Weigl, Wenzel, Flachowsky, Peil, & Hanke, 2015) genes, but the use of dormancy regulators in similar strategies is still to be proposed. A different and potentially game changer strategy would be to modify the expression of dormancy controlling genes by means of state of the art technologies such as CRISPR/Cas9 (Bortesi & Fischer, 2015). This could potentially be done in already grown plants employing viral vectors, a plant-breeding technique with no transmission of genetic modification to the next generation (Yamagishi et al., 2014). In fruit production, the application of chemical compounds is already used as a tool to modulate flowering time and increase yields (see review in Ionescu et al., 2017). Because epigenetic mechanisms participate in gene expression control and hormonal signaling, an alternative for plant breeding approaches is to engineer flowering time by combining efforts on modern genetics and chemicals via the external application of flower-inducing compounds. This is an illustration of how the understanding of flowering time and dormancy control can be important to applied research and to agriculture.

ACKNOWLEDGEMENTS

This work was supported by Embrapa (02.13.05.016.00.02; 02.15.12.001.00.01.001). AMC and TS received a PhD scholarship from 'Coordenação de Aperfeiçoamento de Pessoal de Nível Superior' (CAPES, Ministry of Education). CPS received a postdoctoral fellowship from CAPES. VSF received a grant from the AgreenSkills+ EU fellowship program (FP7-609398).

REFERENCES

Abbott, A. G., Zhebentyayeva, T., Barakat, A., & Liu, Z. (2015). The genetic control of budbreak in trees. *Advances in Botanical Research, 74*, 201–228.

Abe, M., Kobayashi, Y., Yamamoto, S., Daimon, Y., Yamaguchi, A., Ikeda, Y., et al. (2005). FD, a bZIP protein mediating signals from the floral pathway integrator FT at the shoot apex. *Science, 309*(5737), 1052–1056.

Achard, P., Cheng, H., De Grauwe, L., Decat, J., Schoutteten, H., Moritz, T., et al. (2006). Integration of plant responses to environmentally activated phytohormonal signals. *Science, 311*(5757), 91–94.

Achard, P., Baghour, M., Chapple, A., Hedden, P., Van Der, Straeten D, Genschik, P., Moritz, T., & Harberd, N. P. (2007). The plant stress hormone ethylene controls floral transition via DELLA-dependent regulation of floral meristem-identity genes. *Proceedings of the National Academy of Sciences of the United States of America, 104*, 6484–6489.

Achard, P., & Genschik, P. (2009). Releasing the brakes of plant growth: How GAs shutdown della proteins. *Journal of Experimental Botany, 60*, 1085–1092.

Acosta, I. F., Laparra, H., Romero, S. P., Schmelz, E., Hamberg, M., Mottinger, J. P., et al. (2009). tasselseed1 is a lipoxygenase affecting jasmonic acid signaling in sex determination of maize. *Science, 323*(5911), 262–265.

Agger, K., Christensen, J., Cloos, P. A., & Helin, K. (2008). The emerging functions of histone demethylases. *Current Opinion in Genetics & Development, 18*(2), 159–168.

Allard, A., Bink, M. C. A. M., Martinez, S., Kelner, J. J., Legave, J. M., Di Guardo, M., et al. (2016). Detecting QTLs and putative candidate genes involved in budbreak and flowering time in an apple multiparental population. *Journal of Experimental Botany, 67*(9), 2875–2888.

Amasino, R. M. (2005). Vernalization and flowering time. *Current Opinion in Biotechnology, 16*(2), 154–158.

Amasino, R. M. (2010). Seasonal and developmental timing of flowering. *Plant Journal, 61*(6), 1001–1013.

Amasino, R. M., & Michaels, S. D. (2010). The timing of flowering. *Plant Physiology, 154*(2), 516–520.

Andrés, F., & Coupland, G. (2012). The genetic basis of flowering responses to seasonal cues. *Nature Reviews Genetics, 13*(9), 627–639.

Andrés, F., Porri, A., Torti, S., Mateos, J., Romera-Branchat, M., García-Martínez, J. L., et al. (2014). SHORT VEGETATIVE PHASE reduces gibberellin biosynthesis at the *Arabidopsis* shoot apex to regulate the floral transition. *Proceedings of the National Academy of Sciences, 111*(26), 2760–2769.

Angel, A., Song, J., Dean, C., & Howard, M. (2011). A Polycomb-based switch underlying quantitative epigenetic memory. *Nature, 476*(7358), 105–108.

Anh Tuan, P., Bai, S., Saito, T., Imai, T., Ito, A., & Moriguchi, T. (2016). Involvement of *EARLY BUD-BREAK*, an AP2/ERF transcription factor gene, in bud break in Japanese pear (*Pyrus pyrifolia* Nakai) lateral flower buds: Expression, histone modifications and possible target genes. *Plant and Cell Physiology, 57*(5), 1038–1047.

Banerjee, A., Wani, S. H., & Roychoudhury, A. (2017). Epigenetic control of plant cold responses. *Frontiers in Plant Science, 8*, 1643.

Barrero, J. M., Piqueras, P., González-Guzmán, M., Serrano, R., Rodríguez, P. L., Ponce, M. R., et al. (2005). A mutational analysis of the ABA1 gene of *Arabidopsis thaliana* highlights the involvement of ABA in vegetative development. *Journal of Experimental Botany, 56*(418), 2071–2083.

Benjamins, R., Quint, A., Weijers, D., Hooykaas, P., & Offringa, R. (2001). The PINOID protein kinase regulates organ development in Arabidopsis by enhancing polar auxin transport. *Development, 128*(20), 4057–4067.

Benková, E., Michniewicz, M., Sauer, M., Teichmann, T., Seifertová, D., Jürgens, G., et al. (2003). Local, efflux-dependent auxin gradients as a common module for plant organ formation. *Cell, 115*(5), 591–602.

Bennett, S. R. M., Alvarez, J., Bossinger, G., & Smyth, D. R. (1995). Morphogenesis in pinoid mutants of *Arabidopsis thaliana*. *The Plant Journal, 8*(4), 505–520.

Bielenberg, D. G., Wang, Y., Li, Z., Zhebentyayeva, T., Fan, S., Reighard, G. L., et al. (2008). Sequencing and annotation of the evergrowing locus in peach [Prunus persica (L.) Batsch] reveals a cluster of six MADS-box transcription factors as candidate genes for regulation of terminal bud formation. *Tree Genetics and Genomes, 4*(3), 495–507.

Bloomer, R. H., & Dean, C. (2017). Fine-tuning timing: Natural variation informs the mechanistic basis of the switch to flowering in *Arabidopsis thaliana*. *Journal of Experimental Botany, 68*(20), 5439–5452.

Blümel, M., Dally, N., & Jung, C. (2015). Flowering time regulation in crops-what did we learn from Arabidopsis? *Current Opinion in Biotechnology, 32*, 121–129.

Böhlenius, H., Huang, T., Charbonnel-Campaa, L., Brunner, A. M., Jansson, S., Strauss, S. H., et al. (2006). CO/FT regulatory module controls timing of flowering and seasonal growth cessation in trees. *Science, 312*(5776), 1040–1043.

Bortesi, L., & Fischer, R. (2015). The CRISPR/Cas9 system for plant genome editing and beyond. *Biotechnology Advances, 33*(1), 41–52.

Bräutigam, K., Vining, K. J., Lafon-Placette, C., Fossdal, C. G., Mirouze, M., Marcos, J. G., et al. (2013). Epigenetic regulation of adaptive responses of forest tree species to the environment. *Ecology and Evolution, 3*(2), 399–415.

Browse, J. (2009). Jasmonate passes muster: A receptor and targets for the defense hormone. *Annual Review of Plant Biology, 60*(1), 183–205.

Budak, H., Hussain, B., Khan, Z., Ozturk, N. Z., & Ullah, N. (2015). From genetics to functional genomics: Improvement in drought signaling and tolerance in wheat. *Frontiers in Plant Science, 6*, 1012.

Burkle, L., & Irwin, R. (2009). Nectar sugar limits larval growth of solitary bees (Hymenoptera: Megachilidae). *Environmental Entomology, 38*(4), 1293–1300.

Bu, Z., Yu, Y., Li, Z., Liu, Y., Jiang, W., Huang, Y., et al. (2014). Regulation of arabidopsis flowering by the histone mark readers MRG1/2 via interaction with CONSTANS to modulate FT expression. *PLos Genetics, 10*(9), 1–11.

Cai, Q., Yuan, Z., Chen, M., Yin, C., Luo, Z., Zhao, X., et al. (2014). Jasmonic acid regulates spikelet development in rice. *Nature Communications, 5*(1), 3476.

Campos-Rivero, G., Osorio-Montalvo, P., Sánchez-Borges, R., Us-Camas, R., Duarte-Aké, F., & De-la-Peña, C. (2017). Plant hormone signaling in flowering: An epigenetic point of view. *Journal of Plant Physiology, 214*, 16–27.

Campoy, J. A., Ruiz, D., & Egea, J. (2011). Dormancy in temperate fruit trees in a global warming context: A review. *Scientia Horticulturae, 130*(2), 357–372.

Cannell, M. G. R., & Smith, R. I. (1986). Climatic warming, spring budburst and forest damage on trees. *The Journal of Applied Ecology, 23*(1), 177.

Castède, S., Campoy, J. A., Le Dantec, L., Quero-García, J., Barreneche, T., Wenden, B., et al. (2015). Mapping of candidate genes involved in bud dormancy and flowering time in sweet cherry (Prunus avium). *PLos One, 10*(11).

Catterou, M., Dubois, F., Smets, R., Vaniet, S., Kichey, T., Van Onckelen, H., et al. (2002). Hoc: An Arabidopsis mutant overproducing cytokinins and expressing high in vitro organogenic capacity. *The Plant Journal, 30*(3), 273–287.

Celton, J. M. M., Martinez, S., Jammes, M. J. J., Bechti, A., Salvi, S., Legave, J. M. M., et al. (2011). Deciphering the genetic determinism of bud phenology in apple progenies: A new insight into chilling and heat requirement effects on flowering dates and positional candidate genes. *New Phytologist, 192*(2), 378–392.

Chandler, W. H., Kimball, M. H., Philp, G. L., Tufts, W. P., & Weldon, G. P. (1937). Chilling requirements for opening of buds on deciduous orchard trees and some other plants in California. Agricultural Experiment Station, Berkeley, California. *Bulletin, 611*, 3–11.

Chang, H., Jones, M. L., Banowetz, G. M., & Clark, D. G. (2003). Overproduction of cytokinins in petunia flowers transformed with P(SAG12)-IPT delays corolla senescence and decreases sensitivity to ethylene. *Plant Physiology, 132*(4), 2174–2183.

Charron, J. B. F., He, H., Elling, A. A., & Deng, X. W. (2009). Dynamic landscapes of four histone modifications during deetiolation in Arabidopsis. *The Plant Cell, 21*(12), 3732–3748.

Cheng, Y., & Zhao, Y. (2007). A role for auxin in flower development. *Journal of Integrative Plant Biology, 49*(1), 99–104.

Chini, A., Fonseca, S., Fernández, G., Adie, B., Chico, J. M., Lorenzo, O., et al. (2007). The JAZ family of repressors is the missing link in jasmonate signalling. *Nature, 448*(7154), 666–671.

Chinnusamy, V., Gong, Z., & Zhu, J. K. (2008). Abscisic Acid-mediated epigenetic processes in plant development and stress responses. *Journal of Integrative Plant Biology, 50*(10), 1187–1195.

Cho, L. H., Yoon, J., & An, G. (2017). The control of flowering time by environmental factors. *The Plant Journal, 90*(4), 708–719.

Choi, J., Hyun, Y., & Kang, M. J. (2009). Resetting and regulation of *FLOWERING LOCUS C* expression during Arabidopsis reproductive developmentH. Yun, J. Y. Yun, C. Lister, & Y. Choi (Eds.). *The Plant Journal, 57*(5), 918–931.

Chosed, R., Mukherjee, S., Lois, L. M., & Orth, K. (2006). Evolution of a signalling system that incorporates both redundancy and diversity: Arabidopsis SUMOylation. *The Biochemical Journal, 398*(3), 521–529.

Cleland, C. F., & Ajami, A. (1974). Identification of the flower-inducing factor isolated from aphid honeydew as being salicylic acid. *Plant Physiology, 54*(6), 904–906.

Conde, D., González-Melendi, P., & Allona, I. (2013). Poplar stems show opposite epigenetic patterns during winter dormancy and vegetative growth. *Trees, 27*(1), 311–320.

Conde, D., Le Gac, A. L., Perales, M., Dervinis, C., Kirst, M., Maury, S., et al. (2017a). Chilling-responsive DEMETER-LIKE DNA demethylase mediates in poplar bud break. *Plant, Cell and Environment, 40*(10), 2236–2249.

Conde, D., Moreno-Cortés, A., Dervinis, C., Ramos-Sánchez, J. M., Kirst, M., Perales, M., et al. (2017b). Overexpression of DEMETER, a DNA demethylase, promotes early apical bud maturation in poplar. *Plant, Cell and Environment, 40*(11), 2806–2819.

Conti, L. (2017). Hormonal control of the floral transition: Can one catch them all? *Developmental Biology, 430*(2), 288–301.

Cooke, J. E. K., Eriksson, M. E., & Junttila, O. (2012). The dynamic nature of bud dormancy in trees: Environmental control and molecular mechanisms. *Plant, Cell and Environment, 35*(10), 1707–1728.

Coustham, V., Li, P., Strange, A., Lister, C., Song, J., & Dean, C. (2012). Quantitative modulation of polycomb silencing underlies natural variation in vernalization. *Science, 337*(6094), 584–587.

de la Fuente, L., Conesa, A., Lloret, A., Badenes, M. L., & Ríos, G. (2015). Genome-wide changes in histone H3 lysine 27 trimethylation associated with bud dormancy release in peach. *Tree Genetics and Genomes, 11*(3), 45.

de Lucas, M., Davière, J.-M., Rodríguez-Falcón, M., Pontin, M., Iglesias-Pedraz, J. M., Lorrain, S., et al. (2008). A molecular framework for light and gibberellin control of cell elongation. *Nature, 451*(7177), 480–484.

Denardi, F., & Seccon, J. J. (2005). "Castel Gala" - mutação da macieira "Gala" com baixa necessidade de frio e maturação precoce. *Agropecuária Catarinense, 18*(2), 78–82.

Deng, W., Casao, M. C., Wang, P., Sato, K., Hayes, P. M., Finnegan, E. J., et al. (2015). Direct links between the vernalization response and other key traits of cereal crops. *Nature Communications, 6*, 1–8.

Dennis, F. G. (2003). Problems in standardizing methods for evaluating the chilling requirements for the breaking of dormancy in buds of woody plants. *HortScience, 38*, 347–350.

Devoto, A., Nieto-Rostro, M., Xie, D., Ellis, C., Harmston, R., Patrick, E., et al. (2002). COI1 links jasmonate signalling and fertility to the SCF ubiquitin-ligase complex in Arabidopsis. *The Plant Journal, 32*(4), 457–466.

Dill, A., Thomas, S. G., Hu, J., Camille, M., & Steber, T. S. (2004). The arabidopsis F-box protein SLEEPY1 targets gibberellin signaling repressors for gibberellin-induced degradation. *The Plant Cell, 16*(6), 1392–1405.

Dirlewanger, E., Quero-García, J., Le Dantec, L., Lambert, P., Ruiz, D., Dondini, L., et al. (2012). Comparison of the genetic determinism of two key phenological traits, flowering and maturity dates, in three Prunus species: Peach, apricot and sweet cherry. *Heredity, 109*(5), 280–292.

Domagalska, M. A., Schomburg, F. M., Amasino, R. M., Vierstra, R. D., Nagy, F., & Davis, S. J. (2007). Attenuation of brassinosteroid signaling enhances FLC expression and delays flowering. *Development, 134*(15), 2841–2850.

Douhovnikoff, V., & Dodd, R. S. (2014). Epigenetics: A potential mechanism for clonal plant success. *Plant Ecology, 216*(2), 227–233.

El Yaacoubi, A., Malagi, G., Oukabli, A., Hafidi, M., & Legave, J. M. (2014). Global warming impact on floral phenology of fruit trees species in Mediterranean region. *Scientia Horticulturae, 180*, 243–253.

Ensminger, I., Schmidt, L., & Lloyd, J. (2007). Soil temperature and intermittent frost modulate the rate of recovery of photosynthesis in Scots pine under simulated spring conditions. *New Phytologist, 177*(2), 428–442.

EPPO. (1984). EPPO crop growth stage keys: Apple and pear. *Bulletin EPPO, 14*(2), 291–294.

Erez, A. (2000). Bud dormancy; phenomenon, problems and solutions in the tropics and subtropics. *Temperate Fruit Crops in Warm Climates*, (50), 17–48.

Erez, A., Couvillon, G. A., & Hendershott, C. H. (1979). Quantitative chilling enhancement and negation in peach buds by high temperatures in a daily cycle. *Jounal of American Society of Horticultural Science, 104*(4), 536–540.

Erez, A., Fishman, S., Linsley-Noakes, G. C., & Allan, P. (1990). The dynamic model for rest completion in peach buds. *Acta Horticulturae*, (276), 165–174.

Falavigna, V. S., Porto, D. D., Buffon, V., Margis-Pinheiro, M., Pasquali, G., & Revers, L. F. (2014). Differential transcriptional profiles of dormancy-related genes in apple buds. *Plant Molecular Biology Reporter, 32*(4), 796–813.

Fan, S., Bielenberg, D. G., Zhebentyayeva, T. N., Reighard, G. L., Okie, W. R., Holland, D., et al. (2010). Mapping quantitative trait loci associated with chilling requirement, heat requirement and bloom date in peach (Prunus persica). *New Phytologist, 185*(4), 917–930.

Feng, S., Jacobsen, S. E., & Reik, W. (2010). Epigenetic reprogramming in plant and animal development. *Science, 330*(6004), 622–627.

Flachowsky, H., Le Roux, P. M., Peil, A., Patocchi, A., Richter, K., & Hanke, M. V. (2011). Application of a high-speed breeding technology to apple (Malus x domestica) based on transgenic early flowering plants and marker-assisted selection. *New Phytologist, 192*(2), 364–377.

Foster, T., Johnston, R., & Seleznyova, A. (2003). A morphological and quantitative characterization of early floral development in apple (Malus x domestica Borkh.). *Annals of Botany, 92*(2), 199–206.

Freiman, A., Shlizerman, L., Golobovitch, S., Yablovitz, Z., Korchinsky, R., Cohen, Y., et al. (2012). Development of a transgenic early flowering pear (Pyrus communis L.) genotype by RNAi silencing of PcTFL1-1 and PcTFL1-2. *Planta, 235*(6), 1239–1251.

Frewen, B. E., Chen, T. H. H., Howe, G. T., Davis, J., Rohde, A., Boerjan, W., et al. (2000). Quantitative trait loci and candidate gene mapping of bud set and bud flush in populus. *Genetics, 154*(2), 837–845.

Fulneček, J., Matyášek, R., Votruba, I., Holý, A., Křížová, K., & Kovařík, A. (2011). Inhibition of SAH-hydrolase activity during seed germination leads to deregulation of flowering genes and altered flower morphology in tobacco. *Molecular Genetics and Genomics, 285*(3), 225–236.

Furutani, M., Vernoux, T., Traas, J., Kato, T., Tasaka, M., & Aida, M. (2004). PIN-FORMED1 and PINOID regulate boundary formation and cotyledon development in Arabidopsis embryogenesis. *Development, 131*(20), 5021–5030.

Gabay, G., Dahan, Y., Izhaki, Y., Isaacson, T., Elkind, Y., Ben-Ari, G., et al. (2017). Identification of QTLs associated with spring vegetative budbreak time after dormancy release in pear (*Pyrus communis* L.). *Plant Breeding, 136*(5), 749–758.

Gallusci, P., Dai, Z., Génard, M., Gauffretau, A., Leblanc-Fournier, N., Richard-Molard, C., et al. (2017). Epigenetics for plant improvement: Current knowledge and modeling avenues. *Trends in Plant Science, 22*(7), 610–623.

Gälweiler, L., Guan, C., Müller, A., Wisman, E., Mendgen, K., Yephremov, A., et al. (1998). Regulation of polar auxin transport by AtPIN1 in Arabidopsis vascular tissue. *Science, 282*(5397), 2226–2230.

Garcia-Dominguez, M., & Reyes, J. C. (2009). SUMO association with repressor complexes, emerging routes for transcriptional control. *Biochimica et Biophysica Acta (BBA) - Gene Regulatory Mechanisms, 1789*(6–8), 451–459.

Godge, M. R., Kumar, D., & Kumar, P. P. (2008). Arabidopsis HOG1 gene and its petunia homolog PETCBP act as key regulators of yield parameters. *Plant Cell Reports, 27*(9), 1497–1507.

Goldberg-Moeller, R., Shalom, L., Shlizerman, L., Samuels, S., Zur, N., Ophir, R., et al. (2013). Effects of gibberellin treatment during flowering induction period on global gene expression and the transcription of flowering-control genes in Citrus buds. *Plant Science, 198*, 46–57.

Goldfarb, B., Lanz-Garcia, C., Lian, Z., & Whetten, R. (2003). Aux/IAA gene family is conserved in the gymnosperm, loblolly pine (Pinus taeda). *Tree Physiology, 23*(17), 1181–1192.

Graham, J., Hackett, C. A., Smith, K., Woodhead, M., Hein, I., & McCallum, S. (2009). Mapping QTLs for developmental traits in raspberry from bud break to ripe fruit. *Theoretical and Applied Genetics, 118*(6), 1143–1155.

Gregis, V., Andrés, F., Sessa, A., Guerra, R. F., Simonini, S., Mateos, J. L., et al. (2013). Identification of pathways directly regulated by SHORT VEGETATIVE PHASE during vegetative and reproductive development in Arabidopsis. *Genome Biology, 14*(6), 56.

Griffiths, J., Murase, K., Rieu, I., Zentella, R., Zhang, Z.-L., Powers, S. J., et al. (2006). Genetic characterization and functional analysis of the GID1 gibberellin receptors in arabidopsis. *The Plant Cell, 18*(12), 3399–3414.

Guardiola, J. L., Monerri, C., & Agusti, M. (1982). The inhibitory effect of gibberellic acid on flowering in Citrus. *Physiologia Plantarum, 55*(2), 136–142.

Gudesblat, G. E., & Russinova, E. (2011). Plants grow on brassinosteroids. *Current Opinion in Plant Biology, 14*(5), 530–537.

Guilfoyle, T. J., & Hagen, G. (2012). Getting a grasp on domain III/IV responsible for Auxin Response Factor-IAA protein interactions. *Plant Science, 190*, 82–88.

Guo, H., & Ecker, J. R. (2003). Plant responses to ethylene gas are mediated by SCFEBF1/EBF2- dependent proteolysis of EIN3 transcription factor. *Cell, 115*(6), 667–677.

Guo, S., Sun, B., Looi, L.-S., Xu, Y., Gan, E.-S., Huang, J., et al. (2015). Co-ordination of flower development through epigenetic regulation in two model species: Rice and arabidopsis. *Plant and Cell Physiology, 56*(5), 830–842.

Hanke, M. V., Flachowsky, H., Peil, A., & Hättasch, C. (2007). No flower no fruit – genetic potentials to trigger flowering in fruit trees. *Genes, Genomes and Genomics, 1*(1), 1–20.

Hänninen, H., & Tanino, K. (2011). Tree seasonality in a warming climate. *Trends in Plant Science, 16*(8), 412–416.

Harberd, N. P. (2003). Relieving DELLA restraint. *Science, 299*(5614), 1853–1854.

Hardtke, C. S., & Berleth, T. (1998). The Arabidopsis gene MONOPTEROS encodes a transcription factor mediating embryo axis formation and vascular development. *The EMBO Journal, 17*(5), 1405–1411.

Hauagge, R., & Cummins, J. N. (1991). Genetics of length of dormancy period in Malus vegetative buds. *Journal of the American Society for Horticultural Science, 116*(1), 121–126.

He, Y. (2012). Chromatin regulation of flowering. *Trends in Plant Science, 17*(9), 556–562.

He, Y., Michaels, S. D., & Amasino, R. M. (2003). Regulation of flowering time by histone acetylation in arabidopsis. *Science, 302*(5651), 1751–1754.

Heide, O. M. (2011). Temperature rather than photoperiod controls growth cessation and dormancy in Sorbus species. *Journal of Experimental Botany, 62*(15), 5397–5404.

Heide, O. M., & Prestrud, A. K. (2005). Low temperature, but not photoperiod, controls growth cessation and dormancy induction and release in apple and pear. *Tree Physiology, 25*(1), 109–114.

Horvath, D. (2009). Common mechanisms regulate flowering and dormancy. *Plant Science, 177*(6), 523–531.

Horvath, D. (2010). Bud dormancy and growth. *Plant Developmental Biology, 1*, 53–70.

Horvath, D. P., Sung, S., Kim, D., Chao, W., & Anderson, J. (2010). Characterization, expression and function of DORMANCY ASSOCIATED MADS-BOX genes from leafy spurge. *Plant Molecular Biology, 73*(1–2), 169–179.

Howe, G. T., Gardner, G., Hackett, W. P., & Furnier, G. R. (1996). Phytochrome control of short-day-induced bud set in black cottonwood. *Physiologia Plantarum, 97*(1), 95–103.

Hsu, C. Y., Adams, J. P., Kim, H., No, K., Ma, C., Strauss, S. H., et al. (2011). FLOWERING LOCUS T duplication coordinates reproductive and vegetative growth in perennial poplar. *Proceedings of the National Academy of Sciences, 108*(26), 10756–10761.

Hyun, Y., Richter, R., Vincent, C., Martinez-Gallegos, R., Porri, A., & Coupland, G. (2016). Multi-layered regulation of SPL15 and cooperation with SOC1 integrate endogenous flowering pathways at the arabidopsis shoot meristem. *Developmental Cell, 37*(3), 254–266.

Ietswaart, R., Wu, Z., & Dean, C. (2012). Flowering time control: Another window to the connection between antisense RNA and chromatin. *Trends in Genetics, 28*(9), 445–453.

Ingouff, M., Rademacher, S., Holec, S., Šoljić, L., Xin, N., Readshaw, A., et al. (2010). Zygotic resetting of the HISTONE 3 variant repertoire participates in epigenetic reprogramming in arabidopsis. *Current Biology, 20*(23), 2137–2143.

Ionescu, I. A., Møller, B. L., & Sánchez-Pérez, R. (2017). Chemical control of flowering time. *Journal of Experimental Botany, 68*(3), 369–382.

Jackson, S. D., & Hong, Y. (2012). Systemic movement of *FT* mRNA and a possible role in floral induction. *Frontiers in Plant Science, 3*, 1–4.

Jacqmard, A., Gadisseur, I., & Bernier, G. (2003). Cell division and morphological changes in the shoot apex of *Arabidopsis thaliana* during floral transition. *Annals of Botany, 91*(5), 571–576.

Jeong, J. H., Song, H. R., Ko, J. H., Jeong, Y. M., Kwon, Y. E., Seol, J. H., et al. (2009). Repression of FLOWERING LOCUS T chromatin by functionally redundant histone H3 lysine 4 demethylases in Arabidopsis. *PLos One, 4*(11).

Jiang, D., Wang, Y., Wang, Y., & He, Y. (2008). Repression of FLOWERING LOCUS C and FLOWERING LOCUS T by the arabidopsis polycomb repressive complex 2 components. *PLos One, 3*(10).

Jin, J. B., Jin, Y. H., Lee, J., Miura, K., Yoo, C. Y., Kim, W. Y., et al. (2008). The SUMO E3 ligase, AtSIZ1, regulates flowering by controlling a salicylic acid-mediated floral promotion pathway and through affects on FLC chromatin structure. *The Plant Journal, 53*(3), 530–540.

Junttila, O. (1976). Apical growth cessation and shoot tip abscission in Salix. *Physiologia Plantarum, 38*(4), 278–286.

Kamiya, Y., & García-Martínez, J. L. (1999). Regulation of gibberellin biosynthesis by light. *Current Opinion in Plant Biology, 2*(5), 398–403.

Karlberg, A., Englund, M., Petterle, A., Molnar, G., Sjödin, A., Bako, L., et al. (2010). Analysis of global changes in gene expression during activity-dormancy cycle in hybrid aspen apex. *Plant Biotechnology, 27*(1), 1–16.

Kevan, P. G., & Baker, H. G. (1983). Insects as flower visitors and pollinators. *Annual Review of Entomology, 28*(1), 407–453.

Khan, M. R. G., Ai, X. Y., & Zhang, J. Z. (2014). Genetic regulation of flowering time in annual and perennial plants. *Wiley Interdisciplinary Reviews: RNA, 5*(3), 347–359.

Kotoda, N., Hayashi, H., Suzuki, M., Igarashi, M., Hatsuyama, Y., Kidou, S. I., et al. (2010). Molecular characterization of flowering LOCUS t-like genes of apple (Malus × Domestica borkh.). *Plant and Cell Physiology, 51*(4), 561–575.

Kumar, G., Gupta, K., Pathania, S., Swarnkar, M. K., Rattan, U. K., Singh, G., et al. (2017). Chilling affects phytohormone and post-embryonic development pathways during bud break and fruit set in apple (Malus x domestica Borkh.). *Scientific Reports, 7*.

Kumar, G., Rattan, U. K., & Singh, A. K. (2016). Chilling-mediated DNA methylation changes during dormancy and its release reveal the importance of epigenetic regulation during winter dormancy in Apple (Malus x domestica Borkh.). *PLos One, 11*(2).

Kumar, S. V., & Wigge, P. A. (2010). H2A.Z-Containing nucleosomes mediate the thermosensory response in arabidopsis. *Cell, 140*(1), 136–147.

Labuschagné, I. F., Louw, J. H., Schmidt, K., & Sadie, A. (2002). Genetic variation in chilling requirement in apple progeny. *Journal of the American Society for Horticultural Science, 127*(4), 663–672.

Lafon-Placette, C., Faivre-Rampant, P., Delaunay, A., Street, N., Brignolas, F., & Maury, S. (2013). Methylome of DNase I sensitive chromatin in Populus trichocarpa shoot apical meristematic cells: A simplified approach revealing characteristics of gene-body DNA methylation in open chromatin state. *New Phytologist, 197*(2), 416–430.

Lafon-Placette, C., Le Gac, A.-L., Chauveau, D., Segura, V., Delaunay, A., Lesage-Descauses, M.-C., et al. (2018). Changes in the epigenome and transcriptome of the poplar shoot apical meristem in response to water availability affect preferentially hormone pathways. *Journal of Experimental Botany, 69*(3), 537–551.

Lamb, R. C. (1948). Effects of temperature above and below freezing on the breaking of rest in the Latham raspberry. *Journal of the American Society for Horticultural Science, 51*, 313–315.

Lang, G., Early, J., Martin, G., & Darnell, R. (1987). Endo-, para-, and ecodormancy: Physiological terminology and classification for dormancy research. *HortScience, 22*(3), 371–377.

Laurens, F., Aranzana, M. J., Arus, P., Bassi, D., Bink, M., Bonany, J., et al. (2018). An integrated approach for increasing breeding efficiency in apple and peach in Europe. *Horticulture Research, 5*(1), 11.

Lee, S. S., Lee, S. S., Yang, K. Y. Y., Kim, Y. M. M., Park, S. Y. Y., Kim, S. Y., et al. (2006). Overexpression of PRE1 and its homologous genes activates gibberellin-dependent responses in *Arabidopsis thaliana. Plant and Cell Physiology, 47*(5), 591–600.

Lee, T. T., & Skoog, F. (1965). Effects of substituted phenols on bud formation and growth of tobacco tissue cultures. *Physiologia Plantarum, 18*(2), 386–402.

Legave, J. M., Guédon, Y., Malagi, G., El Yaacoubi, A., & Bonhomme, M. (2015). Differentiated responses of apple tree floral phenology to global warming in contrasting climatic regions. *Frontiers in Plant Science, 6*, 1054.

Leida, C., Conesa, A., Llácer, G., Badenes, M. L., & Ríos, G. (2012). Histone modifications and expression of DAM6 gene in peach are modulated during bud dormancy release in a cultivar-dependent manner. *New Phytologist, 193*(1), 67–80.

Li, C., & Cui, Y. (2016). A DNA element that remembers winter. *Nature Genetics, 48*(12), 1451–1452.

Li, C. H., Yu, N., Jiang, S. M., Shangguan, X. X., Wang, L. J., & Chen, X. Y. (2008). Down-regulation of S-adenosyl-l-homocysteine hydrolase reveals a role of cytokinin in promoting transmethylation reactions. *Planta, 228*(1), 125–136.

Li, J. X., Hou, X. J., Zhu, J., Zhou, J. J., Huang, H. B., Yue, J. Q., et al. (2017). Identification of genes associated with lemon floral transition and flower development during floral inductive water deficits: A hypothetical model. *Frontiers in Plant Science, 8*, 1013.

Li, J., Li, Y., Chen, S., & An, L. (2010). Involvement of brassinosteroid signals in the floral-induction network of Arabidopsis. *Journal of Experimental Botany, 61*(15), 4221–4230.

Li, L., Zhao, Y., McCaig, B. C., Wingerd, B. A., Wang, J., Whalon, M. E. B., et al. (2004). The tomato homolog of CORONATINE-INSENSITIVE1 is required for the maternal control of seed maturation, jasmonate-signaled defense responses, and glandular trichome development. *The Plant Cell, 16*(1), 126–143.

Li, P., Filiault, D., Box, M. S., Kerdaffrec, E., van Oosterhout, C., Wilczek, A. M., et al. (2014). Multiple FLC haplotypes defined by independent cis-regulatory variation underpin life history diversity in *Arabidopsis thaliana. Genes & Development, 28*(15), 1635–1640.

Li, Z., Reighard, G. L., Abbott, A. G., & Bielenberg, D. G. (2009). Dormancy-associated MADS genes from the EVG locus of peach [Prunus persica (L.) Batsch] have distinct seasonal and photoperiodic expression patterns. *Journal of Experimental Botany, 60*(12), 3521–3530.

Liu, Y., He, J., Chen, Z., Ren, X., Hong, X., & Gong, Z. (2010). ABA overly-sensitive 5 (ABO5), encoding a pentatricopeptide repeat protein required for cis-splicing of mitochondrial nad2 intron 3, is involved in the abscisic acid response in Arabidopsis. *Plant Journal, 63*(5), 749–765.

Liu, Y., Mu, J., Niklas, K. J., Li, G., & Sun, S. (2012). Global warming reduces plant reproductive output for temperate multi-inflorescence species on the Tibetan plateau. *New Phytologist, 195*(2), 427–436.

Lucas-Reina, E., Ortiz-Marchena, M. I., Romero-Campero, F. J., Calonje, M., Romero, J. M., & Valverde, F. (2016). Evolution of the flowering pathways. *Progress in Botany, 77*, 291–329.

Luo, J., Zhou, J. J., & Zhang, J. Z. (2018). Aux/IAA gene family in plants: Molecular structure, regulation, and function. *International Journal of Molecular Sciences, 19*(1), 259.

Mahajan, R., Delphin, C., Guan, T., Gerace, L., & Melchior, F. (1997). A small ubiquitin-related polypeptide involved in targeting RanGAP1 to nuclear pore complex protein RanBP2. *Cell, 88*(1), 97–107.

Marcelle, R., & Sironval, C. (1963). Effect of gibberellic acid on flowering of apple trees. *Nature, 197*(4865), 405.

March-Díaz, R., & Reyes, J. C. (2009). The beauty of being a variant: H2A.Z and the SWR1 complex in plants. *Molecular Plant, 2*(4), 565–577.

Martínez, C., Pons, E., Prats, G., & León, J. (2004). Salicylic acid regulates flowering time and links defence responses and reproductive development. *The Plant Journal, 37*(2), 209–217.

Mateos, J. L., Madrigal, P., Tsuda, K., Rawat, V., Richter, R., Romera-Branchat, M., et al. (2015). Combinatorial activities of SHORT VEGETATIVE PHASE and FLOWERING LOCUS C define distinct modes of flowering regulation in Arabidopsis. *Genome Biology, 16*(1), 1–23.

Matunis, M. J., Coutavas, E., & Blobel, G. (1996). A novel ubiquitin-like modification modulates the partitioning of the Ran-GTPase-activating protein RanGAP1 between the cytosol and the nuclear pore complex. *The Journal of Cell Biology, 135*(6), 1457–1470.

McKey, D., Elias, M., Pujol, M. E., & Duputié, A. (2010). The evolutionary ecology of clonally propagated domesticated plants. *New Phytologist, 186*(2), 318–332.

Meijón, M., Jesús Cañal, M., Valledor, L., Rodríguez, R., & Feito, I. (2011). Epigenetic and physiological effects of gibberellin inhibitors and chemical pruners on the floral transition of azalea. *Physiologia Plantarum, 141*(3), 276–288.

Melzer, S., Lens, F., Gennen, J., Vanneste, S., Rohde, A., & Beeckman, T. (2008). Flowering-time genes modulate meristem determinacy and growth form in *Arabidopsis thaliana*. *Nature Genetics, 40*(12), 1489–1492.

Michaels, S. D., & Amasino, R. M. (1999). FLOWERING LOCUS C encodes a novel MADS domain protein that acts as a repressor of flowering. *The Plant Cell, 11*(5), 949–956.

Mimida, N., Saito, T., Moriguchi, T., Suzuki, A., Komori, S., & Wada, M. (2015). Expression of DORMANCY-ASSOCIATED MADS-BOX (DAM)-like genes in apple. *Biologia Plantarum, 59*(2), 237–244.

Moon, J., Lee, H., Kim, M., & Lee, I. (2005). Analysis of flowering pathway integrators in arabidopsis. *Plant and Cell Physiology, 46*(2), 292–299.

Moon, J., Suh, S. S., Lee, H., Choi, K. R., Hong, C. B., Paek, N. C., et al. (2003). The SOC1 MADS-box gene integrates vernalization and gibberellin signals for flowering in Arabidopsis. *The Plant Journal, 35*(5), 613–623.

Müller, A., Diener, S., Schnyder, S., Stutz, K., Sedivy, C., & Dorn, S. (2006). Quantitative pollen requirements of solitary bees: Implications for bee conservation and the evolution of bee–flower relationships. *Biological Conservation, 130*(4), 604–615.

Munemasa, S., Hauser, F., Park, J., Waadt, R., Brandt, B., & Schroeder, J. I. (2015). Mechanisms of abscisic acid-mediated control of stomatal aperture. *Current Opinion in Plant Biology, 28*, 154–162.

Murfett, J., Wang, X. J., Hagen, G., & Guilfoyle, T. J. (2001). Identification of Arabidopsis histone deacetylase HDA6 mutants that affect transgene expression. *The Plant Cell, 13*(5), 1047–1061.

Nathan, D., Ingvarsdottir, K., Sterner, D. E., Bylebyl, G. R., Dokmanovic, M., Dorsey, J. A., et al. (2006). Histone sumoylation is a negative regulator in *Saccharomyces cerevisiae* and shows dynamic interplay with positive-acting histone modifications. *Genes & Development, 20*(8), 966–976.

Niu, Q., Li, J., Cai, D., Qian, M., Jia, H., Bai, S., et al. (2016). Dormancy-associated MADS-box genes and microRNAs jointly control dormancy transition in pear (Pyrus pyrifolia white pear group) flower bud. *Journal of Experimental Botany, 67*(1), 239–257.

Noh, B., Lee, S. H., Kim, H. J., Yi, G., Shin, E. A., Lee, M., et al. (2004). Divergent roles of a pair of homologous jumonji/zinc-finger-class transcription factor proteins in the regulation of arabidopsis flowering time. *The Plant Cell, 16*(10), 2601–2613.

Novatchkova, M., Budhiraja, R., Coupland, G., Eisenhaber, F., & Bachmair, A. (2004). SUMO conjugation in plants. *Planta, 220*(1), 1–8.

Ogas, J., Kaufmann, S., Henderson, J., & Somerville, C. (1999). PICKLE is a CHD3 chromatin-remodeling factor that regulates the transition from embryonic to vegetative development in Arabidopsis. *Proceedings of the National Academy of Sciences of the United States of America, 96*(24), 13839–13844.

Olukolu, B. A., Trainin, T., Fan, S., Kole, C., Bielenberg, D. G., Reighard, G. L., et al. (2009). Genetic linkage mapping for molecular dissection of chilling requirement and budbreak in apricot (Prunus armeniaca L.). *Genome, 52*(10), 819–828.

Ong-Abdullah, M., Ordway, J. M., Jiang, N., Ooi, S. E., Kok, S. Y., Sarpan, N., et al. (2015). Loss of Karma transposon methylation underlies the mantled somaclonal variant of oil palm. *Nature, 525*(7570), 533–537.

Palmer, J. L., & Abeles, R. H. (1979). The mechanism of action of S-adenosylhomocysteinase. *The Journal of Biological Chemistry, 254*(4), 1217–1226.

Park, J. H., Halitschke, R., Kim, H. B., Baldwin, I. T., Feldmann, K. A., & Feyereisen, R. (2002). A knock-out mutation in allene oxide synthase results in male sterility and defective wound signal transduction in Arabidopsis due to a block in jasmonic acid biosynthesis. *The Plant Journal, 31*(1), 1–12.

Park, J., Oh, D. H., Dassanayake, M., Nguyen, K. T., Ogas, J., Choi, G., et al. (2017). Gibberellin signaling requires chromatin remodeler PICKLE to promote vegetative growth and phase transitions. *Plant Physiology, 173*(2), 1463–1474.

Pereira, L., Todorova, M., Cai, X., Makaroff, C., Emery, R., & Moffatt, B. (2007). Methyl recycling activities are co-ordinately regulated during plant development. *Journal of Experimental Botany, 58*(5), 1083–1098.

Porto, D. D., Falavigna, V. da S., Arenhart, R. A., Perini, P., Buffon, V., Anzanello, R., et al. (2016). Structural genomics and transcriptional characterization of the Dormancy-Associated MADS-box genes during bud dormancy progression in apple. *Tree Genetics and Genomes, 12*(3), 46.

Przemeck, G. K., Mattsson, J., Hardtke, C. S., Sung, Z. R., & Berleth, T. (1996). Studies on the role of the Arabidopsis gene MONOPTEROS in vascular development and plant cell axialization. *Planta, 200*(2), 229–237.

Putterill, J., & Varkonyi-Gasic, E. (2016). FT and florigen long-distance flowering control in plants. *Current Opinion in Plant Biology, 33*, 77–82.

Qüesta, J. I., Song, J., Geraldo, N., An, H., & Dean, C. (2016). Arabidopsis transcriptional repressor VAL1 triggers Polycomb silencing at FLC during vernalization. *Science, 353*(6298), 485–488.

Quint, M., & Gray, W. M. (2006). Auxin signaling. *Current Opinion in Plant Biology, 9*(5), 448–453.
Ramírez, F., & Davenport, T. L. (2010). Mango (Mangifera indica L.) flowering physiology. *Scientia Horticulturae, 126*(2), 65–72.
Riboni, M., Galbiati, M., Tonelli, C., & Conti, L. (2013). GIGANTEA enables drought escape response via abscisic acid-dependent activation of the florigens and SUPPRESSOR OF OVEREXPRESSION OF CONSTANS1. *Plant Physiology, 162*(3), 1706–1719.
Riboni, M., Test, A. R., Galbiati, M., Tonelli, C., & Conti, L. (2014). Environmental stress and flowering time the photoperiodic connection. *Plant Signaling and Behavior, 9*(7).
Riboni, M., Test, A. R., Galbiati, M., Tonelli, C., & Conti, L. (2016). ABA-dependent control of GIGANTEA signalling enables drought escape via up-regulation of FLOWERING LOCUS T in *Arabidopsis thaliana*. *Journal of Experimental Botany, 67*(22), 6309–6322.
Richards, C. L., Alonso, C., Becker, C., Bossdorf, O., Bucher, E., Colomé-Tatché, M., et al. (2017). Ecological plant epigenetics: Evidence from model and non-model species, and the way forward. *Ecology Letters, 20*(12), 1576–1590.
Richardson, E. A., Seeley, S., & Walker, D. (1974). A model for estimating the completion of rest for Red haven and Elbert peach trees. *HortScience1, 9*(4), 331–332.
Rider, S. D., Henderson, J. T., Jerome, R. E., Edenberg, H. J., Romero-Severson, J., & Ogas, J. (2003). Coordinate repression of regulators of embryonic identity by PICKLE during germination in Arabidopsis. *The Plant Journal, 35*(1), 33–43.
Rinne, P. L. H., Welling, A., Vahala, J., Ripel, L., Ruonala, R., Kangasjärvi, J., et al. (2011). Chilling of dormant buds hyperinduces FLOWERING LOCUS T and recruits GA-inducible 1,3-β-Glucanases to reopen signal conduits and release dormancy in populus. *The Plant Cell, 23*(1), 130–146.
Ríos, G., Leida, C., Conejero, A., & Badenes, M. L. (2014). Epigenetic regulation of bud dormancy events in perennial plants. *Frontiers in Plant Science, 5*, 247.
Rivas-San Vicente, M., & Plasencia, J. (2011). Salicylic acid beyond defence: Its role in plant growth and development. *Journal of Experimental Botany, 62*(10), 3321–3338.
Rocha, P. S. C. F., Sheikh, M., Melchiorre, R., Fagard, M., Boutet, S., Loach, R., et al. (2005). The arabidopsis HOMOLOGY-DEPENDENT GENE SILENCING1 gene codes for an S-Adenosyl-L-Homocysteine hydrolase required for DNA methylation-dependent gene silencing. *The Plant Cell, 17*(2), 404–417.
Rodriguez, P. L., Leube, M. P., & Grill, E. (1998). Molecular cloning in *Arabidopsis thaliana* of a new protein phosphatase 2C (PP2C) with homology to ABI1 and ABI2. *Plant Molecular Biology, 38*(5), 879–883.
Rohde, A., & Bhalerao, R. P. (2007). Plant dormancy in the perennial context. *Trends in Plant Science, 12*(5), 217–223.
Rohde, A., Storme, V., Jorge, V., Gaudet, M., Vitacolonna, N., Fabbrini, F., et al. (2011). Bud set in poplar - genetic dissection of a complex trait in natural and hybrid populations. *New Phytologist, 189*(1), 106–121.
Romeu, J. F., Monforte, A. J., Sánchez, G., Granell, A., García-Brunton, J., Badenes, M. L., et al. (2014). Quantitative trait loci affecting reproductive phenology in peach. *BMC Plant Biology, 14*(1), 52.
Rothkegel, K., Sánchez, E., Montes, C., Greve, M., Tapia, S., Bravo, S., et al. (2017). DNA methylation and small interference RNAs participate in the regulation of MADS-box genes involved in dormancy in sweet cherry (Prunus avium L.). *Tree Physiology, 37*(12), 1739–1751.
Ruiz, D., Campoy, J. A., & Egea, J. (2007). Chilling and heat requirements of apricot cultivars for flowering. *Environmental and Experimental Botany, 61*(3), 254–263.

Saez, A., Rodrigues, A., Santiago, J., Rubio, S., & Rodriguez, P. L. (2008). HAB1-SWI3B interaction reveals a link between abscisic acid signaling and putative SWI/SNF chromatin-remodeling complexes in arabidopsis. *The Plant Cell, 20*(11), 2972–2988.

Saito, T., Bai, S., Imai, T., Ito, A., Nakajima, I., & Moriguchi, T. (2015). Histone modification and signalling cascade of the dormancy-associatedMADS-box gene, PpMADS13-1, in Japanese pear (Pyrus pyrifolia) during endodormancy. *Plant, Cell and Environment, 38*(6), 1157–1166.

Saito, T., Bai, S., Ito, A., Sakamoto, D., Saito, T., Ubi, B. E., et al. (2013). Expression and genomic structure of the dormancy-associated MADS box genes MADS13 in Japanese pears (Pyrus pyrifolia Nakai) that differ in their chilling requirement for endodormancy release. *Tree Physiology, 33*(6), 654–667.

Sánchez-Pérez, R., Dicenta, F., & Martínez-Gómez, P. (2012). Inheritance of chilling and heat requirements for flowering in almond and QTL analysis. *Tree Genetics and Genomes, 8*(2), 379–389.

Santamaría, M., Hasbún, R., Valera, M., Meijón, M., Valledor, L., Rodríguez, J. L., et al. (2009). Acetylated H4 histone and genomic DNA methylation patterns during bud set and bud burst in Castanea sativa. *Journal of Plant Physiology, 166*(13), 1360–1369.

Santamaría, M. E., Rodríguez, R., Cañal, M. J., & Toorop, P. E. (2011). Transcriptome analysis of chestnut (Castanea sativa) tree buds suggests a putative role for epigenetic control of bud dormancy. *Annals of Botany, 108*(3), 485–498.

Sasaki, R., Yamane, H., Ooka, T., Jotatsu, H., Kitamura, Y., Akagi, T., et al. (2011). Functional and expressional analyses of PmDAM genes associated with endodormancy in Japanese apricot. *Plant Physiology, 157*(1), 485–497.

Scaven, V. L., & Rafferty, N. E. (2013). Physiological effects of climate warming on flowering plants and insect pollinators and potential consequences for their interactions. *Current Zoology, 59*(3), 418–426.

Shafiq, S., Berr, A., & Shen, W. H. (2014). Combinatorial functions of diverse histone methylations in *Arabidopsis thaliana* flowering time regulation. *New Phytologist, 201*(1), 312–322.

Shaltout, A. D., Unrath, C. R., & Akademiya, S. (1983). Rest completion prediction model for "Starkrimson Delicious" apples. *Journal of the American Society for Horticultural Science, 108*(6), 957–961.

Shen, J., Xiao, Q., Qiu, H., Chen, C., & Chen, H. (2016). Integrative effect of drought and low temperature on litchi (Litchi chinensis Sonn.) floral initiation revealed by dynamic genome-wide transcriptome analysis. *Scientific Reports, 6*, 32005.

Shiio, Y., & Eisenman, R. N. (2003). Histone sumoylation is associated with transcriptional repression. *Proceedings of the National Academy of Sciences of the United States of America, 100*(23), 13225–13230.

Shinozaki, K., & Yamaguchi-Shinozaki, K. (2007). Gene networks involved in drought stress response and tolerance. *Journal of Experimental Botany, 58*, 221–227.

Shu, K., Chen, Q., Wu, Y., Liu, R., Zhang, H., Wang, S., et al. (2016). ABSCISIC ACID-INSENSITIVE 4 negatively regulates flowering through directly promoting Arabidopsis FLOWERING LOCUS C transcription. *Journal of Experimental Botany, 67*(1), 195–205.

Shu, K., Luo, X., Meng, Y., & Yang, W. (2018). Toward a molecular understanding of abscisic acid actions in floral transition. *Plant and Cell Physiology, 59*(2), 215–221.

Song, J., Angel, A., Howard, M., & Dean, C. (2012a). Vernalization - a cold-induced epigenetic switch. *Journal of Cell Science, 125*(16), 3723–3731.

Song, S., Qi, T., Huang, H., Ren, Q., Wu, D., Chang, C., et al. (2011). The jasmonate-ZIM domain proteins interact with the R2R3-MYB transcription factors MYB21 and MYB24 to affect jasmonate-regulated stamen development in arabidopsis. *The Plant Cell, 23*(3), 1000–1013.

Song, Y., Ma, K., Bo, W., Zhang, Z., & Zhang, D. (2012b). Sex-specific DNA methylation and gene expression in andromonoecious poplar. *Plant Cell Reports, 31*(8), 1393–1405.

Song, Y. H., Ito, S., & Imaizumi, T. (2013). Flowering time regulation: Photoperiod- and temperature-sensing in leaves. *Trends in Plant Science, 18*(10), 575–583.

Southwick, S. M., & Glozer, K. (2000). Reducing flowering with gibberellins to increase fruit size in stone fruit trees: Applications and implications in fruit production. *HortTechnology, 10*, 744–751.

Stintzi, A., & Browse, J. (2000). The Arabidopsis male-sterile mutant, opr3, lacks the 12-oxo-phytodienoic acid reductase required for jasmonate synthesis. *Proceedings of the National Academy of Sciences, 97*(19), 10625–10630.

Stitz, M., Hartl, M., Baldwin, I. T., & Gaquerel, E. (2014). Jasmonoyl-L-isoleucine coordinates metabolic networks required for anthesis and floral attractant emission in wild tobacco (Nicotiana attenuata). *The Plant Cell, 26*(10), 3964–3983.

Suárez-López, P., Wheatley, K., Robson, F., Onouchi, H., Valverde, F., & Coupland, G. (2001). CONSTANS mediates between the circadian clock and the control of flowering in Arabidopsis. *Nature, 410*(6832), 1116–1120.

Sun, C., Chen, D., Fang, J., Wang, P., Deng, X., & Chu, C. (2014). Understanding the genetic and epigenetic architecture in complex network of rice flowering pathways. *Protein and Cell, 5*(12), 889–898.

Sundberg, E., & Østergaard, L. (2009). Distinct and dynamic auxin activities during reproductive development. *Cold Spring Harbor Perspectives in Biology, 1*(6).

Szemenyei, H., Hannon, M., & Long, J. A. (2008). TOPLESS mediates auxin-dependent transcriptional repression during Arabidopsis embryogenesis. *Science, 319*(5868), 1384–1386.

Takeuchi, T., Matsushita, M. C., Nishiyama, S., Yamane, H., Banno, K., & Tao, R. (2018). RNA-sequencing analysis identifies genes associated with chilling-mediated endodormancy release in apple. *Journal of the American Society for Horticultural Science, 143*(3), 194–206.

Tan, F. C., & Swain, S. M. (2006). Genetics of flower initiation and development in annual and perennial plants. *Physiologia Plantarum, 128*(1), 8–17.

Tanaka, H., Masuta, C., Uehara, K., Kataoka, J., Koiwai, A., & Noma, M. (1997). Morphological changes and hypomethylation of DNA in transgenic tobacco expressing antisense RNA of the S-adenosyl-l-homocysteine hydrolase gene. *Plant Molecular Biology, 35*(6), 981–986.

Thirunavukkarasu, N., Hossain, F., Arora, K., Sharma, R., Shiriga, K., Mittal, S., et al. (2014). Functional mechanisms of drought tolerance in subtropical maize (Zea mays L.) identified using genome-wide association mapping. *BMC Genomics, 15*(1), 1182.

Tiwari, S. B., Shen, Y., Chang, H. C., Hou, Y., Harris, A., Ma, S. F., et al. (2010). The flowering time regulator CONSTANS is recruited to the FLOWERING LOCUS T promoter via a unique cis-element. *New Phytologist, 187*(1), 57–66.

Tränkner, C., Lehmann, S., Hoenicka, H., Hanke, M. V., Fladung, M., Lenhardt, D., et al. (2010). Over-expression of an FT-homologous gene of apple induces early xowering in annual and perennial plants. *Planta, 232*(6), 1309–1324.

Turnbull, C. G. N., Anderson, K. L., & Winston, E. C. (1996). Influence of gibberellin treatment on flowering and fruiting patterns in mango. *Australian Journal of Experimental Agriculture, 36*(5), 603–611.

Tylewicz, S., Petterle, A., Marttila, S., Miskolczi, P., Azeez, A., Singh, R. K., et al. (2018). Photoperiodic control of seasonal growth is mediated by ABA acting on cell-cell communication. *Science, 360*(6385), 212–215.

Ubi, B. E., Sakamoto, D., Ban, Y., Shimada, T., Ito, A., Nakajima, I., et al. (2010). Molecular cloning of dormancy-associated MADS-box gene homologs and their characterization during seasonal endodormancy transitional phases of Japanese pear. *Journal of the American Society for Horticultural Science, 135*(2), 174–182.

Upreti, K. K., Reddy, Y. T. N., Prasad, S. R. S., Bindu, G. V., Jayaram, H. L., & Rajan, S. (2013). Hormonal changes in response to paclobutrazol induced early flowering in mango cv. Totapuri. *Scientia Horticulturae, 150*, 414–418.

Urrestarazu, J., Muranty, H., Denancé, C., Leforestier, D., Ravon, E., Guyader, A., et al. (2017). Genome-wide association mapping of flowering and ripening periods in apple. *Frontiers in Plant Science, 8*, 1923.

van Dyk, M. M., Soeker, M. K., Labuschagne, I. F., & Rees, D. J. G. (2010). Identification of a major QTL for time of initial vegetative budbreak in apple (Malus x domestica Borkh.). *Tree Genetics and Genomes, 6*(3), 489–502.

Vitasse, Y., Lenz, A., & Karner, C. (2014). The interaction between freezing tolerance and phenology in temperate deciduous trees. *Frontiers in Plant Science, 5*, 541.

Wang, H., Pan, J., Li, Y., Lou, D., Hu, Y., & Yu, D. (2016). The DELLA-CONSTANS transcription factor cascade integrates gibberellic acid and photoperiod signaling to regulate flowering. *Plant Physiology, 172*(1), 479–488.

Wang, X., Chen, J., Xie, Z., Liu, S., Nolan, T., Ye, H., et al. (2014). Histone lysine methyltransferase SDG8 is involved in brassinosteroid-regulated gene expression in *Arabidopsis thaliana*. *Molecular Plant*, 7(8), 1303–1315.

Wang, Y., Georgi, L. L., Reighard, G. L., Scorza, R., & Abbott, A. G. (2002). Genetic mapping of the evergrowing gene in peach [Prunus persica (L.) Batsch]. *Journal of Heredity, 93*(5), 352–358.

Wang, Y., Li, L., Ye, T., Lu, Y., Chen, X., & Wu, Y. (2013). The inhibitory effect of ABA on floral transition is mediated by ABI5 in Arabidopsis. *Journal of Experimental Botany, 64*(2), 675–684.

Wei, Z., & Li, J. (2016). Brassinosteroids regulate root growth, development, and symbiosis. *Molecular Plant, 9*(1), 86–100.

Weigl, K., Wenzel, S., Flachowsky, H., Peil, A., & Hanke, M. V. (2015). Integration of *BpMADS4* on various linkage groups improves the utilization of the rapid cycle breeding system in apple. *Plant Biotechnology Journal, 13*(2), 246–258.

Weller, J. L., & Ortega, R. (2015). Genetic control of flowering time in legumes. *Frontiers in Plant Science, 6*, 1–13.

Werner, T., Holst, K., Pörs, Y., Guivarc'h, A., Mustroph, A., Chriqui, D., et al. (2008). Cytokinin deficiency causes distinct changes of sink and source parameters in tobacco shoots and roots. *Journal of Experimental Botany, 59*(10), 2659–2672.

Werner, T., & Schmülling, T. (2009). Cytokinin action in plant development. *Current Opinion in Plant Biology, 12*(5), 527–538.

Wickland, D. P., & Hanzawa, Y. (2015). The FLOWERING LOCUS T/TERMINAL FLOWER 1 gene family: Functional evolution and molecular mechanisms. *Molecular Plant, 8*(7), 983–997.

Willige, B. C., Ghosh, S., Nill, C., Zourelidou, M., Dohmann, E. M. N., Maier, A., et al. (2007). The DELLA domain of GA INSENSITIVE mediates the interaction with the GA INSENSITIVE DWARF1A gibberellin receptor of arabidopsis. *The Plant Cell, 19*(4), 1209–1220.

Wu, K., Zhang, L., Zhou, C., Yu, C. W., & Chaikam, V. (2008). HDA6 is required for jasmonate response, senescence and flowering in Arabidopsis. *Journal of Experimental Botany, 59*(2), 225–234.

Wu, M. F., Yamaguchi, N., Xiao, J., Bargmann, B., Estelle, M., Sang, Y., et al. (2015). Auxin-regulated chromatin switch directs acquisition of flower primordium founder fate. *ELife, 4*.

Xiao, Y., Chen, Y., Charnikhova, T., Mulder, P. P. J., Heijmans, J., Hoogenboom, A., et al. (2014). OsJAR1 is required for JA-regulated floret opening and anther dehiscence in rice. *Plant Molecular Biology, 86*(1–2), 19–33.

Xiao, J., & Wagner, D. (2015). Polycomb repression in the regulation of growth and development in Arabidopsis. *Current Opinion in Plant Biology, 23*, 15–24.

Xu, F., Li, T., Xu, P. B., Li, L., Du, S. S., Lian, H. L., et al. (2016). DELLA proteins physically interact with CONSTANS to regulate flowering under long days in Arabidopsis. *FEBS Letters, 590*(4), 541–549.

Xu, L., Zhao, Z., Dong, A., Soubigou-Taconnat, L., Renou, J.-P., Steinmetz, A., et al. (2008). Di- and tri- but not monomethylation on histone H3 lysine 36 marks active transcription of genes involved in flowering time regulation and other processes in *Arabidopsis thaliana*. *Molecular and Cellular Biology, 28*(4), 1348–1360.

Yamagishi, N., Kishigami, R., & Yoshikawa, N. (2014). Reduced generation time of apple seedlings to within a year by means of a plant virus vector: A new plant-breeding technique with no transmission of genetic modification to the next generation. *Plant Biotechnology Journal, 12*(1), 60–68.

Yamaguchi, N., Wu, M. F., Winter, C. M., Berns, M. C., Nole-Wilson, S., Yamaguchi, A., et al. (2013). A molecular framework for auxin-mediated initiation of flower primordia. *Developmental Cell, 24*(3), 271–282.

Yamamuro, C., Zhu, J. K., & Yang, Z. (2016). Epigenetic modifications and plant hormone action. *Molecular Plant, 9*(1), 57–70.

Yamane, H., Kashiwa, Y., Ooka, T., Tao, R., & Yonemori, K. (2008). Suppression subtractive hybridization and differential screening reveals endodormancy-associated expression of an SVP/AGL24-type MADS-box gene in lateral vegetative buds of Japanese apricot. *Journal of the American Society for Horticultural Science, 133*(5), 708–716.

Yan, Y., Christensen, S., Isakeit, T., Engelberth, J., Meeley, R., Hayward, A., et al. (2012). Disruption of OPR7 and OPR8 reveals the versatile functions of jasmonic acid in maize development and defense. *The Plant Cell, 24*(4), 1420–1436.

Yin, H., Zhang, X., Liu, J., Wang, Y., He, J., Yang, T., et al. (2009). Epigenetic regulation, somatic homologous recombination, and abscisic acid signaling are influenced by DNA polymerase epsilon mutation in Arabidopsis. *The Plant Cell, 21*(2), 386–402.

Yin, Y., Vafeados, D., Tao, Y., Yoshida, S., Asami, T., & Chory, J. (2005). A new class of transcription factors mediates brassinosteroid-regulated gene expression in Arabidopsis. *Cell, 120*(2), 249–259.

Yordanov, Y. S., Ma, C., Strauss, S. H., & Busov, V. B. (2014). EARLY BUD-BREAK 1 (*EBB1*) is a regulator of release from seasonal dormancy in poplar trees. *Proceedings of the National Academy of Sciences, 111*(27), 10001–10006.

Young, E. (1992). Timing of high temperature influences chilling negation in dormant apple trees. *Journal of the American Society for Horticultural Science, 117*(2), 271–272.

Yu, S., Galvao, V. C., Zhang, Y. C., Horrer, D., Zhang, T. Q., Hao, Y. H., et al. (2012). Gibberellin regulates the arabidopsis floral transition through mir156-targeted SQUAMOSA PROMOTER BINDING-LIKE transcription factors. *The Plant Cell, 24*(8), 3320–3332.

Yu, X., Li, L., Li, L., Guo, M., Chory, J., & Yin, Y. (2008). Modulation of brassinosteroid-regulated gene expression by jumonji domain-containing proteins ELF6 and REF6 in Arabidopsis. *Proceedings of the National Academy of Sciences, 105*(21), 7618–7623.

Yuan, J. S., Himanen, S. J., Holopainen, J. K., Chen, F., & Stewart, C. N. (2009). Smelling global climate change: Mitigation of function for plant volatile organic compounds. *Trends in Ecology and Evolution, 24*(6), 323–331.

Yun, J., Kim, Y. S., Jung, J. H., Seo, P. J., & Park, C. M. (2012). The AT-hook motif-containing protein AHL22 regulates flowering initiation by modifying FLOWERING LOCUS T chromatin in arabidopsis. *Journal of Biological Chemistry, 287*(19), 15307–15316.

Zhang, C. C., Yuan, W. Y., & Zhang, Q. F. (2012a). RPL1, a gene involved in epigenetic processes regulates phenotypic plasticity in rice. *Molecular Plant, 5*, 482–493.
Zhang, H., Bishop, B., Ringenberg, W., Muir, W. M., & Ogas, J. (2012b). The CHD3 remodeler PICKLE associates with genes enriched for trimethylation of histone H3 lysine 27. *Plant Physiology, 159*(1), 418–432.
Zhang, H., Rider, S. D., Henderson, J. T., Fountain, M., Chuang, K., Kandachar, V., et al. (2008). The CHD3 remodeler PICKLE promotes trimethylation of histone H3 lysine 27. *Journal of Biological Chemistry, 283*(33), 22637–22648.
Zhang, J. Z., Mei, L., Liu, R., Khan, M. R. G., & Hu, C. G. (2014). Possible involvement of locus-specific methylation on expression regulation of LEAFY homologous gene (CiLFY) during precocious trifoliate orange phase change process. *PLos One, 9*(2).
Zhang, L., Wang, Y., Zhang, X., Zhang, M., Han, D., Qiu, C., et al. (2012c). Dynamics of phytohormone and DNA methylation patterns changes during dormancy induction in strawberry (Fragaria × ananassa Duch.). *Plant Cell Reports, 31*(1), 155–165.
Zhang, S., Zhang, D., Fan, S., Du, L., Shen, Y., Xing, L., et al. (2016). Effect of exogenous GA3 and its inhibitor paclobutrazol on floral formation, endogenous hormones, and flowering-associated genes in 'Fuji' apple (Malus x domestica Borkh.). *Plant Physiology and Biochemistry, 107*, 178–186.
Zhebentyayeva, T. N., Fan, S., Chandra, A., Bielenberg, D. G., Reighard, G. L., Okie, W. R., et al. (2014). Dissection of chilling requirement and bloom date QTLs in peach using a whole genome sequencing of sibling trees from an F2mapping population. *Tree Genetics and Genomes, 10*(1), 35–51.
Zhou, C., Zhang, L., Duan, J., Miki, B., & Wu, K. (2005). HISTONE DEACETYLASE19 is involved in jasmonic acid and ethylene signaling of pathogen response in arabidopsis. *The Plant Cell, 17*(4), 1196–1204.

CHAPTER TEN

Epigenetic Regulations of Fleshy Fruit Development and Ripening and Their Potential Applications to Breeding Strategies

Etienne Bucher*, Junhua Kong[§], Emeline Teyssier[§] and Philippe Gallusci[§,1]

*IRHS, Université d'Angers, INRA, AGROCAMPUS-Ouest, Université Bretagne Loire, Angers, France
[§]UMR EGFV, Bordeaux Sciences Agro, INRA, Université de Bordeaux, Villenave d'Ornon, France
[1]Corresponding author: E-mail: philippe.gallusci@inra.fr

Contents

Abstract

Fleshy fruits, which are only found in Angiosperms, are of high economic importance as they represent an essential component of the human daily diet. True fruits derive from carpels, a leaf-like organ, and their development requires major reprogrammings of

Advances in Botanical Research, Volume 88
ISSN 0065-2296
https://doi.org/10.1016/bs.abr.2018.09.015

gene expression patterns of which we are just starting to grasp the importance of epigenetic regulations. Starting at fruit set, fruit development can be subdivided in two main steps, cell division and cell enlargement, that are followed by ripening in the case of fleshy fruits, whereas dry fruits enter a lignification phase before dehiscence occurs. Here, we summarize the state of the art in epigenetic regulation of fruit initiation, development and ripening with a focus on fleshy fruits. We present evidence that stable epigenetic variations, also called epimutations can affect fruit phenotype and discuss the role of DNA methylation and specific histone post translational modifications at the different fruit developmental phases. Indeed most works so far have been performed on tomato, the model for fleshy fruit studies, but recent works also indicate that epigenetic regulations might be of primary importance in other fruits, although the precise mechanisms might differ. Future research directions required to answer key remaining questions are discussed with the idea to develop breeding strategies that will integrate the epigenetic dimension.

1. INTRODUCTION

Flowering plants, also called Angiosperms, have evolved a unique seed-containing organ, the fruit that usually develops from ovaries, although in a few cases fruit tissues derive from accessory tissues such as the receptacle in strawberry or the hypanthium in apple (see Seymour et al, 2013 for a review). In addition to their fundamental function in seed dispersal, fruits have acquired an essential role in human nutrition as they represent an important source of fibers, carbohydrates, antioxidants and vitamins (Klee & Giovannoni, 2011). Fruit development is initiated by pollination and follows a complex pathway that may differ depending on fruit organization and structure (Seymour et al., 2013). As far as fleshy fruits are concerned, two main steps, namely cell division and cell enlargement precede ripening, which is characterized by extensive metabolic changes that result in the accumulation of various compounds including soluble sugars, organic acids, colored compounds resulting in an overall change in fruit texture and appearance (Gapperet al, 2013). In several fleshy fruit species belonging to the Brassicaceae, Cucurbitaceae, Fabaceae, and Solanaceae families, fruits display a high degree of endopolyploidization that occur concomitantly to cell enlargement, a process that has been suggested to promote fruit growth and metabolic activity (Chevalier et al., 2011). In contrast, dry fruits are characterized by a lignification phase followed by fruit dehiscence and seed dispersal (Seymour et al., 2013).

Many studies have described the fundamental role of both genetic and hormonal controls of fleshy fruit development and ripening (recently reviewed in Giovannoni et al, 2017; Gapper et al., 2014), but it is only recently that epigenetic mechanisms were shown to play essential roles in fruits as well (Gallusci et al, 2016). Nowadays, it is widely accepted that epigenetic regulations refer to changes in gene expression that are stable and heritable but do not involve modification of the underlying DNA sequence. Epigenetic marks include DNA methylation at cytosine residues, and histone post translational modifications (HPTMs) that are maintained through DNA replication and cell division (Chapter 2 of this book and references therein). This allows the stable definition and perpetuation of specific gene expression patterns that occur concomitantly to cell differentiation and organ development (Birnbaum & Roudier, 2017; Eichten et al., 2014; Pikaard et al., 2014). So far, several studies in *Arabidopsis thaliana* and in crops have highlighted the relevance of epigenetic mechanisms in the control of various aspects of plant development including reproductive development (Wang & Köhler, 2017), shoot apical meristem functioning (Cao, He, Guo, & Liu, 2015) or root development (Kawakatsu et al., 2016), but also in traits of agronomical importance such as flowering time control (Baulcombe & Dean, 2014), seed development (Kawakatsu et al., 2017; Luo et al., 2009; Xing et al., 2015), heterosis (Dapp et al., 2015) or stress tolerance (Chinnusamy & Zhu, 2009; Crisp et al., 2016; Luo et al., 2012), among others.

As far as fruits are concerned, initial evidence of the role of epigenetic regulations were obtained by Hadfield (1993), who showed that the DNA of genes induced at the onset of tomato fruit ripening became hypomethylated. Since that time, most studies on epigenetic regulations have been performed using tomato fruit as it is an ideal model for the analysis of fleshy fruit development and ripening. However there is evidence accumulating that epigenetic regulations are critical not only for the control of tomato fruit ripening but that they are also involved in the control of fruit set, fruit development, fruit color and maturation in tomato, but also in other fruits (reviewed in; Giovannoni et al., 2017; Farinati et al., 2017; Gallusci et al., 2016), In this chapter, we review the current knowledge of epigenetic regulations with a focus on fleshy fruits and when relevant on the model plant *A. thaliana*. We present evidence that spontaneous or induced epigenetic variations, also called epialleles or epimutations, can generate stable and heritable changes of fruit phenotypes. As fruit set and parthenocarpy are traits of high agronomical value, we discuss the role of epigenetic regulations in these two processes in both Arabidopsis and crops. Indeed, an important aspect concerns the regulation of fruit ripening that was first shown to be under epigenetic control in tomato. The role of epigenetic mechanisms in the regulation of this developmental phase in tomato fruit and in other fleshy fruit species is discussed together with possible applications for fruit crop breeding.

2. EPIALLELES AS EVIDENCE OF THE IMPORTANCE OF EPIGENETIC MECHANISMS IN FRUITS

2.1 Natural and Induced Epigenetic Diversity

Even though epigenetic diversity and how it may affect important traits has been followed with great interest in the plant research field, reports on actual true (*sensu stricto*) "epigenetic mutants" or epialleles are relatively scarce. There may be several reasons for that: either epimutations *per se* are relatively rare, or their detection can be challenging. Indeed, it can be very difficult, for instance in natural populations, to pinpoint traits to genetic and epigenetic changes, because both can fluctuate strongly.

To circumvent this limitation, elegant epigenetic studies have been carried out in the Arabidopsis model plant (Johannes et al., 2009; Reinders et al., 2009). These studies revealed several key findings: epialleles can be artificially induced and stably inherited, they can be associated to specific traits but they can also contribute to increased genetic diversity via the mobilization of transposable elements (Reinders et al., 2009). Another approach was to look at heritable epigenetic changes accumulating during inbreeding (via single seed descent) over 30 generations (Becker et al., 2011). The authors found that tens of thousands of cytosines changed their methylation states in each of the individual lines, showing that the epimutation rate is about an order of magnitude higher than genetic mutation. Quite surprisingly, epimutation rates are not really affected by stress-treatments (Eichten & Springer, 2015). Even though it was observed that certain epimutations were directly linked to the stress treatments, there is currently no indication for a direct adaptive role of these epimutations for the host.

There are several often-mentioned examples of epimutations occurring in nature. These include the peloric epimutation at the *cycloidea* gene in *Linaria vulgaris* which leads to a change in flower symmetry (Cubas, Vincent, & Coen, 1999) and *Cnr* that affects fruit ripening in tomato (discussed below, Manning et al., 2006). In maize, paramutation also contributes to the natural variation of epimutations, in which methylated epialleles impose their methylation state to the other non-methylated allele (Hollick, 2012).

Between deliberately induced and natural epimutations, there are some interesting examples in which epimutations have occurred during clonal propagation (by grafting or *in vitro* culture). These include examples in long-lived trees such as oil palm (Ong-Abdullah et al., 2015) or apple (Daccord et al., 2017; El-Sharkawy, Liang, & Xu, 2015). These epimutations can result in clear changes in economically important traits (oil production and fruit quality respectively). Here we'll discuss in more detail how induced and natural epimutations can influence important fruit traits.

2.2 Epiallelic Variations Contributes to Fruit Phenotypic Diversity

The importance of epigenetic regulations in fruits was initially revealed by the characterization of the tomato *Colourless non ripening (CNR* (Manning et al., 2006)) locus. The *Cnr* allele was isolated as a spontaneous tomato plant variant with fruits showing none of the features normally associated with ripening such as lycopene accumulation, pericarp softening or flavor changes (Eriksson et al., 2004; Thompson et al., 1999). This mutant does show some rare reversion events, suggesting an epigenetic regulation at that locus (Manning et al., 2006). In an elegant study Manning et al. carried out high-resolution genetic mapping of the *Cnr* locus which allowed it to be narrowed down to a 13 kb window. BAC and cosmid sequencing revealed that the mapping interval included four genes (ORF7-10). Interestingly, no genetic difference between *Cnr* and wild-type plants could be identified strengthening the notion that this locus may be under epigenetic control. Among these four genes, ORF7 codes for a SQUAMOSA promoter binding protein-like (SPL) family transcription factor, named *LeSPL-CNR*, that displayed a strong differential gene expression between *Cnr* and wild type fruits (Manning et al., 2006). The role of *LeSPL-CNR* in fruit ripening was then confirmed by virus-induced gene silencing-mediated knockdown. Although, there were no sequence differences between *Cnr* and wild-type genomic sequences at the *CNR* locus, a part of the regulatory region of this gene turned out to be hypermethylated in a 286-bp region located 2.4 kb upstream of the first ATG. This abnormal hypermethylation was only found in lines containing the *Cnr* allele (Manning et al., 2006) and fruit phenotype could be reversed to WT by repressing the *CMT3* gene and to a lower extent the *MET1* or *DRM* genes (Chen et al., 2015). Notably, the

DNA methylation differences between wild-type and non-ripening plants was not only present at the *CNR* locus in fruit, but in all tested tissues (e.g. leaves). This finding is important: Just like genetic mutations, a true epimutation (in contrast to developmentally regulated DNA methylation changes) usually is present in all tissues and is therefore heritable. Hence, *Cnr* presents all characteristics of a spontaneous and stable epimutation and this demonstrates that natural methylation polymorphisms can dramatically affect tomato fruit phenotypes, supporting the potential role of epigenetic variation as a source of phenotypic diversity with potential breeding applications in this species.

Additional epialleles affecting fruit phenotypes have now been identified in tomato and in other plants. Among them, the *VTE3* gene which encodes a 2-methyl-6-phytylquinol methyltransferase underlies a quantitative trait locus (QTL) determining the vitamin E content from the wild tomato species *Solanum pennellii*. This locus was shown to correspond to an expression QTL associated with differential methylation at the *VTE3* promoter region (Quadrana et al., 2014). Methylation occurred at a SINE element located in the *VTE3* promoter region and can vary between tomato varieties thereby impacting gene expression and the Vitamin E content of fruits. Whether specific *VTE3* epialleles can be used for breeding is so far not clear because methylation at the SINE elements located in the *VTE3* promoter region can reverse leading to variable methylation states and expression levels in the progeny (Quadrana et al., 2014). Very recently, additional epialleles were characterized in tomato following methylome analyses in hybrids obtained after crossing the M82 commercial tomato variety with *Solanum penellii*, a wild relative, and in Recombinant Inbred Lines (Gouil & Baulcombe, 2018). Results indicate that interaction between alleles with distinct methylation levels are rather common and present several features of paramutations. These paramutation-like interactions may result in heritable changes in DNA methylation patterns that could lead to the acquisition of novel epialleles useful for tomato improvement provided these newly generated epialleles are stable enough. However, results of Gouil and Baulcombe (Gouil & Baulcombe, 2018) indicate that whereas interactions between epialleles usually led to increased DNA methylation levels, stochastic demethylation events were also observed, limiting possible applications for plant improvement. Clearly additional work is now needed to study the stability and transgenerational inheritance of these new methylation patterns and their impact on plant phenotypes.

Epiallelic variations generating fruit phenotypes have also been identified both in apples and pears. Changes in skin color in fruits of both species were associated with hypermethylation of the *MYB10* gene promoter region resulting in repression of this gene expression and the absence of anthocyanin accumulation (El-Sharkawy et al., 2015; Telias et al., 2011; Wang et al., 2013). Very recently, the methylation of a Karma transposon was shown to be causal to the mantled somaclonal variant of oil palm (*Elaeis guineenis*) fruit (Ong-Abdullah et al., 2015) which presents feminization of flower organs and reduced oil yield. All these examples highlight the relevance DNA methylation polymorphism in shaping phenotypic variations in fleshy fruits.

3. DNA METHYLATION AND HISTONE PTMS ARE INVOLVED IN THE CONTROL OF FRUIT SET

3.1 Hormones Are Essential to the Control of Fruit Set

Fruit set designs the transition from non-growing ovaries to rapidly developing fruits. It normally requires the full completion of fertilization, although in some plants fertilization can be bypassed leading to the formation of a seedless fruit, a process referred to as parthenocarpy (for a recent review Joldersma & Liu, 2018). Fruit set has been the subject of intense research demonstrating that hormones, including auxins, gibberellins (GAs) or cytokinins play essential roles in this process. For example, external auxin application can promote fruit development in citrus and strawberries, whereas GAs have similar effects on rose or apples and cytokinins in water melon, pears or kiwifruit (reviewed in Kumar et al., 2014 and references therein). This is consistent with studies in tomato, Arabidopsis and Eggplant showing that modulation of genes involved in auxin (Rotino, 1997) or GA biosynthesis (De Jong, Mariani, & Vriezen, 2009; Martí et al., 2007) and auxin signal transduction (Mazzucato et al., 2015; Wang et al., 2005) led to parthenocarpic fruit development. Beside the role of hormones in fruit set, evidence is now growing that epigenetic regulation may also play a crucial role is this essential developmental phase.

3.2 Polycomb Proteins Participate to the Control of Fruit Set in Arabidopsis

Early evidence that fruit set is also controlled by epigenetic mechanisms comes from the identification of the Arabidopsis *FIS* and *FIE* mutants.

The *fis* and *fie* mutations are characterized by premature seed development that occurs in the absence of fertilization and by a concomitant elongation of siliques suggesting that mechanisms that allow fertilization-independent seed development also impact fruit set (Chaudhury et al., 1997; Goodrich, et al., 1997). The *FIS* and *FIE* genes encode proteins constitutive of the *Arabidopsis* Polycomb Repressive Complexes 2 (PRC2, for a review (Xiao & Wagner, 2015)). Briefly, PRC2 complexes, initially identified in Drosophila (Schuettengruber et al., 2007), are composed of four core proteins: an Enhancer of Zeste protein (E(z)) containing a SET domain that corresponds to the subunit that catalyzes the trimethylation of lysine 27 on histone H3 (H3K27me3), an EXTRA SEX COMB protein (ESC), a SUPPRESSOR OF ZESTE 12 protein (Su(z)12) and a p55 protein also called Multicopy Supressor of Ira (MSI) or NurF55 (Entrevan et al., 2016). In *Arabidopsis* these proteins are encoded by small multigene families: (1) three genes *SWINGER (SWN)*, *CURLY LEAF (CLF)* and *MEDEA* (also called *FERTILISATION INDEPENDENT SEED DEVELOPMENT 1* (*FIS1*)) encode the Arabidopsis EZ proteins; (2) a unique gene, *FERTILIZATION INDEPENDENT ENDOSPERM DEVELOPMENT* (*FIE*) encodes the ESC orthologue; (3) three genes *VERNALISATION 2*, *EMBRYONIC FLOWER 2* (*EMF2*) and *FERTILISATION INDEPENDENT SEED DEVELOPMENT 2 (FIS2)* encode Su(z)12-like proteins; and (4) five genes, *MSI1* to *5* encode p55 orthologues (for a review: Mozgova et al., 2015). Molecular and genetic evidence allowed defining three main PRC2s in *Arabidopsis* that were named based on their Su(z)12 component, EMF2-PRC2, VRN2-PRC2 and FIS2-PRC2. They play major roles in controlling gene expression at developmental transitions such as the gametophyte to sporophyte, embryo to vegetative, and vegetative to reproductive transitions (Xiao & Wagner, 2015). Among these PRC2s, it is the FIS2-PRC2, which is involved in the repression of seed and fruit development before fertilization. Whether the effects on fruit set are direct or indirect remains, however, to be explored. Noteworthy, Arabidopsis genes involved in auxin biosynthesis and signaling are marked by the H3K27me3 mark specifically in meristems and leaves (Lafos et al., 2011) indicating that Arabidopsis PRC2s participate to the spatial and temporal regulation of auxin related genes. Hence, PRC2s could indirectly regulate fruit set through the control of auxin biosynthesis and signaling. Consistent with this hypothesis the FIS2-PRC2 directly represses the expression of genes of the auxin biosynthesis pathway in the mega-gametophyte thereby leading to a low auxin level that increases in the central cell after fertilization (Figueiredo, Batista, Roszak, & Köhler, 2015). In addition, interactions between hormones and epigenetic regulations are not limited to auxin and polycomb proteins, but also include GAs and cytokinins and several other epigenetic modulators including HDACs, SWI/SNF chromatin remodeling ATPases, or CHD3/mi-2 ATP-dependent chromatin remodeling factors (for a review Yamamuro, Zhu, & Yang, 2016). Whether and how the complex interplay between hormones and epigenetic regulators is involved in the regulation of fruit set remains to be determined.

3.3 The Roles of Polycomb Proteins in Fruit Set Control in Fleshy Fruits

Whether PRC2s have maintained a similar role in the control of seed and fruit development in *Arabidopsis* and in fleshy fruit crops is so far unclear. Indeed, many recent studies have showed the conservation of PRC2 proteins within the plant kingdom and there is evidence that at least in some cases their functional roles have been conserved. For example, the reciprocal complementation of *fie* mutations between *Arabidopsis* and the moss *Physcomitrella patens* is consistent with a functional conservation of PRC2 components during evolution (Mosquna et al., 2009). However genes encoding the different PRC2 subunits have been duplicated in many plant species (Butenko & Ohad, 2011; Lewis, 2017) leading to a diversity of PRC2s that may have acquired specific physiological functions. For example, the *MEDEA* and *FIS2* genes are specific to Arabidopsis and related species and control seed development in these plants (Miyake, Takebayashi, & Wolf, 2009; Spillane et al., 2007). In some cases as in rice (Li et al., 2014; Luo, Platten, Chaudhury, Peacock, & Dennis, 2009), PRC2s with different isoform composition seem to have a similar role, whereas in some others, such as *Hieracium pilosella* evidence for PRC2s controlling early seed development is lacking. In this later case repression of the unique *HpFIE* gene does not lead to proliferation of the megagametophyte central cell in the absence of fertilization but to seed abortion (Rodrigues, Luo, Berger, & Koltunow, 2010). As far as fleshy fruits are concerned, the functional analysis of PRC2 components has only been performed in tomato. The tomato genome contains two functional copies of the *Enhancer of zeste* (*EZ*) genes named *SlEZ1* and *SlEZ2* orthologous respectively to the Arabidopsis *SWN* and *CLF* genes, whereas both FIE and ESC proteins are encoded by unique genes (Boureau, Teyssier, & Drevensek, 2016; Gallusci et al., 2016; How-Kit et al., 2010). As a consequence, tomato PRC2s are characterized by their EZ protagonist rather than by the Su(Z)12 protein as in Arabidopsis. Interestingly, tomato plants with reduced expression of either of the *SlEZ* genes did not show alteration of fruit set (Boureau et al., 2016; How Kit et al., 2010), although ectopic carpel formation was observed with a low frequency in *SlEZ2* RNAi lines (Boureau et al., 2016). Indeed, repression of *SlEZ2* led to additional phenotypes including fasciated stems, abnormal leaves and flowers, and fruits that differed from WT in many aspects: immature fruits were light green to white in contrast with the green WT fruits, ripe fruits were as red as WT but appeared brighter with orange dots and were characterized by modified cutin and epicuticular wax composition as well as by a higher glandular trichome density at maturity (Boureau et al., 2016). Interestingly, repression of the unique gene encoding the tomato FIE resulted in parthenocarpic fruit development, modified flower and fruit shapes, that differed from those observed in *SlEZ* knockdown plants (Liu et al., 2012). As FIE is encoded by a unique gene in the tomato genome, this protein is likely participating to all tomato PRC2s. This may result in effects stronger than those caused by knocking down single *EZ* genes that could partly compensate for each other when co-expressed in tissues or organs. Hence, it is possible that tomato PRC2s participate to the control of fruit set, with possible compensatory effects between the two EZ components (Fig. 1).

3.4 DNA Methylation May Also Be Important for Fruit Set Control

Additional evidence of epigenetic control over fruit set comes from the analysis of Arabidopsis *met1* mutants that present fruits elongating in the absence of fertilization, indicating that maintenance of CG methylation is necessary for the control of fruit set (FitzGerald et al., 2008). The gene networks under methylation control at this developmental transition are so far unknown and their analysis requires tissue specific methylome and transcriptome analyses. It is also unclear whether other DNA methyltransferases are involved.

It is noteworthy that direct interaction between MET1 and MEDEA, a core component of the FIS2-PRC2, has been demonstrated in Arabidopsis and the corresponding genes are co-expressed in the female gametophyte suggesting that they cooperate in the control of endosperm development in this plant (Schmidt et al., 2013). Whether a similar cooperation exists for fruit set has not been investigated so far. Clearly, determining whether

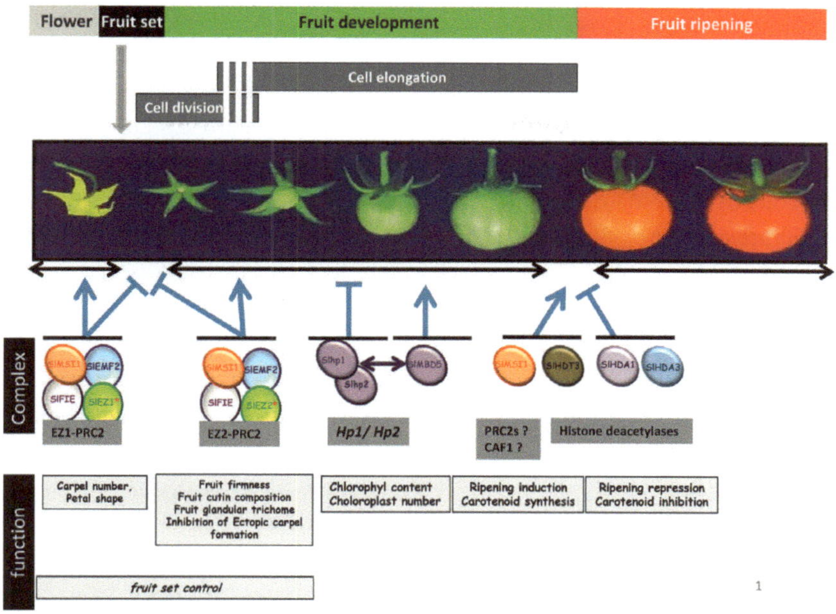

Figure 1 ***Roles of histone PTMs writers, erasers and readers during tomato fruit development.*** Works on tomato genes encoding different putative PRC2 components including the Enhancer of zeste, SlEZ1 and SlEZ2 and the Extra Sex Comb protein FIE have shown that PRC2s are involved in the control of early phases of fruit development and influence flower formation (Boureau et al., 2016; How Kit et al., 2010; Liu et al., 2012). In addition, loss of FIE function, but not of SlEZ1 or SlEZ2, seems to affect fruit set (Liu et al., 2012). As *FIE* is a unique gene in tomato, this is consistent with PRC2s controlling this developmental phase in tomato as well, and suggest functional redundancy between the different PRC2s in the control of fruit set. In addition the EZ2-PRC2 impacts fruit cuticle and cutin composition, limits the number of glandular trichomes in ripe fruits and prevents the formation of ectopic carpels. The HP1 (DDB1, Liu et al., 2004) and HP2 proteins (DET1, Mustilli, 1999), participate to CUL4-DDB1-DET1 complex and control fruit pigmentation during development and ripening. Their activity is antagonized by a MethylCytosine Binding protein, MBD5 that directly interacts with HP2. Recent works have also highlighted the role of various HDAC in the control of fruit ripening that have antagonist functions during ripening and likely play a role *in fine* tuning the induction and kinetic of ripening. EZ1 and 2: ENHANCER of ZESTE 1 and 2; FIE: FERTILISATION INDEPENDENT ENDOSPERM DEVELOPMENT; PRC2: POLYCOMB REPRESSIVE COMPLEX 2; HP1 and 2: HIGH PIGMENT 1 and 2; DDB1: UV-DAMAGED DNA BINDING PROTEIN1; DET1: DE-ETIOLATED-1 PROTEIN; CUL4: CULLIN 4; HDAC: HISTONE DEACETYLASE; MBD5: METHYLCYTOSINE BINDING PROTEIN 5.

PcG and DNA methylation act redundantly or synergistically to control fruit set awaits the study of various simple and double mutants affected in genes encoding for PcG proteins and in those involved in methylation control.

4. FRUIT SHAPE AND FRUIT GROWTH

The main function of fruit development is to facilitate seed dispersal. Much like flower shape and appearance has to be well-adapted to pollinators, fleshy fruit taste, shape, size and color has to be attractive to the animals that consume them. Epigenetic regulation of gene expression can rapidly contribute to the phenotypical diversity thereby increasing selective advantage during environmental changes. Indeed, in the case of flower development, a number of examples have been described in which epigenetic regulation, specifically DNA methylation, can influence flower shape (Cubas et al., 1999) and color (Iida et al., 2004). Fruit development is a complex and highly regulated process involving numerous regulatory pathways contributing to fruit size, shape, secondary metabolites accumulation and texture (Karlova et al., 2014). It is therefore likely that certain of these processes may be subjected to epigenetic regulation.

As of now, we could only find one well-studied example where epigenetic regulation of gene expression directly influences fruit development. In oil palm (*Elaeis guineensis*) somaclonal phenotypic variation in fruit development resulting from *in vitro* propagation has been observed. As described above, the "mantled" developmental defect of the fruit leads to changes in fruit shape up to complete parthenocarpy and thus strongly reduced oil content (Ong-Abdullah et al., 2015).

For fleshy fruits some first observations start to suggest that DNA methylation can contribute to fruit development. In tomato, it was observed that plants defective in DEMETER-like 2 (SlDML2), a DNA demethylase, not only showed delayed ripening (discussed below), but also defects in fruit development (Liu et al., 2015). Some plants developed fruits with multiple carpels. Whether these observations are directly linked to DNA methylation changes still needs to be addressed. It would be very interesting to test if this phenotype is stably inherited after outcrossing the RNAi construct that was used to repress *SlDML2*.

Further support for a potential role for DNA methylation in fruit development comes from studies performed on apple (Daccord et al., 2017). First global DNA methylation comparisons between leaf and young fruit tissues revealed DNA methylation differences, especially in the CHH methylation context. Generally, the trend was going towards an increase of DNA methylation in fruits compared to leaves. Whether these DNA methylation changes have a relevant biological role and if they are truly contributing to regulation of gene expression and that they are not just the result of transcriptional changes remain to be investigated.

The same study also took advantage of the possibility to compare two doubled haploid (homozygous) apple varieties with almost identical genomes (differing just by a few SNPs in protein coding regions). These two varieties produced drastically contrasted apple fruits in term of their sizes. The large fruits had a diameter of 4–5 cm while the small fruits only had a diameter of 2 cm. Histology revealed that the fruit size difference was due to reduced numbers of cell divisions in the smaller apple. In order to test if these phenotypic differences were due to epigenetic differences, DNA methylation patterns were compared between the small and the large fruits. This led to the identification of numerous stable differentially methylated regions (DMRs) close to 294 genes. Several of these genes play a role in fruit development in other species. Of note are an SPL13 orthologue that is differentially methylated between the two apple lines. The SPL pathway that can be regulated by miR156 is interesting because in Arabidopsis it was shown to closely interact with the miR172 pathway. Indeed, overexpression of miR172 in apple leads to a reduced fruit size (Yao et al., 2015). Another gene of interest with a large (>10 kb) DMR was *ACS8* (ACC synthase) which is involved in ethylene signaling. Ethylene can promote cell division in poplar (*Populus*) for instance (Love et al., 2009) and has previously been found to be under epigenetic control (Gapper et al., 2013). While these are interesting leads, the gene responsible for the fruit size difference in the large and small apples still remains to be confirmed. It would then be very interesting to test if fruit-size associated DRMs can also be observed in natural apple tree populations.

5. EPIGENETIC REGULATIONS OF FRUIT RIPENING

5.1 Dynamics of DNA Methylation in Fruits

As mentioned above, early evidence of the dynamics of epigenetic marks during tomato fruit development was provided by Hadfield (1993) who demonstrated changes in DNA methylation patterns at specific loci during fruit ripening. Since, a 30% global decrease in DNA methylation was shown to occur specifically in the pericarp of tomato fruits during ripening (Teyssier et al., 2008). The use of 5-azacytidine, a potent inhibitor of DNA methylation, to treat immature fruits was sufficient to induce premature ripening sectors and was associated with hypomethylation at various loci normally associated with ripening such as the *COLOURLESS NON RIPENING (CNR)*, the *ACC SYNTHASE 2 (ACS2)* and the *PHYTOENE SYNTHASE 1 (PSY1)* genes (Zhong et al., 2013). These results indicate that maintenance of DNA methylation is essential to prevent early ripening. This is also consistent with the observation that genes encoding enzymes involved in methylation maintenance such as *MET1, CMTs* (Chromomethylases) and *DRMs* (Domain Rearranged DNA Methyltransferases) are strongly expressed in the pericarp of young developing fruits, when cell division is highly active and later during fruit growth which is characterized by high endoreduplication rates (Teyssier et al., 2008).

To obtain a more precise understanding of the possible role DNA methylation during fruit ripening, Whole Genome Bisulfite Sequencing (WGBS) was used to describe the dynamics of the methylome over a developmental period starting from immature green to fully ripe tomato fruits (Zhong et al., 2013). The methylome of fruit pericarp was also compared to the one of leaves and of fruits of the *Cnr* and *rin* mutants that are unable to ripen (Manning et al., 2006; Vrebalov et al., 2002). The results confirmed a ripening specific decrease in DNA methylation in the pericarp of tomato fruits whereas fruit pericarp of the *Cnr* and *rin* mutants remained hypermethylated at all stages (Zhong et al., 2013). Demethylation occurred primarily in the CG sequence context and was mainly located in the heterochromatic regions of chromosomes characterized by high transposable element (TE) contents but was also observed in genic regions. Methylation at CG motifs is also the main methylation context in tomato promoters and genes and remained higher in leaves and in immature fruits than in ripe fruits suggesting that it is also responsible for the hypomethylated state of promoters observed during fruit ripening. Indeed, non CG methylation (CHG and CHH) also contributes to the hypomethylated status of promoters in ripening tomato fruits, in contrast to the demethylation at transposons which only occurs in the CG context. Noteworthy, the CHH methylation level is unusually high in tomato ranging from 8.35% in leaves to 13% in immature green fruits and 11% in ripe fruits (Zhong et al., 2013) and differs remarkably from Arabidopsis (1.5% of CHH methylation, Cokus et al., 2008), or rice (4.0% of CHH methylation, Li et al., 2012). The CHH methylation is significantly higher in tomato fruits than in leaves, suggesting that specific regulatory mechanisms control this type of methylation in tomato fruits. Consistent with this view, fruit CHH methylation is associated with miniature TEs and 24 nt small RNAs (Zhong et al., 2013) and several *DRM* genes are expressed in a tissue specific and developmentally regulated manner in tomato fruits (Gallusci et al., 2016; Giovannoni et al., 2017; Teyssier et al., 2008).

Importantly, DNA hypomethylation was found in the promoter region of genes that participate to various aspects of the ripening process, including ethylene synthesis and reception, carotenoid accumulation, cell wall metabolism and fruit softening, and several transcription factors among which the RIPENING INHIBITOR (RIN; Vrebalov et al., 2002), the NON-RIPENING (NOR; Giovannoni, 2004), the COLORLESS NON-RIPENING (CNR; Manning et al., 2006), and the TAGL1 proteins, that together with ethylene coordinate the induction of ripening (Gapper et al., 2013). Reduced methylation in the promoter region of these genes was not random but occurred essentially at or close to the binding sites of the RIN transcription factor. This makes it unlikely that loss of methylation during ripening was primarily due to a passive dilution of methylated cytosines after DNA replication, but rather suggests an active demethylation process.

5.2 Epigenetic Control of Fruit Ripening Requires the Ripening-Induced DNA Demethylase *DML2*

The tomato genome contains four genes encoding putative bifunctional 5-methylcytosine DNA glycosylase-lyases, also called demethylases (Gallusci et al., 2016), for enzymes that have been shown to actively remove methylated cytosines (Zhang et al., 2018). SlDML1 and 2 are both orthologous to the Arabidopsis demethylase ROS1, SlDML3 to DEMETER, whereas SlDML4 appears similarly distant from DML3 and 4 (Liu et al., 2015). Among these four genes, *SlDML2* is the only one strongly induced at the onset of fruit ripening simultaneously to the decrease in genomic DNA methylation observed in the fruit pericarp (Gallusci et al., 2016; Liu et al., 2015; Teyssier et al., 2008). The down-regulation of *SlDML2* using RNAi or VIGS, or its mutation by CRISPR-Cas9 resulted in extremely delayed ripening phenotypes that correlated with the repression of essential transcription factors, including RIN, NOR and CNR that are known to control fruit ripening and the absence of demethylation of their promoter regions. This contrasted with their high expression level and demethylated state in ripening WT fruits (Lang et al., 2017; Liu et al., 2015). WGBS analysis of *SlDML2* mutants generated in the microtome tomato variety indicated that 29,767 regions are hypermethylated in the mutant fruits relative to WT among which 14,704 are located in the proximal promoter of genes. About 600 corresponded to genes induced during fruit ripening which participate to various fruit ripening-related physiological processes such as cell wall metabolisms or carotenoid accumulation or encode TFs including the major regulators of fruit ripening. Additionally, 598 of the hypermethylated loci found in the *dml2* mutant were associated with genes down regulated during ripening, suggesting that DNA demethylation is correlated with their repression. These results suggest that DNA methylation might have a broader role than simply gene repression in plants and that it is required for both transcriptional activation and repression at least in tomato fruits (Fig. 2).

Although Zhong et al. (2013) have investigated DNA methylation in the pericarp only, analyses of the expression of DNA methyltransferase genes, as well as the quantification of the global DNA level in locular tissues, suggest that pericarp and locular tissues present different dynamics of DNA methylation variations in tomato fruits (Teyssier et al., 2008). In a recent study, the transcriptome of the different tomato fruit tissues was investigated. This work indicated that the locular tissue is likely the site of ripening initiation as indicated by the early expression of various major regulators of ripening including *CNR*, *RIN* or *NOR* (Shinozaki et al., 2018). Although no genome wide analysis of DNA methylation variations was performed in this study, the *RIN* promoter was similarly demethylated in the pericarp and locule of ripening tomato fruits, however with different kinetics (Liu et al., 2015; Shinozaki et al., 2018). So far, there is no evidence that a massive demethylation similar to the one observed in the pericarp also occurs in the locular tissues. However, the *SlDML2* gene, necessary for DNA demethylation and ripening induction in the pericarp of tomato fruits as detailed above, is also induced in the locular tissues, earlier though, suggesting that the DML2 enzyme may fulfill similar functions with regards to fruit ripening induction in both tissues. Based on this observation and on the fact that other major regulators of ripening are also induced earlier in the locules than in the pericarp, Giovannoni et al. (2017) proposed a refined ripening model that integrate both genetic, epigenetic and hormonal regulations with spatiotemporal aspects. Briefly, ripening is proposed to occur primarily in the locular tissues and involves the *CNR*, *RIN*, *NOR*, *TAGL1* as well as the *DML2* gene. Ethylene induction occurs later with a positive feedback loop on these different regulators before expanding to the pericarp (Giovannoni et al., 2017).

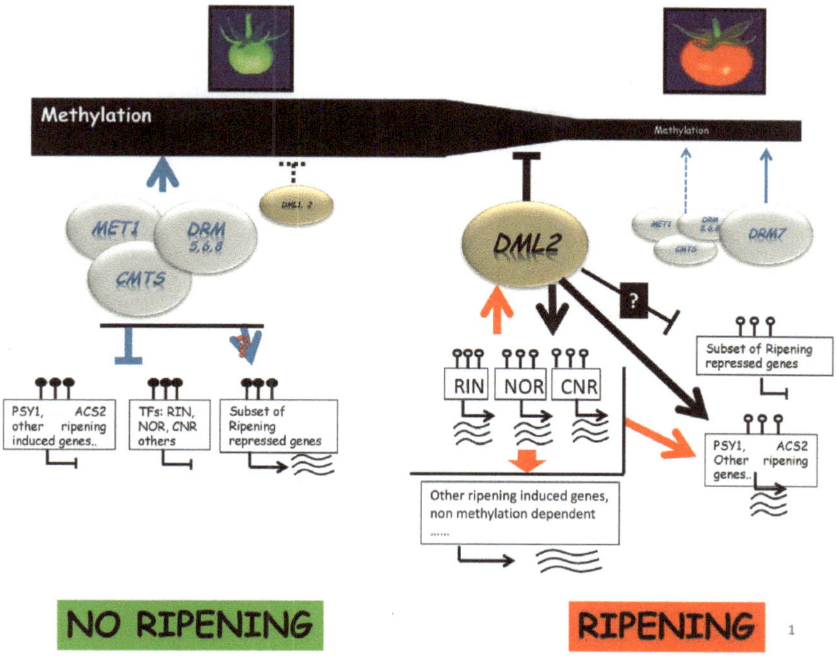

Figure 2 ***Function of genomic DNA methylation in tomato fruit.*** (A) High methylation levels in young fruits correlate with high expression levels of genes involved in maintenance and *de novo* DNA methylation. Master genes controlling fruit ripening (*RIN*, *NOR*, *CNR*) and other genes participating to the ripening process (*PSY1*, Polygalacturonase) and others have their expression repressed in immature green tomato fruits due to high methylation levels in their promoter regions (Lang et al., 2017; Liu et al., 2015), whereas some other genes, including those involved in metabolic processes essential at early developmental phases are expressed although methylated in their promoter region (Lang et al., 2017). (B) At the onset of fruit ripening, DNA methylation is reduced to 70% of the young fruit methylation levels, a process that has been associated with the strong induction of the *SlDML2* gene, whereas those involved in DNA methylation maintenance are barely expressed (Liu et al., 2015; Teyssier et al., 2008). Demethylation occurs in the promoter region of hundreds of genes, including those encoding the master regulators of fruit ripening (*CNR*, *RIN*, *NOR*), the genes involved in ethylene biosynthesis and many others. Demethylation is associated with the induction of these genes and ripening induction (Lang et al., 2017; Liu et al., 2015). In some cases DNA demethylation was suggested to be associated with the repression of gene expression although this has not been demonstrated yet (indicated by question marks) (Lang et al., 2017). Blue arrows indicate direct regulation by DNA methylation. Black arrows and dashes indicate direct regulation by DNA demethylation. Red arrows indicate other types of regulation. SlMET1 (cytosine-DNA-methyltransferase1), CMT (CHROMOMETHYLASE), DRM (DOMAIN, REARANGED METHYLTRASFERASE), DML (DEMETER LIKE DMETHYLASE), PSY1 (PHYTOENE SYNTHASE 1), ACS2 (ACC SYNTHASE 2), RIN (RIPENING INHIBITOR), NOR (NON RIPENING), CNR (COLOURLESS NON RIPENING).

5.3 Histone PTMS Might Have a Conserved Role in the Control of Fruit Ripening in Fleshy and Dry Fruits

Early evidence that histone post-translational modifications (PTMs) are involved in the control of fruit development was provided by the analysis of two high pigment tomato mutants, *hp1* and *hp2*, that are characterized by enhanced color of immature and mature tomato fruits and elevated levels of carotenoids. The *HP1* and *HP2* loci encode the UV-damaged DNA binding protein1 (DDB1, Liu et al., 2004) and the de-etiolated-1 proteins (DET1, Mustilli, 1999), respectively, that form with Cullin 4 (CUL4) the CUL4-DDB1-DET1 complex involved in protein degradation both in Arabidopsis and tomato (Chen, 2006; Tang et al., 2016). In addition, DET1 binds to the amino-terminal tails of histone H2B when non-acetylated thereby mediating transcriptional repression (Benvenuto et al, 2002; Fisher & Franklin, 2011). In *Arabidopsis*, DET1 plays an essential role in modulating H2B ubiquitination homeostasis and RNA PolII activity during photomorphogenesis (Nassrallah et al., 2018) and acts as a transcriptional repressor on light and circadian regulated genes (Lau et al., 2011). Whether the tomato DET1/HP2 protein also shapes the H2Bub landscape in developing fruits has not been established yet. In addition, DDB1/hp1 interacts with *SlMBD5* a methyl CpG binding protein that has an antagonist effect to DDB1/hp1 in fruits (Li et al., 2016). This is also consistent with CUL4-DDB1-DET1 being a transcriptional repressor in contrast to SlMBD5 that binds to hypomethylated DNA and mediate transcriptional activation (Li et al., 2016). These results are consistent with these proteins being involved in the control of fruit development through chromatin remodeling.

Additional evidence that epigenetic regulations are essential for fruit ripening is provided by the recent analysis of both DNA methylation and histone PTMs landscape performed in the frame of the fruitENCODE project (Lü et al., 2017). Using WGBS and ChIP-seq, the fruit epigenomes of more than 12 different fruits that included, in addition to tomato, six climacteric (apple, banana, melon, papaya peach and pear), four non climacteric (cucumber, grape, strawberry and watermelon) fleshy fruit species and as well as two dry fruit species (Arabidopsis and rice) have been determined. In total, 361 RNA-seq profiles, 76 tissue specific ChIP-seq and 43 methylomes were generated in this study to investigate the interactions between genetic and epigenetic regulatory circuits controlling fruit ripening (Lü et al., 2017). Although this work confirmed the prominent role of DNA demethylation in the control of tomato fruit ripening, it also suggested that fleshy fruits have evolved different regulatory mechanisms to control this critical developmental phase. Hence, although DNA demethylation was found to occur in various fleshy fruit species, differentially methylated regions (DMRs) were not always conserved between species and did not always target genes that participate to the regulatory loop controlling fruit ripening. These results question the general function of DNA demethylation in fruit ripening outside tomato (Lü et al., 2017). In contrast, the histone mark H3K27me3 which is set up by PRC2s appears highly conserved and marks genes involved in ethylene biosynthesis, or encoding transcription factors necessary for fleshy fruit ripening control, and their orthologues in the two dry fruit species Arabidopsis and rice (Lü et al., 2017). This led the authors to propose that the mechanisms of fleshy fruit ripening might have evolved from existing regulatory circuits found in ancestral dry fruit plants. However, additional work is now necessary to demonstrate the general function of PRC2s in the control of fleshy fruit ripening. As far as tomato fruits are concerned, RNAi lines affected in various components of the PRC2 complexes did not show alteration of fruit ripening (Boureau et al., 2016; How Kit et al., 2010; Liu et al., 2012). Interestingly, repression of *SlMSI1* gene that is thought to encode the Multisuppressor of Ira component (MSI) of PRC2s, was shown to specifically impact fruit ripening (Liu et al., 2016). This contrasts with the repression of other PRC2 components that only affect early fruit development (Boureau et al., 2016; How Kit et al., 2010; Liu et al., 2012). Whether the *SlMSI1* gene that was investigated by Liu et al. (Liu et al., 2016) encodes a protein that is really part of a PRC2 controlling fruit ripening is however questionable: (1) the authors do not present evidence that H3K27me3 levels were reduced or enhanced when expression of the *MSI1* gene was modified, (2) none of the tomato *EZ* genes is upregulated during fruit ripening in contrast with the *MSI1* gene; (3) there is no evidence that the SlMSI1 interacts with any of the other PRC2 components in ripening fruits; (4) knocking down the unique *FIE* gene in tomato has no impact on fruit ripening, although the corresponding protein is necessarily part of all tomato PRC2s. Altogether, this makes it unlikely that the *SlMSI1* gene is acting through a PRC2 to regulate tomato fruit ripening. In Arabidopsis, MSI1 also participates to the CAF1 complex that controls several aspect of plant development (Mozgova et al., 2018 and references therein). Hence, SlMSI1 action on tomato fruit ripening control could be mediated by CAF1 rather than PRC2s.

Several recent reports also indicate that histone modifiers are likely important for ripening in tomato and other fleshy fruits. First, the expression of genes encoding histone mark writers and erasers including HMTs, HDACs or HATs has been studied in various fruits such as apple (Janssen et al., 2008), banana (Fu et al., 2018), grapevine (Almada et al., 2011; Aquea et al, 2010, 2011), strawberry (Gu, Han, Huang, McAvoy, & Li, 2016) sweet orange (Xu et al., 2015) or tomato (Aiese Cigliano et al., 2013; Zhao et al., 2015) and results indicate fruit specific or preferential expression of a subset of these genes consistent with their recruitment for fruit development control. This was recently confirmed by the demonstration that repressing the tomato HDAC SlHDT3 delays fruit ripening and carotenoid accumulation (Guo et al., 2017), whereas two other tomato putative HDACs, SlHDA3 and SlHDA1 had an opposite role being negative regulators of both ripening and carotenoid synthesis (Guo et al., 2017, 2018). These results suggest complex regulatory networks that involve histone mark writers and erasers and allow tuning the kinetics of fruit ripening. Indeed, more detailed studies are now necessary to unravel the complex interplay between these chromatin based regulations and other essential regulatory of ripening including essential transcription factors and hormones such as ethylene, in tomato but also in other fruit crop species.

5.4 Epigenetic Control of Fruit Pigmentation

Fruit pigmentation is one of the most important commercial quality traits for fruits that are acquired during ripening. Indeed, color change is usually used as an important criterion to determine fruit maturity. While this may seem unfortunate from a scientific point of view, consumers have a strong tendency to select specific fruit colors and color patterns, depending on cultural trends and on what is currently fashionable. Therefore, breeders have long shown a strong interest in obtaining novel colors and/or color patterns on their fruits.

5.4.1 Carotenoid Accumulation Is Controlled by DNA Methylation in Tomato Fruits

As mentioned earlier (parts 1–2 and 4–3), epigenetic regulation of fruit coloration was initially suggested in tomato, following the characterization of the *Cnr* epimutation. In WT tomato, the typical red color of ripe fruit results from a 15-fold increase in carotenoid content, mainly lycopene and to a lower extend β-carotene, associated to the degradation of chlorophylls that occur at the onset of fruit ripening (Fraser et al., 2002). The tomato carotenoid biosynthesis pathway has been studied in great detail and the function of most key enzymes has been identified and confirmed, highlighting the fundamental role of the Phytoene synthase 1 (PSY1), an enzyme that controls the carbon flux entering the carotenoid pathway (Fantini et al., 2013; Fraser et al., 2002). Indeed, the *PSY1* gene and many other ripening induced genes are not expressed in the *Cnr* epimutant, but results suggest a pleiotropic effect of the absence of a functional CNR protein on fruit ripening that indirectly results in the absence of carotenoid accumulation rather that a direct regulation of the carotenoid pathway. Evidence that *PSY1* and other genes of the carotenoid pathway are directly regulated by DNA methylation was provided more recently by silencing the *SlDML2* gene during tomato fruit ripening (Liu et al., 2015). Results indicate an absence of carotenoid accumulation that correlates with *PSY1* gene repression and hypermethylation of its promoter region suggesting a direct link between DNA methylation and *PSY1* gene expression levels, as was also observed for the *NOR* gene (Liu et al., 2015). Interestingly, a conceptual model that used as input *DML2* gene expression profile during fruit development together with the methylation and expression levels of the *NOR* and *PSY1* genes in tomato fruits could successfully simulate methylation at *NOR* and *PSY1* promoters, *NOR* and *PSY1* gene expression profiles and the kinetics of lycopene accumulation during fruit development in WT and RNAi *DML2* fruits (Gallusci et al., 2017). Of course, this model now needs to be validated in tomato varieties that present various kinetics and levels of carotenoid accumulation. However, this work strongly suggests that the control of DNA methylation at key genes is sufficient to predict the accumulation of lycopene, an idea consistent with DNA methylation being a key regulatory pathway for carotenoid accumulation in tomato fruits.

5.4.2 Fruit Color in Fruit Trees: Lessons From Apple

In clonally propagated (grafted) fruit trees, one very common way to obtain color variations is via the spontaneous generation of "mutants" or "sports". Because "mutants" suggest a genetic change and since in most cases the underlying mechanism resulting in the change is unknown, we'll refer to "sports" here. Sports are usually branches on a tree that suddenly display different developmental and physiological properties: Changes in branching architecture, node formation, fruit size, leaf and fruit color just to name a few.

With respect to fruit color, apple is an interesting example. Orchards can contain millions of clones of the same apple tree. Having such a mass of trees combined with the keen eye of a breeder interested in obtaining novel varieties to satisfy the market, numerous sports for fruit coloration have been obtained. Using this approach, numerous novel Gala apple varieties have been obtained in the last decades. These varieties can show different percentages of coloration, stripes, blush types and fruit sizes compared to the Gala variety that is partially blushed. Some examples are "Blondee" or "Délice" which only produce yellow apples or "Galaxy" that produces red stripes. The great advantage of working with sports is that most of the time only one trait is changing, as compared to crosses where all traits are reshuffled after crossing (apple being an obligate outcrosser).

While all these sports greatly contribute to the generation of novel varieties, very little is known about the molecular mechanisms involved in generating those changes in the observed phenotypes. Because all these sports are clones, there are three main factors that may contribute to the phenotypes: Single genetic mutations such as nucleotide polymorphisms (SNPs) or short insertion/deletions (Indels), transposable element (TE) mobility and changes in DNA methylation (epimutations). These events will create genetic (SNPs, TEs) and epigenetic (TEs, DNA methylation) more or less stable somaclonal variability. Because the fruit color quality and quantity can easily be measured, these were the most intensely studied traits in sports. Very commonly, anthocyanins are the red pigments that accumulate in fruits and flowers. One of the most studied key transcription factors controlling anthocyanin accumulation in apple (Telias et al., 2011), pear (Yang et al., 2013) and strawberry (Medina-Puche et al., 2014) is MYB10. This MYB transcription factor regulates the expression of multiple enzymes involved in the anthocyanin biosynthetic pathway. Examples include chalcone synthase (CHS), dihydroxyflavonol-4-reductase (DFR) and UDP-glycose: flavonoid-3-*O*-glycosyltransferase (UFGT) (Takos et al., 2006). Genetic experiments have clearly demonstrated that different alleles of the MYB10 gene exist in red and non-red cultivars explaining the phenotypic differences (Takos et al., 2006).

However, what is happening in sports? Which molecular mechanisms are involved in changing fruit color and patterning? Variable stripe patterns are of particular interest from an epigenetic point of view because they might be caused by variable DNA methylation levels. Petals of morning glory flowers tend to show variable flower pigmentations and stripe patterns. Studies at the molecular level have demonstrated that these stripe patterns can be caused by the transposition or differential methylation of TEs in the promoter region of a *DFR* gene (Iida et al., 2004). Notably, differential methylation of TEs play an important role in regulating pigmentation in plants and animals (Slotkin & Martienssen, 2007) suggesting that highly conserved mechanisms may be acting.

Two studies suggest that epigenetic regulation of gene expression may be involved in these processes in apple skin coloration. The first looked at variable skin patterning in the Honeycrisp variety (Telias et al., 2011). This variety is interesting because it can show two different fruit pigmentation patterns: blushed and striped. To investigate the potential role of DNA methylation in the regulation of MYB10 the authors compared DNA methylation levels at the MYB10 promoter in dissected green and red stripes. The authors find that decreased anthocyanin and *MYB10* transcript accumulation correlate with increased DNA methylation at the *MYB10* promoter. Therefore, mosaicism in DNA methylation levels may explain stripe patterns in apple fruits.

Another study looked at more drastic differences in Gala apple sports where apples contrasted by their overall pigmentation (red and yellow) (El-Sharkawy et al., 2015). Using a transcriptomic approach, the authors identified numerous differentially expressed genes involved in anthocyanin biosynthesis, one of which was *MYB10*. Similar to the previously mentioned report, the DNA methylation level at the *MYB10* promoter inversely correlated with anthocyanin accumulation. Evidence is thus mounting that certain important crop traits are regulated by DNA methylation. Novel tools and drug treatment methods are now being developed in different labs to modify DNA methylation patterns at specific loci and genome-wide in crops (Thakore, Black, Hilton, & Gersbach, 2016; Thieme et al., 2017). These approaches now allow harnessing this additional source of variability for crop breeding.

6. CONCLUSION AND PERSPECTIVES

The understanding of the influence of epigenetic regulation of gene expression on traits of agronomical importance is only in its infancy (Springer & Schmitz, 2017). While interesting examples of epigenetic regulation of certain fruit development traits have been summarized here, it is so far unclear if those are merely exceptions, or if they represent examples of a general rule. For example, the regulation of DNA methylation appears as an essential factor that regulates fruit development and ripening in tomato, but it is unclear whether this epigenetic mechanism will have a similar function in other fleshy fruit species (Lü et al., 2018). This is due to the lack of functional studies of genes that control DNA methylation in fruit crops outside tomato that makes it difficult to determine the precise role of DNA methylation control in various types of fruits and therefore to determine general rules that could be used to develop breeding applications. Modelling carotenoid accumulation in tomato fruits as a function of *SlDML2* expression and DNA methylation level of selected genes provides an example of strategies that can be developed to integrate epigenetic parameters for the specific improvement of traits of high agronomical and economical values (Gallusci et al., 2017).

Additionally, there is also evidence accumulating that histone PTMs play important roles in the control of fruit development in tomato and most likely in other fleshy fruits as well. Undoubtedly the development of high-throughput sequencing and the availability of high-quality genomes will help extending the actual view of epigenome dynamics in various fleshy fruits in addition to tomato. Such genomes are already available for grape (Jaillon et al., 2007), apple (Daccord et al., 2017) or strawberry (Hirakawa et al., 2014; Shulaev et al., 2011) and their use for epigenomic studies should help determining the main epigenetic pathways that are at work in the control of the different phases of fruit development, including fruit set, fruit growth and ripening, but also in determining fruit shelf life. Of course, such approaches are correlative in nature and additional work including the analysis of mutants (natural or induced) or the study of natural diversity will be required to determine the exact role of epigenetic regulators in the different fruit crops. For example the study of Lü et al. (2017) that suggests that the histone PTMs H3K27me3 might be a general mechanism that governs fruit ripening in fleshy fruit species is unlikely to be valid in tomato since tomato mutants impaired in the different proteins that control this mark do not show major ripening alterations in this species (Boureau et al., 2016; How Kit et al., 2010; Liu et al., 2012), unlike those affected in the control of DNA demethylation (Liu et al., 2015). Hence, in addition to high-throughput technologies and epigenome analysis an important effort has to be done to systematically analyze the function of genes encoding histone PTM markers and erasers. Determining the epigenetic regulatory circuits that are important in fleshy fruits will indeed be the first step toward their use in breeding strategies. An additional aspect to consider is the allelic variability of genes encoding epigenetic regulators as this likely contributes to the epigenetic diversity of fruit crops and thereby to fruit phenotypical variations. Of course, epigenetic based phenotypical diversity could be generated by variations in the expression patterns of epigenetic regulators encoding genes or by allelic variants that present differences in their activity. Hence breeding programs that consider epigenetic parameters as phenotypic variables could be initiated to determine links between changes in epigenetic parameters and fruit phenotypes with the aim to identifying favorable alleles of epigenetic regulators. A nice example of such an approach has been provided by (Dubin et al., 2015) who showed that genome-wide CHH methylation variation was strongly associated with SNPs located close to the DNA methyltransferase *CMT2* gene. Whether such association with specific epigenetic regulators can be found for the control of fruit development and quality has never been studied and will require the development of GWAS studies that integrate the analysis of epigenetic parameters, or of epigenomic variations and selected phenotypes (color, kinetic of ripening, shelf life etc.) in a wide range of varieties and closely related wild species of selected fruit crops. As for now such studies do not exist, and are not possible in most species due to a lack of knowledge of the genomes, often impaired by genome size and the important cost for epigenomic studies. Nevertheless, gene candidate approaches could be developed as an alternative, as for example in the case of DNA demethylation in tomato: the expression pattern of the *SlDML2* gene could determine the kinetics and extent of tomato fruit ripening. Hence assessing the impact of the allelic variations and expression patterns at this locus and impact on fruit phenotypes might allow selecting the best *SlDML2* alleles for tomato fruit quality and shelf life.

An important point to consider is that all types of fleshy fruit crops do not share similar propagation strategies. For example, whereas tomato is propagated through sexual reproduction and seeds, apple or grape are clonally propagated and grafted. This adds a new layer of complexity as, depending on the fruit crop under consideration, the heritability of epigenetic marks may differ depending on the mode of propagation. So far, mainly DNA methylation was shown to be in part inherited though meiosis and probably through the female gametohyte whereas histone PTMs are unlikely to be transmitted through generations (reviewed in Gallusci et al., 2017). A totally different situation exists in clonally propagated plants as they are not obtained via meiosis where the epigenetic reprogramming may occur in gametes. In that case epigenetic variations generated in the mother plants are likely maintained in the daughter plants. This somatic inheritance of epigenetic variations is not yet well characterized and could involve both DNA methylation and histone PTMs (see Chapter 5; Gallusci et al., 2017). Also, the stability over years of epigenetic marks in meristems has not been investigated in detail. However, the occurrence of epigenetic somaclonal variations has to be considered together with genetic somaclonal variations, such as SNPs or deletions. Both types of events could contribute to the phenotypical diversification of clones, opening a new research area in fruits crops that are clonally propagated. As discussed above, this point is very elegantly illustrated by studies on fruit color control in apple, a grafted and clonally propagated plant. Indeed, in order to discriminate between the genetic or epigenetic nature of phenotypic variations affecting fruit traits in clonally propagated plant populations the genomes of candidate epimutants have to be compared and scored for mutations and TE activity.

In a more general way, there are still pending questions that need to be addressed to gain a better understanding of the epigenetic regulatory pathways operating during fruit development and ripening:

1. Are DNA methylation modifications the direct cause of changes in gene expression, or could the methylation patterns be the result of changes in transcriptional activities, as was suggested in animal systems (Schübeler, 2015)?
2. What are the perceived signals that result in DNA methylation changes? Are they due to stresses directly or indirectly caused by the mobility of transposable elements elsewhere in the genome? Or are these methylation changes purely stochastic?
3. What is the contribution of other epigenetic regulators in the control of fruit development? Are there common epigenetic processes to all fleshy and/or dry fruits that control the different developmental phases?
4. If epigenetic changes are an important source of phenotypic variability, are there ways to modify the epigenome of fruit trees and other crops in order to obtain epimutants in a more targeted manner?

The recent developments in molecular biology and whole genome sequencing technologies now provide important tools to answer these questions. For instance, the possibility to use CRISPR-Cas9 for targeted DNA methylation changes to create targeted epialleles will open the way to innovative strategies to better understand the role of DNA methylation in regulating gene expression and for fruit crops improvement, without any change in the DNA sequence (Thakore et al., 2016). We have now reached exciting times in which all these issues can be systematically addressed.

REFERENCES

Aiese Cigliano, R., Sanseverino, W., Cremona, G., Ercolano, M. R., Conicella, C., & Consiglio, F. M. (2013). Genome-wide analysis of histone modifiers in tomato: Gaining an insight into their developmental roles. *BMC Genomics, 14*(1), 57. https://doi.org/10.1186/1471-2164-14-57.

Almada, R., Cabrera, N., Casaretto, J. a, Peña-Cortés, H., Ruiz-Lara, S., & González Villanueva, E. (2011). Epigenetic repressor-like genes are differentially regulated during grapevine (*Vitis vinifera* L.) development. *Plant Cell Reports, 30*(10), 1959–1968. https://doi.org/10.1007/s00299-011-1104-0.

Aquea, F., Timmermann, T., & Arce-Johnson, P. (2010). Analysis of histone acetyltransferase and deacetylase families of *Vitis vinifera. Plant Physiology and Biochemistry*. https://doi.org/10.1016/j.plaphy.2009.12.009.

Aquea, F., Vega, A., Timmermann, T., Poupin, M. J., & Arce-Johnson, P. (2011). Genome-wide analysis of the SET DOMAIN GROUP family in Grapevine. *Plant Cell Reports, 30*(6), 1087–1097. https://doi.org/10.1007/s00299-011-1015-0.

Baulcombe, D. C., & Dean, C. (2014). Epigenetic regulation in plant responses to the environment. *Cold Spring Harbor Perspectives in Biology [electronic Resource], 6*(9), a019471. https://doi.org/10.1101/cshperspect.a019471.

Becker, C., Hagmann, J., Müller, J., Koenig, D., Stegle, O., Borgwardt, K., et al. (2011). Spontaneous epigenetic variation in the *Arabidopsis thaliana* methylome. *Nature*. https://doi.org/10.1038/nature10555.

Benvenuto, G., Formiggini, F., Laflamme, P., Malakhov, M., & Bowler, C. (2002). The photomorphogenesis regulator DET1 binds the amino-terminal tail of histone H2B in a nucleosome context. *Current Biology, 12*(17), 1529–1534. https://doi.org/10.1016/S0960-9822(02)01105-3.

Birnbaum, K. D., & Roudier, F. (2017). Epigenetic memory and cell fate reprogramming in plants. *Regeneration, 4*(1), 15–20. https://doi.org/10.1002/reg2.73.

Boureau, L., Teyssier, A. H. E., & Drevensek, S. (2016). A CURLY LEAF homologue controls both vegetative and reproductive development of tomato plants. *Plant Molecular Biology*. https://doi.org/10.1007/s11103-016-0436-0.

Butenko, Y., & Ohad, N. (2011). Polycomb-group mediated epigenetic mechanisms through plant evolution. *BBA - Gene Regulatory Mechanisms, 1809*, 395–406. https://doi.org/10.1016/j.bbagrm.2011.05.013.

Cao, X., He, Z., Guo, L., & Liu, X. (2015). Epigenetic mechanisms are critical for the regulation of *WUSCHEL* expression in floral meristems: Figure 1. *Plant Physiology, 168*(4), 1189–1196. https://doi.org/10.1104/pp.15.00230.

Chaudhury, A. M., Ming, L., Miller, C., Craig, S., Dennis, E. S., & Peacock, W. J. (1997). Fertilization-independent seed development in *Arabidopsis thaliana. Proceedings of the National Academy of Sciences of the United States of America, 94*(8), 4223–4228. https://doi.org/10.1073/pnas.94.8.4223.

Chen, H. (2006). Arabidopsis CULLIN4 forms an E3 ubiquitin ligase with RBX1 and the CDD complex in mediating light control of development. *The Plant Cell Online, 18*(8), 1991–2004. https://doi.org/10.1105/tpc.106.043224.

Chen, W., Kong, J., Qin, C., Yu, S., Tan, J., Chen, Y. R., et al. (2015). Requirement of CHROMOMETHYLASE3 for somatic inheritance of the spontaneous tomato epimutation Colourless non-ripening. *Scientific Reports, 5*, 1–7. https://doi.org/10.1038/srep09192.

Chevalier, C., Nafati, M., Mathieu-Rivet, E., Bourdon, M., Frangne, N., Cheniclet, C., et al. (2011). Elucidating the functional role of endoreduplication in tomato fruit development. *Annals of Botany, 107*(7), 1159–1169. https://doi.org/10.1093/aob/mcq257.

Chinnusamy, V., & Zhu, J. K. (2009). Epigenetic regulation of stress responses in plants. *Current Opinion in Plant Biology, 12*(2), 133–139. https://doi.org/10.1016/j.pbi.2008.12.006.

Cokus, S. J., Feng, S., Zhang, X., Chen, Z., Merriman, B., Haudenschild, C. D., et al. (2008). Shotgun bisulphite sequencing of the Arabidopsis genome reveals DNA methylation patterning. *Nature, 452*(7184), 215–219. https://doi.org/10.1038/nature06745.

Crisp, P. A., Ganguly, D., Eichten, S. R., Borevitz, J. O., & Pogson, B. J. (2016). Reconsidering plant memory: Intersections between stress recovery, RNA turnover, and epigenetics. *Science Advances, 2*(2), e1501340. https://doi.org/10.1126/sciadv.1501340.

Cubas, P., Vincent, C., & Coen, E. (1999). An epigenetic mutation responsible for natural variation in floral symmetry. *Nature*. https://doi.org/10.1038/43657.

Daccord, N., Celton, J. M., Linsmith, G., Becker, C., Choisne, N., Schijlen, E., et al. (2017). High-quality de novo assembly of the apple genome and methylome dynamics of early fruit development. *Nature Genetics, 49*(7), 1099–1106. https://doi.org/10.1038/ng.3886.

Dapp, M., Reinders, J., Bédiée, A., Balsera, C., Bucher, E., Theiler, G., et al. (2015). Heterosis and inbreeding depression of epigenetic Arabidopsis hybrids. *Nature Plants, 1*(7), 15092. https://doi.org/10.1038/nplants.2015.92.

De Jong, M., Mariani, C., & Vriezen, W. H. (2009). The role of auxin and gibberellin in tomato fruit set. *Journal of Experimental Botany, 60*(5), 1523–1532. https://doi.org/10.1093/jxb/erp094.

Dubin, M. J., Zhang, P., Meng, D., Remigereau, M.-S., Osborne, E. J., Paolo Casale, F., et al. (2015). DNA methylation in Arabidopsis has a genetic basis and shows evidence of local adaptation. *ELife, 4*, e05255. https://doi.org/10.7554/eLife.05255.

Eichten, S. R., Schmitz, R. J., & Springer, N. M. (2014). Epigenetics: Beyond chromatin modifications and complex genetic regulation. *Plant Physiology, 165*(3), 933–947. https://doi.org/10.1104/pp.113.234211.

Eichten, S. R., & Springer, N. M. (2015). Minimal evidence for consistent changes in maize DNA methylation patterns following environmental stress. *Frontiers in Plant Science*. https://doi.org/10.3389/fpls.2015.00308.

El-Sharkawy, I., Liang, D., & Xu, K. (2015). Transcriptome analysis of an apple (Malus × domestica) yellow fruit somatic mutation identifies a gene network module highly associated with anthocyanin and epigenetic regulation. *Journal of Experimental Botany, 66*(22), 7359–7376. https://doi.org/10.1093/jxb/erv433.

Entrevan, M., Schuettengruber, B., & Cavalli, G. (2016). Regulation of genome architecture and function by polycomb proteins. *Trends in Cell Biology, 26*(7), 511–525. https://doi.org/10.1016/j.tcb.2016.04.009.

Eriksson, E. M., Bovy, A., Manning, K., Harrison, L., Andrews, J., De Silva, J., et al. (2004). Effect of the Colorless non-ripening mutation on cell wall biochemistry and gene expression during tomato fruit development and ripening. *Plant Physiology, 136*(4), 4184–4197. https://doi.org/10.1104/pp.104.045765.

Fantini, E., Falcone, G., Frusciante, S., Giliberto, L., & Giuliano, G. (2013). Dissection of tomato lycopene biosynthesis through virus-induced gene silencing. *Plant Physiology*. https://doi.org/10.1104/pp.113.224733.

Farinati, S., Rasori, A., Varotto, S., & Bonghi, C. (2017). Rosaceae fruit development, ripening and post-harvest: An epigenetic perspective. *Frontiers in Plant Science, 8*(July), 1–14. https://doi.org/10.3389/fpls.2017.01247.

Figueiredo, D. D., Batista, R. A., Roszak, P. J., & Köhler, C. (2015). Auxin production couples endosperm development to fertilization. *Nature Plants, 1*(November), 1–6. https://doi.org/10.1038/nplants.2015.184.

Fisher, A. J., & Franklin, K. A. (2011). Chromatin remodelling in plant light signalling. *Physiologia Plantarum, 142*(4), 305–313. https://doi.org/10.1111/j.1399-3054.2011.01476.x.

FitzGerald, J., Luo, M., Chaudhury, A., & Berger, F. (2008). DNA methylation causes predominant maternal controls of plant embryo growth. *PLoS One, 3*(5), e2298. https://doi.org/10.1371/journal.pone.0002298.

Fraser, P. D., Romer, S., Shipton, C. A., Mills, P. B., Kiano, J. W., Misawa, N., et al. (2002). Evaluation of transgenic tomato plants expressing an additional phytoene synthase in a fruit-specific manner. *Proceedings of the National Academy of Sciences of the United States of America, 99*(2), 1092–1097. https://doi.org/10.1073/pnas.241374598.

Fu, C. C., Han, Y. C., Guo, Y. F., Kuang, J. F., Chen, J. Y., Shan, W., et al. (2018). Differential expression of histone deacetylases during banana ripening and identification of MaHDA6 in regulating ripening-associated genes. *Postharvest Biology and Technology, 141*(September 2017), 24–32. https://doi.org/10.1016/j.postharvbio.2018.03.010.

Gallusci, P., Dai, Z., Génard, M., Gauffretau, A., Leblanc-Fournier, N., Richard-Molard, C., et al. (2017). Epigenetics for plant improvement: Current knowledge and modeling avenues. *Trends in Plant Science, 22*(7). https://doi.org/10.1016/j.tplants.2017.04.009.

Gallusci, P., Hodgman, T. C., Teyssier, E., & Seymour, G. (2016). DNA methylation and chromatin regulation during fleshy fruit development and ripening. *Frontiers in Plant Science, 7*(June), 807. https://doi.org/10.3389/FPLS.2016.00807.

Gapper, N. E., Giovannoni, J. J., & Watkins, C. B. (2014). Understanding development and ripening of fruit crops in an "omics" era. *Horticulture Research, 1*(May), 1–10. https://doi.org/10.1038/hortres.2014.34.

Gapper, N. E., McQuinn, R. P., & Giovannoni, J. J. (2013). Molecular and genetic regulation of fruit ripening. *Plant Molecular Biology, 82*(6), 575–591. https://doi.org/10.1007/s11103-013-0050-3.

Gapper, N. E., Rudell, D. R., Giovannoni, J. J., & Watkins, C. B. (2013). Biomarker development for external CO_2 injury prediction in apples through exploration of both transcriptome and DNA methylation changes. *AoB Plants, 5*. https://doi.org/10.1093/aobpla/plt021.

Giovannoni, J. J. (2004). Genetic regulation of fruit development and ripening. *The Plant Cell Online, 16*, 170–180. https://doi.org/10.1105/tpc.019158.Fruit.

Giovannoni, J., Nguyen, C., Ampofo, B., Zhong, S., & Fei, Z. (2017). The epigenome and transcriptional dynamics of fruit ripening. *Annual Review of Plant Biology, 68*(1), 61–84. https://doi.org/10.1146/annurev-arplant-042916-040906.

Goodrich, et al. 1997-Nature.pdf. (n.d.).

Gouil, Q., & Baulcombe, D. C. (2018). Paramutation-like features of multiple natural epialleles in tomato. *BMC Genomics, 19*(1), 1–15. https://doi.org/10.1186/s12864-018-4590-4.

Gu, T., Han, Y., Huang, R., McAvoy, R. J., & Li, Y. (2016). Identification and characterization of histone lysine methylation modifiers in Fragaria vesca. *Scientific Reports, 6*(December 2015), 1–13. https://doi.org/10.1038/srep23581.

Guo, J. E., Hu, Z., Yu, X., Li, A., Li, F., Wang, Y., et al. (2018). A histone deacetylase gene, SlHDA3, acts as a negative regulator of fruit ripening and carotenoid accumulation. *Plant Cell Reports, 37*(1), 125–135. https://doi.org/10.1007/s00299-017-2211-3.

Guo, J.-E., Hu, Z., Zhu, M., Li, F., Zhu, Z., Lu, Y., et al. (2017). The tomato histone deacetylase SlHDA1 contributes to the repression of fruit ripening and carotenoid accumulation. *Scientific Reports, 7*(1), 7930. https://doi.org/10.1038/s41598-017-08512-x.

Hadfield 1993.pdf. (n.d.).

Hirakawa, H., Shirasawa, K., Kosugi, S., Tashiro, K., Nakayama, S., Yamada, M., et al. (2014). Dissection of the octoploid strawberry genome by deep sequencing of the genomes of fragaria species. *DNA Research*. https://doi.org/10.1093/dnares/dst049.

Hollick, J. B. (2012). Paramutation: A trans-homolog interaction affecting heritable gene regulation. *Current Opinion in Plant Biology*. https://doi.org/10.1016/j.pbi.2012.09.003.

Iida, S., Morita, Y., Choi, J. D., Park, K. I., & Hoshino, A. (2004). Genetics and epigenetics in flower pigmentation associated with transposable elements in morning glories. *Advances in Biophysics*. https://doi.org/10.1016/S0065-227X(04)80136-9.

Jaillon, O., Aury, J. M., Noel, B., Policriti, A., Clepet, C., Casagrande, A., et al. (2007). The grapevine genome sequence suggests ancestral hexaploidization in major angiosperm phyla. *Nature*. https://doi.org/10.1038/nature06148.

Janssen, B. J., Thodey, K., Schaffer, R. J., Alba, R., Balakrishnan, L., Bishop, R., et al. (2008). Global gene expression analysis of apple fruit development from the floral bud to ripe fruit. *BMC Plant Biology, 8*, 16. https://doi.org/10.1186/1471-2229-8-16.

Johannes, F., Porcher, E., Teixeira, F. K., Saliba-Colombani, V., Simon, M., Agier, N., et al. (2009). Assessing the impact of transgenerational epigenetic variation on complex traits. *PLoS Genetics*. https://doi.org/10.1371/journal.pgen.1000530.

Joldersma, D., & Liu, Z. (2018). The making of virgin fruit: The molecular and genetic basis of parthenocarpy. *Journal of Experimental Botany, 69*(5), 955–962. https://doi.org/10.1093/jxb/erx446.

Karlova, R., Chapman, N., David, K., Angenent, G. C., Seymour, G. B., & De Maagd, R. A. (2014). Transcriptional control of fleshy fruit development and ripening. *Journal of Experimental Botany*. https://doi.org/10.1093/jxb/eru316.

Kawakatsu, T., Nery, J. R., Castanon, R., & Ecker, J. R. (2017). Dynamic DNA methylation reconfiguration during seed development and germination. *Genome Biology, 18*(1), 1–12. https://doi.org/10.1186/s13059-017-1251-x.

Kawakatsu, T., Stuart, T., Valdes, M., Breakfield, N., Schmitz, R. J., Nery, J. R., et al. (2016). Unique cell-type-specific patterns of DNA methylation in the root meristem. *Nature Plants, 2*(May), 16058. https://doi.org/10.1038/nplants.2016.58.

Kit, A. H., Boureau, L., Stammitti-Bert, L., Rolin, D., Teyssier, E., & Gallusci, P. (2010). Functional analysis of SlEZ1 a tomato Enhancer of zeste (E(Z)) gene demonstrates a role in flower development. *Plant Molecular Biology, 74*(3), 201–213. https://doi.org/10.1007/s11103-010-9657-9.

Klee, H. J., & Giovannoni, J. J. (2011). Genetics and control of tomato fruit ripening and quality attributes. *Annual Review of Genetics, 45*, 41–59. https://doi.org/10.1146/annurev-genet-110410-132507.

Kumar, R., Khurana, A., & Sharma, A. K. (2014). Role of plant hormones and their interplay in development and ripening of fleshy fruits. *Journal of Experimental Botany, 65*(16), 4561–4575. https://doi.org/10.1093/jxb/eru277.

Lafos, M., Kroll, P., Hohenstatt, M. L., Thorpe, F. L., Clarenz, O., & Schubert, D. (2011). Dynamic regulation of H3K27 trimethylation during arabidopsis differentiation. *PLoS Genetics, 7*(4). https://doi.org/10.1371/journal.pgen.1002040.

Lang, Z., Wang, Y., Tang, K., Tang, D., Datsenka, T., Cheng, J., et al. (2017). Critical roles of DNA demethylation in the activation of ripening-induced genes and inhibition of ripening-repressed genes in tomato fruit. *Proceedings of the National Academy of Sciences of the United States of America, 114*(22), E4511–E4519. https://doi.org/10.1073/pnas.1705233114.

Lau, O. S., Huang, X., Charron, J. B., Lee, J. H., Li, G., & Deng, X. W. (2011). Interaction of arabidopsis DET1 with CCA1 and LHY in mediating transcriptional repression in the plant circadian clock. *Molecular Cell, 43*(5), 703–712. https://doi.org/10.1016/j.molcel.2011.07.013.

Lewis, Z. A. (2017). Polycomb group systems in fungi: New models for understanding polycomb repressive complex 2. *Trends in Genetics, 33*(3), 220–231. https://doi.org/10.1016/j.tig.2017.01.006.

Li, Y., Deng, H., Miao, M., Li, H., Huang, S., Wang, S., et al. (2016). Tomato MBD5, a methyl CpG binding domain protein, physically interacting with UV-damaged DNA binding protein-1, functions in multiple processes. *New Phytologist, 210*(1), 208–226. https://doi.org/10.1111/nph.13745.

Liu, D.-D., Dong, Q.-L., Fang, M.-J., Chen, K.-Q., & Hao, Y.-J. (2012). Ectopic expression of an apple apomixis-related gene MhFIE induces co-suppression and results in abnormal vegetative and reproductive development in tomato. *Journal of Plant Physiology, 169*(18), 1866–1873. https://doi.org/10.1016/j.jplph.2012.07.018.

Liu, R., How-Kit, A., Stammitti, L., Teyssier, E., Rolin, D., Mortain-Bertrand, A., et al. (2015). A DEMETER-like DNA demethylase governs tomato fruit ripening. *Proceedings of the National Academy of Sciences of the United States of America, 112*(34), 10804–10809. https://doi.org/10.1073/pnas.1503362112.

Liu, Y., Roof, S., Ye, Z., Barry, C., van Tuinen, A., Vrebalov, J., et al. (2004). Manipulation of light signal transduction as a means of modifying fruit nutritional quality in tomato. *Proceedings of the National Academy of Sciences of the United States of America, 101*(26), 9897–9902. https://doi.org/10.1073/pnas.0400935101.

Liu, D. D., Zhou, L. J., Fang, M. J., Dong, Q. L., An, X. H., You, C. X., et al. (2016). Polycomb-group protein SlMSI1 represses the expression of fruit-ripening genes to prolong shelf life in tomato. *Scientific Reports, 6*(July), 1–9. https://doi.org/10.1038/srep31806.

Li, S., Zhou, B., Peng, X., Kuang, Q., Huang, X., Yao, J., et al. (2014). OsFIE2 plays an essential role in the regulation of rice vegetative and reproductive development. *New Phytologist, 201*, 66–79. https://doi.org/10.1111/nph.12472.

Li, X., Zhu, J., Hu, F., Ge, S., Ye, M., Xiang, H., et al. (2012). Single-base resolution maps of cultivated and wild rice methylomes and regulatory roles of DNA methylation in plant gene expression. *BMC Genomics, 13*(1), 300. https://doi.org/10.1186/1471-2164-13-300.

Love, J., Bjorklund, S., Vahala, J., Hertzberg, M., Kangasjarvi, J., & Sundberg, B. (2009). Ethylene is an endogenous stimulator of cell division in the cambial meristem of Populus. *Proceedings of the National Academy of Sciences of the United States of America*. https://doi.org/10.1073/pnas.0811660106.

Luo, M., Liu, X., Singh, P., Cui, Y., Zimmerli, L., & Wu, K. (2012). Chromatin modifications and remodeling in plant abiotic stress responses. *Biochimica et Biophysica Acta - Gene Regulatory Mechanisms, 1819*(2), 129–136. https://doi.org/10.1016/j.bbagrm.2011.06.008.

Luo, M., Platten, D., Chaudhury, A., Peacock, W. J., & Dennis, E. S. (2009). Expression, imprinting, and evolution of rice homologs of the polycomb group genes. *Molecular Plant, 2*(4), 711–723. https://doi.org/10.1093/mp/ssp036.

Lü, P., Yu, S., Zhu, N., Chen, Y.-R., Zhou, B., Pan, Y., et al. (2017). *The fruitENCODE project sheds light on the genetic and epigenetic basis of convergent evolution of climacteric fruit ripening*. https://doi.org/10.1101/231258.

Lü, P., Yu, S., Zhu, N., Chen, Y., Zhou, B., Pan, Y., Tzeng, D., Fabi, J. P., Argyris, J., Garcia-Mas, J., Ye, N., Zhang, J., Grierson, D., Xiang, J., Fei, Z., Giovannoni, J., & Zhong, S. (2018). Genome encode analyses reveal the basis of convergent evolution of fleshy fruit ripening. *Nature Plants, 4*, 784–791.

Manning, K., Tor, M., Poole, M., Hong, Y., Thompson, A. J., King, G. J., et al. (2006). A naturally occurring epigenetic mutation in a gene encoding an SBP-box transcription factor inhibits tomato fruit ripening. *Nature Genetics, 38*(8), 948–952. https://doi.org/10.1038/ng1841.

Martí, C., Orzáez, D., Ellul, P., Moreno, V., Carbonell, J., & Granell, A. (2007). Silencing of DELLA induces facultative parthenocarpy in tomato fruits. *The Plant Journal, 52*(5), 865–876. https://doi.org/10.1111/j.1365-313X.2007.03282.x.

Mazzucato, A., Cellini, F., Bouzayen, M., Zouine, M., Mila, I., Minoia, S., et al. (2015). A TILLING allele of the tomato Aux/IAA9 gene offers new insights into fruit set mechanisms and perspectives for breeding seedless tomatoes. *Molecular Breeding, 35*(1). https://doi.org/10.1007/s11032-015-0222-8.

Medina-Puche, L., Cumplido-Laso, G., Amil-Ruiz, F., Hoffmann, T., Ring, L., Rodríguez-Franco, A., et al. (2014). MYB10 plays a major role in the regulation of flavonoid/phenylpropanoid metabolism during ripening of Fragaria × ananassa fruits. *Journal of Experimental Botany, 65*(2), 401–417. https://doi.org/10.1093/jxb/ert377.

Miyake, T., Takebayashi, N., & Wolf, D. E. (2009). Possible diversifying selection in the imprinted gene, MEDEA, in Arabidopsis. *Molecular Biology and Evolution, 26*(4), 843–857. https://doi.org/10.1093/molbev/msp001.

Mosquna, A., Katz, A., Decker, E. L., Rensing, S. A., Reski, R., & Ohad, N. (2009). Regulation of stem cell maintenance by the Polycomb protein FIE has been conserved during land plant evolution. *Development, 136*(14), 2433–2444. https://doi.org/10.1242/dev.035048.

Mozgova, I., Köhler, C., & Hennig, L. (2015). Keeping the gate closed: Functions of the polycomb repressive complex PRC2 in development. *The Plant Journal, 83*(1), 121–132. https://doi.org/10.1111/tpj.12828.

Mozgova, I., Wildhaber, T., Trejo-Arellano, M. S., Fajkus, J., Roszak, P., Köhler, C., et al. (2018). Transgenerational phenotype aggravation in CAF-1 mutants reveals parent-of-origin specific epigenetic inheritance. *New Phytologist.* https://doi.org/10.1111/nph.15082.

Mustilli, A. C. (1999). Phenotype of the tomato high pigment-2 mutant is caused by a mutation in the tomato homolog of DEETIOLATED1. *The Plant Cell Online, 11*(2), 145–158. https://doi.org/10.1105/tpc.11.2.145.

Nassrallah, A., Rougee, M., Bourbousse, C., Drevensek, S., Fonseca, S., Iniesto, E., et al. (2018). DET1-mediated degradation of a SAGA-like deubiquitination module controls H2Bub homeostasis. *BioRxiv, 309732.* https://doi.org/10.1101/309732.

Ong-Abdullah, M., Ordway, J. M., Jiang, N., Ooi, S. E., Kok, S. Y., Sarpan, N., et al. (2015). Loss of Karma transposon methylation underlies the mantled somaclonal variant of oil palm. *Nature, 525*(7570), 533–537. https://doi.org/10.1038/nature15365.

Pikaard, C. S., Scheid, O. M., Kingston, R. E., Tamkun, J. W., Baulcombe, D. C., & Dean, C. (2014). Epigenetic regulation in plants epigenetic regulation in plants. *Cold Spring Harbor Perspectives in Biology [electronic Resource]*, 1–31. https://doi.org/10.1101/cshperspect.a019315.

Quadrana, L., Almeida, J., Asís, R., Duffy, T., Dominguez, P. G., Bermúdez, L., et al. (2014). Natural occurring epialleles determine vitamin E accumulation in tomato fruits. *Nature Communications, 5*(May), 3027. https://doi.org/10.1038/ncomms5027.

Reinders, J., Wulff, B. B. H., Mirouze, M., Marí-Ordóñez, A., Dapp, M., Rozhon, W., et al. (2009). Compromised stability of DNA methylation and transposon immobilization in mosaic Arabidopsis epigenomes. *Genes & Development, 23*(8), 939–950.

Rodrigues, J. C. M., Luo, M., Berger, F., & Koltunow, A. M. G. (2010). Polycomb group gene function in sexual and asexual seed development in angiosperms. *Sexual Plant Reproduction.* https://doi.org/10.1007/s00497-009-0131-2.

Rotino, G. L., Perri, E., Zottini, M., & Sommer, H. (1997). A Spena Genetic engineering of parthenocarpic plants. *Nature Biotechnology, 15*, 1398–1401.

Schmidt, A., Wöhrmann, H. J. P., Raissig, M. T., Arand, J., Gheyselinck, J., Gagliardini, V., et al. (2013). The Polycomb group protein MEDEA and the DNA methyltransferase MET1 interact to repress autonomous endosperm development in Arabidopsis. *The Plant Journal, 73*(5), 776–787. https://doi.org/10.1111/tpj.12070.

Schübeler, D. (2015). Function and information content of DNA methylation. *Nature, 517*(7534), 321–326. https://doi.org/10.1038/nature14192.

Schuettengruber, B., Chourrout, D., Vervoort, M., Leblanc, B., & Cavalli, G. (2007). Genome regulation by polycomb and trithorax proteins. *Cell, 128*(4), 735–745. https://doi.org/10.1016/j.cell.2007.02.009.

Seymour, G. B., Østergaard, L., Chapman, N. H., Knapp, S., & Martin, C. (2013). Fruit development and ripening. *Annual Review of Plant Biology*. https://doi.org/10.1146/annurev-arplant-050312-120057.

Shinozaki, Y., Nicolas, P., Fernandez-Pozo, N., Ma, Q., Evanich, D. J., Shi, Y., et al. (2018). High-resolution spatiotemporal transcriptome mapping of tomato fruit development and ripening. *Nature Communications, 9*(1). https://doi.org/10.1038/s41467-017-02782-9.

Shulaev, V., Sargent, D. J., Crowhurst, R. N., Mockler, T. C., Folkerts, O., Delcher, A. L., et al. (2011). The genome of woodland strawberry (*Fragaria vesca*). *Nature Genetics*. https://doi.org/10.1038/ng.740.

Slotkin, R. K., & Martienssen, R. (2007). Transposable elements and the epigenetic regulation of the genome. *Nature Reviews. Genetics*. https://doi.org/10.1038/nrg2072.

Spillane, C., Schmid, K. J., Laoueillé-Duprat, S., Pien, S., Escobar-Restrepo, J. M., Baroux, C., et al. (2007). Positive darwinian selection at the imprinted MEDEA locus in plants. *Nature, 448*(7151), 349–352. https://doi.org/10.1038/nature05984.

Springer, N., & Schmitz, R. J. (2017). Exploiting induced and natural epigenetic variation for crop improvement. *Nat Rev Genetics, 18*, 563–575.

Takos, A. M., Jaffe, F. W., Jacob, S. R., Bogs, J., Robinson, S. P., & Walker, A. R. (2006). Light-induced expression of a MYB gene regulates anthocyanin biosynthesis in red apples. *Plant Physiology, 142*(3), 1216–1232. https://doi.org/10.1104/pp.106.088104.

Tang, X., Miao, M., Niu, X., Zhang, D., Cao, X., Jin, X., et al. (2016). Ubiquitin-conjugated degradation of golden 2-like transcription factor is mediated by CUL4-DDB1-based E3 ligase complex in tomato. *New Phytologist, 209*(3), 1028–1039. https://doi.org/10.1111/nph.13635.

Telias, A., Lin-Wang, K., Stevenson, D. E., Cooney, J. M., Hellens, R. P., Allan, A. C., et al. (2011). Apple skin patterning is associated with differential expression of MYb10. *BMC Plant Biology, 11*. https://doi.org/10.1186/1471-2229-11-93.

Teyssier, E., Bernacchia, G., Maury, S., How Kit, A., Stammitti-Bert, L., Rolin, D., et al. (2008). Tissue dependent variations of DNA methylation and endoreduplication levels during tomato fruit development and ripening. *Planta, 228*(3), 391–399. https://doi.org/10.1007/s00425-008-0743-z.

Thakore, P. I., Black, J. B., Hilton, I. B., & Gersbach, C. A. (2016). Editing the epigenome: Technologies for programmable transcription and epigenetic modulation. *Nature Methods*. https://doi.org/10.1038/nmeth.3733.

Thieme, M., Lanciano, S., Balzergue, S., Daccord, N., Mirouze, M., & Bucher, E. (2017). Inhibition of RNA polymerase II allows controlled mobilisation of retrotransposons for plant breeding. *Genome Biology, 18*(1), 1–10. https://doi.org/10.1186/s13059-017-1265-4.

Thompson, A. J., Tor, M., Barry, C. S., Vrebalov, J., Orfila, C., Jarvis, M. C., et al. (1999). Molecular and genetic characterization of a novel pleiotropic tomato-ripening mutant. *Plant Physiology, 120*(2), 383–389. https://doi.org/10.1104/pp.120.2.383.

Vrebalov, J., Ruezinsky, D., Padmanabhan, V., White, R., Medrano, D., Drake, R., et al. (2002). A MADS-box gene necessary for fruit ripening at the tomato *ripening-inhibitor* (*rin*) locus. *Science, 296*(April), 343–346. https://doi.org/10.1126/science.1068181.

Wang, H., Jones, B., Li, Z., Frasse, P., Delalande, C., Regad, F., et al. (2005). The Tomato *Aux/IAA* transcription factor *IAA9 is* involved in fruit development and leaf morphogenesis. *The Plant Cell Online, 17*(October), 2676–2692. https://doi.org/10.1105/tpc.105.033415.1.

Wang, G., & Köhler, C. (2017). Epigenetic processes in flowering plant reproduction. *Journal of Experimental Botany, 68*(4), 797–807. https://doi.org/10.1093/jxb/erw486.

Wang, Z., Meng, D., Wang, A., Li, T., Jiang, S., Cong, P., et al. (2013). The methylation of the PcMYB10 promoter is associated with green-skinned sport in max red bartlett pear. *Plant Physiology, 162*(2), 885–896. https://doi.org/10.1104/pp.113.214700.

Xiao, J., & Wagner, D. (2015). Polycomb repression in the regulation of growth and development in Arabidopsis. *Current Opinion in Plant Biology, 23*, 15–24. https://doi.org/10.1016/j.pbi.2014.10.003.

Xing, M.-Q., Zhang, Y.-J., Zhou, S.-R., Hu, W.-Y., Wu, X.-T., Ye, Y.-J., et al. (2015). Global analysis reveals the crucial roles of DNA methylation during rice seed development. *Plant Physiology, 168*(4), 1417–1432. https://doi.org/10.1104/pp.15.00414.

Xu, J., Xu, H., Liu, Y., Wang, X., Xu, Q., & Deng, X. (2015). Genome-wide identification of sweet orange (*Citrus sinensis*) histone modification gene families and their expression analysis during the fruit development and fruit-blue mold infection process. *Frontiers in Plant Science, 6*(August), 1–16. https://doi.org/10.3389/fpls.2015.00607.

Yamamuro, C., Zhu, J. K., & Yang, Z. (2016). Epigenetic modifications and plant hormone action. *Molecular Plant, 9*(1), 57–70. https://doi.org/10.1016/j.molp.2015.10.008.

Yang, Y., Zhao, G., Yue, W., Zhang, S. ling, Gu, C., & Wu, J. (2013). Molecular cloning and gene expression differences of the anthocyanin biosynthesis-related genes in the red/green skin color mutant of pear (*Pyrus communis* L.). *Tree Genetics and Genomes, 9*(5), 1351–1360. https://doi.org/10.1007/s11295-013-0644-6.

Yao, J. L., Xu, J., Cornille, A., Tomes, S., Karunairetnam, S., Luo, Z., et al. (2015). A microRNA allele that emerged prior to apple domestication may underlie fruit size evolution. *The Plant Journal*. https://doi.org/10.1111/tpj.13021.

Zhang, H., Lang, Z., & Zhu, J. (2018). Dynamics and function of DNA methylation in plants. *Nature Reviews. Molecular Cell Biology*. https://doi.org/10.1038/s41580-018-0016-z.

Zhao, L., Lu, J., Zhang, J., Wu, P.-Y., Yang, S., & Wu, K. (2015). Identification and characterization of histone deacetylases in tomato (*Solanum lycopersicum*). *Frontiers in Plant Science, 5*(January), 1–9. https://doi.org/10.3389/fpls.2014.00760.

Zhong, S., Fei, Z., Chen, Y.-R., Zheng, Y., Huang, M., Vrebalov, J., et al. (2013). Single-base resolution methylomes of tomato fruit development reveal epigenome modifications associated with ripening. *Nature Biotechnology, 31*(2), 154–159. https://doi.org/10.1038/nbt.2462.

CHAPTER ELEVEN

Aspects of Epigenetic Regulation in Cereals

Omar Oltehua-Lopez*, Ana E. Dorantes-Acosta*, Mathieu Ingouff[§], Sophie Lanciano[¶], Olivier Leblanc[§], Daniel Grimanelli[§], Marie Mirouze[¶,1] and Mario A. Arteaga-Vazquez*[,1]

*Laboratory of Epigenetics and Developmental Biology, Instituto de Biotecnología y Ecología Aplicada, Universidad Veracruzana, Xalapa, Mexico
[§]Institut de Recherche pour le Développement (IRD), UMR232, Université de Montpellier, Montpellier, France
[¶]Institut de Recherche pour le Développement (IRD), UMR232, Laboratoire Génome et Développement des Plantes, Perpignan, France
[1]Corresponding authors: E-mail: marie.mirouze@ird.fr, maarteaga@uv.mx

Contents

Abstract

Plants' ability to respond to environmental stimuli and developmental cues depends upon changes in gene expression. In eukaryotes, genetic information encoded by DNA is packed in a highly regulated and dynamic nucleoprotein complex known as chromatin that is subject to epigenetic modifications. Historically, several biological phenomena relying on epigenetic mechanisms were first characterized in plants. Seminal discoveries such as paramutation and silencing of transposable elements were made in maize (*Zea mays*). Later rice (*Oryza sativa*) was selected as a good model for monocotyledons owing to its relatively small genome size and well-annotated sequenced genome. In the past few years an increasing number of epigenetic and epigenomic studies were performed in both maize and rice. In this chapter we will first compare the basic knowledge acquired on epigenetic regulation in rice and maize versus Arabidopsis, then we will describe cereals-specific aspects of epigenetic regulations.

Advances in Botanical Research, Volume 88
ISSN 0065-2296
https://doi.org/10.1016/bs.abr.2018.10.004

1. INTRODUCTION

The term epigenetics has come a long way and over the years it has been employed to describe diverse phenomena ranging from non-Mendelian patterns of inheritance to changes in chromatin or gene expression that are not necessarily mitotically and/or meiotically transmitted. The definition of epigenetics is rooted in the Aristotelian term epigenesis and was conceived by Conrad Waddington from a developmental perspective at a time where the role of genes in the orchestration of developmental programs was still being elucidated and the structure of DNA had not been discovered yet. Waddington defined epigenetics as "the branch of biology that studies the causal interactions between genes and their products which bring the phenotype into being" (Slack, 2002). Epigenetics emerged as a way to address a simple question that has kept curious minds busy for thousands of years: how does life develop? or from a developmental point of view: How can a single fertilized cell (the zygote) give rise to a complex organism harboring specialized cells?

In this chapter, we refer to epigenetics as the study of mitotically and/or meiotically heritable changes in gene expression that occur without changes in DNA sequence and describe the molecular mechanisms (epigenetic marks) underlying those changes in gene expression. Epigenetic regulation is crucial for the dynamic regulation of chromatin states that bring about distinct transcriptional responses and is fundamental to the existence of multicellular organisms as it allows cells with precisely the same genotypes to give rise to a multitude of phenotypes (Heard & Martienssen, 2014). Epigenetic regulation is essential for growth and development of eukaryotic organisms (Henikoff, 2008; Shilatifard, 2006; Suzuki & Bird, 2008) and is also responsible for the maintenance, and reversal of non-genetic cellular memory that records developmental and environmental cues (Bonasio, Tu, & Reinberg, 2010). Epigenetic regulation involves enzyme-mediated chemical modifications of chromatin (DNA and its associated proteins) (Feng, 2010) that include (1) incorporation of histone variants, (2) positioning of nucleosomes, (3) posttranslational modifications of histones (tails and cores) including methylations, acetylations, phosphorylations, ubiquitylations, sumoylations, carbonylations, and glycosylations (reviewed in Berger, 2007; Jenuwein & Allis, 2001; Strahl & Allis, 2000), (4) methylation of cytosine (C) residues in the DNA and (5) the function of some noncoding RNAs (Feng, Jacobsen, & Reik, 2010; Feng & Jacobsen, 2011; Henderson & Jacobsen, 2007; Law & Jacobsen, 2010).

Whether epigenetic variations have contributed to the history of selection and domestication in cereals and could contribute to adaptation in a changing world is an important question in plant biology. In this chapter we will focus on epigenetic regulations unraveled in maize and rice. Maize has the highest production among cereals and is a primary food source around the world (Ranum, Peña-Rosas, & Garcia-Casal, 2014). In addition to its importance as a food crop, maize is a unique research model where classical epigenetic phenomena *i.e.* cycling and silencing of transposable elements (McClintock, 1984; Slotkin & Martienssen, 2007), gene imprinting (Kermicle, 1970; Kermicle & Alleman, 1990; Waters et al., 2013) and paramutation (Arteaga-Vazquez & Chandler, 2010; Brink, 1956; Chandler, Eggleston, & Dorweiler, 2000; Giacopelli & Hollick, 2015) were discovered. Rice is the most important food crop of the developing world and the staple food of more than half of the world's population. The small size of its genome (around 430 megabase pairs), sequenced in 2002 (Goff et al., 2002; IRGSP, 2005), has contributed to establish this crop as a model for cereals (reviewed in Jackson, 2016). Thus, while epigenetic regulations were first dissected in great details in *Arabidopsis thaliana*, today numerous epigenomic studies have been performed in maize and rice. RNAi and CRISPR-Cas tools have further allowed obtaining epigenetic mutants facilitating epigenetic studies in cereals. Here we will discuss the principal mechanisms of epigenetic regulations focusing on maize and rice specificities regarding DNA methylation and will detail some specific epigenetic phenomena in these major crops and their agronomical consequences.

2. IMPORTANCE OF DNA METHYLATION IN CEREALS

One of the best-characterized epigenetic modifications in eukaryotes is the methylation of DNA consisting in the covalent transfer of a methyl (-CH3) group to the fifth carbon of a cytosine residue in the DNA sequence to form 5-methylcytosine (Holliday & Pugh, 1975; Law & Jacobsen, 2010; Zhang, Lang, & Zhu, 2018)(5 mC). In plants, cytosines can be methylated in all symmetrical (CG or CHG, where H can be A, T, or C) and asymmetrical (CHH) sequence contexts. Transposable elements (TEs) and other repetitive sequences are the main targets of DNA methylation in plants (Quadrana & Colot, 2016; Underwood, Henderson, & Martienssen, 2017). Our understanding of the molecular basis and function of DNA methylation in plants comes from remarkable studies in the model system *A. thaliana* that have established a framework for comparative studies (Law & Jacobsen, 2010).

2.1 Establishment of DNA Methylation

The establishment of DNA methylation in plants is mediated by the RNA-dependent DNA methylation (RdDM) pathway (Matzke & Mosher, 2014; Wendte & Pikaard, 2017) and it involves the production of 24 nucleotides (nt) short interfering RNAs (siRNAs) that are able to guide *de novo* cytosine methylation and histone modifications characteristic of heterochromatin in homologous genomic regions (Law & Jacobsen, 2010; Matzke, Kanno, Daxinger, Huettel, & Matzke, 2009) (Fig. 1). In plants, the vast majority of cellular small RNAs correspond to 24 nt siRNAs derived from TEs that direct the establishment of heterochromatin at these repetitive regions (Matzke et al., 2009; Nobuta et al., 2008). RdDM requires the activity of two multisubunit plant specific DNA-dependent RNA polymerases: NUCLEAR RNA POLYMERASE D (NRPD or POL IV) and NRPE (POL V). POL IV generates single-stranded RNAs (ssRNAs) from TEs and repetitive loci that are converted into double stranded RNAs (dsRNAs) by the activity of RNA-DEPENDENT RNA POLYMERASE 2 (RDR2) (Lu et al., 2006; Xie et al., 2004). DICER-LIKE 3 (DCL3) (Gasciolli, Mallory, Bartel, & Vaucheret, 2005; Xie et al., 2004) processes dsRNAs into 24 -nt siRNAs that are stabilized by methylation at their 3′ end terminal OH groups by HUA ENHANCER 1 (HEN1). Members of the ARGONAUTE family, namely AGO4, AGO6 or AGO9 recruit 24-nt siRNAs that recognize and pair with complementary chromatin bound scaffold RNAs that are generated by POL V (from the same loci that gave rise to 24-nt siRNAs) and that will become the substrate for the *de novo* DNA methyltransferase DOMAINS REARRANGED METHYLTRANSFERASE 2 (DRM2). In addition to the interaction mediated by AGO4-loaded 24-nt siRNAs, recruitment of AGO4 and DRM2 to chromatin also occurs through protein-protein interaction between a particular domain in AGO4 termed the "AGO-hook" domain composed of Glycine/Tryptophan (GW/WG) repeats and the Carboxy-Terminal Domain (CTD) of POL V. The great majority of genomic targets recognized by POL V correspond to recent genic and intergenic insertion of TEs in euchromatic regions of the genome that associate with 24 nt siRNAs and DNA methylation and to a lesser extent to pericentromeric heterochromatin where DNA methylation predominantly occurs independent of 24 nt siRNAs. Up to date, the molecular determinants that recruit POL IV and POL V to their chromatin targets

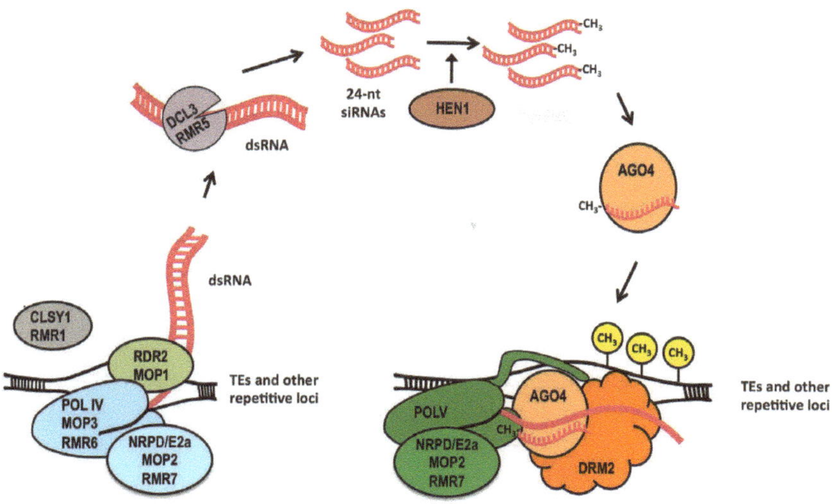

Figure 1 ***RdDM pathway in maize.*** Here we show a simplified version of the RdDM pathway in order to describe the major components in maize. The RdDM pathway targets TEs and other repetitive elements in the genome mainly for *de novo* DNA methylation but it also participates in the maintenance of DNA methylation. TEs and other repetitive loci are transcribed by POL IV (MOP3/RMR6), one of the two plant specific DNA-dependent RNA polymerases, into a single-stranded transcript that by the activity of RDR2 (MOP1) is converted into double-stranded RNA (dsRNA). The RMR1 protein that shares similarity to the CLASSY family is necessary for the production of short interfering RNAs (siRNAs) but its particular role in the pathway in maize is not clear. DCL3 (RMR5) processes the dsRNAs into a duplex of 24-nt siRNAs that are methylated at their 3′-end by the activity of the RNA-methyltransferase HEN1. AGO4 (and other members of the AGO4/6/9 clade) recruits a single stranded 24-nt siRNA and uses it as a guide to recognize a complementary region in the long non coding RNA that is being transcribed by POL V, the second plant specific RNA polymerase involved in silencing. AGO4 interacts with the CTD of POL V through a region known as the "AGO-hook" domain and by a mechanism yet to be described. This complex recruits the *de novo* DNA methyltransferase DRM2 to the loci that will be methylated in all sequence contexts (Alleman et al., 2006; Arteaga-Vazquez & Chandler, 2010; Haag et al., 2012; Hollick, 2012; Matzke & Mosher, 2014; Pikaard & Mittelsten Scheid, 2014; Sidorenko et al., 2009).

are yet to be described. However, di-methylation of Lysine 9 in histone 3 (H3K9me2) plays an important role in attracting POLIV and POLV to loci already harboring this particular chromatin modification (Matzke & Mosher, 2014).

In maize, a functional RdDM pathway is absolutely required for proper development. Mutations in loci encoding components of the RdDM pathway including *NRPD1/mediator of paramutation 3* (*mop3*)/*required to maintain repression 6* (*rmr6*), *NRPD/E2a/Mop2/rmr7* and *mop1*, exhibit pleiotropic defects such as delayed flowering, altered leaf polarity, reduction in plant height, altered sex determination and semi-sterility (Arteaga-Vazquez et al., 2010; Dorweiler et al., 2000; Erhard, Stonaker, Parkinson, & Lim, 2009; Parkinson, Gross, & Hollick, 2007; Sidorenko et al., 2009). Many of the defects observed in these mutants are similar to those observed in mutants that compromise microRNA (miRNA) function in Arabidopsis (Chen, 2009; Jones-Rhoades, Bartel, & Bartel, 2006) and consistent with this observation, a reduction in miRNA levels and disruption of miRNA regulatory circuits have been reported in *mop1* mutants (Arteaga-Vazquez et al., 2010; Hultquist & Dorweiler, 2008). For example, miRNA families showing reduced levels in *mop1-1* mutants include miR159, miR160, miR164, miR168 and miR169. Reduced levels of miR156 can also be observed in *Mop2-1* and *mop3-1* mutants (Arteaga-Vazquez et al., 2010; M.A. Arteaga-Vazquez, unpublished results). Interestingly, inbreeding of *mop3/rmr6* and *mop1* mutants for more than two generations leads to increased sterility and ultimately lethality (Parkinson et al., 2007; Hollick 2010; M.A. Arteaga-Vazquez, unpublished).

In rice, homozygous *Osdrm2* mutants show severe developmental defects. These mutants are completely sterile with a semi-dwarfed phenotype with an abnormal panicle architecture and a reduction in tiller number (Moritoh et al., 2012). The genome-wide DNA methylation level is reduced to 17% in *Osdrm2* compared to the WT suggesting that, in contrast to *Arabidopsis*, OsDRM2 is the major CHH methyltransferase in rice (Tan et al., 2016). Two other *DRM* genes (*DRM1a* and *DRM1b*) are not expressed and might not encode functional DNA methyltransferases (Moritoh et al., 2012). Interestingly, *OsDRM2* is the target of a microRNA miR820 that is encoded in a transposable element (CACTA) revealing the complex regulation between the host epigenetic machinery and TEs (Nosaka et al., 2012).

2.2 Maintenance of DNA Methylation

Different DNA methyltransferases acting in a sequence dependent context are involved in maintaining the patterns of DNA methylation. Following DNA replication, symmetrical methylation in CG can be maintained without the activity of 24-nt siRNAs as the two hemimethylated daughter strands serve as templates to methylate the newly replicated DNA strands. CG methylation is maintained by METHYLTRANSFERASE 1 (MET1) in coordination with VARIANT IN METHYLATION (VIM) proteins. Maintenance of CHG methylation occurs through the activity of the

DNA methyltransferase CHROMOMETHYLASE 3 (CMT3) and at some extent to CMT2 (Bewick et al., 2017; Cao et al., 2003; Finnegan E. J. et al., 1998). In Arabidopsis, a feedback loop is involved in the maintenance of CHG methylation through the ability of CMT3 to bind to dimethylated lysine 9 of histone H3 (H3K9me2) and methylate DNA at CHG sites in the vicinity (Bewick et al., 2017; Stroud et al., 2014). At the same time, methylated CHG can recruit SU(VAR) HOMOLOGUE 4 (SUVH4) SUVH5 and SUVH6 (Ebbs & Bender, 2006; Wendte & Schmitz, 2018) involved in the deposition of H3K9me2, creating this way a CHG-H3K9me2 reinforcement feedback loop. CMT2 can also bind to H3K9me2 but in this case CMT2 exhibits non-CG methylation that contributes to the maintenance of CHH methylation in heterochromatic regions associated with the histone H1 where RDdM is generally inhibited (Zemach et al., 2013). In sharp contrast, asymmetrical CHH methylation requires a constant supply of 24-nt siRNAs to establish DNA methylation after each DNA replication cycle. In summary, the coordinated activity between multiple DNA methyltransferases preserve methylation in all three sequence contexts (Bartee, Malagnac, & Bender, 2001; Bewick et al., 2017; Cao et al., 2003; Zemach et al., 2013). In addition to methyltransferases, DNA methylation is regulated by chromatin factors. DDM1 (Decrease in DNA Methylation 1), a SWI2/SNF2-like chromatin remodelling protein, is necessary to maintain DNA methylation in *Arabidopsis* (Brzeski & Jerzmanowski, 2003; Jeddeloh, Stokes, & Richards, 1999).

Rice possess two putative *MET1* genes: *OsMET1-1* and *OsMET1-2* (Pavlopoulou & Kossida, 2007). *Osmet1-1* mutants do display developmental phenotype while *Osmet1-2* mutants present strong defects in seed development and vegetative growth (Hu et al., 2014). Its global CG methylation level is reduced by 3/4th while it is nearly completely abolished in the *Arabidopsis met1* mutant, suggesting a redundant function between OsMET1-1 and OsMET1-2 (Yamauchi et al., 2008). The *Osmet1-2* mutation also affects gene expression (13% misregulated genes against 2% in *Atmet1*) and transcriptional activity of TEs (Hu et al., 2014). Additionally there are three chromomethylases genes in rice: *OsCMT2*, *OsCMT3a* and *OsCMT3b*. *OsCMT3a* is the only functional *CMT3* orthologue, its loss-of-function affects the expression of genes and TEs and induces transgenerational TE mobility (Cheng et al., 2015). In contrast to *Arabidopsis cmt3*, *Oscmt3a* mutant has several developmental abnormalities: early flowering, nearly sterile plants with short stature. Two genes (*OsDDM1a* and *OsDDM1b*) are orthologous of *DDM1* in rice. The double mutant *Osddm1a*

Osddm1b, with a global DNA methylation level decreased to around 50%, is completely sterile (Higo et al., 2012; Tan et al., 2016).

DNA methylation can be lost passively following DNA replication (passive DNA demethylation) in the absence of DNA methyltransferase activity. DNA methylation can also be enzymatically removed (active DNA demethylation) by the activity of DNA demethylases (Ooi & Bestor, 2008; Smith & Meissner, 2013; Wu & Zhang, 2013, 2017). In plants, DNA demethylases are actually bifunctional enzymes that combine DNA glycosylase activity that allow them to remove 5 mC, cut the DNA and replace 5 mC with C through the base excision repair pathway (Gehring et al., 2006; Li, Kumar, & Qian, 2018; Zhu, Kapoor, Sridhar, Agius, Zhu, 2007; Zhang, Lang, & Zhu 2018). In Arabidopsis, there are four DNA glycosylases involved in active DNA demethylation in all sequence contexts: REPRESSOR OF SILENCING 1 (ROS1), DEMETER (DME), DEMETER LIKE-2 (DML2) AND DML3 (Li et al., 2018; Zhang, Lang, & Zhu, 2018). DNA methylation patterns established by RdDM can be removed by ROS1. In cereals the demethylation pathway is less documented. Rice for example possesses six putative DNA glycosylases: 4 ROS1 (ROS1a-d) and two DML3 (DML3a and b) but no DME homolog (Zemach et al., 2010). *ros1a* mutants display endosperm developmental phenotype, reminiscent of the *dme* mutant phenotype in Arabidopsis, suggesting that ROS1a and DME could share analogous functions in the endosperm.

In summary, the main mechanisms involved in DNA methylation in Arabidopsis are conserved in both maize and rice. However, and in contrast to *Arabidopsis*, rice and maize mutants with altered RdDM pathways display severe developmental phenotypes and some mutants cannot be recovered due to lethality (Li et al., 2014). Of note these three species differ in their DNA methylation level chromosome-wide (Mirouze & Vitte, 2014). While DNA methylation is located mainly at the centromeric and pericentromieric region in *A. thaliana*, it is spread all along the chromosomes in *Oryza sativa* and *Zea mays*, reflecting TE density. DNA methylation mechanisms silencing TEs might play a bigger role in cereal crops with a high TE content. Interestingly, the methylome further depends on the variety. In three different rice varieties, differentially methylated regions (DMRs) were located in the vicinity of genes involved in both epigenetic regulation and abiotic stress response and correlated with a change in gene expression level of these neighbouring genes (Garg, Narayana Chevala, Shankar, & Jain, 2015), showing that DNA methylation affects gene expression in a variety-specific way.

2.3 Dynamics of DNA Methylation

In Arabidopsis genome-wide DNA methylation levels are affected during development notably in the endosperm and during early plant development (Bouyer et al., 2017; Gehring, 2013; Gehring, Bubb, & Henikoff, 2009). In rice the best-documented example is hypomethylation during seed development in the endosperm tissue (Rodrigues et al., 2013). While dicotyledonous plant seeds retain only remnants of the endosperm, the cereals endosperm accumulate starch and storage proteins necessary to germination. This high nutritional value explains the success and the importance of cereals in human diet (Zhou, Yin, & Xue, 2013). Interestingly, while some endosperm-specific genes in rice such as the glutelin genes that produce important storage proteins have hypomethylated promoters in this tissue allowing their overexpression (Zemach et al., 2010), rice endosperm hypomethylation is not associated with a global reactivation of TE mobility. Indeed only one retrotransposon family named *PopRice* was enriched in the extrachromosomal part of the genome in seeds suggesting an active retrotransposition in this tissue (Lanciano et al., 2017). This suggests that most TE families are controlled by multiple epigenetic layers and/or require additional stimuli such as environmental stresses in order to become active.

Some DNA methylation changes are also induced by environmental stresses suggesting that epigenetic regulations could be involved in the Genotype x Environment effect (GxE) and thus be relevant for breeding. Phosphate (Pi) deficiency for instance alters gene transcription and DNA methylation patterns in Arabidopsis but also in rice (Secco et al., 2015). DMRs in response to Pi starvation mainly occur at TEs located near to Pi starvation-induced (PSI) genes. Hypermethylation in response to Pi starvation could facilitate PSI gene transcription. In a transgenerational drought stress experiment, drought-induced epimutations were shown to be non-random and inherited from generation to generation (Zheng et al., 2015).

3. EPIGENETIC REPROGRAMMING AND EPIALLELES

Reprogramming is required to remove epigenetic signatures acquired during development, or imposed by the environment, so that subsequent elaboration of the body plan in the embryo properly reflects the genetic blueprint characteristic of each species (Heard & Martienssen, 2014). Epigenetic marks have the potential to be transmitted meiotically when reprogramming fails. Epigenetic mutations (epimutations) giving rise to epigenetic alleles (epialleles) can be neutral or deleterious and this transgenerational epigenetic inheritance has the potential to be adaptive and in some cases, might even respond to environmental challenges, with major implications for heredity, breeding and evolution (Chandler, 2007; Dickinson, Costa, & Gutierrez-Marcos, 2012; Heard and Martienssen, 2014; Hollick, 2010).

Surprisingly there are few epialleles described yet in cereals (Fig. 2). In rice both *Epi-d1* and *Epi-df* show a dwarf phenotype: DNA hypermethylation in the promoter region of *DWARF1* induces downregulation in the case of *Epi-d1* (Miura et al., 2009) and hypomethylation in the 5′ region of *FERTILIZATION-INDEPENDENT ENDOSPERM1* (*FIE1*) induces ectopic expression or *Epi-df* (Zhang et al., 2012). In contrast, in *Epi-rav6*, the hypomethylation of a MITE (Miniature Inverted-Repeat TE) inserted in the *RELATED TO ABSCISIC ACID INSENSITIVE3 (ABI3)/VIVIPAROUS1 (VP1) 6* (*RAV6*) promoter induces an over-expression of the gene resulting in an increase in leaf angle and smaller grain size (Zhang, Sun, Cao, & Xianwei, 2015). While the phenotypes described above are not of interest for breeding purposes, future epigenomic approaches allowing the identification of epiallelic variants will likely help to expand the repertoire of epialleles and to better appreciate their roles in cereals.

4. IMPRINTING

Genomic imprinting, first discovered in maize at the *R* locus (Kermicle, 1970), is an epigenetic phenomenon responsible for differential allelic expression in a parent-of-origin-dependent manner both in mammals and flowering plants (Feil & Berger, 2007; Gutierrez-Marcos & Dickinson, 2012). Genomic imprinting is required to achieve normal development of the embryo or its nourishing companion tissue, *i.e.* the placenta in mammals and the endosperm in plants. Although several explanations have been proposed, the kindship theory that postulates a parental conflict for nutrient allocation is the most widely accepted for the acquisition of genomic imprinting in such evolutionary divergent organisms (Dilkes, 2004; Haig & Westoby, 1989). Note that the kingship theory is the only hypothesis addressed experimentally so far and that the data obtained in maize provided partial support (Wu et al., 2013). In Arabidopsis and maize, genomic imprinting affects key regulators of endosperm growth (ie, *MEA*, *FIS2*, *PHE1* and *Meg1*, respectively (He, Chen, & Zhu, 2011; Raissig, Baroux,

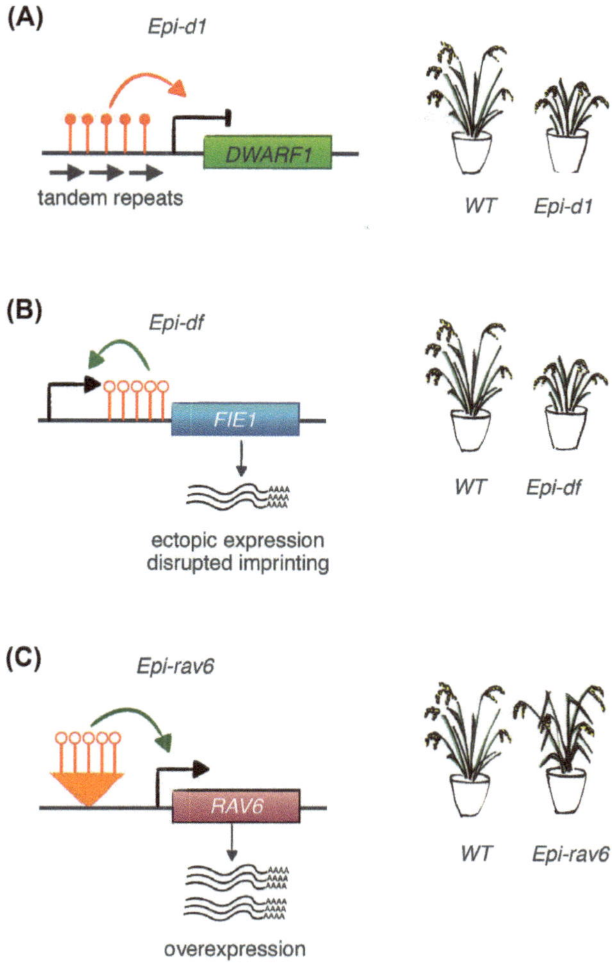

Figure 2 ***Examples of epialleles in rice.*** (A) In the *Epi-d1* epiallele hypermethylation at the *DWARF1* promoter induces metastable silencing of the gene leading to a dwarf plant phenotype (Miura et al., 2009). (B) In the *Epi-df* epiallele hypomethylation in the 5′ region of the *FIE1* gene leads to its ectopic expression and disrupted imprinting associated with a dwarf plant phenotype (Zhang et al., 2012). (C) In the *Epi-rav6* epiallele hypomethylation of a MITE at the *RAV6* promoter induces overexpression of the gene leading to plants with a larger lamina inclination and a smaller grain size (Zhang et al., 2015). *Red lollipops* represent positions with altered methylated patterns (full: hypermethylated compared to wild type, empty: hypomethylated compared to wild type). *Grey arrows* represent tandem repeats. *Black arrows* represent transcription start sites. *Orange triangle* represents a MITE insertion. Rice plants cartoons courtesy of Paco Càrdenas.

& Grossniklaus, 2011) and it is likely involved in the control of parental genomic contributions to the seed (Erilova et al., 2009; Jullien & Berger, 2010; Tiwari et al., 2010), both observations support the expected outcomes postulated by the kindship theory.

Dissection of endosperm imprinting revealed similarities with mammals, a hallmark being the role of DNA methylation in the establishment of gametic, differential DNA methylation either through the antagonistic action of the DNA methylase *MET1* and the DNA-glycosylase *DME* or from cis-elements such as DMRs and repeats or TEs possibly via small RNAs (Borges et al., 2018; He et al., 2011; Wollmann & Berger, 2012). Apart from DNA methylation, additional epigenetic marks might be essential for establishing imprints as allelic repression at the *PHE1* locus requires H3K27me3 (Köhler, Wolff, & Spillane, 2012) or the removal of DNA methylation is not sufficient to reactivate the *MEA* silent allele (Gehring et al., 2006). Intriguingly, Wöhrmann et al. (2012) reported that the ICR (Imprinting Controlling Region) of the *MEA* gene lacks DNA methylation, therefore pinpointing the role of other epigenetic regulators. In support of this, histone H1 is likely involved in allele regulation at imprinted loci in Arabidopsis (Rea et al., 2012) and the inactivated allele of three maternally imprinted genes in maize is repressed by H3K27me2/3 deposition while its active counterpart is enriched in acetylated histones H3 and H4 and H3K4me2 (Haun et al., 2007).

Genome-wide profiling for transcripts and epigenetic marks in both Arabidopsis and cereals (maize, sorghum, and rice) has recently provided insights into allele-specific expression and chromatin organization in seed hybrid tissues (Gehring, Missirian, & Henikoff, 2011; Guo, Rupe, Danilevskaya, Yang, & Hu, 2003; Gutiérrez-Marcos, Costa, & Evans, 2006; Luo et al., 2011; Waters et al., 2011; Wolff et al., 2011; Wyder, Raissig, & Grossniklaus, 2017; Xin et al., 2013; Zang et al., 2011). Genomic imprinting affects the expression of a relatively small number of candidate genes primarily in the endosperm, although imprinting was also reported in the embryo, but to a much lesser extent (Luo et al., 2011). Predicted functions for these candidates are diverse and often irrelevant to nutrient acquisition, except for the maize gene *Meg1* (Becraft & Yi, 2011; Wollmann & Berger, 2012). Finally, these studies revealed a trend for an association of imprinted loci with DMRs while no data is available for additional candidate epigenetic marks as-yet.

The data yet collected have provided considerable information for the molecular factors regulating imprinting of a few genes, but the emerging picture strongly questions the proposed models (Berger, Vu, Li, & Chen, 2012; Borges et al., 2018; Köehler et al., 2011; Wöhrmann et al., 2012; Wollmann & Berger, 2012), particularly regarding the nature and the dynamics of the underlying epigenetic information. In addition, despite strong evolutionary driving forces, genomic imprinting is a highly versatile phenomenon in plants (Raissig et al., 2011), i.e. imprinted genes are poorly conserved across distant species and, to a less extent, within species (Luo et al., 2011; Spillane et al., 2004; Wyder et al., 2017); it can be regulated by different mechanisms, including for paralogs (i.e., *ZmFIE1* and *ZmFIE2*, Dickinson, Costa, & Gutierrez-Marcos, 2012); it can be modulated in interspecific crosses (Gutierrez-Marcos, Pennington, Costa, & Dickinson, 2003; Josefsson, Dilkes, & Comai, 2006) and monoparental reproduction which is frequent in plants. Therefore our understanding of genomic imprinting in plants still remains elusive. Molecular approaches allowing the establishment of detailed epigenetic landscapes and of allele-specific transcriptomic repertoires will likely help to elucidate the regulation of imprinting and its role in seed biology.

5. PARAMUTATION

Paramutation was discovered in maize in the 1950s in specific alleles of the *red1* (*r1*) locus (Brink, 1956). Paramutation was also originally reported for the *sulfurea* (*sulf*) locus in tomato (Hagemann, 1958) and the *booster1* (*b1*) locus in maize (Coe, 1959) and has been documented in other plants and animals (Chandler, 2007). There are four types of alleles that need to be taken into account when studying paramutation: (1) Paramutagenic alleles are alleles capable of promoting a change, (2) Paramutable alleles are alleles prone to be changed, (3) Paramutant alleles represent previously paramutable alleles that when exposed to a paramutagenic allele became paramutagenic (this means a paramutant allele is a previously paramutable allele that has been paramutated) and iv) Neutral alleles (or non paramutagenic) that do not participate in paramutation and therefore are neither capable of inducing a change nor capable of being changed (Chandler & Stam, 2004) (Fig. 3). Crossing a neutral allele with a paramutagenic or a paramutable allele does not change the expression of any of the involved alleles. A unique distinctive feature of paramutation is secondary paramutation and it is observed when a paramutant allele is crossed to a paramutable allele and it generates a new paramutagenic allele. The spreading of paramutation or "infectivity" indicates that

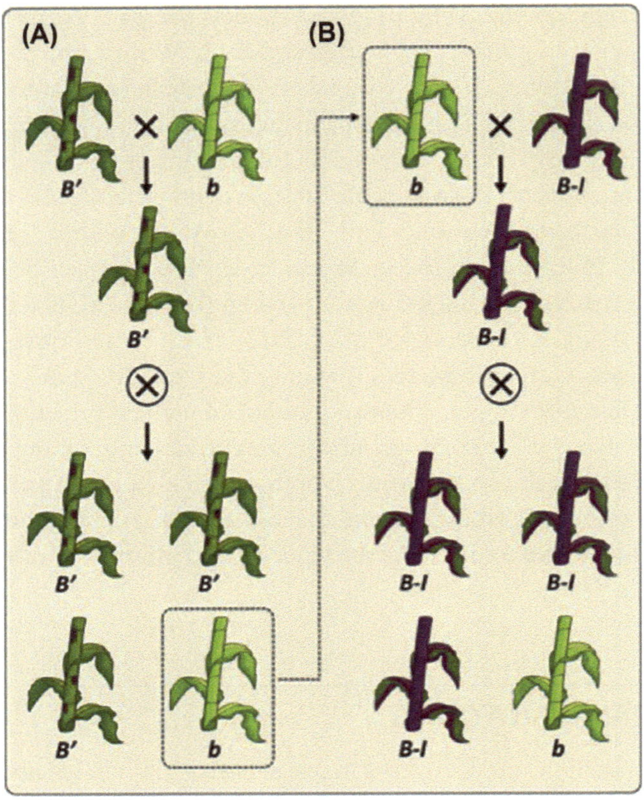

Figure 3 ***Behavior of neutral alleles for paramutation.*** Neutral alleles are not changed nor capable of inducing a change when exposed to paramutagenic or paramutable alleles, respectively. (A) Self-pollination of the F1 progeny of a cross between the neutral allele *b* and the paramutagenic allele *B′* shows segregation of *B′* and *b* plants following a Mendelian pattern. (B) When an F2 *b* plant from (A) is crossed to the paramutable *B-I* allele, the F2 progeny of the self-pollinated F1 population segregates unchanged *B-I* and *b* plants.

paramutation not only changes the expression state of a given allele but it can also alter its regulatory properties (Chandler et al., 2000). In plants, the best-characterized examples of paramutation involve alleles of four different loci that encode transcription factors that activate the biosynthesis of colored flavonoid pigments: *r1*, *b1*, *pl1* and *p1* (Arteaga-Vazquez & Chandler, 2010; Chandler et al., 2000; Chandler & Stam, 2004; Dooner, Robbins, & Jorgensen, 1991; Goff et al., 1990; Hollick, 2010; Stam, Belele, Dorweiler, & Chandler, 2002a). While *r1*, *b1* and *pl1* regulate the production of purple anthocyanin pigment (Dooner et al., 1991; Goff, Cone, & Chandler, 1992), *p1* is required for the synthesis of orange/red phlobaphene pigment

(Grotewold, Athma, & Peterson, 1991). Paramutation at the maize *b1* locus is the most stable and penetrant natural system described to date (Arteaga-Vazquez & Chandler, 2010) (Fig. 4). There are two alleles of the *b1* locus involved in paramutation: (1) *B-I* allele is a highly transcribed, darkly pigmented and paramutable allele, (2) *B′* allele is a weakly expressed, lightly pigmented and paramutagenic allele. When *B-I* and *B′* are crossed together, paramutation always occurs: *B-I* is always changed into *B′* (Chandler & Stam, 2004). *B-I* is unstable and spontaneously can give rise to *B′* at relatively high frequencies ranging from 0.1 to 10% (Chandler, 2010; Coe, 1966; Patterson, Thorpe, & Chandler, 1993). Once *B-I* acquires a *B′* state, it will not naturally turn again into *B-I*. No changes from *B′* to *B-I* have been recorded in over 50 years of research and hundreds of thousands of plants analyzed (Arteaga-Vazquez & Chandler, 2010; Chandler, 2010; Coe, 1966; Hollick, 2017; Patterson & Chandler, 1995). Paramutation at the *b1* locus is one of the best-characterized systems and we will use it as a framework to describe our current understanding of the molecular mechanisms underlying this transgenerational epigenetic phenomenon. The key sequences required for *b1* paramutation are seven direct tandem repeats (*b1TR*) of a 853 bp otherwise unique in the genome, located ~100 kb upstream of the transcription start site of *b1* (Arteaga and Chandler, 2010; Hövel, Pearson, & Stam, 2015; Stam et al., 2002a, 2002b). *B-I* and *B′* carry seven tandem repeats whereas *b1* neutral alleles have a single copy of the repeat unit. Multiple repeats are required for paramutation and the penetrance of paramutation correlates with the number of repeats (Hövel et al., 2015; Stam et al., 2002a). As expected for epialleles, *B-I* and *B′* have identical sequences (throughout a 150 kb including the *b1* gene region) but they show distinct patterns of DNA methylation and chromatin structure within the tandem repeats (Chandler, 2010; Haring et al., 2010; Hövel et al., 2015; Stam et al., 2002a).

6. CONCLUSIONS AND PERSPECTIVES

Rice and maize not only feed the world but are exceptional model organisms proved fundamental for our understanding of genetic and epigenetic mechanisms relevant to all eukaryotes. Epigenetic regulation is particularly important during the formation of seeds and over the last couple of decades we have gained insight into the different mechanisms operating in plants that include dedicated pathways for the "perception" and silencing of TEs and the RNA directed DNA methylation pathways that employs small

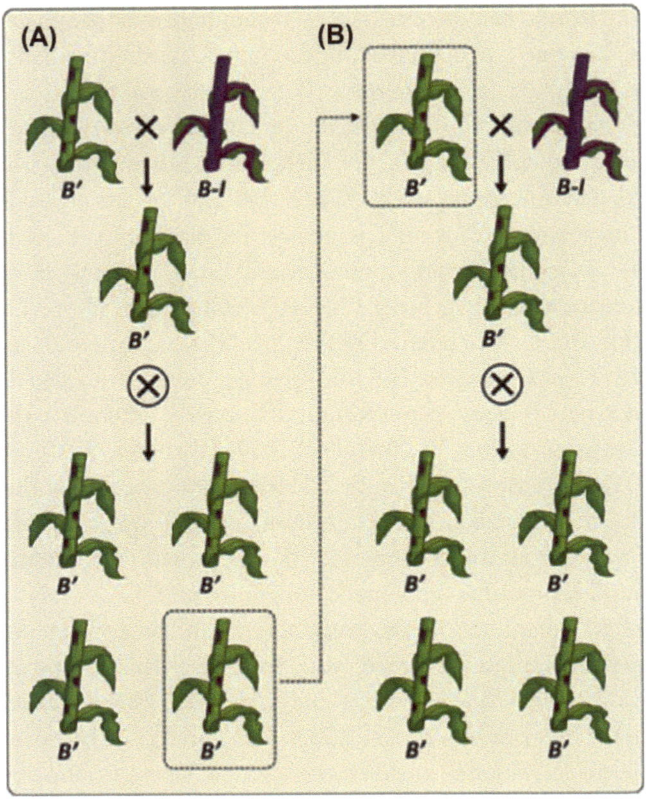

Figure 4 ***Paramutation at the b1 locus.*** Paramutation always occurs when the paramutagenic *B′* and paramutable *B-I* alleles are crossed together. (A) The paramutant F1 progeny (it is important to remark that the newly paramutagenic or paramutant alleles are genetically symbolized as B′* in order to reflect the fact that they were B-I in the previous generation. but for simplicity we refer to all b1 paramutagenic alleles as B′) exhibits both the expression levels of *b1* and the ability to change a *B-I* allele, as those present in the *B′* paramutagenic plants. (B). Secondary paramutation. Paramutant plants from either the F1, F2 or any subsequent filial generation crossed to a naïve *B-I* (that has never been exposed to *B′*) are capable of inducing paramutation. Secondary paramutation is the main characteristics that sets paramutation apart from other epigenetic phenomena. Paramutation has significant evolutionary implications, such as providing an adaptive mechanism for transferring favorable expression states to progeny and a mechanism for establishing functional homozygosity in polyploids (Chandler & Stam, 2004) which is highly relevant for breeding schemes. Identification of additional loci undergoing paramutation in combination with the characterization of additional genes required for paramutation is fundamental to shed light on its molecular basis and to exploit its full potential.

RNAs and long non coding RNAs to establish and maintain distinctive patterns DNA methylation essential for proper development and that in some cases can be transmitted across generations. Even though we are still learning about the molecular basis and extent of the different epigenetic and RNA-based regulatory modules, it is clear that plants have evolved ways to protect their genomes form invasive nucleic acids and at the same time have employed these elements to regulate some of the most intimate aspects of their life such as reproduction.

In the perspective of the future challenges for plant breeding in the context of climate change and increase in human population size, the use of natural or induced epigenetic variation could open new perspectives to improve cereal yield. Most cereal crops have a high TE load and mutants of the epigenetic machinery are difficult to obtain due to an affected fitness. The use of epigenomic editing or the introgression of natural epialleles into elite varieties, although technically feasible, will require a good knowledge of the stability and heritability of these epiallelic forms, as a prerequisite for their use in plant breeding (Springer & Schmitz, 2017). A clear example of the tremendous impact of bringing the power of epigenetics into breeding programs comes from studies in the African oil palm demonstrating that the epiallele termed "Bad Karma" which represents a DNA hypomethylated variant of a LINE retrotransposable element in the intron of the homeotic gene *DEFICIENS* causes abnormalities in the fruit that dramatically affects yield. In sharp contrast, the hypermethylated epiallele (termed "Good Karma") predicts normal fruit set. Based on this key observation, a simple test that can predict the phenotype of the palm trees years before they produce fruits was incorporated into the breeding programs saving this way not only time and money but most important, helping preventing the destruction of a significant number of hectares of rainforest (Ong-Abdullah et al., 2015; Paszkowski, 2015). Increasing the number of characterized epialleles and epigenetic phenomenon in cereal crops would help address these questions in the future.

REFERENCES

Alleman, M., Sidorenko, L., McGinnis, K., Seshadri, V., Dorweiler, J. E., White, J., et al. (July 20, 2006). An RNA-dependent RNA polymerase is required for paramutation in maize. *Nature, 442*(7100), 295–298. https://doi.org/10.1038/nature04884.

Arteaga-Vazquez, M. A., & Chandler, V. L. (2010). Paramutation in maize: RNA mediated trans-generational gene silencing. *Current Opinion in Genetics & Development, 20*, 156–163. https://doi.org/10.1016/j.gde.2010.01.008.

Arteaga-Vazquez, M., Sidorenko, L., Rabanal, F. A., Shrivistava, R., Nobuta, K., Green, P. J., et al. (2010). RNA-mediated trans-communication can establish

paramutation at the b1 locus in maize. *Proceedings of the National Academy of Sciences of the United States of America, 107*, 12986–12991. https://doi.org/10.1073/pnas.1007972107.
Bartee, L., Malagnac, F., & Bender, J. (2001). Arabidopsis cmt3 chromomethylase mutations block non-CG methylation and silencing of an endogenous gene. *Genes & Development, 15*, 1753–1758. https://doi.org/10.1101/gad.905701.
Becraft, P. W., & Yi, G. (2011). Regulation of aleurone development in cereal grains. *Journal of Experimental Botany, 62*, 1669–1675. https://doi.org/10.1093/jxb/erq372.
Berger, S. L. (2007). The complex language of chromatin regulation during transcription. *Nature, 447*, 407–412. https://doi.org/10.1038/nature05915.
Berger, F., Vu, T. M., Li, J., & Chen, B. (2012). Hypothesis: Selection of imprinted genes is driven by silencing deleterious gene activity in somatic tissues. *Cold Spring Harbor Symposia on Quantitative Biology, 77*, 23–29. https://doi.org/10.1101/sqb.2012.77.014514.
Bewick, A. J., Niederhuth, C. E., Ji, L., Rohr, N. A., Griffin, P. T., Leebens-Mack, J., et al. (2017). The evolution of CHROMOMETHYLASES and gene body DNA methylation in plants. *Genome Biology, 18*, 1–13. https://doi.org/10.1186/s13059-017-1195-1.
Bonasio, R., Tu, S., & Reinberg, D. (2010). Molecular signals of epigenetic states. *Science, 330*, 612–616. https://doi.org/10.1126/science.1191078.Molecular.
Borges, F., Parent, J.-S., van Ex, F., Wolff, P., Martínez, G., Köhler, C., et al. (2018). Transposon-derived small RNAs triggered by miR845 mediate genome dosage response in Arabidopsis. *Nature Genetics, 50*, 186–192. https://doi.org/10.1038/s41588-017-0032-5.
Bouyer, D., Kramdi, A., Kassam, M., Heese, M., Schnittger, A., Roudier, F., et al. (2017). DNA methylation dynamics during early plant life. *Genome Biology, 18*, 179. https://doi.org/10.1186/s13059-017-1313-0.
Brink, R. a. (1956). A genetic change associated with the R locus in maize which is directed and potentially reversible. *Genetics, 41*, 872–889.
Brzeski, J., & Jerzmanowski, A. (2003). Deficient in DNA methylation 1 (DDM1) defines a novel family of chromatin-remodeling factors. *Journal of Biological Chemistry, 278*, 823–828. https://doi.org/10.1074/jbc.M209260200.
Cao, X., Aufsatz, W., Zilberman, D., Mette, M. F., Huang, M. S., Matzke, M., et al. (2003). Role of the DRM and CMT3 methyltransferases in RNA-directed DNA methylation. *Current Biology: CB, 13*, 2212–2217. https://doi.org/10.1016/j.cub.2003.11.052.
Chandler, V. L. (2007). Paramutation: From maize to mice. *Cell, 128*, 641–645. https://doi.org/10.1016/j.cell.2007.02.007.
Chandler, V. L. (2010). Paramutation's properties and puzzles. *Science, 330*, 628–629. https://doi.org/10.1126/science.1191044.
Chandler, V. L., Eggleston, W. B., & Dorweiler, J. E. (2000). Paramutation in maize. *Plant Molecular Biology, 43*, 121–145.
Chandler, V. L., & Stam, M. (2004). Chromatin conversations: Mechanisms and implications of paramutation. *Nature Reviews. Genetics, 5*, 532–544. https://doi.org/10.1038/nrg1378.
Chen, X. (2009). Small RNAs and their roles in plant development. *Annual Review of Cell and Developmental Biology, 25*, 21–44. https://doi.org/10.1146/annurev.cellbio.042308.113417.
Cheng, C, Tarutani, Y, Miyao, A, Ito, T, Yamazaki, M, Sakai, H, Fukai, E, & Hirochika, H (2015). Loss of function mutations in the rice chromomethylase OsCMT3a cause a burst of transposition. *The Plant Journal, 83*, 1069–1081. https://doi.org/10.1111/tpj.12952.
Coe, E. (1959). A regular and continuing conversion-type phenomenon at the b locus in maize. *Proceedings of the National Academy of Sciences of the United States of America*, 828–832.
Coe, E. H. (1966). The properties, origin, and mechanism of conversion-type inheritance at the B locus in maize. *Genetics, 53*, 1035–1063.

Dickinson, H., Costa, L., & Gutierrez-Marcos, J. (2012). Epigenetic neofunctionalisation and regulatory gene evolution in grasses. *Trends in Plant Science, 17*, 389–394. https://doi.org/10.1016/j.tplants.2012.04.002.

Dilkes, B. P. (2004). A differential dosage hypothesis for parental effects in seed development. *The Plant Cell Online, 16*, 3174–3180. https://doi.org/10.1105/tpc.104.161230.

Dooner, H. K., Robbins, T. P., & Jorgensen, R. a (1991). *Control of Anthocy an in 1991.*

Dorweiler, J. E., Carey, C. C., Kubo, K. M., Hollick, J. B., Kermicle, J. L., & Chandler, V. L. (2000). Mediator of Paramutation1 is required for establishment and maintenance of paramutation at multiple maize loci. *The Plant Cell Online, 12*, 2101–2118.

Ebbs, M. L., & Bender, J. (2006). Locus-specific control of DNA methylation by the arabidopsis SUVH5 histone. *Methyltransferase, 18*, 1166–1176. https://doi.org/10.1105/tpc.106.041400.1.

Erhard, K., Stonaker, J., Parkinson, S., & Lim, J. (2009). RNA polymerase IV functions in paramutation in Zea mays. *Science, 323*, 1201–1205. https://doi.org/10.1126/science.1164508.

Erilova, A., Brownfield, L., Exner, V., Rosa, M., Twell, D., Scheid, O. M., et al. (2009). Imprinting of the Polycomb group gene MEDEA serves as a ploidy sensor in Arabidopsis. *PLos Genetics, 5*. https://doi.org/10.1371/journal.pgen.1000663.

Feil, R., & Berger, F. (2007). Convergent evolution of genomic imprinting in plants and mammals. *Trends in Genetics, 23*, 192–199. https://doi.org/10.1016/j.tig.2007.02.004.

Feng, S. (2010). Epigenetic reprogramming in plant and animal development. *Science, 330*, 622–627. https://doi.org/10.1126/science.1190614.Epigenetic.

Feng, S., & Jacobsen, S. E. (2011). Epigenetic modifications in plants: An evolutionary perspective. *Current Opinion in Plant Biology, 14*, 179–186. https://doi.org/10.1016/j.pbi.2010.12.002.

Feng, S., Jacobsen, S. E., & Reik, W. (2010). *Epigenetic reprogramming in plant and animal development.*

Finnegan, E. J., Genger, R. K., Peacock, W. J., & Dennis, E. S. (1998). DNA methylation in plants. *Annual Review of Plant Physiology and Plant Molecular Biology, 49*, 223–247. https://doi.org/10.1146/annurev.arplant.49.1.223.

Garg, R., Narayana Chevala, V., Shankar, R., & Jain, M. (2015). Divergent DNA methylation patterns associated with gene expression in rice cultivars with contrasting drought and salinity stress response. *Scientific Reports, 5*, 1–16. https://doi.org/10.1038/srep14922.

Gasciolli, V., Mallory, A. C., Bartel, D. P., & Vaucheret, H. (2005). Partially redundant functions of arabidopsis DICER-like enzymes and a role for DCL4 in producing trans-Acting siRNAs. *Current Biology: CB, 15*, 1494–1500. https://doi.org/10.1016/j.cub.2005.07.024.

Gehring, M. (2013). Genomic imprinting: Insights from plants. *Annual Review of Genetics, 47*, 187–208. https://doi.org/10.1146/annurev-genet-110711-155527.

Gehring, M., Bubb, K. L., & Henikoff, S. (June 12, 2009). Extensive demethylation of repetitive elements during seed development underlies gene imprinting. *Science, 324*(5933), 1447–1451. https://doi.org/10.1126/science.1171609.

Gehring, M., Huh, J. H., Hsieh, T.-F., Penterman, J., Choi, Y., Harada, J. J., et al. (2006). DEMETER DNA glycosylase establishes MEDEA polycomb gene self-imprinting by allele-specific demethylation. *Cell, 124*, 495–506. https://doi.org/10.1016/j.cell.2005.12.034.

Gehring, M., Missirian, V., & Henikoff, S. (2011). Genomic analysis of parent-of-origin allelic expression in arabidopsis thaliana seeds. *PLoS One, 6*. https://doi.org/10.1371/journal.pone.0023687.

Giacopelli, B. J., & Hollick, J. B. (2015). Trans-homolog interactions facilitating paramutation in maize. *Plant Physiology, 168*, 1226–1236. https://doi.org/10.1104/pp.15.00591.

Goff, S. A., Cone, K. C., & Chandler, V. L. (1992). Functional analysis of the transcriptional activator encoded by the maize B gene: Evidence for a direct functional interaction between two classes of regulatory proteins. *Genes & Development, 6*, 864–875. https://doi.org/10.1101/gad.6.5.864.

Goff, S. A., Klein, T. M., Roth, B. a, Fromm, M. E., Cone, K. C., Radicella, J. P., et al. (1990). Transactivation of anthocyanin biosynthetic genes following transfer of B regulatory genes into maize tissues. *The EMBO Journal, 9*, 2517–2522.

Goff, S. A., Ricke, D., Lan, T. H., Presting, G., Wang, R., Dunn, M., et al. (2002). A draft sequence of the rice genome (Oryza sativa L. ssp. japonica). *Science, 296*(5565), 92–100. https://doi.org/10.1126/science.1068275.

Grotewold, E., Athma, P., & Peterson, T. (1991). Alternatively spliced products of the maize P gene encode proteins with homology to the DNA-binding domain of myb-like transcription factors. *Proceedings of the National Academy of Sciences of the United States of America, 88*, 4587–4591. https://doi.org/10.1073/pnas.88.11.4587.

Guo, M., Rupe, M. A., Danilevskaya, O. N., Yang, X., & Hu, Z. (2003). Genome-wide mRNA profiling reveals heterochronic allelic variation and a new imprinted gene in hybrid maize endosperm. *The Plant Journal: for Cell and Molecular Biology, 36*, 30–44. https://doi.org/10.1046/j.1365-313X.2003.01852.x.

Gutiérrez-Marcos, J. F., Costa, L. M., & Evans, M. M. S. (2006). Maternal gametophytic baseless1 is required for development of the central cell and early endosperm patterning in maize (Zea mays). *Genetics, 174*, 317–329. https://doi.org/10.1534/genetics.106.059709.

Gutierrez-Marcos, J. F., & Dickinson, H. G. (2012). Epigenetic reprogramming in plant reproductive lineages. *Plant & Cell Physiology, 53*, 817–823. https://doi.org/10.1093/pcp/pcs052.

Gutierrez-Marcos, J. F., Pennington, P. D., Costa, L. M., & Dickinson, H. G. (2003). Imprinting in the endosperm: A possible role in preventing wide hybridization. *Philosophical Transactions of the Royal Society B, 358*, 1105–1111. https://doi.org/10.1098/rstb.2003.1292.

Haag, J. R., Ream, T. S., Marasco, M., Nicora, C. D., Norbeck, A. D., Pasa-Tolic, L., et al. (2012). In Vitro transcription activities of pol IV, pol V, and RDR2 reveal coupling of pol IV and RDR2 for dsRNA synthesis in plant RNA silencing. *Molecular Cell, 48*, 811–818. https://doi.org/10.1016/j.molcel.2012.09.027.

Hagemann, R. (1958). Somatic conversion in Lycopersicon esculentum mill. *Zeitschrift fuer Vererbungslehre, 89*, 587–613.

Haig, D., & Westoby, M. (1989). Parent-specific gene expression and the triploid endosperm. *The American Naturalist, 134*, 147–155. https://doi.org/10.1086/284971.

Haring, M., Bader, R., Louwers, M., Schwabe, A., Driel, R. Van, Stam, M., et al. (2010). The role of DNA methylation, nucleosome occupancy and histone modifications in paramutation. *The Plant Journal: for Cell and Molecular Biology*, 366–378. https://doi.org/10.1111/j.1365-313X.2010.04245.x.

Haun, W. J., Laoueillé-Duprat, S., O'Connell, M. J., Spillane, C., Grossniklaus, U., Phillips, A. R., et al. (2007). Genomic imprinting, methylation and molecular evolution of maize Enhancer of zeste (Mez) homologs. *The Plant Journal: for Cell and Molecular Biology, 49*, 325–337. https://doi.org/10.1111/j.1365-313X.2006.02965.x.

Heard, E., & Martienssen, R. A. (2014). Transgenerational epigenetic inheritance: Myths and mechanisms. *Cell, 157*, 95–109. https://doi.org/10.1016/j.cell.2014.02.045.

He, X.-J., Chen, T., & Zhu, J.-K. (2011). Regulation and function of DNA methylation in plants and animals. *Cell Research, 21*, 442–465. https://doi.org/10.1038/cr.2011.23.

Henderson, I. R., & Jacobsen, S. E. (2007). Epigenetic inheritance in plants. *Nature, 447*, 418–424. https://doi.org/10.1038/nature05917.

Henikoff, S. (2008). Nucleosome destabilization in the epigenetic regulation of gene expression. *Nature Reviews. Genetics, 9*, 15–26. https://doi.org/10.1038/nrg2206.

Higo, H., Tahir, M., Takashima, K., Miura, A., Watanabe, K., Tagiri, A., et al. (2012). DDM1 (Decrease in DNA Methylation) genes in rice (Oryza sativa). *Molecular Genetics and Genomics, 287*, 785–792. https://doi.org/10.1007/s00438-012-0717-5.

Hollick, J. B. (2010). *Paramutation and Development.* https://doi.org/10.1146/annurev.cellbio.042308.113400.

Hollick, J. B. (2012). Paramutation: A trans-homolog interaction affecting heritable gene regulation. *Current Opinion in Plant Biology, 15*, 536–543. https://doi.org/10.1016/j.pbi.2012.09.003.

Hollick, J. B. (2017). Paramutation and related phenomena in diverse species. *Nature Reviews. Genetics, 18*, 5–23. https://doi.org/10.1038/nrg.2016.115.

Holliday, R., & Pugh, J. E. (1975). DNA modification mechanisms and gene activity during development. *Science, 187*, 226–232. https://doi.org/10.1126/science.1111098.

Hövel, I., Pearson, N. A., & Stam, M. (2015). Cis-acting determinants of paramutation. *Seminars in Cell & Developmental Biology, 44*, 22–32. https://doi.org/10.1016/j.semcdb.2015.08.012.

Hu, L., Li, N., Xu, C., Zhong, S., Lin, X., Yang, J., et al. (2014). Mutation of a major CG methylase in rice causes genome-wide hypomethylation, dysregulated genome expression, and seedling lethality. *Proceedings of the National Academy of Sciences of the United States of America, 111*, 10642–10647. https://doi.org/10.1073/pnas.1410761111.

Hultquist, J. F., & Dorweiler, J. E. (2008). Feminized tassels of maize mop1 and ts1 mutants exhibit altered levels of miR156 and specific SBP-box genes. *Planta, 229*, 99–113. https://doi.org/10.1007/s00425-008-0813-2.

IRGSP International Rice Genome Sequencing Project. (2005). The map-based sequence of the rice genome. *Nature, 436*(7052), 793–800. https://doi.org/10.1038/nature03895.

Jackson, S. A. (2016). Rice: The first crop genome. *Rice (N Y), 9*, 14. https://doi.org/10.1186/s12284-016-0087-4.

Jeddeloh, J. A., Stokes, T. L., & Richards, E. J. (1999). Maintenance of genomic methylation requires a SWI2/SNF2-like protein. *Nature Genetics, 22*, 94–97. https://doi.org/10.1038/8803.

Jenuwein, T., & Allis, C. D. (2001). Allis: Translating the histone code. *Science, 293*, 1074–1080. https://doi.org/10.1126/science.1063127.

Jones-Rhoades, M. W., Bartel, D. P., & Bartel, B. (2006). MicroRNAS and their regulatory roles in plants. *Annual Review of Plant Biology, 57*, 19–53. https://doi.org/10.1146/annurev.arplant.57.032905.105218.

Josefsson, C., Dilkes, B., & Comai, L. (2006). Parent-dependent loss of gene silencing during interspecies hybridization. *Current Biology: CB, 16*, 1322–1328. https://doi.org/10.1016/j.cub.2006.05.045.

Jullien, P. E., & Berger, F. (2010). Parental genome dosage imbalance deregulates imprinting in Arabidopsis. *PLos Genetics, 6*, 1–8. https://doi.org/10.1371/journal.pgen.1000885.

Kermicle, J. L. (1970). Dependence of the R-mottled aleurone phenotype in maize on mode of sexual transmission. *Genetics, 66*, 69–85.

Kermicle, J. L., & Alleman, M. (1990). Gametic imprinting in maize in relation to the angiosperm life cycle. *Development Supplement, 9–14.*

Köhler, C., & Kradolfer, D. (2011). Epigenetic mechanisms in the endosperm and their consequences for the evolution of flowering plants. *Biochim Biophys Acta - Gene Regulatory Mechanisms, 1809*, 438–443. https://doi.org/10.1016/j.bbagrm.2011.04.004.

Köhler, C., Wolff, P., & Spillane, C. (2012). Epigenetic mechanisms underlying genomic imprinting in plants. *Annual Review of Plant Biology, 63*, 331–352. https://doi.org/10.1146/annurev-arplant-042811-105514.

Lanciano, S., Carpentier, M. C., Llauro, C., Jobet, E., Robakowska-Hyzorek, D., Lasserre, E., et al. (2017). Sequencing the extrachromosomal circular mobilome reveals retrotransposon activity in plants. *PLos Genetics, 13*, 1–20. https://doi.org/10.1371/journal.pgen.1006630.

Law, J. a, & Jacobsen, S. E. (2010). Establishing, maintaining and modifying DNA methylation patterns in plants and animals. *Nature Reviews. Genetics, 11*, 204–220. https://doi.org/10.1038/nrg2719.

Li, Q., Eichten, S. R., Hermanson, P. J., Zaunbrecher, V. M., Song, J., Wendt, J., et al. (2014). *Genetic Perturbation of the Maize Methylome, 26*, 4602–4616. https://doi.org/10.1105/tpc.114.133140.

Li, Y., Kumar, S., & Qian, W. (2018). Active DNA demethylation: Mechanism and role in plant development. *Plant Cell Reports, 37*, 77–85. https://doi.org/10.1007/s00299-017-2215-z.

Lu, C., Kulkarni, K., Souret, F. F., MuthuValliappan, R., Tej, S. S., Poethig, R. S., et al. (2006). MicroRNAs and other small RNAs enriched in the Arabidopsis RNA-dependent RNA polymerase-2 mutant. *Genome Research, 16*, 1276–1288. https://doi.org/10.1101/gr.5530106.

Luo, M., Taylor, J. M., Spriggs, A., Zhang, H., Wu, X., Russell, S., et al. (2011). A Genome-Wide survey of imprinted genes in rice seeds reveals imprinting primarily occurs in the endosperm. *PLos Genetics, 7*. https://doi.org/10.1371/journal.pgen.1002125.

Matzke, M., Kanno, T., Daxinger, L., Huettel, B., & Matzke, A. J. M. (2009). RNA-mediated chromatin-based silencing in plants. *Current Opinion in Cell Biology, 21*, 367–376. https://doi.org/10.1016/j.ceb.2009.01.025.

Matzke, M. a, & Mosher, R. a (2014). RNA-directed DNA methylation: An epigenetic pathway of increasing complexity. *Nature Reviews. Genetics, 15*, 394–408. https://doi.org/10.1038/nrg3683.

McClintock, B. (1984). The significance of responses of the genome to challenge. *Science, 226*, 792–801. https://doi.org/10.1126/science.15739260.

Mirouze, M., & Vitte, C. (2014). Transposable elements, a treasure trove to decipher epigenetic variation: Insights from arabidopsis and crop epigenomes. *Journal of Experimental Botany, 65*, 2801–2812. https://doi.org/10.1093/jxb/eru120.

Miura, K., Agetsuma, M., Kitano, H., Yoshimura, A., Matsuoka, M., Jacobsen, S. E., et al. (2009). A metastable DWARF1 epigenetic mutant affecting plant stature in rice. *Proceedings of the National Academy of Sciences of the United States of America, 106*, 11218–11223. https://doi.org/10.1073/pnas.0901942106.

Moritoh, S., Eun, C. H., Ono, A., Asao, H., Okano, Y., Yamaguchi, K., et al. (2012). Targeted disruption of an orthologue of DOMAINS REARRANGED METHYLASE 2, OsDRM2, impairs the growth of rice plants by abnormal DNA methylation. *The Plant Journal: for Cell and Molecular Biology, 71*, 85–98. https://doi.org/10.1111/j.1365-313X.2012.04974.x.

Nobuta, K., Lu, C., Shrivastava, R., Pillay, M., De Paoli, E., Accerbi, M., et al. (2008). Distinct size distribution of endogeneous siRNAs in maize: Evidence from deep sequencing in the mop1-1 mutant. *Proceedings of the National Academy of Sciences of the United States of America, 105*, 14958–14963. https://doi.org/10.1073/pnas.0808066105.

Nosaka, M., Itoh, J. I., Nagato, Y., Ono, A., Ishiwata, A., & Sato, Y. (2012). Role of transposon-derived small RNAs in the interplay between genomes and parasitic DNA in rice. *PLos Genetics, 8*. https://doi.org/10.1371/journal.pgen.1002953.

Ong-Abdullah, M., Ordway, J. M., Jiang, N., Ooi, S. E., Kok, S. Y., & Sarpan, N. (September 24, 2015). Loss of Karma transposon methylation underlies the mantled somaclonal variant of oil palm. *Nature, 24*(7570), 533–537. https://doi.org/10.1038/nature15365.

Ooi, S. K. T., & Bestor, T. H. (2008). The colorful history of active DNA demethylation. *Cell, 133*, 1145–1148. https://doi.org/10.1016/j.cell.2008.06.009.

Parkinson, S. E., Gross, S. M., & Hollick, J. B. (2007). Maize sex determination and abaxial leaf fates are canalized by a factor that maintains repressed epigenetic states. *Developmental Biology, 308*, 462–473. https://doi.org/10.1016/j.ydbio.2007.06.004.

Paszkowski, J. (2015). Epigenetics: The karma of oil palms. *Nature, 525*, 7570. https://doi.org/10.1038/nature15216.

Patterson, G. I., & Chandler, V. L. (1995). Paramutation in maize and related allelic interactions BT. In P. Meyer (Ed.), *Gene silencing in higher plants and related phenomena in other eukaryotes* (pp. 121–141). Berlin, Heidelberg: Springer Berlin Heidelberg. https://doi.org/10.1007/978-3-642-79145-1_9.

Patterson, G. I., Thorpe, C. J., & Chandler, V. L. (1993). Paramutation, an allelic interaction, is associated with a stable and heritable reduction of transcription of the maize b regulatory gene. *Genetics, 135*, 881–894.

Pavlopoulou, A., & Kossida, S. (2007). Plant cytosine-5 DNA methyltransferases: Structure, function, and molecular evolution. *Genomics, 90*, 530–541. https://doi.org/10.1016/j.ygeno.2007.06.011.

Pikaard, C. S., & Mittelsten Scheid, O. (December 1, 2014). Epigenetic regulation in plants. *Cold Spring Harb Perspective in Biology, 6*(12), a019315. https://doi.org/10.1101/cshperspect.a019315.

Quadrana, L., & Colot, V. (2016). Plant transgenerational epigenetics. *Annual Review of Genetics, 50*, 467–491. https://doi.org/10.1146/annurev-genet-120215-035254.

Raissig, M. T., Baroux, C., & Grossniklaus, U. (2011). Regulation and flexibility of genomic imprinting during seed development. *The Plant Cell Online, 23*, 16–26. https://doi.org/10.1105/tpc.110.081018.

Ranum, P., Peña-Rosas, J. P., & Garcia-Casal, M. N. (2014). Global maize production, utilization, and consumption. *Annals of the New York Academy of Sciences, 1312*, 105–112. https://doi.org/10.1111/nyas.12396.

Rea, M., Zheng, W., Chen, M., Braud, C., Bhangu, D., Rognan, T. N., et al. (2012). Histone H1 affects gene imprinting and DNA methylation in Arabidopsis. *The Plant Journal: for Cell and Molecular Biology, 71*, 776–786. https://doi.org/10.1111/j.1365-313X.2012.05028.x.

Rodrigues, J. A., Ruan, R., Nishimura, T., Sharma, M. K., Sharma, R., Ronald, P. C., et al. (2013). Imprinted expression of genes and small RNA is associated with localized hypomethylation of the maternal genome in rice endosperm. *Proceedings of the National Academy of Sciences of the United States of America, 110*, 7934–7939. https://doi.org/10.1073/pnas.1306164110.

Secco, D., Wang, C., Shou, H., Schultz, M. D., Chiarenza, S., Nussaume, L., et al. (2015). Stress induced gene expression drives transient DNA methylation changes at adjacent repetitive elements. *Elife, 4*, 1–26. https://doi.org/10.7554/eLife.09343.001.

Shilatifard, A. (2006). Chromatin modifications by methylation and ubiquitination: Implications in the regulation of gene expression. *Annual Review of Biochemistry, 75*, 243–269. https://doi.org/10.1146/annurev.biochem.75.103004.142422.

Sidorenko, L., Dorweiler, J. E., Cigan, a M., Arteaga-Vazquez, M., Vyas, M., Kermicle, J., et al. (2009). A dominant mutation in mediator of paramutation2, one of three second-largest subunits of a plant-specific RNA polymerase, disrupts multiple siRNA silencing processes. *PLos Genetics, 5*. https://doi.org/10.1371/journal.pgen.1000725. e1000725.

Slack, J. M. W. (2002). Conrad hal Waddington: The last renaissance biologist? *Nature Reviews. Genetics, 3*, 889–895. https://doi.org/10.1038/nrg933.

Slotkin, R. K., & Martienssen, R. (2007). Transposable elements and the epigenetic regulation of the genome. *Nature Reviews. Genetics, 8*, 272–285. https://doi.org/10.1038/nrg2072.

Smith, Z. D., & Meissner, A. (2013). The simplest explanation: Passive DNA demethylation in PGCs. *The EMBO Journal, 32*, 318–321. https://doi.org/10.1038/emboj.2012.349.
Spillane, C., Baroux, C., Escobar-Restrepo, J.-M., Page, D. R., Laoueille, S., & Grossniklaus, U. (2004). Transposons and tandem repeats are not involved in the control of genomic imprinting at the MEDEA locus in arabidopsis. *Cold Spring Harbor Symposia on Quantitative Biology, 69*, 465–476. https://doi.org/10.1101/sqb.2004.69.465.
Springer, N. M., & Schmitz, R. J. (September 2017). Exploiting induced and natural epigenetic variation for crop improvement. *Nature Reviews. Genetics, 18*(9), 563–575. https://doi.org/10.1038/nrg.2017.45.
Stam, M., Belele, C., Dorweiler, J. E., & Chandler, V. L. (2002a). Differential chromatin structure within a tandem array 100 kb upstream of the maize b1 locus is associated with paramutation. *Genes & Development, 16*, 1906–1918. https://doi.org/10.1101/gad.1006702.
Stam, M., Belele, C., Ramakrishna, W., Dorweiler, J. E., Bennetzen, J. L., & Chandler, V. L. (2002b). The regulatory regions required for B' paramutation and expression are located far upstream of the maize b1 transcribed sequences. *Genetics, 162*, 917–930.
Strahl, B. D., & Allis, C. D. (2000). The language of covalent histone modifications. *Nature, 403*, 41–45. https://doi.org/10.1038/47412.
Stroud, H., Do, T., Du, J., Zhong, X., Feng, S., Johnson, L., et al. (2014). Non-CG methylation patterns shape the epigenetic landscape in Arabidopsis. *Nature Structural & Molecular Biology, 21*, 64–72. https://doi.org/10.1038/nsmb.2735.
Suzuki, M. M., & Bird, A. (2008). DNA methylation landscapes: Provocative insights from epigenomics. *Nature Reviews. Genetics, 9*, 465–476. https://doi.org/10.1038/nrg2341.
Tan, F., Zhou, C., Zhou, Q., Zhou, S., Yang, W., Zhao, Y., et al. (2016). Analysis of chromatin regulators reveals specific features of rice DNA methylation pathways. *Plant Physiology, 171*, 2041–2054. https://doi.org/10.1104/pp.16.00393.
Tiwari, S., Spielman, M., Schulz, R., Oakey, R. J., Kelsey, G., Salazar, A., et al. (2010). *Transcriptional profiles underlying parent-of-origin effects in seeds of* Arabidopsis thaliana.
Underwood, C. J., Henderson, I. R., & Martienssen, R. A. (2017). Genetic and epigenetic variation of transposable elements in Arabidopsis. *Current Opinion in Plant Biology, 36*, 135–141. https://doi.org/10.1016/j.pbi.2017.03.002.
Waters, A. J., Bilinski, P., Eichten, S. R., Vaughn, M. W., Ross-Ibarra, J., Gehring, M., et al. (2013). Comprehensive analysis of imprinted genes in maize reveals allelic variation for imprinting and limited conservation with other species. *Proceedings of the National Academy of Sciences of the United States of America, 110*, 19639–19644. https://doi.org/10.1073/pnas.1309182110.
Waters, A. J., Makarevitch, I., Eichten, S. R., Swanson-Wagner, R. A., Yeh, C.-T., Xu, W., et al. (2011). Parent-of-Origin effects on gene expression and DNA methylation in the maize endosperm. *The Plant Cell Online, 23*, 4221–4233. https://doi.org/10.1105/tpc.111.092668.
Wendte, J. M., & Pikaard, C. S. (2017). The RNAs of RNA-directed DNA methylation. *Biochim Biophys Acta - Gene Regulatory Mechanisms, 1860*, 140–148. https://doi.org/10.1016/j.bbagrm.2016.08.004.
Wendte, J. M., & Schmitz, R. J. (2018). Specifications of targeting heterochromatin modifications in plants. *Molecular Plant, 11*, 381–387. https://doi.org/10.1016/j.molp.2017.10.002.
Wöhrmann, H. J. P., Gagliardini, V., Raissig, M. T., Wehrle, W., Arand, J., Schmidt, A., et al. (2012). Identification of a DNA methylation-independent imprinting control region at the *Arabidopsis MEDEA* locus. *Genes & Development, 26*, 1837–1850. https://doi.org/10.1101/gad.195123.112.

Wolff, P., Weinhofer, I., Seguin, J., Roszak, P., Beisel, C., Donoghue, M. T. A., et al. (2011). High-Resolution analysis of parent-of-origin allelic expression in the Arabidopsis endosperm. *PLos Genetics, 7*. https://doi.org/10.1371/journal.pgen.1002126.
Wollmann, H., & Berger, F. (2012). Epigenetic reprogramming during plant reproduction and seed development. *Current Opinion in Plant Biology, 15*, 63–69. https://doi.org/10.1016/j.pbi.2011.10.001.
Wu, B., Peisley, A., Richards, C., Yao, H., Zeng, X., Lin, C., et al. (2013). Structural basis for dsRNA recognition, filament formation, and antiviral signal activation by MDA5. *Cell, 152*, 276–289. https://doi.org/10.1016/j.cell.2012.11.048.
Wu, S. C., & Zhang, Y. (2013). Active DNA demethylation: Many roads lead to rome. *Nature Reviews. Molecular Cell Biology, 11*, 607–620. https://doi.org/10.1038/nrm2950.Active.
Wu, X., & Zhang, Y. (2017). TET-mediated active DNA demethylation: Mechanism, function and beyond. *Nature Reviews. Genetics, 18*, 517–534. https://doi.org/10.1038/nrg.2017.33.
Wyder, S., Raissig, M. T., & Grossniklaus, U. (2017). Consistent reanalysis of genome-wide imprinting studies in plants using generalized linear models increases concordance across datasets. *BioRxiv*.
Xie, Z., Johansen, L. K., Gustafson, A. M., Kasschau, K. D., Lellis, A. D., Zilberman, D., et al. (2004). Genetic and functional diversification of small RNA pathways in plants. *PLos Biology, 2*, 642–652. https://doi.org/10.1371/journal.pbio.0020104.
Xin, M., Yang, R., Li, G., Chen, H., Laurie, J., Ma, C., et al. (2013). Dynamic expression of imprinted genes associates with maternally controlled nutrient allocation during maize endosperm development. *The Plant Cell Online, 25*, 3212–3227. https://doi.org/10.1105/tpc.113.115592.
Yamauchi, T., Moritoh, S., Johzuka-Hisatomi, Y., Ono, A., Terada, R., Nakamura, I., et al. (2008). Alternative splicing of the rice OsMET1 genes encoding maintenance DNA methyltransferase. *Journal of Plant Physiology, 165*, 1774–1782. https://doi.org/10.1016/j.jplph.2007.12.003.
Zemach, A., Kim, M. Y., Hsieh, P.-H., Coleman-Derr, D., Eshed-Williams, L., Thao, K., et al. (2013). The arabidopsis nucleosome remodeler DDM1 allows DNA methyltransferases to access H1-containing heterochromatin. *Cell, 153*, 193–205. https://doi.org/10.1016/j.cell.2013.02.033.
Zemach, A., Kim, M. Y., Silva, P., Rodrigues, J. A., Dotson, B., Brooks, M. D., et al. (2010). Local DNA hypomethylation activates genes in rice endosperm. *Proceedings of the National Academy of Sciences of the United States of America, 107*, 18729–18734. https://doi.org/10.1073/pnas.1009695107.
Zhang, L., Cheng, Z., Qin, R., Qiu, Y., Wang, J.-L., Cui, X., et al. (2012). Identification and characterization of an epi-allele of FIE1 reveals a regulatory linkage between two epigenetic marks in rice. *The Plant Cell Online, 24*, 4407–4421. https://doi.org/10.1105/tpc.112.102269.
Zhang, H., Lang, Z., & Zhu, J.-K. (2018). Dynamics and function of DNA methylation in plants. *Nature Reviews. Molecular Cell Biology, 19*, 489–506. https://doi.org/10.1038/s41580-018-0016-z.
Zhang, X., Sun, J., Cao, X., & Xianwei, S. (2015). Epigenetic mutation of RAV6 affects leaf angle and seed size in rice. *Plant Physiology, 169*. https://doi.org/10.1104/pp.15.00836.
Zhang, M., Zhao, H., Xie, S., Chen, J., Xu, Y., Wang, K., et al. (2011). Extensive, clustered parental imprinting of protein-coding and noncoding RNAs in developing maize endosperm. *Proceedings of the National Academy of Sciences of the United States of America, 108*, 20042–20047. https://doi.org/10.1073/pnas.1112186108.

Zheng, B., Vazin, T., Goodwill, P. W., Conway, A., Verma, A., Ulku Saritas, E., et al. (2015). Magnetic particle imaging tracks the long-term fate of in vivo neural cell implants with high image contrast. *Scientific Reports, 5*, 1–9. https://doi.org/10.1038/srep14055.

Zhou, S. R., Yin, L. L., & Xue, H. W. (2013). Functional genomics based understanding of rice endosperm development. *Current Opinion in Plant Biology, 16*, 236–246. https://doi.org/10.1016/j.pbi.2013.03.001.

Zhu, J., Kapoor, A., Sridhar, V. V., Agius, F., & Zhu, J.-K. (2007). The DNA glycosylase/lyase ROS1 functions in pruning DNA methylation patterns in Arabidopsis. *Current Biology: CB, 17*, 54–59. https://doi.org/10.1016/j.cub.2006.10.059.

CHAPTER TWELVE

Epigenetics in Forest Trees: State of the Art and Potential Implications for Breeding and Management in a Context of Climate Change

Mamadou Dia Sow*, Isabel Allona§, Christophe Ambroise¶, Daniel Conde§, Régis Fichot*, Svetlana Gribkova||, Véronique Jorge#, Grégoire Le-Provost, Luc Pâques#, Christophe Plomion**, Jérôme Salse§§, Léopoldo Sanchez-Rodriguez#, Vincent Segura#, Jörg Tost¶¶ and Stéphane Maury*,[1]**

*Laboratoire de Biologie des Ligneux et des Grandes Cultures, LBLGC, EA 1207 USC 1328, INRA-Université Orléans, Orléans, France
§Centro de Biotecnología y Genómica de Plantas, Universidad Politécnica de Madrid (UPM), Instituto Nacional de Investigación y Tecnología Agraria y Alimentaria (INIA), Madrid, Spain
¶Laboratoire de Mathématiques et Modélisation d'Évry, LaMME, UMR 8071, CNRS-Université d'Evry Val d'Essonne, Evry, France
||Laboratoire de Probabilités et Modèles Aléatoires, LPMA, UMR 8001, CNRS-Université Paris Diderot, Paris, France
#Biologie intégrée pour la valorisation de la diversité des arbres et de la forêt, BioForA, UMR 0588, INRA-ONF, Orléans, France
**Biodiversité, Gènes et Communautés, BIOGECO, INRA, Univ. Bordeaux, Cestas, France
§§Génétique, Diversité et Écophysiologie des Céréales, GDEC, UMR 1095, INRA-Université Clermont Auvergne, Clermont-Ferrand, France
¶¶Laboratoire Épigénétique et Environnement, LEE, CNRGH CEA, Evry, France
[1]Corresponding author: E-mail: stephane.maury@univ-orleans.fr

Contents

Advances in Botanical Research, Volume 88
ISSN 0065-2296
https://doi.org/10.1016/bs.abr.2018.09.003

Abstract

Forest trees are long-lived organisms subject to repeated environmental constraints throughout their long lifetimes. They have developed various mechanisms enabling them to cope with fluctuating environmental conditions during their life span, and to survive to current climate change. Epigenetics has recently emerged as a powerful set of mechanisms regulating various developmental processes, plant growth and responses to environmental variations. Such epigenetic mechanisms, which may remain stable along tree life or across generations, constitute a source of rapid phenotypic variations potentially improving adaptation of the plants in situations in which naturally occurring mutations are very rare.

In this review, we summarize recent advances in forest tree epigenomics. We first draw the particularities of trees and the available (epi) genomics resources and strategies. Then, we discuss the potential contributions of epigenetics to cope with global climate change and regulate various developmental processes, such as developmental transitions during the annual cycle, phenotypic plasticity in response to environmental variations and stress memory, as well as local adaptation. Finally, we propose some challenges for forest management and highlighted the need to take epigenetics into account in forest tree breeding strategies.

1. INTRODUCTION: FEATURES OF FOREST TREES BREEDING

Forest tree breeding programs started in Europe and worldwide only a few decades ago, in the 1950s, boosted by ambitious plantation programs. These programs had two main aims: to guarantee the genetic origin of the forest reproductive material (FRM) used in reforestation and to improve tree characteristics. Biomass production was the main selection criterion for many years, together with overall adaptability to different sites. Further criteria have recently been added, including wood quality traits, stem architecture, and more specific criteria linked to adaptation to abiotic and biotic factors in response to new socioeconomic needs and global changes (Mullin et al. 2011; Pâques, 2013).

Forest tree species are characterized by two key features that render them unique in the domain of plant breeding. Firstly, forest trees were domesticated only recently, so the available gene pools can be considered mostly wide, despite the well-documented introduction of exotic species and/or populations at some sites. Secondly, forest tree species generally cover large areas with contrasting ecological conditions. In Europe, the recolonization routes followed by these species after the most recent period of glaciation have, together with local selection pressures, shaped populations consisting of many different ecotypes. Furthermore, interspecific hybridization is common in forest trees, and several characteristics of the reproductive biology of species (dioecious species, monoecious species with asynchrony between female and male flowers, incompatibility genes, etc.) greatly limit consanguinity. All these factors contribute to the considerable genetic diversity observed in many forest trees (Bagnoli et al. 2011).

Over several decades before the advent of tree breeding programs, many studies focused on how diversity in species is geographically structured. Large-scale provenance trial networks and molecular marker analysis were used as complementary approaches to investigating both adaptive and neutral genetic diversity with two major outcomes. Firstly, it provided evidence in support of the delineation of regions of origin and the selection of local seed sources (selected seed stands). These regions, most of which were defined for native species, were designed to protect local genetic resources and, thus, to restrict the circulation and use of FRM. Secondly, it made possible to set up breeding programs for a limited number of species through identification of the best populations adapted to fit for local zones and to establish breeding populations through the mass-selection of individuals in native stands. In parallel, dedicated plantations (first-generation seed orchards) were established with this material, for mass production of FRM. For many species, these plantations remain the main source of seed for reforestation.

This approach has a major drawback. Indeed, core breeding populations are selected on the basis of their performances across contrasting environments in large common garden networks, but populations are compared on the basis of seed lots collected in a single year, in specific environments. Furthermore, parental contributions to seed lots are probably constrained by a lack of full panmixia and the impact of the environment on seed formation has also been largely ignored although maternal effects on seed weight and subsequent growth are well documented, and evidences for epigenetic memory during embryo formation are also accumulating for forest trees

(Carneros, Yakovlev, Viejo, Olsen, & Fossdal, 2017; Skrøppa, Solvin, & Steffenrem, 2016; Yakovlev et al., 2012; Yakovlev & Fossdal, 2017). Consequently, the genetic differences between populations reported in common garden studies are most likely overestimated.

Due to the length and inherent costs of the breeding cycle in trees, breeding *sensu stricto* has been restricted to a few species of considerable real or potential economic importance and, generally, to those with a short rotation (mostly conifers and poplars). In most cases, breeding programs target reforestation zones outside the native range of the species. Breeding populations are largely exotic and better adapted to reforestation needs. Except in Scandinavia, characterized by strong environmental gradients, breeding programs have mostly aimed to varieties that are not specialized to specific environmental conditions (low GxE (Genetics by Environment) interactions) or traits.

The breeding objectives for forest trees are directed by various expectations from end-users of FRM, such as the reproductive success of base material required by the seed industry, the biomass production required by forest owners (implying adaptation to abiotic and biotic factors), and the stem architecture and properties of wood required by the wood industry for various end uses (Alia & Majada, 2013; Sanchez et al. 2013). Consequently, forest tree varieties actually represent a compromise between a diversity of target traits. There are many natural trade-offs between characters, and the management of antagonistic traits (e.g. growth and wood density) requires alternative approaches, such as the management of breeding populations in sublines, selection procedures for optimizing genetic gains, interspecific hybridization to break down unfavorable correlations and the clonal propagation of ideotypes as sources of FRM (Hallingbäck, Sánchez, & Wu, 2014).

With few exceptions (e.g. interspecific hybridization (poplars, larches) and clonal (poplar, wild cherry) strategies), simple recurrent selection is the most common breeding strategy used for forest trees (e.g. spruces, pines, Douglas fir). This approach is based on recurrent cycles of recombination, testing and selection phases aiming to improve the breeding population over generations and to lead to the release of improved varieties (White, Adams, & Neale, 2007). Simple recurrent selection is based exclusively on additive gene effects, whereas other approaches also exploit non-additive effects (as in reciprocal recurrent selection) or all the available genetic variation (as in clonal strategies). The optimization of mating plans and of selection procedures to increase the frequency of favorable alleles and to discard deleterious alleles whilst maintaining broad genetic variability within the population over breeding cycles is a major challenge that must be met to ensure long-term genetic gains.

Nevertheless, several major morphological and biological constraints slow the whole process, limit breeding possibilities and account for the very small number of forest tree species currently bred beyond a first generation. Space and time are two of the most critical factors. The large size of trees generally limits both the size of breeding populations to a few hundred trees established in clonal banks, and the size and, number of test sites (number of genotypes and replicates tested), decreasing the selection pressure (selection intensity) and the precision of breeding value estimations, thereby restricting genetic gains. Sexual maturity is not reached for a number of years in many tree species, delaying the recombination phase and, thus, the breeding cycle, and the late expression of some traits or their differential expression with age due to ontological effects may postpone the selection phase even further, by several years or even decades. Research efforts to limit these drawbacks and accelerate breeding cycles have focused on flower initiation and stimulation, various techniques to reduce the length of the testing phase (e.g. early selection when there is evidence of strong juvenile—mature correlations for traits, marker-assisted selection) and the timely management of base material establishment for FRM production. Genomic selection, which has only recently been envisaged for some forest trees, shares the same objectives (Muranty et al. 2014; Woolliams, 2013).

The duration of the recombination phase is often limited by estimating the breeding values of genotypes through open-pollinated (or polycross) progenies collected in a single year, as in simple recurrent selection schemes, and, for economic reasons, at a limited number of test sites. Breeding values and other genetic parameters are likely to be biased (overestimated) in cases of strong GxE interactions and/or in the presence of unaccounted for epigenetic effects. As a result, the genetic gains obtained by selection may be smaller than anticipated. Nevertheless, the gains achieved in several breeding programs demonstrate the overall success of this approach (Kimberley, Moore, & Dungey, 2015; Toda & Kurinobu, 2002) at least when GxE interactions are properly accounted for.

Various forms of FRM can be produced, depending on the selection scheme used. Synthetic varieties mass-produced by open-pollination in dedicated plantations (seed orchards) including several tens or even hundreds of genotypes selected on the basis of their GCAs (general combining abilities) are the most common. The hypothetical assumption of panmixia is rarely respected, but these varieties are nevertheless preferred because they guarantee a broad genetic base, together with acceptable levels of genetic gain at relatively low cost. The problem of unequal parental contributions can be overcome (together with the consequences of not accounting for environmental effects during seed formation), by mixing seed crops from several years, but this is seldom done in commercial situations. In more advanced schemes, parent-of-family varieties are produced. This involves the use of a limited number of parental clones and artificial crossing (Nanson, 2004).

Finally, various clonal propagation systems (cuttings, somatic embryogenesis) are also available for some species, for the mass production of a few elite genotypes. The gains from elite genotypes are theoretically greater, as they exploit all the genetic variability available and they can be deployed more rapidly, allowing a faster turnover of varieties in some cases, but at the expense of more limited genetic diversity and frequently higher production costs (Lelu-Walter et al. 2013). The success of clonally propagated trees is conditioned by both genetic and non-genetic effects, such as topophysis and cyclophysis (Nanson, 2004), and more broadly by epigenetic marks during zygotic and somatic embryogenesis, with the potential to modify tree behavior in forests (Kvaalen & Johnsen, 2008).

Here, epigenetics is defined as meiotically or mitotically heritable changes that alter phenotype without any changes in the DNA sequence (Russo, Martienssen, & Riggs, 1996; see Chapter 1 of this book). There are different epigenetic modifications that affect the chromatin structure and therefore gene expression or transposable elements mobilization such as DNA methylation, histones modifications or variants and non-coding RNA (see Chapter 2 of this book). Epigenetics is a fundamental mechanism controlling development and response to environmental variations as well as a tool of choice opening perspectives for forest tree management and breeding notably in the context of actual climate change.

2. FOREST TREE GENOMICS TO SUPPORT EPIGENOMIC RESEARCH

2.1 Sequenced Genomes

All epigenome studies require high-quality genome sequence assemblies. The first forest tree for which a whole-genome sequence was obtained was *Populus trichocarpa* in 2006 (Tuskan et al. 2006), followed, years later by

Eucalyptus grandis (Myburg et al. 2014). Both species have relatively small genomes (<500 Mb). With advances in NGS technologies, the number of whole-genome sequences available for forest trees has steadily increased (Table 1), although the sequences available remain a drop in the ocean of the 60,000 species of trees in the world. Their quality, in terms of completeness (proportion of the genome covered) and scaffold contiguity (up to chromosome arm-level), varies considerably, but new technologies for the sequencing of large single molecules (Deamer et al., 2016) and new approaches improving scaffolding (Jiao et al. 2017), together with new algorithms capable for assembling separate haplotypes in highly heterozygous species, will clearly help to improve forest tree genome assemblies, including those for conifers, which have particularly large genomes (Mackay et al. 2012). Temperate tree species are quite well represented in the list of newly sequenced tree genomes, with representatives of large botanical families, such as Fagaceae (e.g. oak), Salicaceae (e.g. willow); Betulaceae (e.g. birch), Oleaceae (e.g. ash) and Pinaceae (e.g. pine, spruce). By contrast, progress has been slower for other highly diverse tropical forest tree species (but see Silva-Junior, Grattapaglia, Novaes, & Collevatti, 2018), despite their key role in biodiversity and the multiple ecosystem services they provide to humanity, including the supply of food (Daru, Berger, & Wyk, 2016).

Once reference genomes have been established and made publicly available (Neale, Langley, Salzberg, & Wegrzyn, 2013), they can be used to explore a number of fundamental research questions (reviewed in Holliday et al. 2017; Plomion et al. 2016):

- In functional genomics, they help improving our understanding of the ways in which trees develop and respond to biotic interactions and abiotic constraints (plasticity)
- In population genomics, for studies of population diversification, adaptation to local conditions, speciation and to the capacity of species to survive environmental change. In this context, decoding the epigenetic landscape will provide indispensable insight, improving our understanding of forest tree development and adaptation
- In comparative genomics, in which they provide insight into chromosome evolution in terms of past polyploidy events, tandem duplications and the diversification of biological functions. Here, the functional consequences of the macro-evolutionary history of epigenetic signatures has yet to be studied
- In quantitative genetics, to shed light on the consequences of genetic diversity in terms of phenotypic variability. Here, we need to determine

Table 1 Published Forest Tree Genomes With Basic Statistics

Species	Botanical Family	Genome Size (gb/2c)	Assembly Availability (Bioproject Id)	Sequencing Technology	# Scaffolds	Scaffold Cumulative Size (mb)	Scaffold n50 Size (kb)	Organization into Pseudomolecules	References
Gymnosperms									
Picea abies	*Pinacea*	39.2	http://congenie.org (ERP002565)	Illumina	10,253,694	4,300	4.869	No	Nystedt et al. 2013
Picea glauca	*Pinacea*	40	NA (PRJNA83435)	Illumina/ 454FLX	4,300,000	22,500	20.900	No	Birol et al. 2013
Pinus taeda	*Pinacea*	44	NA (PRJNA174450)	Illumina	NA	23,180	66.920	No	Neale et al. 2014
Pinus lambertiana	*Pinacea*	31	http://www. pinegenome.org/ pinerefseq (PRJEB174450)	Illumina	>500bp: 1,089,592	24,700	246.600	No	Stevens et al. 2016
Ginko biloba	*Ginkgoaceae*	10	NA (PRJNA307642)	Illumina	6,459,773	10,600	1358.237	No	Guan et al. 2016
Angiosperms									
Quercus robur	*Fagaceae*	1.5	http://www. oakgenome.fr (PRJEB19898)	Illumina/ 454FLX/ Sanger/ Moleculo	1,409	814	1342.530	Yes	Plomion et al. 2018
Quercus lobata	*Fagaceae*	1.4	https://valleyoak.ucla. edu/ genomicresources/ (PRJNA308314)	Illumina	94,394	1180	278.077	No	Sork et al. 2016
Castanea mollisima	*Fagaceae*	1.6	https:// hardwoodgenomics. org NA	454FLX/ Illumina	41,260	720	39.600	No	https:// hardwood genomics. org

Betula pendula	*Betulaceae*	0.84	https://genomevolution.org/coge/api/v1/genomes/35079/sequence (PRJEB14544)	Illumina, SOLID and PacBio	5,642	430	239.520	Yes	Salojaervi et al. 2017
Betula Nana	*Betulaceae*	0.9	http://www.birchgenome.org/ NA	Illumina	551,923	564	18.689	No	Wang et al. 2013
Juglans regia	*Juglandaceae*	1.2	http://dendrome.ucdavis.edu/ftp/Genome_Data/genome/Reju/ (PRJNA291087)	Illumina/PacBio	186,636	700	464.955	No	Martínez-García et al. 2016
Fraxinus excelsior	*Oleaceae*	1.7	http://www.ashgenome.org/assemblies NA	Illumina/454FLX	89,514	867	103.995	No	Sollars et al. 2017
Eucalyptus grandis	*Myrtaceae*	1.3	http://www.phytozome.net/eucalyptus.php (AUSX00000000)	Sanger	11+4900 unanchored	692	53,900.000	Yes	Myburg et al. 2014
Populus trichocarpa	*Salicaceae*	1	https://genomevolution.org/coge/api/v1/genomes/25127/sequence (AARH00000000)	Sanger	1,446	422	19,465.461	Yes	Tuskan et al. 2006
Populus pruinosa	*Salicaceae*	1.2	NA (PRJNA353148)	Illumina	78,960	479	698.525	No	Yang et al. 2017

(Continued)

Table 1 Published Forest Tree Genomes With Basic Statistics—cont'd

Species	Botanical Family	Genome Size (gb/2c)	Assembly Availability (Bioproject Id)	Sequencing Technology	# Scaffolds	Scaffold Cumulative Size (mb)	Scaffold n50 Size (kb)	Organization into Pseudomolecules	References
Populus tremula x p. Alba	*Salicaceae*	0.9	https://urgi.versailles.inra.fr/Species/Forest-trees/Populus/Clone-INRA-717-1B/NA	Illumina/Ion Torrent	419,969	900	3.850	No	Mader et al. 2016
Salix suchowensis	*Salicaceae*	0.86	NA (AVAC00000000)	Illumina/454FLX	>2,000bp: 7,516	304	925.000	No	Dai et al. 2014
Hevea brasiliensis	*Euohorbiaceae*	1.5	NA (LVXX01000000)	Illumina/Sanger	7,453	1370	1280.000	No	Tang et al. 2016
Macadamian intergrifolia	*Proteaceae*	NA	https://www.ebi.ac.uk/ena/data/view/GCA_900087525 (PRJEB13765)	Illumina	193,493	518	4.745	No	Nock et al. 2016
Handroanth impetiginosus	*Bignoniaceae*	1.1	NA (PRJNA324125)	Illumina	13,206	503	81.316	No	Silva-Junior et al. 2018

the extent to which epigenetic variation accounts for the phenotypic variation of relevant traits and contributes to the debate concerning the causal factors for "*missing heritability*" (i.e. the fact that single genetic variants cannot account for much of the heritability of quantitative traits).

These genomic resources also offer opportunities for converting whole-genome sequences into practical tools for breeding and conservation programs. For instance, high-quality genome sequences can be used to establish an inventory of genetic (SNPs, CNVs, indels) and epigenetic (DNA methylation, chromatin conformation state) variants. These inventories can be used to determine the links between (epi)genotypes and phenotypes, and to develop genomic prediction strategies (Bartholomé et al. 2016; Grattapaglia, 2017) to increase genetic gain per unit time. Another foreseeable impact of enriching the inventory of variants is that it should provide new knowledge concerning neutral and adaptive polymorphisms to facilitate the conservation of threatened species (McMahon, Teeling, & Höglund, 2014), a major endeavor considering the rate at which forest tree species can become extinct (ter Steege et al. 2015).

2.2 Genome Evolution and Potential Role of Epigenetic Modifications

The access to tree genome sequences in recent decades has made it possible to compare tree evolution with the evolution of other plant genomes. However, of the forest tree genomes available and listed in Table 1, only those well assembled into pseudomolecules, such as those of oak, poplar, eucalyptus and birch from the angiosperm (flowering plants) clade are suitable for genome-scale comparative genomics and ancestral genome reconstruction (Fig. 1). Angiosperms arose some 120 to 170 million years ago (mya), and diverged 140–150 mya into monocots and eudicots, the two largest and most diverse groups, accounting for 20% and 75% of modern angiosperm species, respectively (Soltis, Bell, Kim, & Soltis, 2008). It has been suggested that the ancestral angiosperm karyotype consisted of a potential repertoire of 22,899 ancestral genes conserved in present-day species and dating back to 190–238 mya, which evolved into the ancestral monocot karyotype (AMK, with 5 protochromosomes and 6707 ordered protogenes) and the ancestral eudicot karyotype (AEK, with 7 protochromosomes and 6284 ordered protogenes), providing an integrative view of the evolution of angiosperm genomes (Murat, Armero, Pont, Klopp, & Salse, 2017). Oak, poplar, eucalyptus and birch evolved independently from the AEK over the last 150 million years. The AEK experienced an ancestral

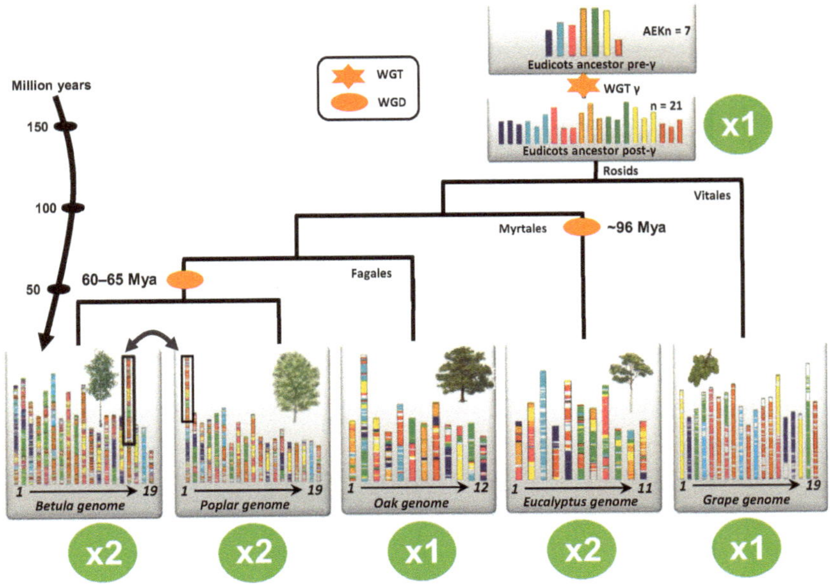

Figure 1 Tree genome evolution from the eudicot ancestor. The present-day tree genomes (birch-poplar-oak-eucalyptus, at the bottom) are represented with color codes to illustrate the evolution of genomic segments from the ancestral eudicot karyotypes (AEKs with 7 and 21 chromosomes, top) over time (time scale on the left in mya). The polyploidization events shaping the structure of the modern tree genomes during their evolution from the AEK are indicated as orange dots (duplication) or stars (triplication). Major angiosperm families are indicated on the tree branches. Green circles show the expected number of modern genes from any ancestral protogene, taking into account the polyploidization events occurring during tree evolution.

whole-genome triplication (γ) common to all modern eudicots, to reach a post-γ AEK of 21 chromosomes (Salse, 2016a). Grape (Vitales subfamily) is the rosid genome most closely resembling the AEK, with 19 modern chromosomes (Murat et al. 2015). Eucalyptus (Myrtales subfamily) experienced a lineage-specific duplication followed by 69 chromosome fissions and 79 chromosome fusions, which shaped the 11 modern chromosomes (Myburg et al. 2014). Poplar (Tuskan et al. 2006) and birch (Salojärvi et al. 2017) emerged from the AEK through a common lineage-specific duplication with a single large chromosomal arrangement involving chromosomes 1 and 16 that differentiates between poplar and birch, consistent with a translocation event between these two genomic fragments (either from chromosome 1 to 16 in birch or from chromosome 16 to 1 in poplar) rather than independent fusions of ancestral chromosomes (on chromosome 1 in poplar or 16 in birch). Finally, it has been suggested that the oak genome has undergone five fissions and 14 fusions since the AEK to achieve its current genome of 12 chromosomes (Plomion et al. 2018). According to this scenario, oak has experienced no lineage-specific whole-genome duplication in addition to the shared ancestral triplication (γ), whereas birch, poplar and eucalypt have experienced an additional duplication. Thus, for any given AEK gene, a single gene would be expected in oak, corresponding to pairs of genes in birch, poplar and eucalyptus (Fig. 1). Such tree species are thus crucial for addressing the role of duplications in promoting the structural and functional partitioning of duplicated genes following polyploidization events (Salse, 2016b).

DNA methylation has been shown to play an important role in duplicate gene evolution in soybean (El Baidouri et al. 2018), wheat (Shaked, Kashkush, Ozkan, Feldman, & Levy, 2001), spartina (Salmon, Ainouche, & Wendel, 2005), rice (Wang et al. 2017c), *Arabidopsis* (Wang et al. 2017b), Brassicaceae (Chen et al. 2015) and cassava (Wang et al. 2015). Differences in the methylation of duplicate genes are involved in the regulation and maintenance of these genes (Keller & Yi, 2014). Greater functional divergence of gene duplicates has been reported in mammals and plants (*i.e.* organisms that have cytosine methylation) as compared to yeast, nematode and flies (*i.e.* organisms that do not have methylated genomes; Rodin & Riggs, 2003). DNA methylation may play a major role in (1) activating transposable elements, leading to insertional mutagenesis and changes in local patterns of gene expression (see Chapter 5), and (2) silencing one copy or driving organ-specific up- or downregulation of one of the duplicated genes. This may result in unequal expression of the duplicates, facilitating rapid genomic reorganization and the stabilization of the newly formed polyploid, contributing to correct meiotic pairing and isolation of the polyploid from its progenitors, increasing evolutionary potential (Paun, Fay, Soltis, & Chase, 2007). Overall, epigenetic regulation of duplicated genes may be a central phenomenon to understand the impact of polyploidization events in plant survival or adaptation to environmental constraints. Research projects targeting post-polyploidization epigenetic plasticity are needed to be accurately used in breeding programs or conservation practices, especially for perennial plant species such as trees that have to face recurrent biotic and abiotic constraints during their long life span.

3. STATE OF THE ART IN FOREST TREE EPIGENOMICS

3.1 Methods and Recent Advances in Epigenomics

Second/next generation sequencing (NGS) has revolutionized the way of interrogating the epigenomes. NGS-based technologies have largely replaced microarrays as the read-out platform for DNA methylation analysis enabling a count-based absolute quantification of DNA methylation levels and current NGS platforms provide robust and high-quality results. The main approaches for the discrimination of methylation are based on three main principles (Table 2):

1) The use of methylation-sensitive restriction endonucleases, i.e., enzymes that are blocked by methylated cytosines in their recognition sequence (Bird & Southern, 1978) are widely used for the analysis of methylation patterns in combination with their methylation-insensitive isoschizomers found within the recognition sequences of restriction enzymes (Fazzari & Greally, 2004). Information complementary to that obtained by methylation-sensitive restriction digests can be obtained by the methylation-dependent restriction enzymes such as *McrBC*, which cleaves between two non-palindromic G/A^{me}C sites (Stewart, Panne, Bickle, & Raleigh, 2000).
2) The methylated fraction of a genome can be enriched by precipitation with a bead-immobilized antibody specific for 5-methylcytosine (Down et al. 2008; Feber et al. 2011) or by affinity purification of methylated DNA with methyl binding domain (MBD) proteins such as MeCP2 (Brinkman et al. 2010) or MBD2 in combination with MBD3L1 (Rauch & Pfeifer, 2005).
3) The most widely used approach consists of the chemical modification of genomic DNA with sodium bisulfite. This chemical reaction induces hydrolytic deamination of non-methylated cytosines to uracils, while methylated cytosines are resistant to the conversion under the chosen reaction conditions (Frommer et al. 1992; Shapiro, DiFate, & Welcher, 1974). This method translates the methylation signal into a sequence difference. After performing PCR, the methylation status at a given position is manifested in the ratio of C (former methylated cytosine) to T (former non-methylated cytosine) and can be analyzed as a virtual C/T polymorphism spanning the entire allele frequency spectrum from 0%–100% in the bisulfite treated DNA. It should be noted that standard bisulfite conversion protocols cannot discriminate between 5-hydoxymethylcytosine and 5-methylcytosine, which are converted

Table 2 Key Characteristics of Most Widely Used Methods for the Genome-wide Analysis of DNA Methylation

Technology	The Technology Behind	Methylated/ Unmethylated Fraction	Single Nucleotide Resolution (Yes/No)	DNA Input in μg	Required Number of Sequencing Reads per Sample in m	Coverage of an Average Genome	Comments	References
WGBS/ MethylC-seq	Adaptor ligation before or after bisulfite treatment of genomic DNA prior to sequencing	U+M	Yes	0.1–5	500	95%	Comprehensive coverage of all cytosines in a genome, several variations of library preparation exist, cost-intensive, but will with decreasing sequencing costs become the gold standard for the discovery technologies, no distinction between 5 mC and 5 hmC	Urich et al. 2015, Daviaud et al. 2018
RRBS	Restriction digest allows for size selection of small fragments, used for library construction, bisulfite conversion and sequencing	U+M	Yes	0.01–1	20	4%	85% of CpG islands are covered, cost-effective sequencing approach, low coverage in CpG poor regions, no distinction between 5 mC and 5 hmC	Meissner et al. 2005, Gu et al. 2011

(*Continued*)

Table 2 Key Characteristics of Most Widely Used Methods for the Genome-wide Analysis of DNA Methylation—cont'd

Technology	The Technology Behind	Methylated/ Unmethylated Fraction	Single Nucleotide Resolution (Yes/No)	DNA Input in µg	Required Number of Sequencing Reads per Sample in m	Coverage of an Average Genome	Comments	References
Agilent SureSelect^{XT} methylation capture	Libraries are prepared form genomic DNA fragments, hybridized to the capture probes, eluted, bisulfite converted, and amplified before being sequenced by a NGS	U+M	Yes	3	50	2.8%	Off the shelf product, captures all known promoters and upstream regions, uses the procedure used for exome sequencing with an additional bisulfite conversion step, relatively expensive, custom design possible	Borno et al. 2012, Walker et al. 2015
SeqCapEpi CpGiant enrichment	Libraries are prepared form genomic DNA fragments, bisulfite converted, pre-amplified hybridized to the capture probes, eluted, b and amplified before being sequenced by a NGS	U+M	Yes	1	40–50	2.7%	Less input compared to the SureSelect capture as capture is performed on bisulfite treated libraries after pre-amplification, relatively expensive, custom design possible	Li et al. 2015, Walker et al. 2015

MBD-seq/ MethylCap-seq	Enrichment of methylated DNA using a Methyl-binding domain protein before or after a library construction followed by sequencing	M	No	0.2–1	30–40	95%	Also suitable for FFPE, amenable to high-throughput and analysis of large cohorts, absolute quantification of DNA methylation difficult, specific for 5 mC	Brinkman et al. 2010
MeDIP-seq	Library construction, Enrichment of methylated DNA using an antibody against 5 mC followed by sequencing	M	No	0.3–0.5	50–60	95%	Captures and allows to analyze also the transposons of the human genome, specific for 5 mC, absolute quantification of DNA methylation difficult, lower signal to noise compared to MBD- and MIRA-seq, better coverage in CpG poor regions	Down et al. 2008, Taiwo et al. 2012
MIRA-seq	Enrichment of methylated DNA using a Methyl-binding domain protein before or after a library construction followed by sequencing	M	No	0.1–1	40–50	95%	Higher specificity compared to MBD-seq, absolute quantification of DNA methylation difficult, specific for 5 mC	Jung et al. 2015

(Continued)

Table 2 Key Characteristics of Most Widely Used Methods for the Genome-wide Analysis of DNA Methylation—cont'd

Technology	The Technology Behind	Methylated/ Unmethylated Fraction	Single Nucleotide Resolution (Yes/No)	DNA Input in µg	Required Number of Sequencing Reads per Sample in m	Coverage of an Average Genome	Comments	References
MRE-SEQ	Digestion of genomic DNA in parallel with several methylation-sensitive restriction enzymes, size selection, library construction, sequencing	U	Yes, at restriction site	1.5–2.5	30	4%	Analyses preferentially unmethylated CpGs in high density CpG regions, in combination with MeDIP comprehensive genome coverage is achieved	Maunakea et al. 2010
HELP	Digestion of genomic DNA in parallel with a methylation-sensitive restriction enzymes and its methylation-insensitive isoschizomer, size selection, library construction, sequencing	U+M	Yes, at restriction site	0.3	30	4%	Normalization of the counts signals from the two digestions allows for quantitative analysis at restriction sites, limited coverage due to the use of restriction enzymes, variation allows for the parallel analysis of 5 hmC	Suzuki et al. 2010, Greally, 2018

MSAP	Digestion of genomic DNA in parallel with a methylation-sensitive restriction enzymes and its methylation-insensitive isoschizomer in parallel in combination with a frequently cutting enzyme, adaptor ligation, size selection, library construction, sequencing	U+M	Yes at restriction site	0.5	15	< 1%	One of the most widely used methods in plants and trees, MSAP has mostly been performed using PAGE electrophoresis as read-out platform using a fluorescently or radioactive labeled amplification primer, rarely by sequencing, most diff. methylated regions are located in intergenic regions and repetitive elements	Fulnecek & Kovarik, 2014, Chwialkowska, Korotko, Kosinska, Szarejko, & Kwasniewski, 2017

5 mC: 5-methylcytosine; 5 hmC; 5-hydroxymethylcytosine; WGBS: Whole Genome Bisulfite Sequencing; RRBS: Reduced Representation Bisulfite Sequencing; MBD: Methyl-Binding Domain; MeDIP: Methylated DNA ImmunoPrecipitation; MIRA: methylated-CpG island recovery assay; MRE: Methylation-sensitive Restriction Enzyme sequencing; HELP: *Hpa*II tiny fragment Enrichment by Ligation-mediated PCR; MSAP: Methylation-Sensitive Amplification Polymorphism.

with a similar efficiency requiring specialized protocols such as oxidative bisulfite sequencing (Booth et al. 2012) or TAB-seq (Yu et al. 2012).

Whole-genome bisulfite sequencing (WGBS) or MethylC-seq can be considered as the current gold standard for the genome-wide identification of differentially methylated cytosines (DMCs) and differentially methylated regions (DMRs) at single nucleotide resolution (Lister et al. 2008; Urich, Nery, Lister, Schmitz, & Ecker, 2015). The whole genome read-out after bisulfite conversion overcomes the limitations of cloning and Sanger sequencing, in which the quantitative resolution was limited by the number of clones analyzed (in most studies < 20), or pyrosequencing limited to a small number of loci of interest. Several protocols have been devised to reduce the loss of material, adding sequencing adaptors through transposition rather than ligation (Adey & Shendure, 2012) or adding these only after bisulfite treatment, which leads to a substantial loss of library complexity using standard protocols (Miura, Enomoto, Dairiki, & Ito, 2012).

These and other technical improvements allow now the comprehensive methylome analysis from limited amounts of cells (Daviaud, Renault, Mauger, Deleuze, & Tost, 2018), or even single cells (Kelsey, Stegle, & Reik, 2017). However, the unprecedented quantitative and spatial resolution that is currently transforming DNA methylation analysis still comes at a high cost requiring substantial sequencing to obtain a proper and even coverage, and requires specialized bioinformatic expertise and resources. Noteworthy, as only part of the analyzed sequences will contain potentially methylated cytosines, a number of strategies have been devised to concentrate the analysis on parts of the genome including Reduced Representation Bisulfite Sequencing (RRBS, Meissner et al. 2005), methylation-sensitive enzymes (MRE-seq; Maunakea et al. 2010) or affinity based approaches using methyl-binding proteins or antibodies (Brinkman et al., 2010; Jung et al., 2015; Taiwo et al., 2012; see also Tost, 2016 for a detailed review). However, these do not necessarily allow the analysis of all regions of interest, due to the absence of restriction-enzyme recognition sites in the size range which is commonly sequenced (RRBS, MRE-seq) or do not provide base-resolution data and only relative quantification of DNA methylation levels (MeDIP/MBD).

Alternative approaches have been developed to focus DNA methylation analysis together with a supposed biological function (i.e. expressed or silenced regions of the genome) making use of the presence of other epigenetic modifications such as histone modifications or open chromatin (Lafon-Placette et al. 2013; Statham et al. 2012). Another now widely used approach captures regions of interest with complementary oligonucleotides similarly to standard exome sequencing analyzing up to 8 M CpGs (Walker et al. 2015; Wendt, Rosenbaum, Richmond, Jeddeloh, & Burgess, 2018). Capture can be made prior or after the bisulfite conversion with the first approach making use of the increased complexity of the genome prior to bisulfite conversion, but prohibiting amplification prior to the capture, while the second approach needs to accommodate a much larger number of probes to capture molecules with different methylation patterns but permitting DNA amplification prior to capture. If fewer regions of interest need to be interrogated, a number of multiplexed amplification systems have been devised enabling the amplification of up to several hundred regions of interest in parallel (Diep et al. 2012; Paul et al. 2014), which are subsequently analyzed on benchtop NGS sequencers. For locus-specific DNA methylation analysis Sanger sequencing (Frommer et al. 1992), Pyrosequencing (Tost & Gut, 2007), MALDI mass spectrometry (Ehrich et al. 2005), methylation-specific PCR (Herman, Graff, Myohanen, Nelkin, & Baylin, 1996) and its real time variations (Eads et al. 2000; Tost, 2016) have been most-widely used methods, but are also more and more replaced by a NGS-based amplicon sequencing. Benchtop sequencers (MiSeq or PGM) allow generating high levels of coverage (100's-1000's x) that yield precise measurements of the quantitative levels of cytosine methylation. In addition, due to the sequencing of clonal clusters, these methods provide co-methylation patterns on individual molecules within the limits of the length of the reads. With a current output of ~50 M reads for the MiSeq, yielding between 3.8 and 15 GB of sequence depending on the used sequencing kit, several tens to hundreds of target regions can be analyzed simultaneously depending on the desired coverage and number of samples analyzed in parallel. Their short run time, relatively low running costs and wide availability make them a valuable alternative for targeted DNA methylation analysis. A recent large-scale evaluation study including the most widely used methods for locus-specific DNA methylation analysis demonstrated that technologies have reached a level of technical maturity yielding reproducible and accurate results, with pyrosequencing and bisulfite amplicon-sequencing showing slightly improved performance (Blueprint Consortium, 2016; http://www.blueprint-epigenome.eu/).

New read-out technologies enabling a direct-readout of DNA methylation patterns without the need for chemical conversion and/or amplification such as single-molecule real-time (SMRT) and Nanopore sequencing have been devised. SMRT sequencing is based on different kinetics during polymerization by an immobilized polymerase when encountering methylated bases while for nanopore sequencing single DNA molecules are pulled through a protein or synthetic pore and nucleobase dependent changes in the electric potential of the pore are measured (Clarke et al. 2009; Flusberg et al. 2010; Song et al. 2012). Despite recent improvements in the base calling algorithms for nanopore sequencing (Rand et al. 2017; Simpson et al. 2017), these need to be further optimized, the high error rate associated with the sequencing, which currently requires sequencing with very deep coverage, needs to be reduced before these technologies can be routinely applied to DNA methylation analysis.

In contrast, few methods are available for the analysis of histone modifications now commonly analyzed by chromatin immunoprecipitation (ChIP (Gilmour & Lis, 1984)) followed by qPCR for the detection and quantification of the analyzed modification at a locus of interest using specific primers or NGS (ChIP-seq) for the genome-wide mapping of an histone modification at very high resolution (Landt et al. 2012).

3.2 Epigenomics in Forest Trees: Role in the Response to Environmental Variations

These last years, the available epigenomic data in forest trees are steadily increasing. These studies focused especially on DNA methylation, histones modification and small RNAs. Table 3 summarizes available epigenomic data in forest tree species for DNA methylation, histones modification and small RNAs. For DNA methylation we chose to not include studies only focusing on global DNA methylation level (Plomion et al. 2016). Although miRNAs are mitotically and meiotically hereditable factors that control gene expression without changes in the DNA sequence, they are classically not included in epigenetic mechanisms. However, growing evidences, on cancer studies, show their substantial role in the control of several epigenetic mechanisms. Specifically, miRNAs regulate at the post-transcriptional level many epigenetic-related-genes, a subgroup of miRNAs, epi-miRNAs, can regulate the expression of effectors of the epigenetic mechanisms by directly or indirectly targeting epigenetic-modifying enzymes and molecules or can act in the nucleus by stimulating or repressing genes transcription in a manner strictly correlated to the chromatin state (Ramassone, Pagotto, Veronese, & Visone, 2018; Xia, Guo, & Deng, 2014). According to this, it is possible to consider miRNAs in an enlarged view of mechanisms associated to epigenetic processes. For these reasons, we referred to some miRNAs studies in trees in the Table 3. Few of the epigenomic studies

Table 3 Available Epigenomics Data in Forest Tree Species

Species	Family	Approach	Main Conclusion	References
DNA Methylation				
Picea abies	*PINACEAE*	WGBS	The Norway spruce genome is heavily methylated because of high transposon content. Somatic embryogenesis cultures used in the industry show altered DNA methylation patterning.	Ausin et al. 2016
Pinus pinaster	*PINACEAE*	MSAP	DNA methylation varies during aging of embryogenic cultures and in response to 5-azacytidine treatment	Klimaszewska et al. 2009
Pinus radiata	*PINACEAE*	MSAP	DNA methylation variations during needle maturation in relation to the loss of *in vitro* organogenic capability	Valledor, Meijón, Hasbún, Cañal, & Rodríguez, 2010
Pinus pinea	*PINACEAE*	MSAP	Variable epigenetic markers discriminate individuals and differentiates two well represented populations contrary to genetic variation	Sáez-Laguna et al., 2014
Quercus lobata	*FAGACEAE*	RRBS	While CHG methyl polymorphisms are not playing a significant role and would make poor targets for natural selection, our findings suggest that CpG methyl polymorphisms as a whole are involved in local adaptation, either directly or through linkage to regions under selection.	Platt et al. 2015

(Continued)

Table 3 Available Epigenomics Data in Forest Tree Species—cont'd

Species	Family	Approach	Main Conclusion	References
Quercus lobata	*FAGACEAE*	RRBS	Climate and spatial variables explain more overall variance in CG-SMV among individuals than in SNPs, CHG-SMVs or CHH-SMVs. These results from natural oak populations provide initial evidence for a role of CG methylation in locally adaptive evolution or plasticity in plant response	Gugger et al. 2016
Quercus ilex	*FAGACEAE*	MSAP	DNA methylation variations under drought stress	Rico, Ogaya, Barbeta, & Peñuelas, 2014
Eucalyptus globulus	*MYRTACEAE*	MeDIP-seq MRE-seq	Locus-specific methylation could be major regulators of vegetative phase change	Hasbún, Iturra, Bravo, Rebolledo-Jaramillo, & Valledor, 2016
Populus trichocarpa	*SALICACEAE*	WGBS	If patterns of methylation are very similar in flowering plants, CHG methylation levels in transposons and repeats were much higher in poplar	Feng et al. 2010
Populus trichocarpa	*SALICACEAE*	MeDIP-SEQ	DNA in recombination hotspots is less methylated	Slavov et al. 2012
Populus trichocarpa	*SALICACEAE*	MeDIP-SEQ	DNA methylation is tissue specific and gene-body DNA methylation has a more repressive effect on transcription than promoter methylation	Vining et al. 2012

Populus trichocarpa	*SALICACEAE*	MeDIP-SEQ	DNA methylation varies in a highly gene- and chromosome-differential manner during *in vitro* differentiation and regeneration. Hypermethylation of gene bodies may serve a protective role against activation of abundant transposable elements	Vining et al. 2013
Populus trichocarpa	*SALICACEAE*	WGBS	Differential methylation was associated with site-dependent, different total amounts of miRNA, with few miRNAs sequences directly targeted by differential methylation. This may explain habitat or seasonal memory in perennials and site-dependent growth performance	Schönberger et al. 2016
Populus trichocarpa	*SALICACEAE*	WGBS	DNA methylation in response to stress regulates genes by methylating TEs in promoters and gene body of transcription factors	Liang et al. 2014
Populus tremula x alba	*SALICACEAE*	RNAi strategy	First report for a DNA methylation modified-tree. The phenotypic consequences of reduced *DDM1* activity and DNA methylation appears to increase with cumulative plant propagation and growth	Zhu et al. 2013
Populus tremula × alba	*SALICACEAE*	Immunofluorescence	CsDML gene induces bud formation needed for the survival of the apical meristem under the harsh conditions of winter	Conde et al. 2017a, 2017b
Populus tremula × alba	*SALICACEAE*	WGBS	A chilling-dependent DEMETER-like DNA demethylase mechanisms being involved in the shift from winter dormancy to a condition that precedes shoot apical vegetative growth in poplar	Conde et al. 2017a, 2017b

(Continued)

Table 3 Available Epigenomics Data in Forest Tree Species—cont'd

Species	Family	Approach	Main Conclusion	References
Populus alba	*SALICACEAE*	MSAP	DNA methylation would adapt gene expression in response to biotic and abiotic stress	Cicatelli et al. 2014
Populus alba	*SALICACEAE*	MSAP	The genetic biodiversity of poplars is quite limited but it is counterbalanced by epigenetic inter-population molecular variability. Plant biodiversity should no longer be restricted to genetic aspects, especially in the case of vegetatively propagated plant species	Guarino et al. 2015
Populus tomentosa	*SALICACEAE*	MSAP	DNA methylation sites have the potential to regulate the genes' transcript levels	Song et al. 2012
Poplar latin	*SALICACEAE*	MSAP	DNA methylation is nonlinearly related to the ploidy level	Li et al. 2011
Populus deltoides	*SALICACEAE*	MeDIP-SEQ	The methylation patterns of the parents both partially and dynamically passed onto their hybrids and F1 hybrids has a non-additive methylation level	Gao et al. 2014
Populus simonii	*SALICACEAE*	MSAP, Degradome Bisulfite-seq	DNA methylation probably regulates the expression of miRNA genes, thus affecting expression of their target genes, likely through the gene-silencing function of miRNAs, to maintain cell survival under abiotic stress conditions	Ci et al. 2016
P. pseudo-simonii × *P. nigra and P. beijingensis*	*SALICACEAE*	MSAP	Both hybridization and polyploidization contributed to cytosine methylation variation.	Suo, Dong, & Kang, 2015

Populus. simonii	*SALICACEAE*	HPLC MSAP	The SDMR162 region, consisting of Psi-MIR396e and PsiLNCRNA00268512, is regulated by epigenetic pathways and we speculate that PsiLNCRNA00268512 regulates miR396e levels by acting as a target mimic.	Song, Ci, Tian, & Zhang, 2016
Populus nigra	*SALICACEAE*	MeDIP-seq	Methylated genes were prevalent in the poplar genome, but that only a few of these participated in diurnal gene expression regulation	Ding, Liang, Diao, Su, & Zhang, 2018
Populus×euramericana (P. deltoides batr.×P. nigra l.)	*SALICACEAE*	MeDIP-chip	Shoot apical meristem responses to changes in water availability involved coordinated variations in DNA methylation, as well as in gene expression, with a specific targeting of genes involved in hormone pathways, a factor that may enable phenotypic plasticity.	Lafon-Placette et al. 2018
Jatropha curcas	*EUPHORBIACEAE*	MSAP	DNA methylation polymorphism is observed between individuals and in response to gamma-irradiation	Kanchanaketu, Sangduen, Toojinda, & Hongtrakul, 2012
Acacia mangium	*FABACEAE*	MSAP	Differences in DNA methylation exist between juvenile and mature plant materials	Baurens, Nicolleau, Legavre, Verdeil, & Monteuuis, 2004

(Continued)

Table 3 Available Epigenomics Data in Forest Tree Species—cont'd

Species	Family	Approach	Main Conclusion	References
Laguncularia racemosa	*COMBRETACEAE*	MSAP	Higher DNA methylation differentiation than genetic one between adapted populations to distinct environment	Lira-Medeiros et al. 2010
Elaeis guineensis jacq.	*ARECACEAE*	MSAP	DNA methylation polymorphism discriminates between the two phenotypes (normal and mantled) only when they were from the same genetic origin	Jaligot, Beulé, Baurens, Billotte, & Rival, 2004
Histones Modifications				
Picea asperata	*PINACEAE*	antibody-enrichment, MS	Functional characterization of the acetylated proteins revealed that in the desiccated somatic embryos, LysAc is mainly involved in the response to stress and central metabolism	Xia et al. 2016
Quercus suber l.	*FAGACEAE*	HPCE MS-RAPD Protein Gel Blot	Epigenetic mechanisms such as DNA methylation and histone H3 acetylation have opposite and particular dynamics that can be crucial for the stepwise establishment of cork Oak into high stress, allowing its acclimation and survival	Correia et al. 2013
Eucalyptus grandis	*MYRTACEAE*	ChIP-seq	In developing *E. grandis* xylem, H3K4me3 enrichment is an indicator of active transcription, consistent with its known role in sustaining pre-initiation complex formation in yeast	Hussey, Mizrachi, Groover, Berger, & Myburg, 2015

Populus trichocarpa	*SALICACEAE*	HDAC Colorimetric Assay	HDACs were required for de novo organogenesis and normal growth of populus roots.	Ma, Zhang, Zhang, Yang, & Li, 2016
Populus simonii x Populus nigra	*SALICACEAE*	ChIP assay	Histone acetylation positively regulates the tissue dependent expression pattern of the poplar homologs of C4 homologous genes. This regulatory mechanism seems to be conserved among the C3 and C4 species	Li et al. 2017
Small RNAS (miRNA & siRNA)				
Picea abies	*PINACEAE*	sRNA-seq	Norway spruce contains a set of conserved miRNAs as well as a large proportion of novel non-conserved miRNAs. The differentially expression of specific miRNAs indicate their putative participation in the epigenetic regulation	Yakovlev et al. 2010
Picea abies	*PINACEAE*	sRNA-seq RNA-seq	A significant number of epigenetic regulators were differentially expressed during embryogenesis at different epitype-inducing conditions. This support that methylation of DNA and histones, as well as sRNAs, are pivotal for the establishment of the epigenetic memory	Yakovlev et al. 2016
Picea abies	*PINACEAE*	Rt-qPCR	Epigenetic memory affects the timing of bud burst phenology and the expression of bud burst related genes in genetically identical Norway spruce epitypes in a manner usually associated with ecotypes.	Carneros et al. 2017

(Continued)

Table 3 Available Epigenomics Data in Forest Tree Species—cont'd

Species	Family	Approach	Main Conclusion	References
Picea abies	*PINACEAE*	sRNA-seq	Fine-tuning of the miRNA production likely participates in both developmental regulation and epigenetic memory formation in Norway spruce	Yakovlev & Fossdal, 2017
Picea glauca	*PINACEAE*	sRNA-seq	Lacking of 24-nt sRNAs at the late conifer seed developmental phase may result in less constraint in TE activities, thus contributing to the massive expansion of genome size.	Liu & El-Kassaby, 2017
Pinus taeda	*PINACEAE*	miRNA-array	conserved miRNAs are expressed in mature and germinated loblolly pine pollen	Quinn, Iriyama, & Fernando, 2014
Pinus tabuliformis	*PINACEAE*	sRNA-seq PARE-seq	The sRNA pathways have higher activity in female than in male cones, and the miRNA pathways are the main sRNA pathways in *P. tabuliformis*	Niu et al. 2015
Ginkgo biloba l.	*Ginkgoaceae*	sRNA-seq	Identification a large number of miRNAs in mature female and male *G. biloba* leaves	Wang et al. 2015
Populus trichocarpa	*SALICACEAE*	sRNA-seq, Degradome sequencing	Five upregulated miRNAs and seven downregulated miRNAs in response to drought stress discovered. This will promote the understanding of miRNA functions during the drought response.	Shuai et al. 2013

Populus balsamifera	*SALICACEAE*	sRNA-seq	A large fraction of miRNAs vary among species. The non-conserved miRNAs may regulate cellular, physiological or developmental processes specific to the taxa that produce them, as appears likely to be the case for those miRNAs that have only been observed in Populus	Barakat, Wall, DiLoreto, dePamphilis, & Carlson, 2007
Populus. tomentosa	*SALICACEAE*	sRNA-seq qRT-PCR	Significant changes in the expression of 17 conserved miRNA families and nine novel miRNAs were observed in response to drought stress, and in seven conserved miRNA families and five novel miRNAs in response to flooding stress.	Ren et al. 2012
Populus tomentosa	*SALICACEAE*	Bisulfite sequencing	miRNA 172b might play an important role in the regulation of bisexual flower development-related gene expression in andromonoecious poplar, via modification of methylation. Hyper-methylation in andromonoecious and gynomonoecious poplar might function as an important regulator in bisexual flower development	Song, Tian, Ci, & Zhang, 2015
Populus euphratica	*SALICACEAE*	sRNA-seq	Expression changes of miRNAs were inversely correlated with the expression profiles of their putative targets and might be involved in some biological process related stress tolerance	Duan et al. 2016

(*Continued*)

Table 3 Available Epigenomics Data in Forest Tree Species—cont'd

Species	Family	Approach	Main Conclusion	References
Cunninghamia lanceolata	*CUPRESSACÉES*	sRNA-seq	Identification of a complex population of sRNAs in Chinese fir through high throughput sequencing	Wan et al. 2012
Cryptomeria japonica	*CUPRESSACÉES*	sRNA-seq	Both conserved and species-specific sRNAs contribute to the development of male strobili in *C. japonica*	Ujino-Ihara, Ueno, Uchiyama, & Futamura, 2018
Acer palmatum thunb.	*SAPINDACEAE*	RNA-seq sRNA-seq	RNA-seq and sRNA-seq data show that gene differentiation associates with *Acer palmatum* leaf coloration in different light conditions	Li et al. 2015
Phoenix dactylifera l.	*ARECACEAE*	sRNA-seq	Date palm contains a large population of conserved and non-conserved miRNAs that function at the post-transcriptional level, and are important for adaptation to salinity	Yaish et al. 2015

WGBS: Whole Genome Bisulfite Sequencing; RRBS: Reduced Representation Bisulfite Sequencing; MeDIP: Methylated DNA ImmunoPrecipitation; MRE-seq: Methylation-sensitive Restriction Enzyme sequencing; MSAP: Methylation-Sensitive Amplification Polymorphism; sRNA-seq: small RNA sequencing; ChIP: Chromatin ImmunoPrecipitation; HDAC: Histone DeACetylase; MS-RAPD: Methylation Sensitive - Random Amplified Polymorphic DNA; HPCE: High-Performance Capillary Electrophoresis; HPLC: High-Performance Liquid Chromatography.

are developed thereafter as examples showing the growing evidences for the role of epigenetics in the response of trees to environmental variations.

3.2.1 Epigenetics and Developmental Transitions During the Annual Cycle

Many studies have reported changes in the developmental stages of cells, tissues or organs from different organisms accompanied by global changes in chromatin epigenetic marks (Kumar, Kumari, Sharma, & Sharma, 2013). These observations suggest that switches in the epigenetic state of chromatin generate a new landscape for gene expression that is required to promote entry into a new developmental stage and has led to research to characterize the role of chromatin remodeling in plant development over the entire life cycle (Liu et al. 2015).

In cold and temperate regions, the annual growth cycles of trees are tightly linked to the succession of environmental conditions during the course of the seasons. The decrease in day length during the fall causes a cessation of growth and induces bud set in several tree species, including poplar. The first stage of this process is called ecodormancy. Greater exposure to short days and low temperatures later in the year induces endodormancy. In several tree species, such as apple, low temperatures alone are sufficient to induce dormancy. Once in endodormancy, the apical meristem and leaf primordia, which are already protected by buds, become insensitive to growth-promoting signals and cell division is stopped by endogenous factors that remain poorly understood. Many species have a chilling requirement to break dormancy, a need for exposure to low, non-lethal temperatures. Once dormancy is released, growth is halted purely by environmental factors, such as low winter temperatures, in particular. An increase in temperature during the spring promotes bud burst and restores growth. Changes in the epigenetic profile of chromatin over the course of the annual cycle have been observed in several tree species. In chestnut, for example, total DNA methylation increases during bud set in the fall and decreases during bud break in the spring (Santamaría et al. 2009). Furthermore, acetylated histone 4 (AcH4) levels in the apical bud are higher during bud break than during bud set (Santamaría et al. 2009). In peach, the dormancy regulators DORMANCY-ASSOCIATED MADS-BOX (DAM), DAM1, DAM4, DAM5 and DAM6, are significantly enriched in H3K27me3 during dormancy release relative to the dormant state (de la Fuente, Conesa, Lloret, Badenes, & Ríos, 2015). H3K27me3 is a chromatin mark associated with stable gene silencing in this situation. In poplar stems, DNA methylation levels are higher and AcK8H4 levels are lower during winter dormancy than during active growth (Conde, González-Melendi, & Allona, 2013). Moreover, DNA methylation is highly restricted to the central zone of the poplar SAM (shoot apical meristem) during the active growth period, whereas methylation is widespread throughout the SAM and in other bud regions during growth cessation under short days (Conde et al. 2017b). In poplar, the transition from dormancy to the reactivation of cell division in the SAM is accompanied by an initial increase, followed by a gradual decrease in global DNA methylation level (Conde et al. 2017a). In apple (*Malus* x *domestica*), Kumar, Rattan, and Singh (2016) found that DNA methylation levels decreased gradually over four developmental stages, from bud dormancy to fruit set: dormant bud, silver tip, green tip and initial fruit set. This dynamic in apple buds occurs only when apple trees are grown in environmental conditions satisfying the chilling requirement for winter dormancy release, as reported for poplar bud break (Conde et al. 2017a; Kumar et al. 2016), highlighting the importance of environmental conditions for the developmental reprogramming of DNA methylation (see Chapters 10 and 11 of this book).

In addition to these changes in epigenetic profile during annual growth cycle of trees, several transcriptomic analyses have highlighted differential patterns of expression for genes involved in chromatin remodeling over the various stages of tree development during the course of the year: in poplar, FERTILIZATION-INDEPENDENT ENDOSPERM (FIE) is rapidly upregulated following the perception of short days (Ruttink et al. 2007). FIE is part of t POLYCOMB REPRESSION COMPLEX 2 (PRC2), which is involved in the H3K27m3 methylation. Tylewicz et al. (2018) used a transgenic approach to analyze the function in poplar of PICKLE (PKL), a chromatin remodeling antagonist of PRC2 identified in *Arabidopsis*. They showed that PKL was downregulated by abscisic acid (ABA) under short days, and required for plasmodesmata closure and bud dormancy establishment in poplar (Fig. 2). Shim et al. (2014) identified five genes involved in DNA methylation that were differentially expressed at different developmental stages of winter dormancy in poplar stems: two poplar homologs of *Arabidopsis* METHYLTRANSFERASE 1 (MET1), a homolog of CHROMOMETHYLASE 3 (CMT3) and two DEMETER-like (DML) genes. One poplar DML gene was upregulated when growth ceased at the onset of winter dormancy, whereas the other was more strongly expressed during the exit from dormancy after the chilling requirement had been fulfilled.

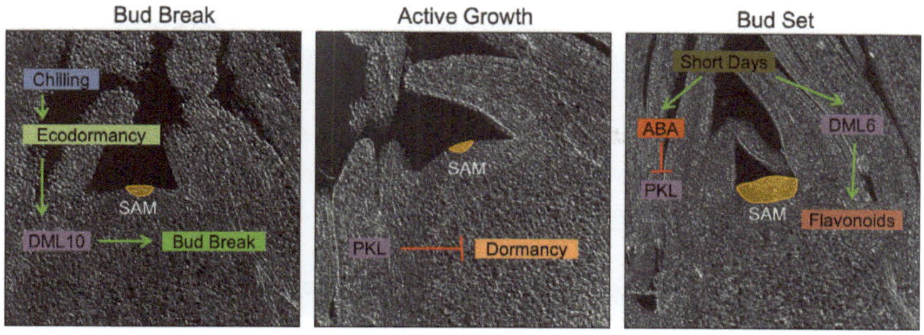

Figure 2 Schematic model of the role of chromatin remodeling in poplar bud dormancy. During bud break, DML10-mediated active demethylation promotes the induction of growth-promoting genes require to re-establish shoot elongation. During active growth, DNA methylation is tightly restricted to the central zone of the SAM (yellow patch). During the growing season the PKL chromatin remodeling factor represses the induction of bud dormancy. During bud set, DNA methylation spreads to the rest of the SAM (yellow patch). Short days induce ABA, which represses PKL, promoting dormancy. Short days and cold temperatures also induce DML6, a DNA demethylase promoting the expression of flavonoid biosynthetic enzymes. Flavonoids accumulate in the SAM and bud scales during apical bud formation.

Based on the changes in DNA methylation and in the expression of genes involved in the establishment or removal of this epigenetic mark over the annual cycle in poplar, Conde et al. (2017a and 2017b) identified and characterized the role of DMLs in trees. One DML induced by short days and low temperatures in poplar and chestnut participates in apical bud formation during autumn, by activating flavonoid biosynthesis enzymes. Flavonoids accumulate in bud scales and the SAM during the onset of winter dormancy, and are required for the survival of the apical meristem under harsh winter conditions (Conde et al. 2017b) (Fig. 2). Poplar transgenic lines with a reduced expression of PtaDML10 showed a delay in bud burst after winter dormancy release, compared to wild type. Genome-wide transcriptome and methylome analysis and data mining have identified gene targets for the active DML-dependent DNA demethylation associated with bud break (Conde et al. 2017a). These data suggest that DEMETER-like mediated DNA demethylation is involved in an environmentally induced developmental stage transition, during the shift from winter dormancy to vegetative growth of the shoot apex in poplar (Fig. 2).

In summary, there is growing evidence to suggest that chromatin remodeling changes are important during the transitions between developmental stages in the annual growth cycle of trees annual developmental stages transitions. The small number of functional characterizations carried out to date has led to the identification of chromatin remodelers involved in both the establishment of bud dormancy in the fall and bud break in the spring. Further studies are required to determine the mechanisms by which the epigenetic state of chromatin senses the environmental conditions under which the tree is growing, throughout its life cycle.

3.2.2 Epigenetics as a Mediator of Phenotypic Plasticity in Response to Environmental Variations

Phenotypic plasticity is defined as the ability of a genotype to express different phenotypes in different environments (Bradshaw, 2006). Significant progress has been made towards understanding the molecular mechanisms underlying phenotypic plasticity (Kelly, Tami, & Andrew, 2012), but the role of epigenetics in this process remains unclear, despite recent findings suggesting that it is of great importance (Bossdorf, Richards, & Pigliucci, 2008; Bräutigam et al. 2013; Herrera & Bazaga, 2013; Lira-Medeiros et al. 2010; Zhang, Fischer, Colot, & Bossdorf, 2013). Using *Arabidopsis thaliana* epigenetic recombinant inbred lines (i.e. differing only in their level of DNA methylation), Kooke et al. (2015) showed that traits such as leaf area, flowering time, and rosette branching display variable plasticity under neutral and saline conditions, suggesting that the variation of DNA methylation may play a key role in phenotypic plasticity. The capacity of the epigenetic marks (DNA methylation, histone modifications, etc.) to alter genes expression and transposable elements mobility could lead to deep physiological modulation affecting plant fitness.

In forest trees, epigenetic mechanisms may play a crucial role in survival and adaptation under stressful conditions, by mediating phenotypic modifications with beneficial effects in response to the environment. However, it remains difficult to separate the genetic and epigenetic components underlying phenotypic modifications. Many tree species propagate easily through clones, and this feature can be exploited in artificial conditions, such as breeding programs, for dissemination. Clonal propagation opens up possibilities for unravelling the epigenetic basis of phenotypic plasticity. Only a few studies investigating the role of epigenetics in tree phenotypic plasticity have been published, almost certainly due to the lack of fully sequenced genome for trees. In 2010, Gourcilleau et al. investigated a possible link between epigenetics and the plasticity to drought stress response in poplar, and were able to establish correlations between morphological and epigenetic variables. In particular, they found a positive correlation between DNA methylation levels and biomass productivity under well-watered conditions only. Recently, correlations between DNA methylation levels and biomass productivity have been confirmed on a collection of poplar genotypes grown outdoors in distinct water availability conditions (Le Gac et al. 2018). In the same model species, Raj et al. (2011) found differences in transcript abundance levels under drought stress that were probably linked to the geographic origin and correlated with DNA methylation level, suggesting that DNA methylation may modulate gene expression, leading to phenotypic variation under environmental stress. Recently, Lafon-Placette et al. (2018) have shown when comparing the genome wide distribution of differentially methylated regions (DMRs) to the differentially expressed genes (DEGs) that variations in soil water availability induced changes in DNA methylation levels preferentially at genes involved in phytohormones metabolism and signaling, potentially promoting phenotypic plasticity. Sáez-Laguna et al. (2014) showed epigenetic variability in stone pine, which is characterized by very low levels of genetic variation and high levels of plasticity for many traits. Using epigenetic markers, they were able to discriminate between two populations that were indistinguishable with classical molecular markers. They concluded that the high level of phenotypic plasticity found in stone pine trees, was probably linked to epigenetic variations between individuals or populations.

However, despite the many observations made in these studies, the relationship between epigenetics and phenotypic plasticity in trees remains underexplored. The main challenges for the coming years relate to the role of epigenetics in modulating the phenotype of forest trees under ongoing climate change and the stability of this environmentally-induced plasticity between individuals and across generations.

3.2.3 Epigenetics and Environmental (Stress) Memory

The concept of memory can be defined as the capacity of the organism to benefit from its past experience (Tulving, 1985). In plants, memory, has been linked to epigenetic modifications (Crisp, Ganguly, Eichten, Borevitz, & Pogson, 2016; D'Urso & Brickner, 2016; Latzel, Rendina González, & Rosenthal, 2016; Schönberger, Chen, Mager, & Ludewig, 2016). Indeed, several studies in recent decades have focused on the memory effect in plants and have shown that it is not exclusively dependent on genotype (in terms of the alleles present), but that it results from physiological processes dependent on gene activity (Lauria & Rossi, 2011). In addition to memory, the priming event is followed by a period of stress memory which involves storing information on the priming stress, potentially through an epigenetic phenomenon, and results in a modified response upon a recurring stress or in a sustained response after the priming stress (Lämke & Bäurle, 2017). This memory may last from several days to weeks for somatic stress memory, and in some cases, may even extend to offspring.

In forest trees species, which are characterized by a long life time and complex life cycle, phenological behavior, such as growth, bud phenology and frost hardiness, may be influenced by climatic conditions (Conde et al. 2017a; Johnsen et al. 2005b, 2005a; Johnsen & Skrøppa, 2001; Skrøppa, Tollefsrud, Sperisen, & Johnsen, 2010; Skrøppa & Johnsen, 2000). The best known epigenetic memory in trees is that reported for Norway spruce. In 2005, Johnsen et al. established that differences in day length and temperature during female meiosis affect progeny performance. Cold and short days environment advances bud burst during the spring, whereas high temperature and short days delay bud set during the summer. Yakovlev et al. (2014) and Yakovlev, Carneros, Lee, Olsen and Fossdal (2016) have identified a temperature-dependent epigenetic memory affecting the timing of bud burst and bud set in trees generated by temperature changes during somatic embryogenesis (Carneros et al. 2017; Yakovlev & Fossdal, 2017). Yakovlev, Fossdal, and Johnsen (2010), Yakovlev et al. (2012 and 2016) subsequently showed that this temperature-dependent "epigenetic memory" may be accounted for in part by the differential expression of specific microRNAs (miRNAs). As already mentioned (see Section 3.2), if miRNAs are not strictly classified in epigenetic mechanisms, growing evidences of their interactions with epigenetic mechanisms propose that it could be relevant to include them in an enlarged view of mechanisms controlling epigenetic phenomenon (Ramassone et al., 2018; Xia et al. 2014).

Similar epigenetic effects have been found in other species, such as white spruce, Scots pine, longleaf pine and *Larix* spp (Dormling & Johnsen, 1992; Greenwood & Hutchison, 1996; Schmidtling & Hipkins, 2004; Stoehr, L'Hirondelle, Binder, & Webber, 1998; Webber, Ott, Owens, & Binder, 2005). In poplar, Schönberger et al. (2016) showed that, despite their clonal origin, stem cuttings derived from different sites (with different amounts of phosphorus) established differently in a common environment. Cuttings derived from the site with the lowest amount of phosphorus grew in a manner suggesting a memory of the original environment. These differences in behavior were associated with different levels of methylation and amounts of miRNAs as a function of previous habitat, modifying the roots in a habitat-dependent manner under conditions of phosphate starvation, suggesting that these epigenetic mechanisms may account for habitat or seasonal memory in perennials. Le Gac et al. (2018) reported the existence of an environmental epigenetic memory in shoot apical meristems (SAMs) of *Populus* × *euramericana* trees in field conditions. SAMs are the center of shoot morphogenesis, undergoing constant cell division. It therefore seems likely that these meristematic cells are the best candidates for the transmission of environmental variants to the newly formed organs (leafs, stems, etc.), leading to phenotypic plasticity (Fig. 3) (Gourcilleau et al. 2010; Lafon-Placette et al. 2013, 2018). Using trees grown in common gardens under drought or from different sites with a pedoclimatic gradient, Le Gac et al. (2018) showed that winter-dormant SAMs express an epigenetic memory of the drought stress they experienced during vegetative growth.

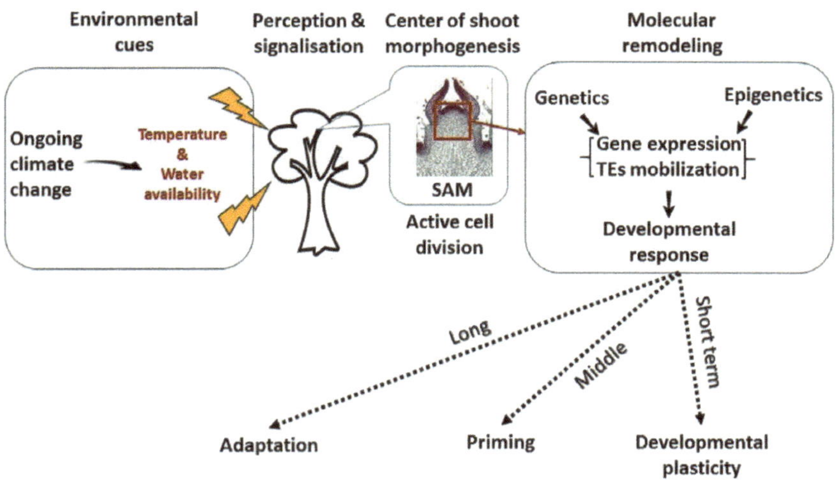

Figure 3 Schematic model of the link between environmental cues and epigenetic regulation in the shoot apical meristem (SAM). SAMs are the center of shoot morphogenesis and undergo constant cell division. During environmental constraints (heat, cold, drought), the signals perceived by the plant induce dynamic changes in the SAM affecting gene expression and TE mobilization. These signals are mostly linked to changes in epigenetic state, but also, to a lesser extent, to genetic changes. At short term, they may affect plant development and promote stress tolerance, leading to developmental plasticity. Once the constraint is released, the plant may develop the ability to remember the stress episode (priming, middle term), allowing it to react more efficiently if subsequently exposed to the same stress, thereby resulting in higher resistance. In cases of acute and long-lasting environmental pressure (long term), these changes may be transmitted to subsequent generations, potentially promoting adaptation to the new environmental conditions.

They identified DMRs conserved in different experimental set-ups (greenhouse, common garden, field conditions) with similar environmental constraints (water availability), and overlapping with genes most strongly related to the stress response, suggesting a potential role in priming.

Despite these important studies, there is still a long way to go before the underlying mechanisms of epigenetic memory in long-lived forest trees species are fully understood. However, progress in NGS technologies and CRISPR-cas9 should rapidly enhance future investigations. The elucidation of epigenetic memory in forest trees species will provide a valuable resources for the adaptation of trees to local environments, because this memory effect can persist for several years under field conditions (Skrøppa, Kohmann, Johnsen, Steffenrem, & Edvardsen, 2007). Indeed, memory could be a good tool for successful dispersion of genetically heterogeneous cohorts in stable conditions, in the sense that imprint could prepare and "correct" new seed for the conditions that prevail at the parents' place, smoothing out different allelic setups. In that sense, memory would prevent natural selection from weeding out all genetic variation, and preserving extra amounts that may become useful in future conditions. It is important to note that the stability of epigenetic modifications, which is still under investigation, is a crucial issue for breeding perspectives (see part 4. of the present chapter; Achour et al. 2017).

Most studies on local adaptation have focused on genetic variation (SNP, CNV, etc.). There is currently no concrete evidence for an involvement of epigenetic mechanisms in this process of local adaptation, but several studies have suggested that epigenetics may explain some of the phenotypic differentiation between populations that cannot otherwise be explained by DNA polymorphisms (Gugger, Fitz-Gibbon, PellEgrini, & Sork, 2016; Platt, Gugger, Pellegrini, & Sork, 2015; Skrøppa et al. 2010; Sork, 2016, 2017; Verhoeven, von Holdt, & Sork, 2016). It has been suggested that epigenetics can modulate plant fitness under environmental constraints (Bewick & Schmitz, 2017; Bräutigam et al. 2013; Kooke et al., 2015; Kawakatsu et al. 2016; Lande, 2009; Meyer, 2015; Seymour & Becker, 2017). This was interesting suggesting that genetics (allelic variation) is not the only source of phenotypic variation between individuals in a population or species. There is also evidence for a link between genetic and epigenetic mechanisms or for genetic control over epigenetic variation (Becker et al. 2011; Dubin et al. 2015). Epigenetic modifications may also generate genetic variants, such as SNPs (deamination of 5 mC to yield T) or CNV (reactivation of transposable elements (TEs) leading to new insertions elsewhere in the genome). Platt et al. (2015) recently showed, in a comparison of SNPs (single-nucleotide polymorphisms) and SMPs (single-methylation polymorphisms), that population differentiation levels were lower for SNPs in loci without SMPs than for those in loci with methylation polymorphisms. One explanation could be stabilizing selection against locally driven methylations. Plant adaptation is frequently associated with changes in DNA sequence, but it is becoming increasingly evident that changes in epigenetic state (mostly changes in DNA methylation level) under environmental constraints can lead to the reactivation of certain TEs, which may then insert next to coding regions, causing a stable silencing of the genes concerned that is transmissible across generations (Horváth, Merenciano, & González, 2017; Wang et al. 2017a; Zheng et al. 2017). Such epigenetic mechanisms represent a rapid source of adaptation since naturally occurring mutations are very rare (Becker et al. 2011). Indeed, epimutation rate has been estimated at 3.10^{-4} (van der Graaf et al., 2015) which is four times higher than the substitution rate estimated at 7.10^{-9} in *A. thaliana* for example.

Interestingly, Platt et al. (2015) showed, in natural oak tree populations in southern California, that changes in DNA methylation, particularly at CpG sites, are strongly associated with local adaptation, either directly or through linkage to regions under selection. However, no relationship was found between methylation in a CHG context and adaptation. Methylation in the CpG context seems to be more frequent than non-CG methylation in the genes of diverse species, suggesting that CpG methylation may have a higher adaptation potential, due to its tight association with the coding regions subjected to selection. Indeed, Gugger et al. (2016) identified several climate-associated methylation variants that tend to occur in or near genes, highlighting the potential importance of DNA methylation in the CpG context as a determinant process for the plant response to environment leading to local adaptation in natural populations.

Further studies are required to characterize clearly the role of epigenetics in local adaptation. Improvements in our understanding of the role of epigenetics in phenotypic plasticity and memory will be essential for this. Yakovlev et al. (2010, 2012 and 2016) demonstrated the existence of a temperature-dependent "epigenetic memory" during embryo development in Norway spruce that regulates bud set and cold acclimation. This epigenetic memory was found to be of ecological significance, as it reproduced the differences observed between latitudes in natural conditions (Skrøppa et al. 2010).

4. POTENTIAL USES OF EPIGENETICS IN FOREST TREE BREEDING

4.1 Epigenetic Diversity (Natural or Induced) in Tree Clonally-Propagated or Seed Conservation

Plants have two types of propagation strategy: sexual and asexual. Sexual reproduction is a natural process generating seeds that give rise to plants not genetically identical to the parent plant. By contrast, asexual propagation (division, grafting, air-layering, cuttings, somatic embryogenesis, etc.) is frequently used for forest trees, as it maintains the identity and performance of the parent tree in the context of tree breeding. The clonal propagation of some forest trees may limit their genetic diversity, potentially increasing the importance of epigenetic diversity for adaptation. However, many questions remain unanswered and will be of increasing relevance in the coming years. Is epigenetic diversity fine-tuned by environmental cues? Will it be sufficient to cope with ongoing climate changes?

Forest decline due to drought in conditions of heat stress has already been observed in various regions of the world (Allen et al. 2010). Epigenetics is emerging as a key driver of phenotypic plasticity and adaptation, because epigenetic mechanisms can be induced in a reversible manner by environmental conditions. Several studies have investigated the role of epigenetics in clonally propagated trees (Ausin et al. 2016; Carneros et al. 2017; Guarino, Cicatelli, Brundu, Heinze, & Castiglione, 2015; Klimaszewska et al. 2009; Liang et al. 2014; Schönberger et al. 2016; Teyssier et al. 2014; Vining et al. 2013; Yakovlev, Carneros, Lee, Olsen, & Fossdal, 2016). Epigenetics has been shown to provide an additional layer of regulation for various biological process in clonally propagated tree, which may help clonal trees with limited genetic diversity to modulate their phenotype under changing conditions (Cicatelli et al. 2014; Fresnedo-Ramírez, Chan, Parfitt, Crisosto, & Gradziel, 2017; Guarino et al. 2015; Lafon-Placette et al. 2018; Liang et al. 2014; Schönberger et al. 2016; Shuai, Liang, Zhang, Yin, & Xia, 2013). There is also finding suggesting that embryo development and seed maturation may be the sensitive periods for stable epigenetic modification (Carneros et al. 2017; Yakovlev et al. 2016).

Somatic embryogenesis is an emerging technique for obtaining embryos from somatic or embryonic cells. It is emerging as a powerful tool for studying molecular and physiological mechanisms during embryo development (Lelu-Walter et al. 2013, 2006; Trontin, Klimaszewska, Morel, Hargreaves, & Lelu-Walter, 2016). It is also widely used for the multiplication of elite individuals as part of forest improvement and deployment programs. Plant regeneration via somatic embryogenesis occurs in five steps: initiation, multiplication, maturation, germination and acclimation. Given the suggested sensitivity of embryo development as a period for stable modification, this technique of clonal propagation is a valuable tool for studies of the role of environmental constraints during early plant life. The modification of the annual growth cycle reported in Norway spruce (see Section 3.2.3) due to stable epigenetic changes in the early life of the plant opens up new possibilities for tree deployment. Indeed, somatic embryogenesis could be used to produce trees better able to acclimate to the anticipated global warming. Another advantage of somatic embryogenesis is that the embryogenic masses obtained after the multiplication phase can be cryopreserved in liquid nitrogen (Lelu-Water et al. 2006). This makes it possible to store veritable banks of individuals for years, a major advantage over seed banking.

Seed banking and conservation are used to prevent the total extinction of species due to climate change. Seeds are stored at low temperatures, to maintain them in a dormant state until their use for replanting. However, seed conservation has been reported to lead to yield losses. Little is known about the possible effects of epigenetics on seed storage (Michalak et al. 2013, 2015; Plitta et al. 2014; Plitta-Michalak et al. 2018). For instance, in *Pyrus communis*, Michalak et al. (2013) detected epigenetic changes in severely desiccated seeds, potentially reflecting a specific response of orthodox seeds to drying. Senescence and aging in recalcitrant *Quercus robur* seeds were also correlated with a decrease in global DNA methylation levels, suggesting that epigenetic mechanisms may play a key role in seed viability and should be assessed to improve seed storage practices (Michalak et al. 2015; Plitta-Michalak et al. 2018).

4.2 Epigenetic Component of Heterosis and Hybrid Homeostasis

Hybrid vigor or heterosis is defined as a superior performance of hybrids, for many traits, relative to their parental inbred lines (see Chapter 8 of this book). Hybrids have been widely used in crop and tree breeding since ancient times, although the molecular mechanisms underlying heterosis have yet to be fully elucidated. Various genetic concepts have been put forward to account for this phenomenon: dominance, overdominance and epistasis. Genetic differences between parental inbred lines have long been considered as a determinant of performance when creating hybrids, but it has recently been shown that hybrids may outperform their parents even if the parental genomes are almost identical (Groszmann et al. 2011; Meyer, Törjék, Becher, & Altmann, 2004). This finding raises the interesting question of the origin of this heterosis. In theory, heterosis involves the creation of new genotypic combinations, altering the gene expression profile of the progeny. However, epigenetic mechanisms can alter gene expression without altering the DNA sequence, and there is growing evidence for a role of epigenetic factors in heterosis (Dapp et al. 2015; Groszmann, Greaves, Fujimoto, James Peacock, & Dennis, 2013; Schnable & Springer, 2013). DNA methylation and small RNA-mediated mechanisms are the most epigenetic processes most frequently associated to heterosis (Chodavarapu et al. 2012; Gao et al. 2014; Kawanabe et al. 2016; Lauss et al. 2018; Shen et al. 2017; Zhang et al. 2016).

The importance of these epigenetic mechanisms for heterosis was demonstrated in *Arabidopsis,* in which the inhibition of DNA methylation (5-azacytidine) or the disruption of an RNA-dependent DNA methylation process neutralized heterosis in hybrids (Shen et al. 2012). Lauss et al. (2018) recently suggested that locus-specific epigenetic divergence between the parental lines can directly or indirectly trigger heterosis independently of genetic changes. Very few studies have been published investigating the role of epigenetics in the heterosis of forest trees. In 2014, Gao et al. investigated the role of DNA methylation changes in *Populus deltoides* hybrids from intraspecific parental lines (i.e. with similar genetic backgrounds and different levels of DNA methylation). They found that hybrids outperformed the parental lines for most of the traits studied, and had higher levels of DNA methylation. They suggested that the methylation patterns of the parental inbred lines were passed on, at least in part, to the progeny, and that these hybrids had non-additive levels of methylation, suggesting a possible key role of DNA methylation in the heterosis of forest trees. Further studies are required to characterize hybrid vigor in trees in more detail. For the moment, it is clear that epigenetic variations, such as decreases in DNA methylation during homologous recombination, can lead to the reactivation of TEs, which can then insert into new regions of the genome, creating epialleles. Environmental stresses may trigger the generation of these epialleles, which create epigenetic diversity, with the potential to improve the performance of phenotypic traits in the progeny. Thus, in the face of ongoing change, heterosis is a tool of choice for the survival of forest trees species in new environmental conditions, at least for those trees for which hybrid breeding programs have been established. The identification of epigenetic markers for trait improvement will help tree breeders to produce plants with higher performances capable of acclimating rapidly to climate change.

4.3 Epigenetic Markers for Trait Improvement and Breeding

Plant breeding programs to date have focused almost exclusively on genetic variation between individuals and populations. Quantitative genetics is generally used merely to explain the genetic basis of variation in complex traits. However, as described here, it is now evident that epigenetics plays a fundamental role in gene expression and transposon silencing leading to phenotypic variation and that epigenetic marks can be inherited by future generations (Law & Jacobsen, 2010; Richards et al. 2017). These epigenetic factors also provide the genome with a certain degree of plasticity, as they can be induced reversibly by environmental conditions. Their extensive involvements in various biological processes and their impact on complex traits have made them essential for population and quantitative genomic studies. In 2014, Cortijo et al. showed that DMRs could account for up to 90% of the heritability of complex traits in *Arabidopsis* epi-mutant lines, suggesting that DMRs could act as quantitative trait loci (epiQTL) potentially used for artificial selection. Indeed, the traditional quantitative genetics methods used to estimate the genetic component of the phenotype could also be used to assess natural epigenetic variations. DNA methylation could therefore be considered as traits, and the estimation of narrow-sense heritability (h^2) or genetic differentiation index (epiQ_{ST}), for these marks could help to shed light on the possible relationship between genetic and epigenetic variation and more precisely, on the genetic control of epigenetic variation (Dubin et al. 2015; Sow et al. 2018). Conversely, epigenetic polymorphisms, such as SMPs (single-methylation polymorphisms) or DMRs could be used as epigenetic markers in genome scan based approaches, to assess the role of epigenetics into population structure and phenotypic variation (epiFST, epiQTL) (Fig. 4). Only few studies based on these epigenetic mapping approaches have been published (Cortijo et al. 2014; Li, Eichten, Hermanson, & Springer, 2014; Long et al. 2011; Zou et al. 2017), and no study has yet reported the use of epigenetic markers for association mapping analyses in forest tree species. However, there is growing interest in the use of these methods in selection. Indeed, marker-assisted selection should no longer focus purely on genetics, but should also consider the interaction between genetics, epigenetics and the environment (Fig. 4). The success of these new integrative approaches will depend principally on the development of powerful statistical models capable of dealing with different variables in multivariate analysis.

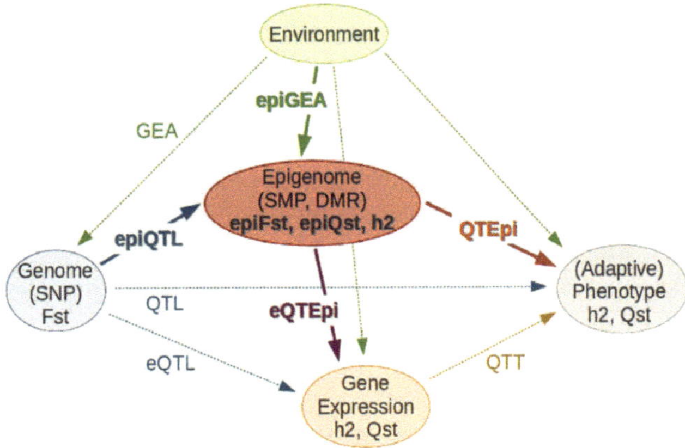

Figure 4 Quantitative and population genomics analyses for the integrative analysis of genetic, epigenetic and environmental effects. GEA: gene–environment association; epiGEA: epigenetics–environment association; QTL: quantitative trait locus; epiQTL: epigenetic QT; eQTL: expression QTL; QTEpi: quantitative trait epilocus; eQTEpi: expression QTEpi; QTT: quantitative trait transcript; h^2: heritability; Fst: genetic differentiation between populations; epiFst: epigenetic differentiation between populations; Qst: quantitative differentiation between populations; epiQst: epigenetic quantitative differentiation between populations; SNP: single-nucleotide polymorphism; SMP: single-methylation polymorphism; DMR: differentially methylated region.

4.4 Biotechnology and Epigenome Editing

In recent decades, the development of new tools for genome engineering has opened up new perspectives and helped to decipher the molecular mechanisms underlying different biological process. In *Arabidopsis* and crop plants, the success of genome modification has been well documented and such techniques have been incorporated into breeding programs (Podevin, Davies, Hartung, Nogué, & Casacuberta, 2013). Various tools have recently been proposed for targeting the epigenome. Most of these techniques target genes or mRNAs involved in epigenetic process (readers and erasers of epigenetic marks, chromatin remodelers, etc.). Several studies on *Arabidopsis* have demonstrated the utility of such methods for highlighting the role of epigenetics in important biological process. This is clearly illustrated by the epiRILs (epigenetic recombinant inbred lines) obtained by crossing *A. thaliana ddm1* mutant lines with a wild-type line, leading to the creation of lines differing only in terms of their DNA methylation levels (Johannes, Colot, & Jansen, 2008; see Chapter 4 of this book).

Very few studies on epigenomic engineering in forest tree species have been published. The first hypomethylated trees were generated in 2013.

Using an RNAi strategy, Zhu et al. (2013) targeted the homology of the *A. thaliana DDM1* gene in hybrid poplar (*Populus tremula x alba*). This gene encodes a chromatin remodeler that provides DNA methyltransferases with to access to the DNA sequence. It is therefore required for the maintenance of DNA methylation and heterochromatin assembly. Poplar *ddm1* RNAi lines had lower levels of DNA methylation and a mottled leaf phenotype after a cycle of dormancy. Conde et al. (2017a) used this RNAi strategy in poplar to decrease the expression of *DEMETER-LIKE (DML)* genes. They found that transgenic poplars with *DML8/10* downregulation had significantly higher levels of DNA methylation, which delayed bud break. Conversely, the overexpression of a *DML* from chestnut (*Castanea sativa*) in poplar hybrids resulted in transgenic lines in which bud formation was accelerated (Conde et al. 2017b). No data have yet been published for the knockout of genes implicated in epigenetic processes in forest tree species.

The recent development of new genome editing technologies has opened up new avenues for investigating the role of epigenetics in gene expression, transposon silencing and genome stability. The CRISPR/Cas9 system is emerging as a major technique due to its simplicity, design flexibility and high degree of efficiency and it has been used for the editing of genomes in diverse taxa, from bacteria to eukaryotes (see Chapter 6 of this book). This system has a number of different uses including the suppression or direct insertion at specific loci of a gene, or of a chromosome or gene rearrangement. Few studies involving the use of this technique in forest trees have been published, and none of the existing studies focused on epigenetic marks (Fan et al. 2015; Zhou, Jacobs, Xue, Harding, & Tsai, 2015). Much remains to be done to develop these new advanced technologies for application in forest tree species. The biology of such trees (with their complex life cycle and long generation time) and the strategies used for their propagation (clonal propagation) impose additional constraints, limiting the routine use of these techniques. Although there are still some technical difficulties to be solved, the development of new propagation strategies, such as somatic embryogenesis, should help to overcome some of these limitations, by making it possible to work with embryonic cells, resulting in the correct segregation of transgenes in the final product. The development of these strategies should also make it possible to gain insight into molecular mechanisms for forest tree breeding.

4.5 Integrative Approach, Systems Biology, Statistics, and Modeling

Understanding how tree phenotype is controlled at the molecular level in response to environmental conditions is an important and challenging task involving studies of the complex interplay between different genetic and epigenetic mechanisms. Modern "-omics" technologies can provide genomic, transcriptomic, and epigenomic data for species growing under different environmental conditions, and opportunities for studying phenotypic plasticity from several perspectives. The integrative statistical analysis of "-omics" data, combining several heterogeneous data sets obtained within a given species, increases study power and provides insight into the interactions between the various mechanisms of molecular regulation determining the phenotype. Studies of the role of epigenetic factors in the context of their relationship to other mechanisms are particularly useful in this respect.

Such analyses are challenging methodologically, because they involve studies of statistical interactions between several high-dimensional "-omics" data sets and the construction of mathematical model for predicting phenotypes from these explanatory data sets taking into account their large numbers of dimensions, different origins and statistical correlation. A recurrent problem in such analyses is that there are large numbers and diverse types of -omics variables for a relatively small number of observations. The following strategies are classically adopted to make it possible to use multivariate statistical models in this context: use of additional knowledge (regularization, Bayesian priors), limitation of the number of variables (variable selection, exploration of candidate models) or modification of data representation through the use of kernel methods or variable compression.

The mathematical tool used depends on the goal of the study and the context. One of the main methods for studying relationships between heterogeneous data sets, such as DNA sequencing, RNA sequencing, and methylation data, for example, is based on mathematical network models. This approach is based on the representation of large biological datasets as networks, with biological entities (e.g. genes) as nodes and edges corresponding to links between these biological entities. If only one type of link is considered (e.g. a transcriptomic link), the resulting network diagram summarizes the system of links of this type between genes (e.g. correlation of expression levels). If several links of different types are considered simultaneously, an integrative network is constructed, representing several layers of cellular information and their interplay in the same network diagram (Fabres, Collins, Cavagnaro, & Rodríguez López, 2017; Wong & Matus, 2017).

Another strategy is required if the goal is to study the link between one or several of the explanatory data sets and the phenotype. For instance, the statistical analysis of epigenetic variation and its relationship to environmental phenotypic plasticity can be performed according to a classical strategy in which the association of each methylation site or region with the considered phenotype is tested separately (Bräutigam et al. 2013; Nicotra et al. 2010). If other explanatory data sets are available, this classical approach may be extended by taking into account influential covariates, such as transcriptomic, genetic, or environmental variables.

Kernels and kernel-based methods are another promising tool for data integration for classification and phenotype prediction. Kernel methods were first proposed as a technique for data integration by Lanckriet, De Bie, Cristianini, Jordan, and Noble (2004). Each dataset is represented by a kernel function, which defines generalized similarity relationships between pairs of entities, such as genes or proteins. The kernel representation is both flexible and efficient, and can be applied to many different types of data. Moreover, several computational algorithms have been developed, making it possible to combine kernel functions derived from different types of data efficiently. Kernel-based methods provide a natural framework for combining variables of different types, as they make it possible to take into account the nature of each type of "-omics" variable and to combine them in multiple kernel methods (Alam, Lin, Calhoun, & Wang, 2017).

5. CONCLUSION

Trees are outstanding organisms in terms of their extreme size, long lifespan, complex life cycles and wood production and forests, play a major role in Earth's ecology, by supplying a range of ecosystem services. In the past decades, widespread forest die-off due to drought and/or heat constraints has been observed around the world for all forest biomes, and is predicted to increase with ongoing climate (Allen et al. 2010). In recent years, epigenetics, that is sensitive to environmental variations, has emerged as a major regulator of various biological processes conferring plasticity to the genome that can be transmitted across generations. However, in forest tree species, there is still a long way to go in to decipher the role of epigenetics in ecological responses particularly in the frame of climate change. Here, we have reviewed the increasing genomics resources that will support the development of forest tree epigenomics. We have also synthesized the major evidences actually available in forest trees supporting a role of epigenetics in phenotypic plasticity, stress memory and adaptation. Further investments are required to clarify the role and applications of epigenetics in genetic resource conservation, breeding and forest management as for other plants (see other chapters of this book; Achour et al. 2017). Here, we reviewed several potential uses of epigenetics for trees with sexual or clonal propagation (Fig. 5) in the frame of the production of elite individuals, assisted migration and forest deployment programs. Thus, it is now urgent to inform and assist tree breeders and forest managers to include a questioning towards epigenetics

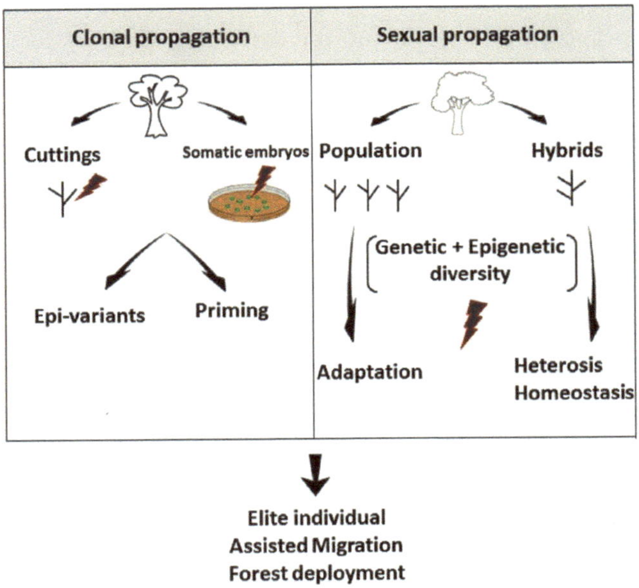

Figure 5 Recommendations for forest management and tree breeding in the face of climate changes. Some trees are vegetatively propagated, which may lead to limited genetic diversity. However, it is now evident that epigenetic diversity plays a key role in adaptation, due to its sensitivity to environmental conditions. During clonal propagation (cuttings or somatic embryos), epigenetic variants (epi-variants) or priming effect may be induced by the application of controlled stresses (heat, drought) early in the life of the plant. The early-induced epigenetic changes may promote developmental plasticity and, subsequently, local adaptation, if these changes are transmitted across generations. Sexually propagated trees tend to display greater genetic and epigenetic diversity, which may be crucial for population adaptation under environmental constraints. Another way of combating forest decline is to produce progenies with higher performances. In this case, the concept of heterosis should not be restricted to genetic aspects, but should also take into account the complex interactions between genetic and epigenetic diversity, because the expression of a trait of interest in forestry may be dependent on its epigenetic state. The trees with the highest performance generated by such strategies could be used in assisted migration and forest deployment programs.

in their future program. This is one of the concerns as well as some of the research questions and proposed avenues indicated above that will be addressed within the research project **EPITREE** (**E**volutionary and functional impact of **EPI**genetic variation in forest **TREE**s; ANR-17-CE32–0009–01, 2018–21, https://www6.inra.fr/epitree-project/Le-projet-EPITREE). EPITREE focuses on the evolutionary and functional impact of epigenetic variation in forest trees. Two emblematic trees, poplar and oak, are used in an innovative integrative analysis to assess how genetic and epigenetic variations contribute to phenotypic plasticity and adaptation to local environment.

REFERENCES

Achour, Z., Archipiano, M., Barneche, F., Baurens, C., Beckert, M., Ben, C., et al. (2017). *Epigenetics in plant breeding.* Article de positionnement du Groupement d'intérêt scientifique Biotechnologies vertes et de l'Alliance nationale de recherche pour l'environnement http://www.gisbiotechnologiesvertes.com/fr/publications/position-paper-epigenetics-in-plant-breeding.

Adey, A., & Shendure, J. (2012). Ultra-low-input, tagmentation-based whole-genome bisulfite sequencing. *Genome Research, 22*(6), 1139–1143.

Alam, M., Lin, H.-Y., Calhoun, V., & Wang, Y.-P. (2017). *Kernel method for detecting higher order interactions in multi-view data: An application to imaging, genetics, and epigenetics.* arXiv preprint arXiv:1707.04368.

Alia, R., & Majada, J. (2013). Phenotyping for the future and the future of phenotyping. In S. Lee, & J. Woolliams (Eds.), *Novel tree breeding* (pp. 53–62). Madrid, Spain: PublINIA. http://libros.inia.es/libros/download.php?id=709&pdf=products_pdfpreview.

Allen, C. D., Macalady, A. K., Chenchouni, H., Bachelet, D., McDowell, N., Vennetier, M., et al. (2010). A global overview of drought and heat-induced tree mortality reveals emerging climate change risks for forests. *Forest Ecology and Management, 259*(4), 660–684. https://doi.org/10.1016/j.foreco.2009.09.001.

Ausin, I., Feng, S., Yu, C., Liu, W., Kuo, H. Y., Jacobsen, E. L., et al. (2016). DNA methylome of the 20-gigabase Norway spruce genome. *Proceedings of the National Academy of Sciences of the United States of America, 113*(50), E8106–E8113. https://doi.org/10.1073/pnas.1618019113.

Bagnoli, F., Fady, B., Fineschi, S., Oddou-Muratorio, S., Piotti, A., Sebastiani, F., et al. (2011). Neutral patterns of genetic variation and applications to conservation in conifer species. In C. Plomion, J. Bousquet, & C. Kole (Eds.), *Genetics, genomics and breeding of crop plants science publ.* Enfield, NH, USA: CRC Press.

Barakat, A., Wall, P. K., DiLoreto, S., dePamphilis, C. W., & Carlson, J. E. (2007). Conservation and divergence of microRNAs in Populus. *BMC Genomics, 8*(1), 481. https://doi.org/10.1186/1471-2164-8-481.

Bartholomé, J., Van Heerwaarden, J., Isik, F., Boury, C., Vidal, M., Plomion, C., et al. (2016). Performance of genomic prediction within and across generations in maritime pine. *BMC Genomics, 17*(1). https://doi.org/10.1186/s12864-016-2879-8.

Baurens, F.-C., Nicolleau, J., Legavre, T., Verdeil, J.-L., & Monteuuis, O. (2004). Genomic DNA methylation of juvenile and mature Acacia mangium micropropagated in vitro with reference to leaf morphology as a phase change marker. *Tree Physiology, 24*(4), 401–407. https://doi.org/10.1093/treephys/24.4.401.

Becker, C., Hagmann, J., Müller, J., Koenig, D., Stegle, O., Borgwardt, K., et al. (2011). Spontaneous epigenetic variation in the *Arabidopsis thaliana* methylome. *Nature, 480*(7376), 245–249. https://doi.org/10.1038/nature10555.

Bewick, A. J., & Schmitz, R. J. (2017). Gene body DNA methylation in plants. *Current Opinion in Plant Biology, 36*, 103–110. https://doi.org/10.1016/j.pbi.2016.12.007.

Bird, A. P., & Southern, E. M. (1978). Use of restriction enzymes to study eukaryotic DNA methylation: I. The methylation pattern in ribosomal DNA from *Xenopus laevis*. *Journal of Molecular Biology, 118*(1), 27–47.

Birol, I., Raymond, A., Jackman, S. D., Pleasance, S., Coope, R., Taylor, G. A., et al. (2013). Assembling the 20 Gb white spruce (Picea glauca) genome from whole-genome shotgun sequencing data. *Bioinformatics, 29*(12), 1492–1497. https://doi.org/10.1093/bioinformatics/btt178.

Blueprint Consortium. (2016). Quantitative comparison of DNA methylation assays for biomarker development and clinical applications. *Nature Biotechnology, 34*(7), 726–737.

Booth, M. J., Branco, M. R., Ficz, G., Oxley, D., Krueger, F., Reik, W., et al. (2012). Quantitative sequencing of 5-methylcytosine and 5-hydroxymethylcytosine at single-base resolution. *Science, 336*(6083), 934–937.

Borno, S. T., Fischer, A., Kerick, M., Falth, M., Laible, M., Brase, J. C., et al. (2012). Genome-wide DNA methylation events in TMPRSS2-ERG fusion-negative prostate cancers implicate an EZH2-dependent mechanism with miR-26a hypermethylation. *Cancer Discovery, 2*(11), 1024–1035.

Bossdorf, O., Richards, C. L., & Pigliucci, M. (2008). Epigenetics for ecologists. *Ecology Letters, 0*(0). https://doi.org/10.1111/j.1461-0248.2007.01130.x, 071117033013002–???

Bradshaw, A. D. (2006). Unravelling phenotypic plasticity – why should we bother? *New Phytologist, 170*(4), 644–648. https://doi.org/10.1111/j.1469-8137.2006.01761.x.

Bräutigam, K., Vining, K. J., Lafon-Placette, C., Fossdal, C. G., Mirouze, M., Marcos, J. G., et al. (2013). Epigenetic regulation of adaptive responses of forest tree species to the environment. *Ecology and Evolution, 3*(2), 399–415. https://doi.org/10.1002/ece3.461.

Brinkman, A. B., Simmer, F., Ma, K., Kaan, A., Zhu, J., & Stunnenberg, H. G. (2010). Whole-genome DNA methylation profiling using MethylCap-seq. *Methods, 52*(3), 232–236.

Carneros, E., Yakovlev, I., Viejo, M., Olsen, J. E., & Fossdal, C. G. (2017). The epigenetic memory of temperature during embryogenesis modifies the expression of bud burst-related genes in Norway spruce epitypes. *Planta, 246*(3), 553–566. https://doi.org/10.1007/s00425-017-2713-9.

Chen, X., Ge, X., Wang, J., Tan, C., King, G. J., & Liu, K. (2015). Genome-wide DNA methylation profiling by modified reduced representation bisulfite sequencing in Brassica rapa suggests that epigenetic modifications play a key role in polyploid genome evolution. *Frontiers in Plant Science, 6*. https://doi.org/10.3389/fpls.2015.00836.

Chodavarapu, R. K., Feng, S., Ding, B., Simon, S. A., Lopez, D., Jia, Y., et al. (2012). Transcriptome and methylome interactions in rice hybrids. *Proceedings of the National Academy of Sciences of the United States of America, 109*(30), 12040–12045. https://doi.org/10.1073/pnas.1209297109.

Chwialkowska, K., Korotko, U., Kosinska, J., Szarejko, I., & Kwasniewski, M. (2017). Methylation sensitive amplification polymorphism sequencing (MSAP-Seq)-A method for high-throughput analysis of differentially methylated CCGG sites in plants with large genomes. *Front Plant Science, 8*, 2056.

Cicatelli, A., Todeschini, V., Lingua, G., Biondi, S., Torrigiani, P., & Castiglione, S. (2014). Epigenetic control of heavy metal stress response in mycorrhizal versus non-mycorrhizal poplar plants. *Environmental Science and Pollution Research, 21*(3), 1723–1737. https://doi.org/10.1007/s11356-013-2072-4.

Ci, D., Song, Y., Du, Q., Tian, M., Han, S., & Zhang, D. (2016). Variation in genomic methylation in natural populations of *Populus simonii* is associated with leaf shape and photosynthetic traits. *Journal of Experimental Botany, 67*(3), 723–737. https://doi.org/10.1093/jxb/erv485.

Clarke, J., Wu, H. C., Jayasinghe, L., Patel, A., Reid, S., & Bayley, H. (2009). Continuous base identification for single-molecule nanopore DNA sequencing. *Nature Nanotechnology, 4*(4), 265–270.

Conde, D., González-Melendi, P., & Allona, I. (2013). Poplar stems show opposite epigenetic patterns during winter dormancy and vegetative growth. *Trees, 27*(1), 311–320. https://doi.org/10.1007/s00468-012-0800-x.

Conde, D., Le Gac, A.-L., Perales, M., Dervinis, C., Kirst, M., Maury, S., et al. (2017a). Chilling-responsive DEMETER-LIKE DNA demethylase mediates in poplar bud break: Role of active DNA demethylase in trees' bud break. *Plant, Cell and Environment, 40*(10), 2236–2249. https://doi.org/10.1111/pce.13019.

Conde, D., Moreno-Cortés, A., Dervinis, C., Ramos-Sánchez, J. M., Kirst, M., Perales, M., et al. (2017b). Overexpression of DEMETER, a DNA demethylase, promotes early apical bud maturation in poplar. *Plant, Cell and Environment, 40*(11), 2806–2819. https://doi.org/10.1111/pce.13056.

Correia, B., Valledor, L., Meijón, M., Rodriguez, J. L., Dias, M. C., Santos, C., et al. (2013). Is the interplay between epigenetic markers related to the acclimation of cork oak plants to high temperatures? *Plos One, 8*(1), e53543. https://doi.org/10.1371/journal.pone.0053543.

Cortijo, S., Wardenaar, R., Colome-Tatche, M., Gilly, A., Etcheverry, M., Labadie, K., et al. (2014). Mapping the epigenetic basis of complex traits. *Science, 343*(6175), 1145–1148. https://doi.org/10.1126/science.1248127.

Crisp, P. A., Ganguly, D., Eichten, S. R., Borevitz, J. O., & Pogson, B. J. (2016). Reconsidering plant memory: Intersections between stress recovery, RNA turnover, and epigenetics. *Science Advances, 2*(2), e1501340. https://doi.org/10.1126/sciadv.1501340.

Dai, X., Hu, Q., Cai, Q., Feng, K., Ye, N., Tuskan, G. A., et al. (2014). The willow genome and divergent evolution from poplar after the common genome duplication. *Cell Research, 24*(10), 1274.

Dapp, M., Reinders, J., Bédiée, A., Balsera, C., Bucher, E., Theiler, G., et al. (2015). Heterosis and inbreeding depression of epigenetic Arabidopsis hybrids. *Nature Plants, 1*(7). https://doi.org/10.1038/nplants.2015.92.

Daru, B. H., Berger, D. K., & Wyk, A. E. (2016). Opportunities for unlocking the potential of genomics for African trees. *New Phytologist, 210*(3), 772–778.

Daviaud, C., Renault, V., Mauger, F., Deleuze, J. F., & Tost, J. (2018). Whole-genome bisulfite sequencing using the ovation(R) ultralow methyl-seq protocol. *Methods in Molecular Biology, 1708*, 83–104.

Deamer, D., Akeson, M., & Branton, D. (2016). Three decades of nanopore sequencing. *Nature Biotechnology, 34*, 518–524.

de la Fuente, L., Conesa, A., Lloret, A., Badenes, M. L., & Ríos, G. (2015). Genome-wide changes in histone H3 lysine 27 trimethylation associated with bud dormancy release in peach. *Tree Genetics & Genomes, 11*(3). https://doi.org/10.1007/s11295-015-0869-7.

Diep, D., Plongthongkum, N., Gore, A., Fung, H. L., Shoemaker, R., & Zhang, K. (2012). Library-free methylation sequencing with bisulfite padlock probes. *Nature Methods, 9*(3), 270–272.

Ding, C.-J., Liang, L.-X., Diao, S., Su, X.-H., & Zhang, B.-Y. (2018). Genome-wide analysis of day/night DNA methylation differences in Populus nigra. *Plos One, 13*(1), e0190299. https://doi.org/10.1371/journal.pone.0190299.

Dormling, I., & Johnsen, Ø. (1992). Effects of the parental environment on full-sib families of *Pinus sylvestris*. *Canadian Journal of Forest Research, 22*(1), 88–100. https://doi.org/10.1139/x92-013.

Down, T. A., Rakyan, V. K., Turner, D. J., Flicek, P., Li, H., Kulesha, E., et al. (2008). A Bayesian deconvolution strategy for immunoprecipitation-based DNA methylome analysis. *Nature Biotechnology, 26*(7), 779–785.

Duan, H., Lu, X., Lian, C., An, Y., Xia, X., & Yin, W. (2016). Genome-wide analysis of MicroRNA responses to the phytohormone abscisic acid in populus euphratica. *Frontiers in Plant Science, 7*. https://doi.org/10.3389/fpls.2016.01184.

Dubin, M. J., Zhang, P., Meng, D., Remigereau, M.-S., Osborne, E. J., Casale, F. P., et al. (2015). DNA methylation in Arabidopsis has a genetic basis and shows evidence of local adaptation. *Elife, 4*, e05255.

D'Urso, A., & Brickner, J. H. (2016). Epigenetic transcriptional memory. *Current Genetics, 63*(3), 435–439. https://doi.org/10.1007/s00294-016-0661-8.

Eads, C. A., Danenberg, K. D., Kawakami, K., Saltz, L. B., Blake, C., Shibata, D., et al. (2000). MethyLight: A high-throughput assay to measure DNA methylation. *Nucleic Acids Research, 28*(8), e32.

Ehrich, M., Nelson, M. R., Stanssens, P., Zabeau, M., Liloglou, T., Xinarianos, G., et al. (2005). Quantitative high-throughput analysis of DNA methylation patterns by base-specific cleavage and mass spectrometry. *Proceedings of the National Academy of Sciences of the United States of America, 102*(44), 15785–15790.

El Baidouri, M., Kim, K. D., Abernathy, B., Li, Y.-H., Qiu, L.-J., & Jackson, S. A. (2018). Genic C-methylation in soybean is associated with gene paralogs relocated to transposable element-rich pericentromeres. *Molecular Plant, 11*(3), 485–495. https://doi.org/10.1016/j.molp.2018.02.006.

Fabres, P. J., Collins, C., Cavagnaro, T. R., & Rodríguez López, C. M. (2017). A concise review on multi-omics data integration for terroir analysis in Vitis vinifera. *Frontiers in Plant Science, 8*. https://doi.org/10.3389/fpls.2017.01065.

Fan, D., Liu, T., Li, C., Jiao, B., Li, S., Hou, Y., et al. (2015). Efficient CRISPR/Cas9-mediated targeted mutagenesis in populus in the first generation. *Scientific Reports, 5*. https://doi.org/10.1038/srep12217.

Fazzari, M. J., & Greally, J. M. (2004). Epigenomics: Beyond CpG islands. *Nature Reviews. Genetics, 5*(6), 446–455.

Feber, A., Wilson, G., Zhang, L., Presneau, N., Idowu, B., Down, T. A., et al. (2011). Comparative methylome analysis of benign and malignant peripheral nerve sheath tumours. *Genome Research, 21*(4), 515–524. https://doi.org/10.1101/gr.109678.110.

Feng, S., Cokus, S. J., Zhang, X., Chen, P.-Y., Bostick, M., Goll, M. G., et al. (2010). Conservation and divergence of methylation patterning in plants and animals. *Proceedings of the National Academy of Sciences of the United States of America, 107*(19), 8689–8694. https://doi.org/10.1073/pnas.1002720107.

Flusberg, B. A., Webster, D. R., Lee, J. H., Travers, K. J., Olivares, E. C., Clark, T. A., et al. (2010). Direct detection of DNA methylation during single-molecule, real-time sequencing. *Nature Methods, 7*(6), 461–465.

Fresnedo-Ramírez, J., Chan, H. M., Parfitt, D. E., Crisosto, C. H., & Gradziel, T. M. (2017). Genome-wide DNA-(de)methylation is associated with Noninfectious Bud-failure exhibition in Almond (*Prunus dulcis* [Mill.] D.A.Webb). *Scientific Reports, 7*, 42686. https://doi.org/10.1038/srep42686.

Frommer, M., McDonald, L. E., Millar, D. S., Collis, C. M., Watt, F., Grigg, G. W., et al. (1992). A genomic sequencing protocol that yields a positive display of 5-methylcytosine residues in individual DNA strands. *Proceedings of the National Academy of Sciences of the United States of America, 89*(5), 1827–1831.

Fulnecek, J., & Kovarik, A. (2014). How to interpret methylation sensitive amplified polymorphism (MSAP) profiles? *BMC Genetics, 15*, 2.

Gao, M., Huang, Q., Chu, Y., Ding, C., Zhang, B., & Su, X. (2014). Analysis of the leaf methylomes of parents and their hybrids provides new insight into hybrid vigor in Populus deltoides. *BMC Genetics, 15*(Suppl 1), S8. https://doi.org/10.1186/1471-2156-15-S1-S8.

Gilmour, D. S., & Lis, J. T. (1984). Detecting protein-DNA interactions in vivo: Distribution of RNA polymerase on specific bacterial genes. *Proceedings of the National Academy of Sciences of the United States of America, 81*(14), 4275–4279.

Gourcilleau, D., Bogeat-Triboulot, M.-B., Le Thiec, D., Lafon-Placette, C., Delaunay, A., El-Soud, W. A., et al. (2010). DNA methylation and histone acetylation: Genotypic variations in hybrid poplars, impact of water deficit and relationships with productivity. *Annals of Forest Science, 67*(2), 208.

Grattapaglia, D. (2017). Status and perspectives of genomic selection in forest tree breeding. In M. E. Sorrells (Ed.), *Genomic selection for crop improvement* (pp. 199–257). Cham: Springer.

Greally, J. M. (2018). The HELP-based DNA methylation assays. *Methods in Molecular Biology, 1708*, 191–207.

Greenwood, M. S., & Hutchison, K. W. (1996). Genetic after effects of increased temperature in Larix. In J. Hom, R. Birdsey, & K. O'Brian (Eds.), *Proceedings of the 1995 meeting of the northern global change program* (pp. 56–62). Radnor: USDA Forest Service Report.

Groszmann, M., Greaves, I. K., Albertyn, Z. I., Scofield, G. N., Peacock, W. J., & Dennis, E. S. (2011). Changes in 24-nt siRNA levels in Arabidopsis hybrids suggest an epigenetic contribution to hybrid vigor. *Proceedings of the National Academy of Sciences of the United States of America, 108*(6), 2617–2622. https://doi.org/10.1073/pnas.1019217108.

Groszmann, M., Greaves, I. K., Fujimoto, R., James Peacock, W., & Dennis, E. S. (2013). The role of epigenetics in hybrid vigour. *Trends in Genetics, 29*(12), 684–690. https://doi.org/10.1016/j.tig.2013.07.004.

Guan, R., Zhao, Y., Zhang, H., Fan, G., Liu, X., Zhou, W., et al. (2016). Draft genome of the living fossil Ginkgo biloba. *GigaScience, 5*(1). https://doi.org/10.1186/s13742-016-0154-1.

Guarino, F., Cicatelli, A., Brundu, G., Heinze, B., & Castiglione, S. (2015). Epigenetic diversity of clonal white poplar (*Populus alba* L.) populations: Could methylation support the success of vegetative reproduction strategy? *Plos One, 10*(7). https://doi.org/10.1371/journal.pone.0131480.

Gugger, P. F., Fitz-Gibbon, S., PellEgrini, M., & Sork, V. L. (2016). Species-wide patterns of DNA methylation variation in *Quercus lobata* and their association with climate gradients. *Molecular Ecology, 25*(8), 1665–1680. https://doi.org/10.1111/mec.13563.

Gu, H., Smith, Z. D., Bock, C., Boyle, P., Gnirke, A., & Meissner, A. (2011). Preparation of reduced representation bisulfite sequencing libraries for genome-scale DNA methylation profiling. *Nature Protocols Other Titles: Protocols, 6*(4), 468–481.

Hallingbäck, H. R., Sánchez, L., & Wu, H. X. (2014). Single versus subdivided population strategies in breeding against an adverse genetic correlation. *Tree Genetics & Genomes, 10*(3), 605–617. https://doi.org/10.1007/s11295-014-0707-3.

Hasbún, R., Iturra, C., Bravo, S., Rebolledo-Jaramillo, B., & Valledor, L. (2016). Differential methylation of genomic regions associated with heteroblasty detected by M&M algorithm in the nonmodel species *Eucalyptus globulus* Labill. *International Journal of Genomics, 2016*. https://doi.org/10.1155/2016/4395153.

Herman, J. G., Graff, J. R., Myohanen, S., Nelkin, B. D., & Baylin, S. B. (1996). Methylation-specific PCR: A novel PCR assay for methylation status of CpG islands. *Proceedings of the National Academy of Sciences of the United States of America, 93*(18), 9821–9826.

Herrera, C. M., & Bazaga, P. (2013). Epigenetic correlates of plant phenotypic plasticity: DNA methylation differs between prickly and nonprickly leaves in heterophyllous *Ilex aquifolium* (Aquifoliaceae) trees: Heterophylly and DNA methylation. *Botanical Journal of the Linnean Society, 171*(3), 441–452. https://doi.org/10.1111/boj.12007.

Holliday, J. A., Aitken, S. N., Cooke, J. E., Fady, B., González-Martínez, S. C., Heuertz, M., et al. (2017). Advances in ecological genomics in forest trees and applications to genetic resources conservation and breeding. *Molecular Ecology, 26*(3), 706–717.

Horváth, V., Merenciano, M., & González, J. (2017). Revisiting the relationship between transposable elements and the eukaryotic stress response. *Trends in Genetics, 33*(11), 832–841. https://doi.org/10.1016/j.tig.2017.08.007.

Hussey, S. G., Mizrachi, E., Groover, A., Berger, D. K., & Myburg, A. A. (2015). Genome-wide mapping of histone H3 lysine 4 trimethylation in Eucalyptus grandis developing xylem. *BMC Plant Biology, 15*(1). https://doi.org/10.1186/s12870-015-0499-0.

Jaligot, E., Beulé, T., Baurens, F.-C., Billotte, N., & Rival, A. (2004). Search for methylation-sensitive amplification polymorphisms associated with the « mantled » variant phenotype in oil palm (*Elaeis guineensis* Jacq.). *Genome, 47*(1), 224–228. https://doi.org/10.1139/g03-085.

Jiao, W.-B., Accinelli, G. G., Hartwig, B., Kiefer, C., Baker, D., Severing, E., et al. (2017). Improving and correcting the contiguity of long-read genome assemblies of three plant species using optical mapping and chromosome conformation capture data. *Genome Research, 27*(5), 778–786. https://doi.org/10.1101/gr.213652.116.

Johannes, F., Colot, V., & Jansen, R. C. (2008). Epigenome dynamics: A quantitative genetics perspective. *Nature Reviews. Genetics, 9*(11), 883–890. https://doi.org/10.1038/nrg2467.

Johnsen, Ø., Dæhlen, O. G., Østreng, G., & Skrøppa, T. (2005a). Daylength and temperature during seed production interactively affect adaptive performance of *Picea abies* progenies. *New Phytologist, 168*(3), 589–596. https://doi.org/10.1111/j.1469-8137.2005.01538.x.

Johnsen, Ø., Fossdal, C. G., Nagy, N., Mølmann, J., DæHLEN, O. G., & Skrøppa, T. (2005b). Climatic adaptation in *Picea abies* progenies is affected by the temperature during zygotic embryogenesis and seed maturation. *Plant, Cell and Environment, 28*(9), 1090–1102. https://doi.org/10.1111/j.1365-3040.2005.01356.x.

Johnsen, Ø., & Skrøppa, T. (2001). The influence of the reproductive environment during sexual reproduction on adaptations of conifers along latitudinal and altitudinal gradients. In S. Huttunen, H. Heikkilä, J. Bucher, B. Sundberg, P. Jarvis, & R. Matyssek (Eds.), *Trends in European forest tree physiology research, cost action E6 EUROSILVA* (pp. 207–221). Dordrecht, the Netherlands: Kluwer Academic Publishers.

Jung, M., Kadam, S., Xiong, W., Rauch, T. A., Jin, S. G., & Pfeifer, G. P. (2015). MIRA-seq for DNA methylation analysis of CpG islands. *Epigenomics, 7*(5), 695–706.

Kanchanaketu, T., Sangduen, N., Toojinda, T., & Hongtrakul, V. (2012). Genetic diversity analysis of *Jatropha curcas* L. (Euphorbiaceae) based on methylation-sensitive amplification polymorphism. *Genetics and Molecular Research, 11*(2), 944–955. https://doi.org/10.4238/2012.April.13.2.

Kawakatsu, T., Huang, S. C., Jupe, F., Sasaki, E., Schmitz, R. J., Urich, M. A., et al. (2016). Epigenomic diversity in a global collection of *Arabidopsis thaliana* accessions. *Cell, 166*(2), 492–505. https://doi.org/10.1016/j.cell.2016.06.044.

Kawanabe, T., Ishikura, S., Miyaji, N., Sasaki, T., Wu, L. M., Itabashi, E., et al. (2016). Role of DNA methylation in hybrid vigor in *Arabidopsis thaliana*. *Proceedings of the National Academy of Sciences of the United States of America, 113*(43), E6704–E6711. https://doi.org/10.1073/pnas.1613372113.

Keller, T. E., & Yi, S. V. (2014). DNA methylation and evolution of duplicate genes. *Proceedings of the National Academy of Sciences of the United States of America, 111*(16), 5932–5937. https://doi.org/10.1073/pnas.1321420111.

Kelly, S. A., Tami, M. P., & Andrew, M. S. (2012). Phenotypic plasticity: Molecular mechanisms and adaptive significance. In R. Terjung (Ed.), *Comprehensive physiology*. Hoboken, NJ, USA: John Wiley & Sons, Inc. https://doi.org/10.1002/cphy.c110008.

Kelsey, G., Stegle, O., & Reik, W. (2017). Single-cell epigenomics: Recording the past and predicting the future. *Science, 358*(6359), 69–75.

Kimberley, M. O., Moore, J. R., & Dungey, H. S. (2015). Quantification of realised genetic gain in radiata pine and its incorporation into growth and yield modelling systems. *Canadian Journal of Forest Research, 45*(12), 1676–1687. https://doi.org/10.1139/cjfr-2015-0191.

Klimaszewska, K., Noceda, C., Pelletier, G., Label, P., Rodriguez, R., & Lelu-Walter, M. A. (2009). Biological characterization of young and aged embryogenic cultures of *Pinus pinaster* (Ait.). *In Vitro Cellular & Developmental Biology - Plant, 45*(1), 20–33. https://doi.org/10.1007/s11627-008-9158-6.

Kooke, R., Johannes, F., Wardenaar, R., Becker, F., Etcheverry, M., Colot, V., et al. (2015). Epigenetic basis of morphological variation and phenotypic plasticity in *Arabidopsis thaliana*. *The Plant Cell Online, 27*(2), 337–348. https://doi.org/10.1105/tpc.114.133025.

Kumar, S., Kumari, R., Sharma, V., & Sharma, V. (2013). Roles, and establishment, maintenance and erasing of the epigenetic cytosine methylation marks in plants. *Journal of Genetics, 92*(3), 629–666.

Kumar, G., Rattan, U. K., & Singh, A. K. (2016). Chilling-mediated DNA methylation changes during dormancy and its release reveal the importance of epigenetic regulation during winter dormancy in apple (Malus x domestica Borkh.). *Plos One, 11*(2), e0149934. https://doi.org/10.1371/journal.pone.0149934.

Kvaalen, H., & Johnsen, Ø. (2008). Timing of bud set in Picea abies is regulated by a memory of temperature during zygotic and somatic embryogenesis. *New Phytologist, 0*(0). https://doi.org/10.1111/j.1469-8137.2007.02222.x, 071009191540003–???

Lafon-Placette, C., Faivre-Rampant, P., Delaunay, A., Street, N., Brignolas, F., & Maury, S. (2013). Methylome of DNase I sensitive chromatin in *Populus trichocarpa* shoot apical meristematic cells: A simplified approach revealing characteristics of gene-body DNA methylation in open chromatin state. *New Phytologist, 197*(2), 416–430. https://doi.org/10.1111/nph.12026.

Lafon-Placette, C., Le Gac, A.-L., Chauveau, D., Segura, V., Delaunay, A., Lesage-Descauses, M.-C., et al. (2018). Changes in the epigenome and transcriptome of the poplar shoot apical meristem in response to water availability affect preferentially hormone pathways. *Journal of Experimental Botany, 69*(3), 537–551. https://doi.org/10.1093/jxb/erx409.

Lämke, J., & Bäurle, I. (2017). Epigenetic and chromatin-based mechanisms in environmental stress adaptation and stress memory in plants. *Genome Biology, 18*, 124.

Lanckriet, G. R. G., De Bie, T., Cristianini, N., Jordan, M. I., & Noble, W. S. (2004). A statistical framework for genomic data fusion. *Bioinformatics, 20*(16), 2626–2635. https://doi.org/10.1093/bioinformatics/bth294.

Lande, R. (2009). Adaptation to an extraordinary environment by evolution of phenotypic plasticity and genetic assimilation. *Journal of Evolutionary Biology, 22*(7), 1435–1446. https://doi.org/10.1111/j.1420-9101.2009.01754.x.

Landt, S. G., Marinov, G. K., Kundaje, A., Kheradpour, P., Pauli, F., Batzoglou, S., et al. (2012). ChIP-seq guidelines and practices of the ENCODE and modENCODE consortia. *Genome Research, 22*(9), 1813–1831.

Latzel, V., Rendina González, A. P., & Rosenthal, J. (2016). Epigenetic memory as a basis for intelligent behavior in clonal plants. *Frontiers in Plant Science, 7*. https://doi.org/10.3389/fpls.2016.01354.

Lauria, M., & Rossi, V. (2011). Epigenetic control of gene regulation in plants. *Biochimica et Biophysica Acta (BBA) - Gene Regulatory Mechanisms, 1809*(8), 369–378. https://doi.org/10.1016/j.bbagrm.2011.03.002.

Lauss, K., Wardenaar, R., Oka, R., van Hulten, M. H. A., Guryev, V., Keurentjes, J. J. B., et al. (2018). Parental DNA methylation states are associated with heterosis in epigenetic hybrids. *Plant Physiology, 176*(2), 1627–1645. https://doi.org/10.1104/pp.17.01054.

Law, J. A., & Jacobsen, S. E. (2010). Establishing, maintaining and modifying DNA methylation patterns in plants and animals. *Nature Reviews. Genetics, 11*(3), 204–220. https://doi.org/10.1038/nrg2719.

Le Gac, A.-L., Lafon-Placette, C., Chauveau, D., Segura, V., Delaunay, A., Fichot, R., et al. (2018). Winter-dormant shoot apical meristem in poplar trees shows environmental epigenetic memory. *Journal of Experimental Botany, 69*(20), 4821–4837. https://doi.org/10.1093/jxb/ery271.

Lelu-Walter, M.-A., Bernier-Cardou, M., & Klimaszewska, K. (2006). Simplified and improved somatic embryogenesis for clonal propagation of *Pinus pinaster* (Ait.). *Plant Cell Reports, 25*(8), 767–776. https://doi.org/10.1007/s00299-006-0115-8.

Lelu-Walter, M.-A., Thompson, D., Harvengt, L., Sanchez, L., Toribio, M., & Pâques, L. E. (2013). Somatic embryogenesis in forestry with a focus on Europe: State-of-the-art, benefits, challenges and future direction. *Tree Genetics & Genomes, 9*(4), 883–899. https://doi.org/10.1007/s11295-013-0620-1.

Liang, D., Zhang, Z., Wu, H., Huang, C., Shuai, P., Ye, C.-Y., et al. (2014). Single-base-resolution methylomes of populus trichocarpa reveal the association between DNA methylation and drought stress. *BMC Genetics, 15*(Suppl 1), S9. https://doi.org/10.1186/1471-2156-15-S1-S9.

Li, Y., Dong, X.-M., Jin, F., Shen, Z., Chao, Q., & Wang, B.-C. (2017). Histone acetylation modifications affect tissue-dependent expression of poplar homologs of C4 photosynthetic enzyme genes. *Frontiers in Plant Science, 8*. https://doi.org/10.3389/fpls.2017.00950.

Li, Q., Eichten, S. R., Hermanson, P. J., & Springer, N. M. (2014). Inheritance patterns and stability of DNA methylation variation in maize near-isogenic lines. *Genetics, 196*(3), 667–676. https://doi.org/10.1534/genetics.113.158980.

Li, A., Hu, B.-Q., Xue, Z.-Y., Chen, L., Wang, W.-X., Song, W.-Q., et al. (2011). DNA methylation in genomes of several annual herbaceous and woody perennial plants of varying ploidy as detected by MSAP. *Plant Molecular Biology Reporter, 29*(4), 784–793. https://doi.org/10.1007/s11105-010-0280-3.

Lira-Medeiros, C. F., Parisod, C., Fernandes, R. A., Mata, C. S., Cardoso, M. A., & Ferreira, P. C. G. (2010). Epigenetic variation in mangrove plants occurring in contrasting natural environment. *Plos One, 5*(4), e10326. https://doi.org/10.1371/journal.pone.0010326.

Lister, R., O'Malley, R. C., Tonti-Filippini, J., Gregory, B. D., Berry, C. C., Millar, A. H., et al. (2008). Highly integrated single-base resolution maps of the epigenome in Arabidopsis. *Cell, 133*(3), 523–536.

Li, Q., Suzuki, M., Wendt, J., Patterson, N., Eichten, S. R., Hermanson, P. J., et al. (2015). Post-conversion targeted capture of modified cytosines in mammalian and plant genomes. *Nucleic Acids Research, 43*(12), e81.

Liu, Y., & El-Kassaby, Y. A. (2017). Landscape of fluid sets of hairpin-derived 21-/24-nt-long small RNAs at seed set uncovers special epigenetic features in Picea glauca. *Genome Biology and Evolution*. https://doi.org/10.1093/gbe/evw283. evw283.

Liu, R., How-Kit, A., Stammitti, L., Teyssier, E., Rolin, D., Mortain-Bertrand, A., et al. (2015). A DEMETER-like DNA demethylase governs tomato fruit ripening. *Proceedings of the National Academy of Sciences of the United States of America, 112*(34), 10804–10809. https://doi.org/10.1073/pnas.1503362112.

Long, Y., Xia, W., Li, R., Wang, J., Shao, M., Feng, J., et al. (2011). Epigenetic QTL mapping in Brassica napus. *Genetics, 189*(3), 1093–1102. https://doi.org/10.1534/genetics.111.131615.

Mackay, J., Dean, J. F. D., Plomion, C., Peterson, D. G., Cánovas, F. M., Pavy, N., et al. (2012). Towards decoding the conifer giga-genome. *Plant Molecular Biology, 80*(6), 555–569. https://doi.org/10.1007/s11103-012-9961-7.

Mader, M., Le Paslier, M.-C., Bounon, R., Bérard, A., Rampant, P. F., Fladung, M., et al. (2016). Whole-genome draft assembly of *Populus tremula* x *P. alba* clone INRA 717-1B4. *Silvae Genetica, 65*(2). https://doi.org/10.1515/sg-2016-0019.

Martínez-García, P. J., Crepeau, M. W., Puiu, D., Gonzalez-Ibeas, D., Whalen, J., Stevens, K. A., et al. (2016). The walnut (*Juglans regia*) genome sequence reveals diversity in genes coding for the biosynthesis of non-structural polyphenols. *The Plant Journal, 87*(5), 507–532. https://doi.org/10.1111/tpj.13207.

Maunakea, A. K., Nagarajan, R. P., Bilenky, M., Ballinger, T. J., D'Souza, C., Fouse, S. D., et al. (2010). Conserved role of intragenic DNA methylation in regulating alternative promoters. *Nature, 466*(7303), 253–257.

Ma, X., Zhang, C., Zhang, B., Yang, C., & Li, S. (2016). Identification of genes regulated by histone acetylation during root development in Populus trichocarpa. *BMC Genomics, 17*(1). https://doi.org/10.1186/s12864-016-2407-x.

McMahon, B. J., Teeling, E. C., & Höglund, J. (2014). How and why should we implement genomics into conservation? *Evolutionary Applications, 7*(9), 999–1007. https://doi.org/10.1111/eva.12193.

Meissner, A., Gnirke, A., Bell, G. W., Ramsahoye, B., Lander, E. S., & Jaenisch, R. (2005). Reduced representation bisulfite sequencing for comparative high-resolution DNA methylation analysis. *Nucleic Acids Research, 33*(18), 5868–5877.

Meyer, P. (2015). Epigenetic variation and environmental change: Fig. 1. *Journal of Experimental Botany, 66*(12), 3541–3548. https://doi.org/10.1093/jxb/eru502.

Meyer, R. C., Törjék, O., Becher, M., & Altmann, T. (2004). Heterosis of biomass production in Arabidopsis. Establishment during early development. *Plant Physiology, 134*(4), 1813–1823. https://doi.org/10.1104/pp.103.033001.

Michalak, M., Barciszewska, M. Z., Barciszewski, J., Plitta, B. P., & Chmielarz, P. (2013). Global changes in DNA methylation in seeds and seedlings of *Pyrus communis* after seed desiccation and storage. *Plos One, 8*(8), e70693. https://doi.org/10.1371/journal.pone.0070693.

Michalak, M., Plitta-Michalak, B. P., Naskręt-Barciszewska, M., Barciszewski, J., Bujarska-Borkowska, B., & Chmielarz, P. (2015). Global 5-methylcytosine alterations in DNA during ageing of *Quercus robur* seeds. *Annals of Botany, 116*(3), 369–376. https://doi.org/10.1093/aob/mcv104.

Miura, F., Enomoto, Y., Dairiki, R., & Ito, T. (2012). Amplification-free whole-genome bisulfite sequencing by post-bisulfite adaptor tagging. *Nucleic Acids Research, 40*(17), e136.

Mullin, T. J., Andersson, B., Bastien, J. C., Beaulieu, J., Burdon, R. D., Dvorak, W. S., et al. (2011). Economic importance, breeding objectives and achievements. In C. Plomion, J. Bousquet, & C. Kole (Eds.), *Genetics, genomics and breeding of crop plants science publ.* Enfield, NH, USA: CRC Press.

Muranty, H., Jorge, V., Bastien, C., Lepoittevin, C., Bouffier, L., & Sanchez, L. (2014). Potential for marker-assisted selection for forest tree breeding: Lessons from 20 years

of MAS in crops. *Tree Genetics & Genomes, 10*(6), 1491–1510. https://doi.org/10.1007/s11295-014-0790-5.
Murat, F., Armero, A., Pont, C., Klopp, C., & Salse, J. (2017). Reconstructing the genome of the most recent common ancestor of flowering plants. *Nature Genetics, 49*(4), 490–496.
Murat, F., Zhang, R., Guizard, S., Gavranović, H., Flores, R., Steinbach, D., et al. (2015). Karyotype and gene order evolution from reconstructed extinct ancestors highlight contrasts in genome plasticity of modern rosid crops. *Genome Biology and Evolution, 7*(3), 735–749. https://doi.org/10.1093/gbe/evv014.
Myburg, A. A., Grattapaglia, D., Tuskan, G. A., Hellsten, U., Hayes, R. D., Grimwood, J., et al. (2014). The genome of *Eucalyptus grandis*. *Nature, 510*(7505), 356–362. https://doi.org/10.1038/nature13308.
Nanson, A. (2004). *Génétique et amelioration des arbres forestiers*. Gembloux: Presses Agronomiques de Gembloux, 712 p., 103 ill., 25 tab.
Neale, D. B., Langley, C. H., Salzberg, S. L., & Wegrzyn, J. L. (2013). Open access to tree genomes: The path to a better forest. *Genome Biology, 14*(6), 120.
Neale, D. B., Wegrzyn, J. L., Stevens, K. A., Zimin, A. V., Puiu, D., Crepeau, M. W., et al. (2014). Decoding the massive genome of loblolly pine using haploid DNA and novel assembly strategies. *Genome Biology, 15*(3), R59.
Nicotra, A. B., Atkin, O. K., Bonser, S. P., Davidson, A. M., Finnegan, E. J., Mathesius, U., et al. (2010). Plant phenotypic plasticity in a changing climate. *Trends in Plant Science, 15*(12), 684–692. https://doi.org/10.1016/j.tplants.2010.09.008.
Niu, S.-H., Liu, C., Yuan, H.-W., Li, P., Li, Y., & Li, W. (2015). Identification and expression profiles of sRNAs and their biogenesis and action-related genes in male and female cones of *Pinus tabuliformis*. *BMC Genomics, 16*(1). https://doi.org/10.1186/s12864-015-1885-6.
Nock, C. J., Baten, A., Barkla, B. J., Furtado, A., Henry, R. J., & King, G. J. (2016). Genome and transcriptome sequencing characterises the gene space of *Macadamia integrifolia* (Proteaceae). *BMC Genomics, 17*(1). https://doi.org/10.1186/s12864-016-3272-3.
Nystedt, B., Street, N. R., Wetterbom, A., Zuccolo, A., Lin, Y.-C., Scofield, D. G., et al. (2013). The Norway spruce genome sequence and conifer genome evolution. *Nature, 497*(7451), 579–584. https://doi.org/10.1038/nature12211.
Pâques, L. E. (2013). In L. E. Pâques (Ed.), *Forest tree breeding in Europe. Current state-of-the-art and perspectives* (p. 25). Springer. https://doi.org/10.1007/978-94-007-6146-9.
Paul, D. S., Guilhamon, P., Karpathakis, A., Butcher, L. M., Thirlwell, C., Feber, A., et al. (2014). Assessment of RainDrop BS-seq as a method for large-scale, targeted bisulfite sequencing. *Epigenetics, 9*(5), 678–684.
Paun, O., Fay, M. F., Soltis, D. E., & Chase, M. W. (2007). Genetic and epigenetic alterations after hybridization and genome doubling. *Taxon, 56*(3), 649. https://doi.org/10.2307/25065850.
Platt, A., Gugger, P. F., Pellegrini, M., & Sork, V. L. (2015). Genome-wide signature of local adaptation linked to variable CpG methylation in oak populations. *Molecular Ecology, 24*(15), 3823–3830. https://doi.org/10.1111/mec.13230.
Plitta-Michalak, B. P., Naskręt-Barciszewska, M. Z., Kotlarski, S., Tomaszewski, D., Tylkowski, T., Barciszewski, J., et al. (2018). Changes in genomic 5-methylcytosine level mirror the response of orthodox (*Acer platanoides* L.) and recalcitrant (*Acer pseudoplatanus* L.) seeds to severe desiccation. *Tree Physiology, 38*(4), 617–629. https://doi.org/10.1093/treephys/tpx134.
Plitta, B. P., Michalak, M., Bujarska-Borkowska, B., Barciszewska, M. Z., Barciszewski, J., & Chmielarz, P. (2014). Effect of desiccation on the dynamics of genome-wide DNA methylation in orthodox seeds of *Acer platanoides* L. *Plant Physiology and Biochemistry, 85*, 71–77. https://doi.org/10.1016/j.plaphy.2014.10.014.

Plomion, C., Aury, J. M., Amselem, J., Leroy, T., Murat, F., Duplessis, S., et al. (2018). Oak genome reveals facets of long lifespan. *Nature Plants, 4*, 440–452.

Plomion, C., Bastien, C., Bogeat-Triboulot, M. B., Bouffier, L., Déjardin, A., Duplessis, S., et al. (2016). Forest tree genomics: 10 achievements from the past 10 years and future prospects. *Annals of Forest Science, 73*(1), 77–103. https://doi.org/10.1007/s13595-015-0488-3.

Podevin, N., Davies, H. V., Hartung, F., Nogué, F., & Casacuberta, J. M. (2013). Site-directed nucleases: A paradigm shift in predictable, knowledge-based plant breeding. *Trends in Biotechnology, 31*(6), 375–383. https://doi.org/10.1016/j.tibtech.2013.03.004.

Quinn, C. R., Iriyama, R., & Fernando, D. D. (2014). Expression patterns of conserved microRNAs in the male gametophyte of loblolly pine (*Pinus taeda*). *Plant Reproduction, 27*(2), 69–78. https://doi.org/10.1007/s00497-014-0241-3.

Raj, S., Bräutigam, K., Hamanishi, E. T., Wilkins, O., Thomas, B. R., Schroeder, W., et al. (2011). Clone history shapes Populus drought responses. *Proceedings of the National Academy of Sciences of the United States of America, 108*(30), 12521–12526.

Ramassone, A., Pagotto, S., Veronese, A., & Visone, R. (2018). Epigenetics and MicroRNAs in cancer. *International Journal of Molecular Sciences, 19*, 459. https://doi.org/10.3390/ijms19020459.

Rand, A. C., Jain, M., Eizenga, J. M., Musselman-Brown, A., Olsen, H. E., Akeson, M., et al. (2017). Mapping DNA methylation with high-throughput nanopore sequencing. *Nature Methods, 14*(4), 411–413.

Rauch, T., & Pfeifer, G. P. (2005). Methylated-CpG island recovery assay: A new technique for the rapid detection of methylated-CpG islands in cancer. *Laboratory Investigation; a Journal of Technical Methods and Pathology, 85*(9), 1172–1180.

Ren, Y., Chen, L., Zhang, Y., Kang, X., Zhang, Z., & Wang, Y. (2012). Identification of novel and conserved Populus tomentosa microRNA as components of a response to water stress. *Functional & Integrative Genomics, 12*(2), 327–339. https://doi.org/10.1007/s10142-012-0271-6.

Richards, C. L., Alonso, C., Becker, C., Bossdorf, O., Bucher, E., Colomé-Tatché, M., et al. (2017). Ecological plant epigenetics: Evidence from model and non-model species, and the way forward. *Ecology Letters, 20*(12), 1576–1590. https://doi.org/10.1111/ele.12858.

Rico, L., Ogaya, R., Barbeta, A., & Peñuelas, J. (2014). Changes in DNA methylation fingerprint of *Quercus ilex* trees in response to experimental field drought simulating projected climate change. *Plant Biology, 16*(2), 419–427. https://doi.org/10.1111/plb.12049.

Rodin, S. N., & Riggs, A. D. (2003). Epigenetic silencing may aid evolution by gene duplication. *Journal of Molecular Evolution, 56*(6), 718–729. https://doi.org/10.1007/s00239-002-2446-6.

Russo, V. E. A., Martienssen, R. A., & Riggs, A. D. (1996). *Epigenetic mechanisms of gene regulation* (Vol. 32). Cold Spring Harbor, NY: Cold Spring Harbor Laboratory Press. i–xii.

Ruttink, T., Arend, M., Morreel, K., Storme, V., Rombauts, S., Fromm, J., et al. (2007). A molecular timetable for apical bud formation and dormancy induction in poplar. *The Plant Cell Online, 19*(8), 2370–2390. https://doi.org/10.1105/tpc.107.052811.

Sáez-Laguna, E., Guevara, M.-Á., Díaz, L.-M., Sánchez-Gómez, D., Collada, C., Aranda, I., et al. (2014). Epigenetic variability in the genetically uniform forest tree species *Pinus pinea* L. *Plos One, 9*(8). https://doi.org/10.1371/journal.pone.0103145.

Salmon, A., Ainouche, M. L., & Wendel, J. F. (2005). Genetic and epigenetic consequences of recent hybridization and polyploidy in spartina (poaceae): Genetic and epigenetic changes in spartina. *Molecular Ecology, 14*(4), 1163–1175. https://doi.org/10.1111/j.1365-294X.2005.02488.x.

Salojärvi, J., Smolander, O.-P., Nieminen, K., Rajaraman, S., Safronov, O., Safdari, P., et al. (2017). Genome sequencing and population genomic analyses provide insights into the adaptive landscape of silver birch. *Nature Genetics, 49*(6), 904–912. https://doi.org/10.1038/ng.3862.

Salse, J. (2016a). Ancestors of modern plant crops. *Current Opinion in Plant Biology, 30*, 134–142. https://doi.org/10.1016/j.pbi.2016.02.005.

Salse, J. (2016b). Deciphering the evolutionary interplay between subgenomes following polyploidy: A paleogenomics approach in grasses. *American Journal of Botany, 103*(7), 1167–1174. https://doi.org/10.3732/ajb.1500459.

Sanchez, L., Bastien, C., Rozenberg, P., Lee, S., Woolliams, J., Lagercrantz, U., et al. (2013). Novel tree breeding. In S. J. Lee, & J. Woolliams (Eds.), *Monografias INIA, monographies, forestry series*.

Santamaría, M., Hasbún, R., Valera, M., Meijón, M., Valledor, L., Rodríguez, J. L., et al. (2009). Acetylated H4 histone and genomic DNA methylation patterns during bud set and bud burst in *Castanea sativa*. *Journal of Plant Physiology, 166*(13), 1360–1369. https://doi.org/10.1016/j.jplph.2009.02.014.

Schmidtling, R. C., & Hipkins, V. (2004). The after-effects of reproductive environment in shortleaf pine. *Forestry, 77*(4), 287–295. https://doi.org/10.1093/forestry/77.4.287.

Schnable, P. S., & Springer, N. M. (2013). Progress toward understanding heterosis in crop plants. *Annual Review of Plant Biology, 64*(1), 71–88. https://doi.org/10.1146/annurev-arplant-042110-103827.

Schönberger, B., Chen, X., Mager, S., & Ludewig, U. (2016). Site-dependent differences in DNA methylation and their impact on plant establishment and phosphorus nutrition in *Populus trichocarpa*. *Plos One, 11*(12), e0168623. https://doi.org/10.1371/journal.pone.0168623.

Seymour, D. K., & Becker, C. (2017). The causes and consequences of DNA methylome variation in plants. *Current Opinion in Plant Biology, 36*, 56–63. https://doi.org/10.1016/j.pbi.2017.01.005.

Shaked, H., Kashkush, K., Ozkan, H., Feldman, M., & Levy, A. A. (2001). Sequence elimination and cytosine methylation are rapid and reproducible responses of the genome to wide hybridization and allopolyploidy in wheat. *The Plant Cell Online, 13*(8), 1749–1759.

Shapiro, R., DiFate, V., & Welcher, M. (1974). Deamination of cytosine derivatives by bisulfite. Mechanism of the reaction. *Journal of the American Chemical Society, 96*(3), 906–912.

Shen, H., He, H., Li, J., Chen, W., Wang, X., Guo, L., et al. (2012). Genome-wide analysis of DNA methylation and gene expression changes in two Arabidopsis ecotypes and their reciprocal hybrids. *The Plant Cell Online, 24*(3), 875–892. https://doi.org/10.1105/tpc.111.094870.

Shen, Y., Sun, S., Hua, S., Shen, E., Ye, C.-Y., Cai, D., et al. (2017). Analysis of transcriptional and epigenetic changes in hybrid vigor of allopolyploid *Brassica napus* uncovers key roles for small RNAs. *The Plant Journal, 91*(5), 874–893. https://doi.org/10.1111/tpj.13605.

Shim, D., Ko, J.-H., Kim, W.-C., Wang, Q., Keathley, D. E., & Han, K.-H. (2014). A molecular framework for seasonal growth-dormancy regulation in perennial plants. *Horticulture Research, 1*(1). https://doi.org/10.1038/hortres.2014.59.

Shuai, P., Liang, D., Zhang, Z., Yin, W., & Xia, X. (2013). Identification of drought-responsive and novel *Populus trichocarpa* microRNAs by high-throughput sequencing and their targets using degradome analysis. *BMC Genomics, 14*(1), 233. https://doi.org/10.1186/1471-2164-14-233.

Silva-Junior, O. B., Grattapaglia, D., Novaes, E., & Collevatti, R. G. (2018). Genome assembly of the Pink Ipê (*Handroanthus impetiginosus*, Bignoniaceae), a highly valued, ecologically keystone Neotropical timber forest tree. *GigaScience, 7*(1). https://doi.org/10.1093/gigascience/gix125.

Simpson, J. T., Workman, R. E., Zuzarte, P. C., David, M., Dursi, L. J., & Timp, W. (2017). Detecting DNA cytosine methylation using nanopore sequencing. *Nature Methods, 14*(4), 407–410.

Skrøppa, T., & Johnsen, Ø. (2000). Patterns of adaptive genetic variation in forest tree species; the reproductive environment as an evolutionary force in *Picea abies*. In C. Mátyás (Ed.), *Forest genetics and sustainability* (pp. 49–58). Dordrecht, the Netherlands: Kluwer Academic Publications.

Skrøppa, T., Kohmann, K., Johnsen, Ø., Steffenrem, A., & Edvardsen, Ø. M. (2007). Field performance and early test results of offspring from two Norway spruce seed orchards containing clones transferred to warmer climates. *Canadian Journal of Forest Research, 37*(3), 515–522. https://doi.org/10.1139/X06-253.

Skrøppa, T., Solvin, T., & Steffenrem, A. (2016). Epigenetic response to climate conditions during the reproductive phase; an evolutionary factor of northern Norway spruce populations?. In *Final meeting of cost action FP1202 MaP-FGR and conference on "marginal and peripheral tree populations: A key genetic resource for European forests" Arezzo, Italy, 26–29 September 2016.*

Skrøppa, T., Tollefsrud, M. M., Sperisen, C., & Johnsen, Ø. (2010). Rapid change in adaptive performance from one generation to the next in *Picea abies*—central European trees in a Nordic environment. *Tree Genetics & Genomes, 6*(1), 93–99. https://doi.org/10.1007/s11295-009-0231-z.

Slavov, G. T., DiFazio, S. P., Martin, J., Schackwitz, W., Muchero, W., Rodgers-Melnick, E., et al. (2012). Genome resequencing reveals multiscale geographic structure and extensive linkage disequilibrium in the forest tree *Populus trichocarpa*. *New Phytologist, 196*(3), 713–725. https://doi.org/10.1111/j.1469-8137.2012.04258.x.

Sollars, E. S. A., Harper, A. L., Kelly, L. J., Sambles, C. M., Ramirez-Gonzalez, R. H., Swarbreck, D., et al. (2017). Genome sequence and genetic diversity of European ash trees. *Nature, 541*(7636), 212–216. https://doi.org/10.1038/nature20786.

Soltis, D. E., Bell, C. D., Kim, S., & Soltis, P. S. (2008). Origin and early evolution of angiosperms. *Annals of the New York Academy of Sciences, 1133*(1), 3–25. https://doi.org/10.1196/annals.1438.005.

Song, Y., Ci, D., Tian, M., & Zhang, D. (2016). Stable methylation of a non-coding RNA gene regulates gene expression in response to abiotic stress in *Populus simonii*. *Journal of Experimental Botany, 67*(5), 1477–1492. https://doi.org/10.1093/jxb/erv543.

Song, C. X., Clark, T. A., Lu, X. Y., Kislyuk, A., Dai, Q., Turner, S. W., et al. (2012). Sensitive and specific single-molecule sequencing of 5-hydroxymethylcytosine. *Nature Methods, 9*(1), 75–77.

Song, Y., Tian, M., Ci, D., & Zhang, D. (2015). Methylation of microRNA genes regulates gene expression in bisexual flower development in andromonoecious poplar. *Journal of Experimental Botany, 66*(7), 1891–1905. https://doi.org/10.1093/jxb/eru531.

Sork, V. L. (2016). Gene flow and natural selection shape spatial patterns of genes in tree populations: Implications for evolutionary processes and applications. *Evolutionary Applications, 9*(1), 291–310. https://doi.org/10.1111/eva.12316.

Sork, V. L. (2017). Genomic studies of local adaptation in natural plant populations. *Journal of Heredity, 109*(1), 3–15. https://doi.org/10.1093/jhered/esx091.

Sork, V. L., Fitz-Gibbon, S. T., Puiu, D., Crepeau, M., Gugger, P. F., Sherman, R., et al. (2016). First draft assembly and annotation of the genome of a California endemic oak Quercus lobata nee (Fagaceae). *Genes|Genomes|Genetics*. https://doi.org/10.1534/g3.116.030411.

Sow, M. D., Segura, V., Chamaillard, S., Jorge, V., Delaunay, A., Lafon-Placette, C., et al. (2018). Narrow-sense heritability and PST estimates of DNA methylation in three *Populus nigra L.* populations under contrasting water availability. *Tree Genetics and Genomes, 14*(5). https://doi.org/10.1007/s11295-018-1293-6.

Statham, A. L., Robinson, M. D., Song, J. Z., Coolen, M. W., Stirzaker, C., & Clark, S. J. (2012). Bisulfite sequencing of chromatin immunoprecipitated DNA (BisChIP-seq) directly informs methylation status of histone-modified DNA. *Genome Research, 22*(6), 1120–1127.

Stevens, K. A., Wegrzyn, J. L., Zimin, A., Puiu, D., Crepeau, M., Cardeno, C., et al. (2016). Sequence of the sugar pine megagenome. *Genetics, 204*(4), 1613–1626. https://doi.org/10.1534/genetics.116.193227.

Stewart, F. J., Panne, D., Bickle, T. A., & Raleigh, E. A. (2000). Methyl-specific DNA binding by McrBC, a modification-dependent restriction enzyme. *Journal of Molecular Biology, 298*(4), 611–622.

Stoehr, M. U., L'Hirondelle, S. J., Binder, W. D., & Webber, J. E. (1998). Parental environment aftereffects on germination, growth, and adaptive traits in selected white spruce families. *Canadian Journal of Forest Research, 28*(3), 418–426. https://doi.org/10.1139/x98-012.

Suo, Y., Dong, C., & Kang, X. (2015). Inheritance and Variation of Cytosine Methylation in Three Populus Allotriploid Populations with Different Heterozygosity. *Plos One, 10*(4), e0126491. https://doi.org/10.1371/journal.pone.0126491.

Suzuki, M., Jing, Q., Lia, D., Pascual, M., McLellan, A., & Greally, J. M. (2010). Optimized design and data analysis of tag-based cytosine methylation assays. *Genome Biology, 11*, R36. https://doi.org/10.1186/gb-2010-11-4-r36.

Taiwo, O., Wilson, G. A., Morris, T., Seisenberger, S., Reik, W., Pearce, D., et al. (2012). Methylome analysis using MeDIP-seq with low DNA concentrations. *Nature Protocols Other Titles: Protocols*, 7(4), 617–636.

Tang, C., Yang, M., Fang, Y., Luo, Y., Gao, S., Xiao, X., et al. (2016). The rubber tree genome reveals new insights into rubber production and species adaptation. *Nature Plants, 2*(6), 16073. https://doi.org/10.1038/nplants.2016.73.

ter Steege, H., Pitman, N. C. A., Killeen, T. J., Laurance, W. F., Peres, C. A., Guevara, J. E., et al. (2015). Estimating the global conservation status of more than 15,000 Amazonian tree species. *Science Advances, 1*(10), e1500936. https://doi.org/10.1126/sciadv.1500936.

Teyssier, C., Maury, S., Beaufour, M., Grondin, C., Delaunay, A., Le Metté, C., et al. (2014). In search of markers for somatic embryo maturation in hybrid larch (*Larix × eurolepis*): Global DNA methylation and proteomic analyses. *Physiologia Plantarum, 150*(2), 271–291. https://doi.org/10.1111/ppl.12081.

Toda, T., & Kurinobu, S. (2002). Realized genetic gains observed in progeny tolerance of selected red pine (*Pinus densiflora*) and black pine (*P. thunbergii*) to pine wilt disease. *Silvae Genetica, 51*(1), 42–44.

Tost, J. (2016). Current and emerging technologies for the analysis of the genome-wide and locus-specific DNA methylation patterns. *Advances in Experimental Medicine and Biology, 945*, 343–430.

Tost, J., & Gut, I. G. (2007). DNA methylation analysis by pyrosequencing. *Nature Protocols Other Titles: Protocols, 2*(9), 2265–2275.

Trontin, J.-F., Klimaszewska, K., Morel, A., Hargreaves, C., & Lelu-Walter, M.-A. (2016). Molecular aspects of conifer zygotic and somatic embryo development: A review of genome-wide approaches and recent insights. In M. A. Germana, & M. Lambardi (Eds.), *In vitro embryogenesis in higher plants* (Vol. 1359, pp. 167–207). New York, NY: Springer New York. https://doi.org/10.1007/978-1-4939-3061-6_8.

Tulving, E. (1985). Memory and consciousness. *Canadian Psychology/Psychologie Canadienne, 26*(1), 1–12. https://doi.org/10.1037/h0080017.

Tuskan, G. A., DiFazio, S., Jansson, S., Bohlmann, J., Grigoriev, I., Hellsten, U., et al. (2006). The genome of black cottonwood, *Populus trichocarpa* (Torr. & Gray). *Science, 313*(5793), 1596–1604. https://doi.org/10.1126/science.1128691.

Tylewicz, S., Petterle, A., Marttila, S., Miskolczi, P., Azeez, A., Singh, R. K., et al. (2018). Photoperiodic control of seasonal growth is mediated by ABA acting on cell-cell communication. *Science, 360*(6385), 212–215. https://doi.org/10.1126/science.aan8576.

Ujino-Ihara, T., Ueno, S., Uchiyama, K., & Futamura, N. (2018). Comprehensive analysis of small RNAs expressed in developing male strobili of *Cryptomeria japonica*. *Plos One, 13*(3), e0193665. https://doi.org/10.1371/journal.pone.0193665.

Urich, M. A., Nery, J. R., Lister, R., Schmitz, R. J., & Ecker, J. R. (2015). MethylC-seq library preparation for base-resolution whole-genome bisulfite sequencing. *Nature Protocols Other Titles: Protocols, 10*(3), 475–483.

Valledor, L., Meijón, M., Hasbún, R., Cañal, M. J., & Rodríguez, R. (2010). Variations in DNA methylation, acetylated histone H4, and methylated histone H3 during Pinus radiata needle maturation in relation to the loss of in vitro organogenic capability. *Journal of Plant Physiology, 167*(5), 351–357. https://doi.org/10.1016/j.jplph.2009.09.018.

van der Graaf, A., Wardenaar, R., Neumann, D. A., Taudt, A., Shaw, R. G., Jansen, R. C., et al. (2015). Rate, spectrum, and evolutionary dynamics of spontaneous epimutations. *Proceedings of the National Academy of Sciences of the United States of America, 112*(21), 6676–6681.

Verhoeven, K. J. F., von Holdt, B. M., & Sork, V. L. (2016). Epigenetics in ecology and evolution: What we know and what we need to know. *Molecular Ecology, 25*(8), 1631–1638. https://doi.org/10.1111/mec.13617.

Vining, K., Pomraning, K. R., Wilhelm, L. J., Ma, C., Pellegrini, M., Di, Y., et al. (2013). Methylome reorganization during in vitro dedifferentiation and regeneration of *Populus trichocarpa*. *BMC Plant Biology, 13*(1), 92. https://doi.org/10.1186/1471-2229-13-92.

Vining, K. J., Pomraning, K. R., Wilhelm, L. J., Priest, H. D., Pellegrini, M., Mockler, T. C., et al. (2012). Dynamic DNA cytosine methylation in the Populus trichocarpa genome: Tissue-level variation and relationship to gene expression. *BMC Genomics, 13*(1), 27.

Walker, D. L., Bhagwate, A. V., Baheti, S., Smalley, R. L., Hilker, C. A., Sun, Z., et al. (2015). DNA methylation profiling: Comparison of genome-wide sequencing methods and the infinium human methylation 450 bead chip. *Epigenomics, 7*(8), 1287–1302.

Wang, J., Blevins, T., Podicheti, R., Haag, J. R., Tan, E. H., Wang, F., et al. (2017b). Mutation of Arabidopsis SMC4 identifies condensin as a corepressor of pericentromeric transposons and conditionally expressed genes. *Genes & Development, 31*(15), 1601–1614. https://doi.org/10.1101/gad.301499.117.

Wang, D., Qu, Z., Yang, L., Zhang, Q., Liu, Z.-H., Do, T., et al. (2017a). Transposable elements (TEs) contribute to stress-related long intergenic noncoding RNAs in plants. *The Plant Journal, 90*(1), 133–146. https://doi.org/10.1111/tpj.13481.

Wang, N., Thomson, M., Bodles, W. J. A., Crawford, R. M. M., Hunt, H. V., Featherstone, A. W., et al. (2013). Genome sequence of dwarf birch (*Betula nana*) and cross-species RAD markers. *Molecular Ecology, 22*(11), 3098–3111. https://doi.org/10.1111/mec.12131.

Wang, X., Zhang, Z., Fu, T., Hu, L., Xu, C., Gong, L., et al. (2017c). Gene-body CG methylation and divergent expression of duplicate genes in rice. *Scientific Reports, 7*(1). https://doi.org/10.1038/s41598-017-02860-4.

Wang, L., Zhao, J., Zhang, M., Li, W., Luo, K., Lu, Z., et al. (2015). Identification and characterization of microRNA expression in *Ginkgo biloba* L. leaves. *Tree Genetics & Genomes, 11*(4). https://doi.org/10.1007/s11295-015-0897-3.

Wan, L.-C., Wang, F., Guo, X., Lu, S., Qiu, Z., Zhao, Y., et al. (2012). Identification and characterization of small non-coding RNAs from Chinese fir by high throughput sequencing. *BMC Plant Biology, 12*(1), 146. https://doi.org/10.1186/1471-2229-12-146.

Webber, J., Ott, P., Owens, J., & Binder, W. (2005). Elevated temperature during reproductive development affects cone traits and progeny performance in *Picea glauca* ×

engelmannii complex. *Tree Physiology, 25*(10), 1219–1227. https://doi.org/10.1093/treephys/25.10.1219.

Wendt, J., Rosenbaum, H., Richmond, T. A., Jeddeloh, J. A., & Burgess, D. L. (2018). Targeted bisulfite sequencing using the SeqCap epi enrichment system. *Methods in Molecular Biology, 1708*, 383–405.

White, T. L., Adams, W. T., & Neale, D. B. (2007). *Forest genetics*. Wallingford and Cambridge: CABI Publishing. p x + 682.

Wong, D. C. J., & Matus, J. T. (2017). Constructing integrated networks for identifying new secondary metabolic pathway regulators in grapevine: Recent applications and future opportunities. *Frontiers in Plant Science, 8*. https://doi.org/10.3389/fpls.2017.00505.

Woolliams, J. (2013). Genome wide selection: A radical re-think or more of the same? In S. J. Lee, & J. Woolliams (Eds.), *Novel tree breeding*. Monografias INIA: serie Forestal no 24.

Xia, J., Guo, X., & Deng, K. (2014). Epigenetics, MicroRNAs and human cancer. In S. Babashah (Ed.), *MicroRNAs: Key regulators of oncogenesis* (pp. 29–57). Cham: Springer International Publishing. https://doi.org/10.1007/978-3-319-03725-7_2.

Xia, Y., Jing, D., Kong, L., Zhang, J., OuYang, F., Zhang, H., et al. (2016). Global lysine acetylome analysis of desiccated somatic embryos of picea asperata. *Frontiers in Plant Science, 7*. https://doi.org/10.3389/fpls.2016.01927.

Yaish, M. W., Sunkar, R., Zheng, Y., Ji, B., Al-Yahyai, R., & Farooq, S. A. (2015). A genome-wide identification of the miRNAome in response to salinity stress in date palm (*Phoenix dactylifera* L.). *Frontiers in Plant Science, 6*. https://doi.org/10.3389/fpls.2015.00946.

Yakovlev, I. A., Carneros, E., Lee, Y., Olsen, J. E., & Fossdal, C. G. (2016). Transcriptional profiling of epigenetic regulators in somatic embryos during temperature induced formation of an epigenetic memory in Norway spruce. *Planta, 243*(5), 1237–1249. https://doi.org/10.1007/s00425-016-2484-8.

Yakovlev, I. A., & Fossdal, C. G. (2017). In silico analysis of small RNAs suggest roles for novel and conserved miRNAs in the formation of epigenetic memory in somatic embryos of Norway spruce. *Frontiers in Physiology, 8*. https://doi.org/10.3389/fphys.2017.00674.

Yakovlev, I. A., Fossdal, C. G., & Johnsen, Ø. (2010). MicroRNAs, the epigenetic memory and climatic adaptation in Norway spruce. *New Phytologist, 187*(4), 1154–1169. https://doi.org/10.1111/j.1469-8137.2010.03341.x.

Yakovlev, I., Fossdal, C. G., Skrøppa, T., Olsen, J. E., Jahren, A. H., & Johnsen, Ø. (2012). An adaptive epigenetic memory in conifers with important implications for seed production. *Seed Science Research, 22*(02), 63–76. https://doi.org/10.1017/S0960258511000535.

Yakovlev, I. A., Lee, Y., Rotter, B., Olsen, J. E., Skrøppa, T., Johnsen, Ø., et al. (2014). Temperature-dependent differential transcriptomes during formation of an epigenetic memory in Norway spruce embryogenesis. *Tree Genetics & Genomes, 10*(2), 355–366. https://doi.org/10.1007/s11295-013-0691-z.

Yang, W., Wang, K., Zhang, J., Ma, J., Liu, J., & Ma, T. (2017). The draft genome sequence of a desert tree *Populus pruinosa*. *GigaScience, 6*(9), 1–7. https://doi.org/10.1093/gigascience/gix075.

Yu, M., Hon, G. C., Szulwach, K. E., Song, C. X., Jin, P., Ren, B., et al. (2012). Tet-assisted bisulfite sequencing of 5-hydroxymethylcytosine. *Nature Protocols Other Titles: Protocols, 7*(12), 2159–2170.

Zhang, Y.-Y., Fischer, M., Colot, V., & Bossdorf, O. (2013). Epigenetic variation creates potential for evolution of plant phenotypic plasticity. *New Phytologist, 197*(1), 314–322. https://doi.org/10.1111/nph.12010.

Zhang, Q., Li, Y., Xu, T., Srivastava, A. K., Wang, D., Zeng, L., et al. (2016). The chromatin remodeler DDM1 promotes hybrid vigor by regulating salicylic acid metabolism. *Cell Discovery, 2*, 16027. https://doi.org/10.1038/celldisc.2016.27.
Zheng, X., Chen, L., Xia, H., Wei, H., Lou, Q., Li, M., et al. (2017). Transgenerational epimutations induced by multi-generation drought imposition mediate rice plant's adaptation to drought condition. *Scientific Reports, 7*, 39843. https://doi.org/10.1038/srep39843.
Zhou, X., Jacobs, T. B., Xue, L.-J., Harding, S. A., & Tsai, C. J. (2015). Exploiting SNPs for biallelic CRISPR mutations in the outcrossing woody perennial *Populus* reveals 4-coumarate:CoA ligase specificity and redundancy. *New Phytologist, 208*(2), 298–301. https://doi.org/10.1111/nph.13470.
Zhu, R., Shevchenko, O., Ma, C., Maury, S., Freitag, M., & Strauss, S. H. (2013). Poplars with a PtDDM1-RNAi transgene have reduced DNA methylation and show aberrant post-dormancy morphology. *Planta, 237*(6), 1483–1493. https://doi.org/10.1007/s00425-013-1858-4.
Zou, M., Lu, C., Zhang, S., Chen, Q., Sun, X., Ma, P., et al. (2017). Epigenetic map and genetic map basis of complex traits in cassava population. *Scientific Reports, 7*, 41232. https://doi.org/10.1038/srep41232.

FSC
www.fsc.org
MIX
Papier | Fördert gute Waldnutzung
FSC® C083411

Kolibri 360 GmbH
Ferdinand-Jühlke-Straße 7
99095 Erfurt, Deutschland
produktsicherheit@kolibri360.de